Laser Spectroscopy
and Its Applications

OPTICAL ENGINEERING

Series Editor

Brian J. Thompson

William F. May Professor of Engineering
and Provost, University of Rochester
Rochester, New York

Laser Engineering Editor:
Peter K. Cheo
United Technologies Research Center
Hartford, Connecticut

Laser Advances Editor:
Leon J. Radziemski
New Mexico State University
Las Cruces, New Mexico

Optical Materials Editor:
Solomon Musikant
Paoli, Pennsylvania

Other Volumes in Preparation

.

Laser Spectroscopy and Its Applications

edited by

Leon J. Radziemski
Department of Physics
New Mexico State University
Las Cruces, New Mexico

Richard W. Solarz
Jeffrey A. Paisner
Lawrence Livermore National Laboratory
Livermore, California

CRC Press
Taylor & Francis Group
Boca Raton London New York

CRC Press is an imprint of the
Taylor & Francis Group, an **informa** business

First published 1987 by Marcel Dekker, Inc.

Published 2019 by CRC Press
Taylor & Francis Group
6000 Broken Sound Parkway NW, Suite 300
Boca Raton, FL 33487-2742

© 1987 by Taylor & Francis Group, LLC
CRC Press is an imprint of Taylor & Francis Group, an Informa business

First issued in paperback 2019

No claim to original U.S. Government works

ISBN 13: 978-0-367-45156-1 (pbk)
ISBN 13: 978-0-8247-7525-4 (hbk)

**Visit the Taylor & Francis Web site at
http://www.taylorandfrancis.com**

**and the CRC Press Web site at
http://www.crcpress.com**

Library of Congress Cataloging-in-Publication Data

Laser spectroscopy and its applications.

   (Optical engineering ; 11)
   Includes bibliographies and index.
   1. Laser spectroscopy.   2. Atomic spectroscopy.
3. Molecular spectroscopy.   I. Radziemski, Leon J.
II. Solarz, Richard W.   III. Paisner, Jeffrey A.
IV. Series: Optical engineering (Marcel Dekker, Inc.) ;
v. 11.
QC454.L3L343   1987   535.5'8   86-16638
ISBN 0-8247-7525-2

About the Series

Optical science, engineering, and technology have grown rapidly in the last decade so that today optical engineering has emerged as an important discipline in its own right. This series is devoted to discussing topics in optical engineering at a level that will be useful to those working in the field or attempting to design systems that are based on optical techniques or that have significant optical subsystems. The philosophy is not to provide detailed monographs on narrow subject areas but to deal with the material at a level that makes it immediately useful to the practicing scientist and engineer. These are not research monographs, although we expect that workers in optical research will find them extremely valuable.

Volumes in this series cover those topics that have been a part of the rapid expansion of optical engineering. The developments that have led to this expansion include the laser and its many commercial and industrial applications, the new optical materials, gradient index optics, electro- and acousto-optics, fiber optics and communications, optical computing and pattern recognition, optical data reading, recording and storage, biomedical instrumentation, industrial robotics, integrated optics, infrared and ultraviolet systems, etc. Since the optical industry is currently one of the major growth industries this list will surely become even more extensive.

Brian J. Thompson
University of Rochester
Rochester, New York

Foreword

Twenty-five years have passed since the laser was created based on
fundamental work on quantum electronics by scientists from the
United States and the USSR. During this period the laser has become
not only a scientific tool generating coherent light for research goals
but also an instrument revolutionizing extensive areas of science and
engineering, from communications to medicine, where light had almost
not been used before. Laser light has also dramatically changed the
areas traditionally connected with light, particularly optics, spectroscopy, and photochemistry. Different types of tunable lasers were developed in the 1960s and 1970s that provoked rapid progress resulting
in new methods for laser spectroscopy and selective action of laser
light on atoms and molecules. A significant contribution to the elabora-
tion of these trends was made by the scientists of two leading U.S.
laboratories—the Lawrence Livermore National Laboratory and the Los
Alamos National Laboratory. While solving the practical problems of
laser isotope separation, they obtained numerous scientific results
exceeding its limits and of interest to a wider circle of scientists. So
one can only welcome the idea conceived by the editors of this volume,
Lee Radziemski, Richard Solarz, and Jeff Paisner, active participants
in the above-mentioned laser programs, to compose the present
monograph on the basis of "the laser experience" of these laboratories.
The monograph offers to readers' attention detailed and easy-to-read
descriptions of the main techniques of laser spectroscopy and its
applications written by highly skilled experts. The book will be useful to a vast group of readers, beginning with students, who come
across the application of laser spectroscopy methods.

Professor V. S. Letokhov
Institute of Spectroscopy
Troitzk, Moscow, USSR

Preface

Laser spectroscopy is a discipline of great breadth that has applications in diverse fields ranging from combustion studies and trace-sample detection to biological research. These innovative applications have been strongly coupled to the remarkable progress made in the last 25 years in lasers and electro-optic devices. In turn, laser spectroscopy has contributed greatly to the discovery of the hundreds of lasers now available to the user. This symbiotic relationship has allowed the field to expand very rapidly so that individuals now must specialize to maintain their expertise. The purpose of this volume is to familiarize the reader with the tools and concepts used by practitioners in the field and provide extensive, but not exhaustive, reference material for further study. The contributors to the book are all in the forefront of their respective fields: resonance photoionization, laser absorption, laser-induced breakdown, photodissociation, Raman scattering, remote sensing, and laser-induced fluorescence. The authors of each chapter directed their material to a wide audience: engineers, technical managers, nonlaser scientists who require an introduction to the field, graduate students who seek a primer, and laser spectroscopists who wish to expand their knowledge.

An attempt to provide a current, complete, and comprehensive review of laser spectroscopy and its applications is a formidable undertaking and perhaps an impossible task in light of rapid developments in the field. The editors have selected the seven topics listed above for inclusion in this volume. These provide excellent coverage of the field. Other applications of laser spectroscopy merit study, in particular, metrology and photobiology. Novel experiments in these fields have generated great excitement in the scientific community during the last few years. Excellent references in metrology include J. L. Hall, *Science,* Vol. 202, pp. 147 to 156 (1979), and V. S. Letokhov and V. P. Chebotayev, *Nonlinear Laser Spectroscopy* (Springer-Verlag, NY, 1977). The reader is also referred to the continuing Springer-Verlag series on the proceedings from the International Conferences on Laser Spectroscopy. The most recent meeting in this series was held in the summer of 1985 in Maui, HI [*Laser Spectroscopy VII*, T. W. Hansch and Y. R. Shen, Eds. (Springer-Verlag, NY, 1985), Vol. 49 in the Springer Series in Optical Sciences]. A useful text on photobiology is *Lasers in Photomedicine and Photobiology*, R. Pratesi and C. A. Sacchi, Eds. (Springer-Verlag, NY, 1980), Vol. 22 in the Springer Series in Optical Sciences.

The current volume is arranged as follows: Chapter 1 contains an overview of the semiclassical theory of atomic and molecular spectra. The last part of the chapter contains new information on superfine structure in polyatomic molecules. Chapter 2 represents a break with the overall philosophy of the book. The fundamentals of lasers are so well covered elsewhere that it seems pointless to review them here. Rather the author has discussed the state-of-the-art lasers and those of the next generation that are likely to be useful in spectroscopy. He has also included references to fundamental texts on laser theory for those who need a more fundamental understanding.

At this point a chapter on the general subject of resonant interaction of laser light with atoms and molecules could have been included. However, this subject is treated well in the references given in Chapters 3 and 4. In addition, some of the necessary theory is covered where appropriate. Chapter 3 treats resonant absorption and detection by photoionization, the technique recently chosen as a commercially viable method for atomic vapor laser isotope separation for the enrichment of uranium and other metals. Chapter 4 treats laser absorption spectroscopy, actually quite a mature field. Chapters 5, 6, and 7 deal primarily, though not exclusively, with nonresonant techniques. Chapter 5 treats the use of laser-created plasmas in spectrochemical analysis. The subject of molecular photodissociation is discussed in Chapter 6. This scheme also has been demonstrated to be useful for isotope separation. Chapter 7 discusses Raman scattering techniques, both resonant and nonresonant. Chapter 8 describes uses of most of these methods for remote sensing of the atmosphere. Finally, Chapter 9 treats applications of laser-induced fluorescence with particular emphasis on combustion diagnostics.

It is our hope that this book can take both the expert and those studying laser spectroscopy for the first time to the point where they will be stimulated to continue their interest into the current journal literature. The vitality and richness of laser spectroscopy and its applications after more than 25 years of laser research are proof of the importance of the field and its value to society. We hope the readers share the feelings of the authors that the subject matter of this text is both elegant and useful.

Leon J. Radziemski
Richard W. Solarz
Jeffrey A. Paisner

Acknowledgments

The editors would like to thank their colleagues at the Lawrence Livermore National Laboratory, especially Drs. E. F. Worden, J. R. Morris, I. P. Herman, B. W. Shore, J. K. Crane, and C. A. Haynam for their valuable comments, and L. J. Evans and R. E. Hendrickson for their expertise in preparing the manuscript. We also wish to thank Dr. R. R. Jacobs of Spectra Physics, San Jose, California, Dr. D. F. Heller of Allied Corporation, Corporate Technology, Mt. Bethel, New Jersey, Dr. L. A. Rahn of Sandia National Laboratory, California, Dr. J. Tiee of Los Alamos National Laboratory, and Dr. R. Measures, University of Toronto, for reading portions of the manuscript.

Contents

Contributors

Hao-Lin Chen Laser Isotope Separation Program, Lawrence Livermore National Laboratory, Livermore, California

David A. Cremers Chemistry Division, Los Alamos National Laboratory, Los Alamos, New Mexico

William B. Grant Earth and Space Sciences Division, Jet Propulsion Laboratory, California Institute of Technology, Pasadena, California

John L. Lyman Chemistry Division, Los Alamos National Laboratory, Los Alamos, New Mexico

Robert P. Lucht Combustion Research Facility, Sandia National Laboratories, Livermore, California

John R. Murray Advanced Laser Development Program, Lawrence Livermore National Laboratory, Livermore, California

Jeffrey A. Paisner Laser Isotope Separation Program, Lawrence Livermore National Laboratory, Livermore, California

Chris W. Patterson Theoretical Division, Los Alamos National Laboratory, Los Alamos, New Mexico

Leon J. Radziemski Department of Physics, New Mexico State University, Las Cruces, New Mexico

Richard W. Solarz Laser Isotope Separation Program, Lawrence Livermore National Laboratory, Livermore, California

James J. Valentini Department of Chemistry, University of California, Irvine, California

Laser Spectroscopy
and Its Applications

1
Semiclassical Principles of Atomic and Molecular Spectra

CHRIS W. PATTERSON *Los Alamos National Laboratory, Los Alamos, New Mexico*

1.1 INTRODUCTION

The basic principles of atomic and molecular spectra have not changed since the advent of quantum mechanics in 1925. However, with the use of lasers in the last 20 years there has been a remarkable improvement in both the intensity and resolution of the spectral light source. With increased light intensity we can now probe highly excited states of atoms and molecules and with increased spectral resolution we can observe more structural detail. With few exceptions atomic and molecular theory has had to evolve to keep pace with the increase in observed structural detail of highly excited states. We shall try to emphasize in this chapter those aspects of atomic and molecular theory which are important to our understanding of high-resolution laser spectroscopy of highly excited states.

Curiously, this understanding of atomic and molecular fine structure has revived the use of semiclassical techniques that predate the quantum mechanics of Schrödinger and Heisenberg. Physicists and chemists are rediscovering the semiclassical techniques of Planck, Bohr, and Sommerfeld and applying them to gain both a qualitative and quantitative understanding of the fine structure of highly excited levels of atoms and molecules.

The semiclassical picture has been useful for determining the fine structure of highly excited levels such as electronic Rydberg levels, vibrational local-mode levels, and rotational cluster levels—all of which are important to recent atomic and molecular schemes for isotope separation. These schemes require the absorption of many photons before the ionization of an atom or the dissociation of a molecule. For atoms the excitation can occur for just one electron, in which case the Rydberg levels are hydrogenic with high principal quantum number. For molecules the excitation can occur in just one normal or local mode to high vibrational and rotational quanta. In this chapter we are concerned with high-angular-momentum quanta carried by the electrons in atoms or the vibrations and rotations in molecules which give rise to fine-structure spectra. In these cases the electronic, vibrational, and rotational moton can be accurately described with a semiclassical rather than a quantum picture. Our exposition in this chapter will be in terms of this almost forgotten semiclassical picture which is just now gaining popularity as a useful tool in understanding high-resolution spectroscopy.

In this chapter we also discuss another useful but often ignored tool for understanding the combinatorics of both the electrons in atoms and the nuclei in molecules. When treating more than two identical electrons in atoms or two identical nuclei in molecules, it is very useful to introduce a tableau notation to handle the permutational intricacies of the Pauli principle. This notation allows for the quick enumeration of the possible atomic and molecular fine-structure levels.

We will begin this chapter by discussing in Sec. 1.2 the semiclassical principles of absorption and emission of radiation where the radiation field is treated as continuous and the atomic and molecular energies are treated as discrete. Quantum transition probabilities are determined from semiclassical arguments for electric dipole radiation. The selection rules developed throughout this chapter are for atomic and molecular electric dipole transitions only. We also discuss in this section line shapes, cross sections, and saturation of transitions appropriate to high-resolution spectroscopy. Atomic spectroscopy is treated in Sec. 1.3, and vibrational and rotational spectroscopy of molecules is treated in Secs. 1.4 and 1.5, respectively.

Because of space limitations, we do not treat the spectra of atoms and molecules in static electric fields (Stark effect) or magnetic fields (Zeeman effect), nor do we treat the electronic spectra of molecules except for analogies between vibrational and electronic molecular structure.

1.2 ABSORPTION AND EMISSION OF RADIATION

The rate-equation approach to the absorption and emission of radiation given here is valid only for experimental time scales long compared to natural lifetimes and collisional effects. The rate-equation approach also ignores quantum optics phenomena such as multiphoton absorption and Rabi oscillations, which are developed later in the book. A good review may be found in the book by Loudon [Loudon, 1983].

1.2.1 Transition Probabilities

The total power radiated by a classical electric dipole oscillator of frequency ν is given in cgs units by

$$\frac{64\pi^4 \nu^2 |\underline{J}_0|^2}{3c^3} \tag{1.1}$$

where $\underline{J}_0$ is the dipole current vector

$$\underline{J}_0 = \int \underline{J} d^3 r$$

The corresponding quantum expression results from the relation

$$|\underline{J}_0| \rightarrow 2\pi i \nu \mu_{12}^* \tag{1.2}$$

where $\nu = \nu_1 - \nu_2$ is the frequency of the transition from the initial state 2 to the final state 1 and where the dipole moment of the transition is

$$\underline{\mu}_{12} \equiv \langle 1|\underline{\mu}|2\rangle = \exp\left(\int \Psi_1^* \underline{r} \Psi_2 \, d^3 r\right) \tag{1.3}$$

in terms of the initial and final quantum wavefunctions Ψ_2 and Ψ_1 of states 2 and 1, respectively (see Fig. 1.1) and charge e.

The power radiated for spontaneous electric dipole emission from upper level 2 to lower level 1 is obtained by substituting Eq. (1.2) into (1.1). Dividing by $h\nu$ then gives the transition probability per unit time for *spontaneous emission*:

$$A_{21} = \frac{64\pi^4 \nu^3 \mu^2 g_1}{3hc^3 g_2} \tag{1.4}$$

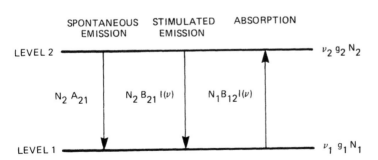

FIG. 1.1 Transition rates in a two-level system. For a given level 1, N_1 is the number of atoms or molecules occupying that level, g_1 is the level degeneracy, and ν_1 is the level frequency.

which is called the Einstein spontaneous emission coefficient. We define g_1 and g_2 to be the number of states (or degeneracy) of levels 1 and 2, respectively. All such states have the same transition dipole moment μ_{12}. Consequently, the degeneracy factor in Eq. (1.4) is simply a result of summing over the final states of level 1 and averaging over the initial ones of level 2. The transition dipole moment μ is defined such that $\mu^2 = g_2\mu_{12}^2$. The inverse of the coefficient A_{21} is the *natural lifetime* τ_{21} of a given level 2 before a transition to level 1 in which a photon of frequency ν is emitted. The sum over all such final levels of different energy in Eq. (1.4) results in the total transition rate for spontaneous emission from level 2,

$$A_2 = \sum_f A_{2f} \tag{1.5}$$

with total natural lifetime $\tau_2 = 1/A_2$. The *branching ratio* for a transition from level 2 to level 1 is simply $A_{21}/A_2 = \tau_2/\tau_{21}$.

The unit of the dipole moment μ is the debye (D; 10^{-18} esu-cm), for which the natural lifetime is given in Table 1.1. For visible wavelengths, such as $\lambda = 589.6$ nm for the sodium D_1 lines, we have $\mu = 6.3$ D ($g_1 = g_2 = 2$) and $\tau = 16.5$ ns. Such radiative lifetimes due to spontaneous emission are typical of electronic transitions in both atoms and molecules. However, because natural lifetimes are proportional to the cube of the wavelength of emitted radiation, the lifetimes are considerably longer for infrared and microwave transitions. For a molecular ν_3 vibrational transition of SF_6 with infrared wavelength of 10.55 μm and $\mu = 0.39$ D ($g_1 = 1$ and $g_2 = 3$), the natural lifetime is 74 ms. For a molecular rotational $J = 1$ transition with a microwave wavelength of 18.6 cm and $\mu = 1.50$ D ($g_1 = g_2 = 3$), the natural lifetime is about 300 years! For molecular vibrations and rotations an excited state will usually decay due to nonradiative effects before it has a chance to radiate.

It is also possible for an atom or molecule to radiate due to stimulated emission. For example, in the two-level system shown in Fig. 1.1, a light source with intensity per unit frequency $I(\nu)$ (in W/cm^2-s) can have its intensity increased by both spontaneous and stimulated transitions from level 2 to level 1 or its intensity decreased by transitions from level 1 to level 2. The equations for the transition rates of the two-level system are given by

$$\frac{dN_1}{dt} = -\frac{dN_2}{dt} = N_2A_{21} + N_2B_{21}I(\nu) - N_1B_{12}I(\nu) \tag{1.6}$$

where N_1 and N_2 are the number of atoms or molecules in levels 1 and 2, respectively. The relation above defines the Einstein B coefficients for stimulated emission, B_{21}, and absorption, B_{12}.

TABLE 1.1

Natural lifetime

$$\tau_{21}(\text{seconds}) = \frac{3.19 \times 10^6 \lambda^3(\text{cm}^3)g_2}{\mu^2(\text{debye}^2)g_1}$$

Oscillator strength

$$f = \frac{8\pi^2(\text{cm})\mu^2}{3he^2\lambda}$$

$$= \frac{1.50\lambda^2(\text{cm}^2)g_2}{\tau_{21}(\text{seconds})g_1} \qquad \text{where} \quad \tau_{21} = \frac{1}{A_{21}}$$

or

$$\mu(\text{debye}) = 842\sqrt{\lambda(\text{cm})f}$$

Rabi width

$$\Omega = \frac{\mu E}{\sqrt{3}h} = \mu\sqrt{\frac{8\pi I}{3ch^2}}$$

or

$$\Omega(\text{cm}^{-1}) = \frac{\Omega(\text{Hz})}{2\pi c}$$

$$= 2.66 \times 10^{-4}\mu(\text{debye})\sqrt{I(\text{watts}/\text{cm}^2)}$$

In thermal equilibrium, the transition rates must be zero or

$$N_2 A_{21} + N_2 B_{21} I(\nu) = N_1 B_{12} I(\nu) \tag{1.7}$$

Also, in thermal equilibrium, the intensity of the radiation source is that of a blackbody given by Planck's law,

$$I(\nu) = \frac{8\pi h\nu^3}{c^2}\left(e^{h\nu/kT} - 1\right)^{-1} \tag{1.8}$$

and the populations must have a Boltzmann distribution,

$$\frac{N_2}{N_1} = \frac{g_2}{g_1} e^{-h\nu/kT}$$ (1.9)

Substituting Eqs. (1.8) and (1.9) into Eq. (1.7) and solving for B_{12}, we find

$$g_1 B_{12} = g_2 B_{21}$$ (1.10a)

and

$$B_{12} = \frac{8\pi^3}{3ch^2} \mu^2$$ (1.10b)

The question arises: When is the stimulated emission due to thermal radiation equal to the spontaneous emission? We find that

$$\frac{B_{21} I(\nu)}{A_{21}} = \bar{n}(\nu) = \left(e^{h\nu/kT} - 1 \right)^{-1}$$ (1.11)

where $\bar{n}$ is the average number of photons at frequency ν. The thermal stimulate emission will be less than the spontaneous emission whenever

$$h\nu \geqslant kT$$

or   (1.12)

$$\lambda \leqslant 50 \ \mu m \ (\text{at room temperature})$$

Thus for most electronic transitions we may ignore the lifetime due to thermally stimulated transitions compared to the natural lifetime of a state.

1.2.2 Line Shapes and Cross Sections

Let $f(\nu)$ be the fraction of the transitions that occur at frequency ν per unit frequency. Thus $f(\nu)$ represents a typical line-shape function. We may write $I(\nu) = If(\nu)$, where I is the total intensity of a source in the range $d\nu$ about frequency ν. The attenuation or gain of the source intensity I is related to the number of molecules in volume $V = A \ dz$ which undergo a transition, where A is the area of the source beam and dz is its distance of propagation. Let us assume that all the population is in level 1, so that $N = N_1$. For each transition from level 1 to level 2 the molecule will absorb energy $h\nu$ from the source or

$$A \ dI = - \frac{NB_{12}If(\nu)h\nu A \ dz}{V}$$

If we define $n = N/V$ to be the population density, then

$$dI = -nB_{12}If(\nu)h\nu \ dz \qquad (1.13)$$

Integrating, we have Beer's law,

$$I = I_0 e^{-kz} \qquad (1.14)$$

Thus the intensity is attenuated exponentially with *absorption coefficient* k given by

$$k(cm^{-1}) = nB_{12}f(\nu)h\nu \qquad (1.15)$$

We may relate the absorption coefficient to the cross section of absorption using Eq. (1.10b),

$$\sigma(cm^2) = \frac{k}{n} = \frac{8\pi^3 \nu f(\nu)\mu^2}{3ch} \qquad (1.16)$$

We have assumed that all the population is in level 1. In thermal equilibrium, the population in level 1 will be the total population times a Boltzmann factor resulting from the thermal distribution. Integrating both sides of Eq. (1.16) over frequency and assuming that ν is essentially constant over the line-shape function $f(\nu)$, we find

$$\int \sigma(\nu) \ d\nu = h\nu B_{12} \qquad (1.17)$$

From Eqs. (1.17) and (1.10) it is clear that the integrated cross section depends linearly on the frequency. This dependence typifies both absorption and stimulated emission, in contrast to spontaneous emission.

The actual peak absorption cross section depends on the line-shape function $f(\nu)$. Broadening of lines can arise from the different Doppler frequency shifts of transitions. Such broadening gives rise to a Gaussian line-shape function due to the Maxwellian distribution of particle velocities. The broadening is *inhomogeneous* because different particles can have different transition frequencies. The Doppler line-shape function is given by the Gaussian

$$f_G(\nu) = \frac{(4 \ln 2/\pi)^{1/2}}{\gamma_D} e^{-4 \ln 2 (\nu - \nu_0)^2 / \gamma_D^2} \qquad (1.18)$$

where the Doppler full width at half-maximum (FWHM) γ_D is given by

$$\gamma_D (\text{cm}^{-1}) = \left(\frac{8R \ln 2\, T}{Mc^2} \right)^{1/2} \frac{1}{\lambda}$$

$$= 7.16 \times 10^{-7} \left(\frac{T}{M} \right)^{1/2} \frac{1}{\lambda(\text{cm})} \qquad (1.19)$$

where M is the atomic mass number and T the temperature in K.

For visible transitions in sodium ($\lambda = 589.6$ nm) the Doppler width at room temperature is 0.04 cm^{-1}. Assuming that all the population is in the ground state, the sodium D_1 transition with a dipole moment strength of 6.3 D and the line width noted above would have a peak cross section at line center ν_0 of $\sigma_0 \sim 7 \times 10^{-12}$ cm^2.

For infrared transitions ($\lambda = 10.55$ µm) in heavy molecules such as SF$_6$, the Doppler broadening at room temperature is quite small, with $\gamma_D \cong 0.001$ cm^{-1}. Consequently, the instrumental or collisional broadening many dominate.

Broadening of lines can also arise from the finite collisional or natural lifetime of a state. Such broadening gives rise to a Lorentzian line-shape function due to the effective damping of the classically os-cillating dipole. The broadening is *homogeneous* since each particle has the same distribution of transition frequencies. The Lorentzian line-shape function is given by

$$f_L(\nu) = \frac{\gamma_L / 2\pi}{(\nu - \nu_0)^2 + \gamma_L^2 / 4} \qquad (1.20)$$

where γ_L is the Lorentzian line width (FWHM). Examples of Gaussian and Lorentzian line-shape functions are given in Fig. 1.2.

For collisional broadening the line width is proportional to the pressure or

$$\gamma_L (\text{cm}^{-1}) = pC_p \qquad (1.21)$$

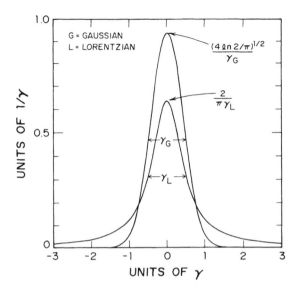

FIG. 1.2 Gaussian (inhomogeneous) and Lorentzian (homogeneous) line functions. The area under each curve is normalized to unity. The line width γ is the full width at half maximum (FWHM).

The constant of proportionality C_p is called the pressure broadening coefficient. For gas kinetic collisions rates this constant is typically ~ 0.0002 cm^{-1}/torr. For broadening due to the natural lifetimes of spontaneous emission, we have

$$\gamma_L (\text{cm}^{-1}) = \frac{1}{2\pi c \tau} \tag{1.22}$$

Even for atomic transitions the natural broadening is very small. For our example of the sodium D lines, the natural line width would be $\gamma_L \sim 8 \times 10^{-4}$ cm^{-1}. Since we are using units of cm^{-1} for γ_L and γ_G, they must be converted to frequency units when using $f(\nu)$ in Eqs. (1.15) and (1.16).

For small pressures, where the Doppler line broadening dominates, the absorption coefficient k_0 at line center will be linearly dependent on the number density n and hence the pressure. For high pressures, where the collisional line broadening dominates, the absorption coefficient will be independent of pressure, as seen from Eqs. (1.15), (1.20), and (1.21).

1.2.3 Saturation of Transitions

For high-resolution spectroscopy it is important to overcome line-broadening effects, which often obscure the detailed atomic and molecular substructure which is hidden beneath the line profile in ordinary absorption spectroscopy.

Atomic and molecular beam spectroscopy allows one to reduce both the collisional and Doppler broadening by using a well-collimated beam of particles at very low pressure. A light source excites the particles perpendicular to their trajectories, where the velocity distribution is narrow and the Doppler broadening small. This spectroscopic technique is limited by the small number density of atoms or molecules in the beam and by the short path length. Thus only absorbers with high cross sections can be detected and only in spectral regions where the detectors are very sensitive.

Saturation spectroscopy allows for very long path lengths but requires the high radiation intensity of a laser. An intense laser at frequency ν_L is tuned across a Doppler-broadened absorption line as shown in Fig. 1.3. At the laser frequency the transition is saturated so that there are nearly equal populations in the upper and lower states of the transition.

We can see this saturation effect by referring to Eq. (1.6). We can write the rate of population increase in level 2 in terms of the photon flux $\phi = I/h\nu$ in units of photons/cm^2-s and the homogeneous cross section σ_L of the transition. We have, ignoring the spontaneous emission for high laser intensities and assuming that the levels have equal degeneracy such that $B_{12} = B_{21} = B$,

$$\frac{dN_2}{dt} = (N_1 - N_2)BIf_L(\nu)$$
$$= (N_1 - N_2)\sigma_L \phi \tag{1.23}$$

Solving with initial conditions $N = N_1 + N_2 = N_1$, we have

$$N_2 = \frac{N}{2}(1 - e^{-2\sigma_L \phi t}) \tag{1.24}$$

For sufficiently long laser pulses t_p, where

$$t_p \gg \frac{1}{\sigma_L \phi} \tag{1.25}$$

the population of level 2 saturates at half the population initially in level 1. It is important to realize that the population N refers to a subset of the total population with velocity ν_x, so that the Doppler-shifted frequency is resonant, as shown in Fig. 1.3.

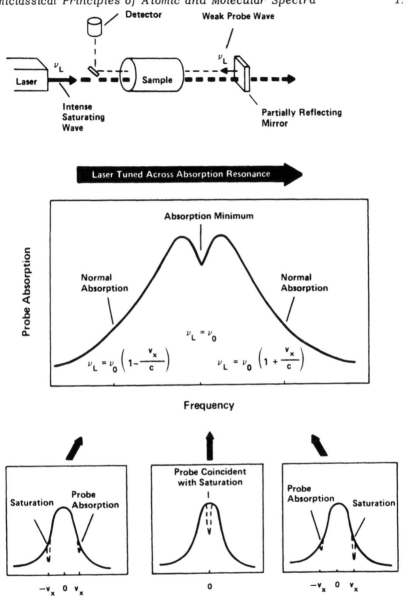

FIG. 1.3 Saturation spectroscopy. An intense laser pump at frequency ν_L is tuned across a Doppler-broadened absorption feature with a small part reflected back as a probe. If $\nu_L = \nu_0$ the pump and probe see the same velocity subset ($v_x = 0$) and the probe absorption is reduced. The result is a sharp absorption minimum at the resonant frequency [McDowell, 1982].

$$\nu_L = \begin{cases} \nu_0 \left(1 - \dfrac{v_x}{c} \right) & \nu_L < \nu_0 \\[2ex] \nu_0 \left(1 + \dfrac{v_x}{c} \right) & \nu_L > \nu_0 \end{cases} \qquad (1.26)$$

The cross section must be adjusted for this limited population according to Eq. (1.16) and indeed corresponds to the cross section σ_L for the Lorentzian homogeneous line width.

In saturation spectroscopy a small part of the laser beam is reflected back as shown in Fig. 1.3 and acts as a probe. For the probe and pump beam opposite velocity components will be on resonance. Only if $\nu_L = \nu_0$ will the two laser beams see the same velocity subset, namely $\nu_x = 0$. However, this velocity subset has been saturated by the pump so that the probe absorption is small. In Fig. 1.3 we see the absorption of the *probe* beam as the laser is tuned across the Doppler line profile. The result is a sharp absorption minimum at the resonant frequency corresponding to the much smaller Lorentzian homogeneous line width.

We may write Eq. (1.23) in terms of the "Rabi width"

$$\Omega = \frac{\mu}{\sqrt{3}} \frac{E}{\hbar} \qquad (1.27)$$

The factor $1/\sqrt{3}$ is a result of the spatial average of $\underline{\mu} \cdot \underline{E}$ over all possible orientations of $\underline{\mu}$ with respect to the applied electric field vector $\underline{E}$. One must be careful since this factor is sometimes included in the definition of the dipole moment μ itself. For electronic transitions the oscillator strength f is commonly used instead of the dipole moment μ. In Table 1.1 we show some common relations between the oscillator strengths f, the dipole moment μ, and the Rabi width Ω. The light intensity of a sinusoidally oscillating electric field is

$$I = \frac{c}{8\pi} E^2 \qquad (1.28)$$

At line center Eq. (1.23) becomes

$$\frac{dN_2}{dt} = \frac{(N_2 - N_1)\Omega^2}{2\pi\gamma_L} \qquad (1.29)$$

The rate equations we have used assume that the damping γ_L is large compared to the Rabi width Ω. Otherwise, there can be population oscillations during the pulse at frequency Ω and multiphoton transitions can dominate.

1.3 ATOMIC STRUCTURE

We shall first develop the semiclassical physics of the hydrogen atom thoroughly since an intuitive understanding of this simple system will aid in the understanding of more complicated atoms. The spectral lines of the hydrogen atom follow a pattern that can often be discerned with some modification in the spectra of other atoms, such as the alkali metals and highly excited Rydberg states. For these the semiclassical picture can be both qualitatively and quantitatively accurate. For a good review, see the book by White (White, 1934).

1.3.1 The Hydrogen Atom

1.3.1A. Bohr Quantization

To find the quantized energy levels of hydrogen we may use the Bohr semiclassical quantization conditions for circular orbits of the electron about the nucleus. The angular momentum of the electron orbiting the nucleus is, in polar coordinates (r, ϕ),

$$p_\phi = mr^2 \dot{\phi} = mrv \qquad (1.30)$$

The semiclassical quantization condition stipulates that the action for an orbit of the classical trajectory is an integral number of h or

$$\int p_\phi \, d\phi = p_\phi \int_0^{2\pi} d\phi = nh \qquad (1.31)$$

Since the angular momentum is a constant of the motion, we have

$$p_\phi = \frac{nh}{2\pi} = n\hbar \quad n = 1, 2, \ldots \qquad (1.32)$$

or

$$v = \frac{n\hbar}{mr} \qquad n = 1, 2, \ldots \qquad (1.33)$$

We may equate the Coulomb force of attraction to the centrifugal force,

$$\frac{mv^2}{r} = \frac{e^2 Z}{r^2} \qquad (1.34)$$

where Ze is the charge at the nucleus for single-electron atoms. The atomic number Z applies, for example, to singly ionized He and doubly ionized Li, where Z = 2 and Z = 3, respectively. From Eq. (1.33) we find the radius of the quantized orbits,

$$r \equiv a_n = \frac{n^2 \hbar^2}{me^2 Z} = n^2 a_1 \quad n = 1, 2, \ldots \tag{1.35}$$

and velocity

$$v = \frac{e^2 Z}{n \hbar} \quad n = 1, 2, \ldots \tag{1.36}$$

The smallest orbital radius, when n = 1, is called the Bohr radius,

$$a_1 = 0.0529 \text{ nm}$$

and is the basic unit of atomic dimensions. Recently, orbits have been observed with $n \cong 100$ in atomic beams for which the radius is nearly 1 μm [Kleppner, 1981]. Atoms with such a large *principal* quantum number n are called Rydberg atoms and in many respects they behave classically, as we shall see below.

The total energy E depends solely on the principal quantum number. For circular orbits, the sum of the kinetic and potential energy is

$$E = \frac{1}{2} mv^2 - \frac{e^2 Z}{r}$$

or from Eqs. (1.35) and (1.36),

$$E_n = \frac{me^4 Z^2}{2n^2 \hbar^2} \quad n = 1, 2, \ldots \tag{1.37}$$

Since the nucleus is not of infinite mass, m should be replaced by the reduced mass $\mu = mM/(m + M)$, where M is the nuclear mass. These energy levels are drawn on the right of Fig. 1.4. There are an infinite number of discrete energy levels which become more closely spaced as the principal quantum number becomes large. The n = 0 orbit with r = 0 is not allowed since it would require infinite energy.

It is customary to write the energy in wave numbers by dividing E_n by hc:

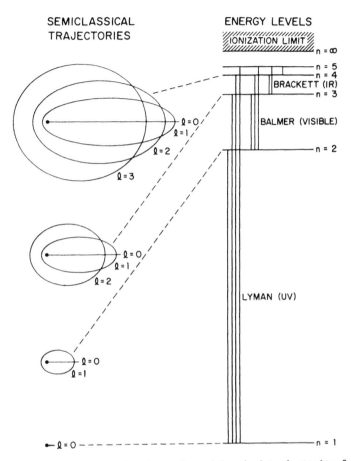

FIG. 1.4 Energy levels and semiclassical trajectories for the hydrogen atom. For a given principal quantum number, n, the eccentricity of the semiclassical elliptical orbits are determined by the angular momentum quantum number of the electron, ℓ. The semimajor axis, a, has a n^2 dependence. Possible electric dipole transitions are shown.

$$\frac{E_n}{hc} = -\frac{RZ^2}{n^2} \qquad (1.38)$$

where R, the Rydberg, is

$$R_H = \frac{\mu e^4}{4\pi c\hbar^3} = 109,677.58 \ \mathrm{cm}^{-1} \ \text{for hydrogen} \qquad (1.39)$$

For large n the energy approaches zero, so that Rydberg states are very weakly bound. The ionization energy from the ground state (n = 1) is simply equal to 13.6 eV.

The electron is moving the fastest when it is closest to the nucleus with n = 1. Comparing the speed of the electron to the speed of light for n = 1, we find

$$\alpha = \frac{v}{c} = \frac{e^2}{c\hbar} = \frac{1}{137} \tag{1.40}$$

The constant α is called the fine-structure constant. We shall see later that fine-structure corrections to the energy E_n are of the order α^2.

1.3.1B. Bohr-Sommerfeld Quantization

In general there is one semiclassical quantization condition for each degree of freedom of a classical system. Since the electron orbiting the nucleus has three degrees of freedom, we have ignored two quantization conditions. It is indeed surprising that we have arrived at the correct expression for the energy.

From classical physics we know that the electron will in general orbit the nucleus in an ellipse, not a circle, so that the radius r is not a constant and accounts for another degree of freedom. We also know that the ellipse will be in a plane perpendicular to the angular momentum vector $\underline{\ell}$. The orientation of this plane with respect to the laboratory z axis accounts for another degree of freedom, θ. As shown in Fig. 1.5, the three coordinates r, θ, and ϕ comprise the three degrees of freedom of the hydrogen atom. The radius r is assumed to be measured from the center of mass.

The three quantization conditions become

$$\int p_r \, dr = n_r h \qquad n_r = 1, 2, \ldots \tag{1.41a}$$

$$\int p_\theta \, d\theta = n_\theta h \qquad n_\theta = 0, 1, 2, \ldots \tag{1.41b}$$

$$\int p_\phi \, d\phi = mh \qquad m = 0, 1, 2, \ldots \tag{1.41c}$$

As shown in Fig. 1.5, p_ϕ is a constant of motion since it is just the projection of the total angular momentum $\underline{\ell}$ on the laboratory z axis. Thus we may write

$$\ell_z = p_\phi = m\hbar \tag{1.42}$$

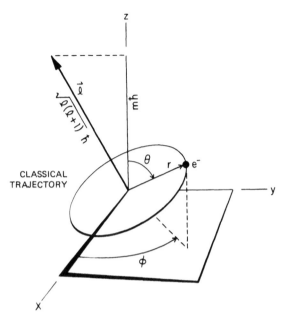

FIG. 1.5 Spherical coordinates (r, θ, ϕ) for the semiclassical quantization of the hydrogen atom. The classical trajectory for the electron
is an ellipse with the orbital angular momentum $\underline{\ell}$ perpendicular to the
plane of the ellipse. Both $\underline{\ell}$ and $\underline{\ell}_z = m\hbar$ are constants of the motion.

One calls m the *spatial* (or magnetic) quantum number and it determines the orientation of the plane of the ellipse.
      The remaining Bohr-Sommerfeld conditions lead to the quantized
elliptical orbits as shown clearly by Pauling [Pauling, 1935]. The
semimajor axis, a, of the ellipse depends only on the principal quantum number n and is equal to the Bohr radius for circular orbits,
given by Eq. (1.35) with

$$n = n_r + n_\theta + m = 1, \, 2, \, \ldots \tag{1.43}$$

The semiminor axis, b, of the ellipse is given by

$$b = \frac{\sqrt{\ell^2}}{n} \, a \tag{1.44}$$

where

$$\ell = n_\theta + m = 0, \, 1, \, 2, \, \ldots, n \tag{1.45}$$

is the *azimuthal* quantum number such that the total angular momentum is $|\underline{\ell}| = \hbar\sqrt{\ell^2}$. The azimuthal quantum number cannot be greater than n because of the condition $b \leqslant a$ for an ellipse. Also, the magnitude of the spatial quantum number $|m|$ cannot be greater than ℓ because $\ell_z \leqslant |\underline{\ell}|$. Finally, the quantized energy is the same as for circular orbits given by Eq. (1.37).

By considering all the degrees of freedom, we have many semi-classical elliptical orbits with the same energy. For a given energy specified by n we have the *degenerate* orbits with

$$m = -\ell, \; -\ell + 1, \ldots, \ell \tag{1.46a}$$

and

$$\ell = 0, \; 1, \ldots, n \tag{1.46b}$$

Although Sommerfeld ruled out the $\ell = 0$ orbit because it would touch the nucleus, it turns out that only the $\ell = n$ orbit is ruled out from quantum mechanical considerations. That is, circular orbits are not really allowed! Also, the quantum solution of Schrödinger's equation for the hydrogen atom results in the replacement $\ell^2 \to \ell(\ell + 1)$ above. Accordingly, the allowed ℓ orbits for a given value of n are shown on the left of Fig. 1.4.

The energy of the hydrogen atom is independent of m since it cannot depend on the orientation of the ellipse. The fact that at this stage the energy is independent of ℓ also is more intriguing. The ℓ degeneracy is a consequence of the fixed semimajor axis (i.e., the ellipse does not precess). We shall see later that when the ellipse precesses the energy will depend on ℓ.

1.3.1C. Wavefunctions

The solution to Schrödinger's equation gives the quantum mechanical wavefunction. For any spherically symmetric interaction, as in the case of the Coulomb potential, the angular dependence of the wavefunction is given by the spherical harmonics $\psi_{\ell m}(\theta, \phi)$ (see, e.g., the book by Cowan [Cowan, 1981]). The total wavefunction is the product of the spherical harmonics and the radial function:

$$\Psi_{n\ell m}(r, \theta, \phi) = R_{n\ell}(r)\psi_{\ell m}(\theta, \phi) \tag{1.47}$$

The probability of finding the electron at coordinates r, θ, ϕ is given by the square of the total wavefunction $\Psi^2_{n\ell m}(r, \theta, \phi)$. This probability looks nothing like the ellipses shown in Fig. 1.4. However, for large n, it has been recently shown [Snider, 1983] that one can find linear combinations of the $\Psi_{n\ell m}$ for a given n which look very similar to the classical ellipses. Thus the semiclassical ellipses give accurate representations of quantum Rydberg states where n is large.

We shall not discuss the properties of the spherical harmonics, which are considered at length in many standard texts, since they arise for any spherically symmetric potential; instead, we shall explore the properties of the radial wavefunctions $R_{n\ell}(r)$ for the Coulomb potential. In Fig. 1.6 we compare the quantum mechanical probability of finding the electron at distance r with the corresponding classical probability. The classical probability is proportional to the inverse of the radial velocity and so is infinite at the classical turning points located at the pericenter and apocenter of the ellipse, as shown in this figure. The classical turning points give roughly the limits of the quantum probability. For large n the classical distribution is the average of the quantum mechanical distribution, as can be seen in Fig. 1.6. Note that the most radially localized states occur for $\ell = n - 1$, where the classical orbits are nearly circular.

1.3.1D. Transitions

The ith component of the dipole moment vector μ_i for i = x, y, z for a transition from initial state nℓm to final state n'ℓ'm' is given by

$$\mu_i = \int \Psi_{n\ell m}(r,\theta,\phi) r_i \Psi_{n'\ell'm'}(r,\theta,\phi) \, d^3r \qquad i = x, y, z \qquad (1.48)$$

Evaluating this integral, we find

$$m' - m = \Delta m = \begin{cases} 0 & i = z \\ \\ \pm 1 & i = x \text{ or } y \end{cases} \qquad (1.49)$$

and

$$\ell' - \ell = \Delta \ell = \pm 1 \qquad (1.50)$$

These are the selection rules for hydrogenic transitions.

Since there are no conditions on n, we can have any of the transitions shown in Fig. 1.4. Each set of transitions shown in Fig. 1.4 to the same final lower level n_f results in a spectroscopic series with wavenumbers

$$\frac{1}{\lambda_{n_i}} = RZ^2\left(\frac{1}{n_f^2} - \frac{1}{n_i^2}\right) \qquad n_i = n_f + 1, \, n_f + 2, \, \ldots \qquad (1.51)$$

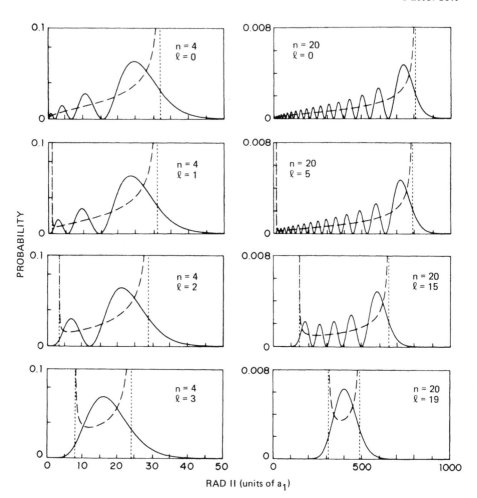

FIG. 1.6 Comparison of the quantum (solid curve) with the classical (dashed curve) probability for finding the hydrogen electron at a given radius from the nucleus. For high n the classical probability distribution is the average of the quantum distribution. The classical turning points at the pericenter and apocenter of the respective ellipses (see Fig. 1.4) determine the limits of the distribution as shown by the vertical dashed lines.

For $\Delta n = 1$ transitions, the dipole moments are roughly the electron charge (4.8×10^{-10} esu) times the Bohr radius (5.28×10^{-9} cm), so that dipole moments of a debye (10^{-18} esu-cm) are typical. For $\Delta n > 1$ the spatial overlap of the initial and final wavefunctions is small, as seen from Fig. 1.4, resulting in a small dipole moment.

For large n the frequency of $\Delta n = 1$ transitions is the same as the classical frequency of the orbiting electron. This is known as Bohr's correspondence theorem and we shall see later similar examples for vibrations and rotations in molecules.

Letting $n_i \approx n_f \approx n \gg 1$ in the equation above, we have transition frequencies

$$v_n(Hz) = \frac{2RcZ^2}{n^3} \Delta n$$

$$= f \quad \text{for } \Delta n = 1 \tag{1.52}$$

where f is the classical orbital frequency. Note that from Eq. (1.38) we may write

$$v_n = \frac{1}{h} \frac{dE_n}{dn} = f \tag{1.53}$$

to relate the classical frequencies to the quantum energies.

1.3.2 The Atomic Shell Model

The hydrogenic wavefunctions are the basis for the atomic shell model. For other elements one replaces the Coulomb potential with the nuclear and electron "effective" spherically symmetric core potential.

We have seen that the characteristic radii or semimajor axes of the Bohr-Sommerfeld elliptical orbits depend only on n. Thus the principal quantum number n is said to specify the *shell* of a given electron. The number of electrons that can be accommodated in a given shell is the number of possible states within the shell. This is a consequence of the *Pauli exclusion principle*, which specifies that no two electrons can be in exactly the same state. For a given azimuthal quantum number ℓ there are $2\ell + 1$ possible spatial states labeled by m, as seen from Eq. (1.46a).

However, an electron also has a spin angular momentum of magnitude $\sqrt{s(s+1)}\ \hbar$, where $s = 1/2$. There are two possible spin orientations corresponding to spin spatial quantum numbers $m_s = 1/2$ or $-1/2$ for spin-up (↑) or spin-down (↓) electrons, respectively. Therefore, the total number of electron states for each shell is

$$2 \sum_{\ell=0}^{n-1} (2\ell + 1) = 2n^2 \tag{1.54}$$

The $2(2\ell + 1)$ electrons (including spin) for a given azimuthal quantum number ℓ are said to form a *subshell*. We shall show later that when a subshell is filled with $2(2\ell + 1)$ electrons it forms an electron core with a spherically symmetric core potential. Because an electron with low ℓ spends more time near the nucleus than an electron with high ℓ (see Fig. 1.6), it will have less screening by the other electrons from the nucleus and feel more of the charge Z of the nucleus. Since the potential energy is proportional to $-Z/r$, the electrons with low ℓ will have *lower* potential energy on the average. Thus, within a shell, the electrons will first fill the lower ℓ subshells. In effect, the electron screening in multielectron atoms causes a single electron trajectory to precess and its energy will depend on ℓ. The subshells are labeled by s, p, d, and f for $\ell = 0, 1, 2$, and 3. They are filled in order of increasing energy $E_{n,\ell}$ as follows:

n, ℓ label:	1s	2s	2p	3s	3p	4s	3d	4p	5s	4d	5p	6s	4f
$2(2\ell + 1)$:	2	2	6	2	6	2	10	6	2	10	6	2	14

The 4s subshell is filled before the 3d subshell because eventually the large Z in the potential near the nucleus for low ℓ becomes more important than the $1/n^2$ energy dependence for high n. Full shells form extremely stable atoms, as in the case of the noble gases. The alkali metals correspond to the elements with an electron in an s subshell outside a filled shell. Alkali metals are hydrogenic since only the outer s electron is generally optically excited to higher orbitals. Filling the d subshells gives rise to the transition metals, while filling the f subshell gives rise to the rare earth metals. Atoms with the same partially filled subshells are in the same *group* and exhibit similar spectral features and chemistry. Thus the group of elements Li, Na, K, Rb, and Cs have partially filled subshells 2s, 3s, 4s, 5s, and 6s, while the group B, Al, Ga, In, and Tl have partially filled subshells 2p, 3p, 4p, 5p, and 6p.

1.3.3 Spectra of Alkali Metals

The single electron of an alkali metal which is outside the core of filled subshells will feel a spherically symmetric potential. However, the elliptical trajectories of this electron will precess due to penetration into the electron core. This precession is greater for lower ℓ, where the penetration is deepest. The penetrating orbits now have an energy which is dependent on ℓ and is given by

$$E_{n\ell} = \frac{-RZ^2}{(n - \delta_\ell)^2} \qquad (1.55)$$

where δ_ℓ is called the *quantum defect* and

$$n_\ell = n - \delta_\ell \qquad (1.56)$$

is the *effective* principal quantum number. It is possible to calculate δ_ℓ classically on the basis of a precessing elliptical trajectory due to core penetration and core polarization [White, 1934]. For large ℓ, where the electron orbits are more circular, the core penetration is small and δ_ℓ approaches zero. For example, in sodium, the quantum defects are $\delta_0 = 0.35$ and $\delta_3 = -0.01$ with a slight n dependence.

A sodium energy-level diagram is shown in Fig. 1.7. Because of the selection rules $\Delta\ell = \pm 1$, the number of possible transitions are limited as shown. Thus only transitions from p orbitals are allowed to the ground state, giving rise to the *principal* series shown. The first term in this series at 590 nm is the yellow sodium D line first seen in the sun's emission spectra by Fraunhofer in 1814. Transitions from both s and d orbitals are allowed to the 3p first excited states, giving rise to the *sharp* and *diffuse* series, respectively. The next excited state is a 4s orbital. Although transitions from higher-lying p orbitals are allowed to this state, they occur more frequently to the ground state in emission spectra because of the ν^3 dependence for spontaneous emission. Transitions from both the p and f orbitals are allowed to the 3d excited state. Again, transitions from the p orbital will preferably go to the ground state. Transitions from the f orbitals give rise to the *fundamental* series. The orbital labels s, p, d, and f were named for the spectral series shown in Fig. 1.7. Similar spectra occur for the other alkali metals.

1.3.4 Fine Structure for One Optically Active Electron

We have seen how the alkali metals exhibit spectra similar to those of a one-electron atom. We now discuss the observed fine structure seen at higher resolution. This fine structure is due to the intrinsic electron angular momentum or spin, which gives rise to an electron magnetic momentum. We first consider the magnetic moment produced by the electron's *orbital* angular momentum, however.

The angular momentum of the electron is given by

$$\underline{\ell} = m\underline{r} \times \underline{v}$$

The magnetic moment of the current loop produced by a circular orbiting electron is simply

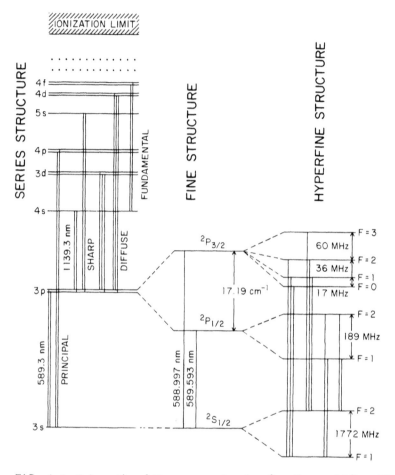

FIG. 1.7 Schematic of the energy levels of sodium. (After Allen and Eberly [Allen, 1975].)

$$\underline{\mu}_\ell = \frac{e}{2c} \, \underline{r} \times \underline{v} = \frac{e}{2mc} \, \underline{\ell}$$

$$= -\frac{g_\ell \mu_B}{\hbar} \, \underline{\ell} \tag{1.57}$$

where the Bohr magneton μ_B is given by

$$\mu_B = -\frac{e\hbar}{2mc} = 0.927 \times 10^{-20} \text{ erg/G} \tag{1.58}$$

and $g_\ell = 1$ is the orbital g factor, which is positive for the negatively charged electron. The relations above hold, as well, for elliptical orbits. The magnetic moment of the electron due to the intrinsic electron spin angular momentum $\underline{s}$ is given by a similar equation.

$$\underline{\mu}_s = \frac{g_s \mu_B}{\hbar} \, \underline{s} \qquad (1.59)$$

where $g_s = 2$ has been experimentally determined. The projection of ℓ about the spaced fixed z axis has been shown to be $\ell_z = m\hbar$. Similarly, the electron spin can have values $s_z = m_s\hbar$ with $m_s = -1/2, 1/2$.

Now the electron moving in the electronic field of the nucleus will experience a magnetic field given by

$$\underline{H} = -\frac{1}{c}\underline{v} \times \underline{E} = -\frac{Ze}{mcr^3}\,\underline{\ell}$$

where the electric field at the electron due to the nuclear charge is

$$\underline{E} = \frac{Ze\underline{r}}{r^3}$$

The energy of the electron's magnetic dipole moment in this magnetic field is simply

$$E(\text{spin-orbit}) = -\underline{\mu}_s \cdot \underline{H} = -\frac{g_s \mu_B}{\hbar}\,\underline{s} \cdot \underline{H}$$

or

$$E(\text{spin-orbit}) = \frac{g_s \mu_B}{mc\hbar r^3}\,Ze^2\underline{s} \cdot \underline{\ell}$$

This is the spin-orbit energy in the reference frame of the moving electron. In the inertial frame of the nucleus there is a relativistic correction due to the Thomas precession [White, 1934] of the electron relative to the nucleus. The result is that the spin-orbit energy is reduced by a factor of 2. Using the expression Eq. (1.58) for the Bohr magneton, the final result is

$$E(\text{spin-orbit}) = \frac{Ze^2}{2m^2c^2r^3}\,\underline{s} \cdot \underline{\ell}$$

Let the total angular momentum $\underline{j}$ be the sum of the orbital angular momentum $\underline{\ell}$ and the spin angular momentum $\underline{s}$ or

$$\underline{j} = \underline{\ell} + \underline{s}$$

Then

$$\underline{\ell} \cdot \underline{s} = \frac{1}{2} [\, |\underline{j}|^2 - |\underline{\ell}|^2 - |\underline{s}|^2]$$

$$= \frac{1}{2} \left[j(j + 1) - \ell(\ell + 1) - \frac{3}{4} \right] \hbar^2$$

where $j = \ell \pm 1/2$ depending on whether the spin $\underline{s}$ is in the same or opposite direction to $\underline{\ell}$. When $\ell = 0$, then $j = s = 1/2$. Thus except for $\ell = 0$ s orbitals there are two values for the spin-orbit energy for each ℓ. We have

$$E(\text{spin-orbit}) = \frac{Ze^2 \hbar^2}{4m^2 c^2 r^3} \times \left\{ \begin{array}{c} \ell \\ -(\ell + 1) \end{array} \right\} \quad \begin{array}{c} \text{for } j = \ell + \dfrac{1}{2} \\[2mm] \text{for } j = \ell - \dfrac{1}{2} \end{array} \quad (\ell \neq 0)$$

$$(1.62)$$

This splitting of the ℓ levels into doublets is called fine-structure splitting. The degeneracy of the upper level is $(2j + 1) = 2\ell + 2$ while the degeneracy of the lower level is $2j + 1 = 2\ell$. Notice that the spin-orbit energy times the degeneracy of the upper and lower levels cancel, so that the energy "center of gravity" is zero. Also, the smaller j value is lower in energy. This is shown in Fig. 1.7 for sodium.

We may evaluate the average value of $1/r^3$ for the trajectory assuming hydrogenic wavefunctions:

$$\left\langle \frac{1}{r^3} \right\rangle = \int R_{n\ell}^*(r) \frac{1}{r^3} R_{n\ell}(r) r^2 \, dr$$

$$= \frac{Z^3}{a_1^3 n^3 \ell(\ell + 1)(\ell + 1/2)}$$

$$(1.63)$$

Using the value of E_n for hydrogen, we find

$$E(\text{spin-orbit}) = \frac{Z^2 |E_n| \alpha^2}{n(2\ell + 1)(\ell + 1)\ell} \times \left\{ \begin{array}{c} \ell \\ -(\ell + 1) \end{array} \right\} \quad \begin{array}{c} \text{for } j = \ell + \dfrac{1}{2} \\[2mm] \text{for } j = \ell - \dfrac{1}{2} \end{array} \quad (\ell \neq 0)$$

$$(1.64)$$

in terms of fine-structure constant $\alpha = e^2/\hbar c$. Thus the spin-orbit splitting of a given energy level is of the order of 1 part in $1/\alpha^2 \sim 10^4$. For high n and ℓ where the hydrogenic wavefunction will be appropriate to the alkaline metals, this can be 1 cm^{-1} or less. For example, the spin-orbit splitting of the 3p level of sodium is 17.2 cm^{-1}, whereas for the 6p level it is 1.4 cm^{-1}. For alkali atoms where n and ℓ are not large, we should replace Ze^2/r^3 by $(1/r)(dV/dr)$, where V is the Coulomb potential.

For hydrogen there is an additional relativistic fine-structure splitting due to the high velocity of the electron at the pericenter of the ellipse for low ℓ. This splitting for hydrogen is comparable to the spin-orbit splitting. For alkaline metals the velocity of the optically active electron is not relativistically important, especially for high n. The fine-structure doublets are labeled according to the *terms* $^{2s+1}\ell_j$ as shown in Fig. 1.7. The superscript 2s + 1 denotes the spin degeneracy, which, except for $\ell = 0$, is the number of fine-structure levels.

Besides the previously mentioned one-electron selection rule $\Delta \ell = \pm 1$, the fine-structure transitions also obey the following selection rule for atoms with one optically active electron:

$$\Delta j = 0, \pm 1 \tag{1.65}$$

The rules above lead to only two or three allowed fine-structure transitions for each member of a series, as shown in Fig. 1.7. The sodium D line is a doublet split by some 17.2 cm^{-1} corresponding to the spin-orbit splitting of the 3p configuration.

The boron group of elements also exhibit doublet and triplet fine-structure transitions. This group has one p electron outside a filled s subshell, so that the ground state will be a fine-structure split pair of levels.

1.3.5 Fine Structure for Two Optically Active Electrons

The carbon group of elements C, Si, Ge, Sn, and Pb have two optically active electrons in p subshells with n = 2, 3, 4, 5, and 6, respectively. Since for a given carbon group element the two electrons have identical quantum numbers n and ℓ, the electrons can only be distinguished by their spatial and spin quantum numbers m and m_s. The total wavefunction for the two electrons is some linear combination of the product of the wavefunctions for each electron,

$$\psi_{n\ell m}(1)\psi_{n\ell m'}(2)\phi_{m_s}(1)\phi_{m_s'}(2)$$

where $\psi(i)$ and $\phi(i)$ are the orbital and spin wavefunctions for the ith electron.

The proper linear combination or "coupling" of wavefunctions depends on the relative strength of the electrostatic or Coulomb interactions between electrons and the spin-orbit interactions. For a low-Z atom such as carbon the electrostatic interaction will be larger than the spin-orbit interaction in the ground-state configuration, resulting in a coupling scheme called Russell-Saunders or *LS coupling*. For a high-Z atom such as lead, the spin-orbit interaction dominates, resulting in the *jj-coupling* scheme. Both of these are described below. In both schemes one must be careful not to violate the Pauli exclusion principle.

1.3.5A. LS coupling

Let us ignore the spin-orbit interaction for the moment and consider only the energy due to the electrostatic interaction between electrons. The electrostatic interaction is independent of the rotation of all the electrons by the same angle. As a result the total angular momentum of the electrons $\underline{L}$ will be a constant of the motion. States with the same $\underline{L}$ must have the same energy; that is, $\underline{L}$ is a good quantum number. For two electrons with individual angular moments $\underline{\ell}_1$ and $\underline{\ell}_2$, we have

$$\underline{L} = \underline{\ell}_1 + \underline{\ell}_2 \qquad (1.66)$$

and

$$L_z = \ell_{1z} + \ell_{2z} = M_L \hbar \qquad (1.67)$$

where

$$M_L = -L, \; -L + 1, \ldots, L \qquad (1.68)$$

The spatial quantum numbers for total angular momentum obey the same conditions as before. The classical limits on $|L|$ are determined by whether $\underline{\ell}_1$ and $\underline{\ell}_2$ point in the same or opposite directions or

$$|\ell_1 + \ell_2| > L > |\ell_1 - \ell_2|$$

From Eqs. (1.67) and (1.68), the quantum limits on L are similar to the classical ones. We have

$$L = |\ell_1 + \ell_2|, \; |\ell_1 + \ell_2 - 1|, \; \ldots, \; |\ell_1 - \ell_2|$$

for the quantum addition of two angular momenta.

The same rules apply for the coupling of two spins $\underline{s}_1$ and $\underline{s}_2$, so that

$$\underline{S} = \underline{s}_1 + \underline{s}_2 \tag{1.69}$$

or

$$S = 0 \text{ or } 1 \tag{1.70}$$

and

$$M_S = -S, \; -S + 1, \; \ldots, \; S \tag{1.71}$$

For example, with two $\ell = 1$ p electrons as in the case of carbon, the total orbital angular momentum, according to the rules above, can be $L = 0$, 1, or 2. It is customary to label the states according to the convention ^{2S+1}L. Thus for two p orbitals we can have ^1S, ^1P, and ^1D states for $S = 0$ and ^3S, ^3P, and ^3D states for $S = 1$.

However, not all of these states will be allowed according to the Pauli exclusion principle. The fact that no two electrons can be in the same state means that the total wavefunctions must be antisymmetric with respect to interchange of particles. This introduces combinatorial complications which will exclude some L for electrons in the same subshell. We shall introduce a tableau notation for treating electrons in subshells which greatly simplifies the combinatorics resulting from the Pauli exclusion principle and is especially useful for the treatment of complex spectra. Underlying this notation is the fundamental structure of permutational and unitary transformation theory [Harter, 1976].

For two electrons there are only two ways to antisymmetrize the total wavefunction and still produce states of total angular momentum L and S. We may symmetrize the orbital wavefunctions and antisymmetrize the spin wavefunctions with respect to particle interchange, and vice versa. We define a tableau notation such that rows represent symmetrization and columns represent antisymmetrization. We write

$$\boxed{m\,|\,m'} \equiv \frac{1}{\sqrt{2}} [\psi_{n\ell m}(1)\psi_{n\ell m'}(2) + \psi_{n\ell m}(2)\psi_{n\ell m'}(1)]$$

$$\boxed{\begin{array}{c} m \\ m' \end{array}} \equiv \frac{1}{\sqrt{2}} [\psi_{n\ell m}(1)\psi_{n\ell m'}(2) - \psi_{n\ell m}(2)\psi_{n\ell m'}(1)]$$

for the symmetrized and antisymmetrized orbital states for a given $n\ell$ and similarly

$$\boxed{\text{m}_\text{s}\,|\,\text{m}'_\text{s}} \equiv \frac{1}{\sqrt{2}} [\phi_{m_s}(1)\phi_{m'_s}(2) + \phi_{m_s}(2)\phi_{m'_s}(1)]$$

$$\boxed{\begin{array}{c}\text{m}_\text{s}\\ \hline \text{m}'_\text{s}\end{array}} \equiv \frac{1}{\sqrt{2}}[\phi_{m_s}(1)\phi_{m'_s}(2) - \phi_{m_s}(2)\phi_{m'_s}(1)]$$

for the symmetrized and antisymmetrized spin states.

Note that if two quantum numbers in a column are the same (m = m' or $m_s = m'_s$), then the antisymmetrized wavefunction is zero. That is, we cannot antisymmetrize particles with identical quantum numbers. Thus we have the rule that *quantum numbers in tableau columns must be different.* There is no such restriction for tableau rows. However, if we interchange the quantum numbers in the rows or column we obtain the same wavefunction (although the overall sign can be reversed). To avoid repetition we *order quantum numbers in the columns decreasing to the right and in the rows decreasing downward.*

Following these transparent conventions we show the possible total spin and orbital angular momentum tableaus in Fig. 1.8 for a p^2 configuration. The total M_L or M_S may be found simply by adding the m or m_s in the tableau boxes. For the antisymmetric S = 0 spin tableau there are six possible symmetric orbital tableaus. From the M_L of these six tableau we see that they correspond to L = 2 and L = 0, although the linear combinations for the M_L = 0 tableaus are not specified. The total wavefunction is simply a product of these antisymmetrized spin and symmetrized orbital tableaus giving rise to states ^{1}D and ^{1}S. Similarly, for the three symmetric S = 1 spin tableaus there are three possible antisymmetric orbital tableaus with L = 1 giving rise to a ^{3}P state. Thus the states ^{1}P, ^{3}S, and ^{3}D are not allowed by the Pauli exclusion principle for a p^2 configuration.

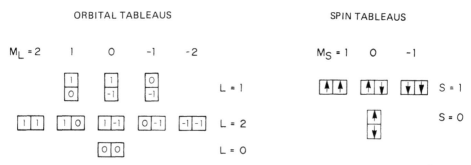

FIG. 1.8 Spin tableaus and the conjugate orbital tableus for an np^2 configuration. The M_S and M_L of the tableaus determine the possible S and L, respectively.

A schematic for the first five energy levels of the carbon atom is shown on the left of Fig. 1.9. In the LS-coupling limit the ordering of the energy levels is determined by Hund's rules, which state:

1. States with the higher total spin S are lower in energy.
2. For a given S, states with higher total L are lower in energy.

Hund's rules are usually applicable to the ground-state configuration, although they are often violated for excited-state configurations. Hund's first rule is a direct result of the Pauli principle. In our case, the S = 1 spin states correspond to antisymmetrized orbital states, where the probability is small for two electrons to occupy nearby positions. Thus for S = 1 states the electrons are farther apart from each other, where the Coulomb interaction is weaker and the energy is lower. The fact that electrons tend to be more isolated from each other with high S is referred to as the correlation or residual Coulomb interaction.

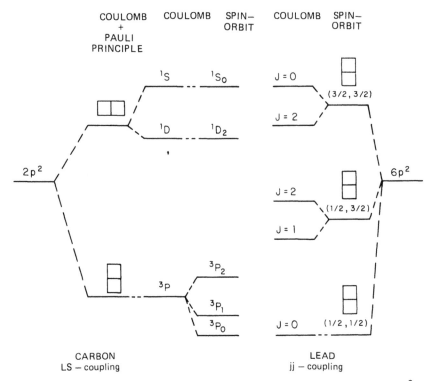

FIG. 1.9 Schematic of the fine-structure energy levels of the $2p^2$ and $6p^2$ configuration in the LS-coupling (carbon) and jj-coupling (lead) limits, respectively.

The spin-orbit interaction will split a given LS level into *multiplets* labeled by their total angular momentum $\underline{J}$, where

$$\underline{J} = \underline{L} + \underline{S} \tag{1.72}$$

as shown in Fig. 1.9 for the carbon atom. The possible values of J are determined again by the coupling relations:

$$J = |L + S|, \; |L + S - 1|, \; \ldots, \; |L - S| \tag{1.73}$$

Thus for $L = 1$ and $S = 1$ we have three *terms* $^{2S+1}L_J = {}^3P_0, {}^3P_1,$ and $^3P_2$. Thus $2S + 1$ is the number of terms in a multiplet (for $S \leqslant L$).
The spin-orbit energy is proportional to

$$E(\text{spin-orbit}) \propto \sum_i \underline{s}(i) \cdot \underline{\ell}(i) \tag{1.74}$$

where the sum is over the number of electrons and $\underline{s}(i)$ or $\underline{\ell}(i)$ are the spin and angular momentum for the ith electron. As in the case of one optically active electron, this leads to an energy dependence given by

$$E(\text{spin-orbit}) \propto [J(J + 1) - L(L + 1) - S(S + 1)] \tag{1.75}$$

for a given L and S for which the spacing of adjacent levels is proportional to the J of the higher level. This is called the Landé interval rule. Also, because the constant of proportionality in Eq. (1.75) is positive, as in the one-electron case, the lower J in the multiplet has the lower energy.

The LS-coupling selection rules for electric dipole transitions are the same as the one-electron rules. Since the electric dipole is independent of spin, we have

$$\Delta S = 0 \tag{1.76}$$

The foregoing condition is rigorous for higher-order transition moments as well (magnetic dipole, quadrapole, etc.). Thus transitions from singlet to triplet levels are strictly forbidden to the extent that the LS-coupling scheme is valid. States for which transitions to lower levels are spin forbidden are called metastable states since they are long lived. However, for large spin-orbit interactions as in the jj-coupling scheme to the right of Fig. 1.9, there will be an appreciable mixing of the total electron spin S within each energy level. The remaining selection rules for LS-coupling transitions are

$$\Delta L = 0, \pm 1 \tag{1.77a}$$

$$\Delta J = 0, \pm 1 \quad \text{not } J = 0 \rightarrow J = 0 \tag{1.77b}$$

Electric dipole transitions must occur between configurations of opposite parity $\Pi(-1)^{\ell_i}$. In general, transitions cannot occur within the same configuration.

1.3.5B. jj coupling

We have seen that as Z increases so does the interaction between the electron spin and orbital angular momentum. For lower-energy configurations of heavy atoms, such as lead, and higher-energy configurations of light atoms, the spin-orbit interaction is stronger than the electrostatic interaction between electrons. When this happens the terms $\underline{s}(i) \cdot \underline{\ell}(i)$ in Eq. (1.74) cause the spin and orbital angular momentum of each electron to couple, giving total angular momentum j_i for the ith electron. For two electrons we have $j_1 = \ell_1 \pm 1/2$ and $j_2 = \ell_2 \pm 1/2$ with energy

$$E(\text{spin-orbit}) = \alpha \left[j_1(j_1 + 1) - \ell_1(\ell_1 + 1) - \frac{3}{4} \right]$$

$$+ \left[\beta\ j_2(j_2 + 1) - \ell_2(\ell_2 + 1) - \frac{3}{4} \right] \tag{1.78}$$

The spin-orbit states may be labeled by (j_1, j_2) and, according to Eq. (1.78), the states with lower (j_1, j_2) will have lower energy, where in general α and β are positive for less-than-half-filled subshells.

On the right of Fig. 1.9 is shown a schematic of the spin-orbit energy levels for a heavy atom such as lead. The spin-orbit levels will be split by the Coulomb interacts to form levels labeled by total angular momentum $\underline{J} = \underline{J}_1 + \underline{J}_2$, where

$$J = |j_1 + j_2|, \ |j_1 + j_2 - 1|, \ldots, \ |j_1 - j_2| \tag{1.79}$$

subject to the constraint of the Pauli exclusion principle that the resultant wavefunction be antisymmetric with respect to exchange of particles (denoted by tableau columns in Fig. 1.9). When $j_1 = j_2$ ($\ell_1 = \ell_2$) and the particles are identical, not all of the J values above will be allowed by the exclusion principle. The splitting of the (j_1, j_2) spin-orbit pair into J levels is shown on the right of Fig. 1.9.

The selection rules for transitions in the jj-coupling limit are

$$\Delta j_1 = 0, \pm 1$$

$$\Delta j_2 = 0, \pm 1 \tag{1.80}$$

$$\Delta J = 0, \pm 1 \ \text{not} \ J = 0 \rightarrow J = 0$$

where as in the LS-coupling limit, electric dipole transitions must occur between configurations with opposite parity. We should note that in configurations of both excited- and ground-state atoms neither the LS-coupling nor the jj-coupling limits are appropriate when the spin-orbit interaction and electrostatic interaction are nearly equal. There also exist intermediate coupling schemes [Sobelman, 1979].

The $\Delta J = 0, \pm 1$ selection rule is valid for either coupling scheme since J is always a good quantum number for a free atom. Since J is always known for the ground electronic state of a neutral atom, this selection rule may be used to probe excited states with several lasers to determine the structure of states that are pumped by a laser.

1.3.6 Fine Structure for More Than Two Optically Active Electrons

When there are more than two electrons in an open subshell the tableau notation greatly simplifies the enumeration of states allowed by the Pauli exclusion principle. Since the LS-coupling scheme is more often encountered, especially in the ground-state subshell, we shall develop only this limit here.

We have seen how tableau rows represent symmetrization and tableau columns represent antisymmetrization. When the number of electrons N in a subshell is greater than two, it is possible to have permutational characteristics other than symmetrization or antisymmetrization, which again can be represented by tableaus with N boxes, as shown in Fig. 1.10. Since we are not allowed to have identical quantum numbers in tableau columns, we cannot have columns longer than the number of quantum numbers that differentiate a state. For spin states where $2s + 1 = 2$ we are limited to tableaus with two rows or less. For orbital states we are limited to tableaus with $2\ell + 1$ rows or less. The last characterization of tableaus with N boxes is that the number of boxes in successive rows be monotonically decreasing. These rules are obeyed by the spin and orbital tableaus illustrated in Fig. 1.10.

The tableaus uniquely characterize the permutational characteristics of the spin and orbital wavefunctions under interchange of particles. In order for the product of the spin and orbital wavefunctions to be antisymmetric, the *tableau frame for the orbital wavefunction must be conjugate to the tableau frame for the spin wavefunction* as in Fig. 1.10. The conjugate orbital frame is obtained by exchanging the rows with the columns of the spin frame as shown in Fig. 1.10. Note that in Fig. 1.9 the two electron spin and orbital tableau frames are conjugate to each other. The number of columns of the spin tableau is the number of rows of the orbital tableau, and vice versa. Thus orbital tableaus conjugate to spin tableaus are limited to two columns. This is a statement of the Pauli exclusion principle in tableau notation for electrons.

(a)
Orbital
Frame

(b)
Spin
Frame

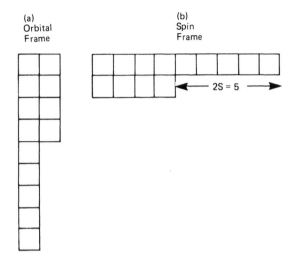

FIG. 1.10 The orbital frame (a) must be conjugate to the spin frame (b) by exchanging rows and columns. The number of unpaired boxes in the columns of the spin frame determines the total spin quantum number S as shown.

The number of unpaired boxes in the spin tableau columns corresponds to the total number of spin states $2S + 1$ as seen in Figs. 1.10 and 1.8. Thus each spin tableau frame uniquely specifies the spin S.

Let us now consider the possible total spin and orbital angular momenta quantum numbers S and L for the np^3 configuration of the nitrogen group. On the right of Fig. 1.11 we see there are two possible spin tableau frames, corresponding to the $S = 3/2$ and $S = 1/2$ quartet and doublet states, for which we have enumerated the possible tableau and M_S for each spin frame. On the left of this figure are the conjugate orbital frames with the possible orbital tableau and M_L for each frame. From the values of M_L we determine the total angular momenta $L = 2$ or 1 for $S = 1/2$ and $L = 0$ for $S = 3/2$, giving rise to terms ^4S, ^2D, and ^2P in order of increasing energy according to Hund's rules. As shown in Fig. 1.12, the spin-orbit interaction then results in terms $^4S_{3/2}$, $^2D_{3/2}$, $^2D_{5/2}$, $^2P_{1/2}$, and $^2P_{3/2}$ in order of increasing energy.

So far we have considered the lithium, boron, carbon, and nitrogen groups, which have half- or less-than-half-filled subshells in s and p orbitals. The tableau notation allows the determination of the spectroscopic terms for more than half-filled subshells by considering them as electron "holes" in the less than half-filled subshell.

The orbital and spin tableau for a filled d subshell is shown in Fig. 1.13. There are 10 boxes in these tableaus, corresponding to

ORBITAL TABLEAUS   SPIN TABLEAUS

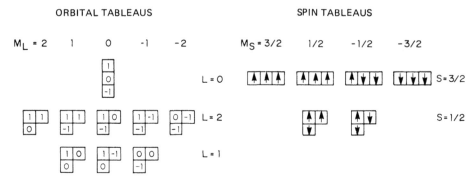

FIG. 1.11 Spin tableaus and the conjugate orbital tableaus for an np^3 configuration. The M_S and M_L of the tableaus determine the possible S and L, respectively.

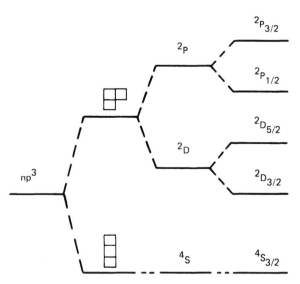

FIG. 1.12 Schematic of the fine-structure energy levels of an np^3 configuration in the LS-coupling limit. The residual Coulomb splitting is indicated with tableaus. The spin-orbit splitting of the ^{2S+1}L levels leads to the $^{2S+1}L_J$ terms shown on the far right.

ORBITAL TABLEAU   SPIN TABLEAU

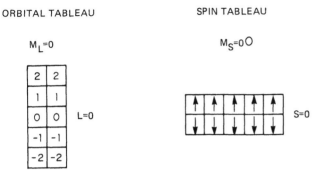

FIG. 1.13 Spin tableau and conjugate orbital tableau for a filled d subshell.

the $2(2\ell + 1) = 10$ electrons. Since there are no unpaired electrons, there is only one allowed tableau (with $M_S = 0$), which corresponds to total spin angular momentum $S = 0$. Similarly, the conjugate orbital frame has only one tableau (with $M_L = 0$), which corresponds to total orbital angular momentum $L = 0$. For $L = 0$ the total orbital wavefunction must be spherically symmetric. Hence the term arising from this configuration is $^1S_0$. In general, electron cores made up of filled subshells give rise to spherically symmetric potentials, as indicated previously.

The tableau notation greatly simplifies the combinatorics of the Pauli principle for more-than-half-filled subshells by means of the *associated tableau* as defined in Fig. 1.14. We see that the associated orbital tableau has $L_z = -M_L\hbar$, which is negative that of the original tableau. Similarly, the associated spin tableau has $S_z = -M_S\hbar$, which is negative that of the original tableau. Thus the associated tableaus contain the same spin and angular momentum terms as the original tableaus. This means that a more-than-half-filled subshell with N electrons (where $N > 2\ell + 1$) has the same terms as the less-than-half-filled subshell with $4\ell + 2 - N$ electrons.

The fluorine group has the same terms in their np^5 configuration as the np configuration of the lithium group, and the oxygen group has the same terms in their np^4 configuration as the np^2 configuration of the carbon group. The ordering of L and S levels will follow Hund's rules described above. However, the ordering of the J levels will be reversed, so that atoms with more-than-half-filled subshells have inverted fine structure. This is because an electron hole acts like a positively charged electron, changing the sign for spin-orbit coupling energies in Eq. (1.75). As a result, the oxygen ground state has terms $^3P_2$, $^3P_1$, $^3P_0$ in inverse order to carbon in Fig. 1.9.

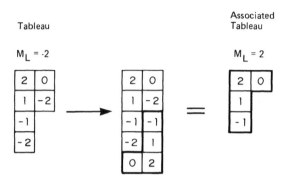

FIG. 1.14 The associated tableaus for the half-filled d^6 configuration may be found by subtraction from a filled subshell as shown. The M_L of the associated tableau has the opposite sign of the original tableau and corresponds to electron "holes" in the less-than-half-filled d^4 configuration for which the spectroscopic terms are more easily calculated.

1.3.7 Hyperfine Structure

So far, J has been a good quantum number independent of coupling scheme. This may no longer be true when one considers the intrinsic spin $\underline{I}$ of the atomic nucleus. The interaction of the nuclear magnetic moment with the magnetic field produced by orbiting electrons gives rise to atomic hyperfine structure. Since the magnetic moment of the nucleus is much smaller than the magnetic moment of the electron, the hyperfine structure splitting J levels is much less than the fine-structure splitting of L levels.

The nuclear magnetic moment $\underline{\mu}_I$ due to the nuclear spin angular momentum $\underline{I}$ is given by

$$\underline{\mu}_I = \frac{g_I u_N}{\hbar} \, \underline{I} \tag{1.81}$$

where the nuclear magneton μ_N is defined by

$$\mu_N = \frac{e\hbar}{2Mc} = \frac{1}{1836} \frac{e\hbar}{2mc} = \frac{1}{1836} \mu_B \tag{1.82}$$

The factor $1/1836$ is ratio of the mass of the electron to the mass of a nucleon (proton or neutron).

The nuclear spin-orbit interaction is analogous to the electron spin-orbit interaction given by Eq. (1.75), although it is roughly 2000 times weaker. Typically, hyperfine splittings due to nuclear magnetic dipole interactions are of the order of the Doppler line widths or less, and in these cases sub-Doppler spectroscopic techniques must be used to resolve them completely.

In analogy to Eq. (1.75), the hyperfine interaction couples the nuclear spin angular momentum $\underline{I}$ and the electron angular momentum $\underline{J}$ to give total angular momentum $\underline{F}$, where $\underline{F} = \underline{J} + \underline{I}$ with energy.

$$E(\text{nuclear spin-orbit}) = A[F(F + 1) - J(J + 1) - I(I + 1)] \quad (1.83)$$

We see that the hyperfine structure levels obey Landé interval rules, similar to the fine-structure levels. Some of the hyperfine structure levels are shown for sodium, where $I = 3/2$ in Fig. 1.7 and have been observed in saturation spectra [Hänsch, 1971].

There is also a smaller hyperfine splitting in atoms when $I \geqslant 1$ due to the nuclear quadrupole moment interaction with the electric field gradient of the electrons. For sodium this interaction is negligible and the Landé interval rule is obeyed.

The selection rules for hyperfine transitions for both LS and jj coupling are

$$\Delta J = 0, \pm 1$$

$$\Delta F = 0, \pm 1 \quad \text{not } F = 0 \to F = 0 \quad\quad\quad\quad (1.84)$$

These rules result in four possible hyperfine transitions, comprising the sodium fine-structure D "doublet" shown in Fig. 1.7.

Since laser bandwidths are very narrow, a laser can often access one hyperfine F component of a given J multiplet. The selection rules above are consequently very important when several lasers are used to ionize an atom by multistep excitation.

1.3.8 Isotope Effects

A change in the mass of the nucleus due to the number of neutrons will change the electrostatic energy, causing different isotopes to have "mass-shifted" levels. Shifting of levels also results from the fact that the nucleus is not a point source and the Coulomb potential will change in the neighborhood of the nucleus. The different nuclear volumes of different isotopes cause isotopic-level shifts. The volume effect is largest for s and p electrons, which spend appreciable time near the nucleus. Isotopic shifts in light elements are due primarily to mass effects, while those of heavy elements are due primarily to volume effects.

As we have seen, the hyperfine splitting depends on the nuclear magnetic and quadrupole moments, both of which can change with isotopes. Even-even (proton-neutron numbers) nuclei have zero nuclear magnetic moment. Thus uranium ^{238}U, for example, has I = 0 and no quadrupole moment, while ^{235}U has I = 7/2 with septuplet hyperfine splittings (for J $\geqslant$ I).

1.4 VIBRATIONAL STRUCTURE OF POLYATOMIC MOLECULES

The motion of electrons in a molecule are so rapid that in studying the electronic properties of molecules the nuclei may be regarded as fixed. One may then, in principle, calculate the electronic potential for any given internuclear configuration during the vibration to see what forces the nuclei experience. Such a vibrational potential is a result of the *Born-Oppenheimer* approximation, which assumes that the electrons have time to equilibrate for any internuclear configuration.

This vibrational potential will be proportional to the internuclear displacements from the nuclear equilibrium configuration of the molecule, so that low-amplitude vibrational motion will be harmonic. We discuss below the properties of a one-, two-, and three-dimensional harmonic oscillator and apply these properties to vibrational excitation of molecules with intense radiation fields—or multiple photon excitation. As in the atomic case, we use the semiclassical quantization conditions to determine the discrete energy levels of the harmonic oscillator. The classical picture, again, is most appropriate for the high vibrational excitations that will be considered here. For a review, see the book by Herzberg [Herzberg, 1945].

1.4.1 The Harmonic Oscillator

1.4.1A Semiclassical Quantization

For a one-dimensional harmonic oscillator, the force on a particle is proportional to the displacement x according to Hooke's law, or

$$F = m\ddot{x} = -kx$$

where k is Hooke's constant. Solving for the displacement, we find

$$x = a \sin 2\pi\nu t \tag{1.85a}$$

$$\nu = 2\pi \sqrt{\frac{k}{m}} \tag{1.85b}$$

where a is the amplitude of the oscillation and ν is its frequency. Integrating over one period $t = 1/\nu$, the semiclassical quantization condition becomes

$$\int p_x dx = n_x h \quad n_x = 0, 1, 2\ldots$$

or

$$m \int_0^t (\dot{x})^2 dt = 2\pi^2 m \nu a_{n_x}^2 = n_x h \tag{1.86}$$

The energy of the harmonic oscillator becomes

$$E_{n_x} = \frac{1}{2} m \dot{x}^2 + \frac{1}{2} kx^2 = 2\pi^2 m \nu^2 a_{n_x}^2$$

$$= n_x h \nu \quad n_x = 0, 1, 2, \ldots \tag{1.87}$$

Thus the quantum energy levels for an harmonic oscillator are equally spaced, forming a vibrational "ladder." The amplitude, a, depends only on the square root of the quantum number n_x.

For a three-dimensional harmonic oscillator in Cartesian coordinates we have

$$F_x = m\ddot{x} = -k_x x$$

$$F_y = m\ddot{y} = -k_y y \tag{1.88}$$

$$F_z = m\ddot{z} = -k_z z$$

The three semiclassical quantum conditions are

$$\int p_x dx = n_x h \quad n_x = 0, 1, 2, \ldots$$

$$\int p_y dy = n_y h \quad n_y = 0, 1, 2, \ldots \tag{1.89}$$

$$\int p_z dz = n_z h \quad n_z = 0, 1, 2, \ldots$$

If the three-dimensional oscillator is isotropic such that $k_x = k_y = k_z$, the energy is

$$E_n = \frac{1}{2}m(\dot{x}^2 + \dot{y}^2 + \dot{z}) + \frac{1}{2}k(x^2 + y^2 + z^2)$$

$$= (n_x + n_y + n_z)h\nu$$

$$= nh\nu \qquad\qquad (1.90)$$

which follows from Eqs. (1.89) in analogy to the one-dimensional case. Again, the quantum energy levels are equally spaced, as shown on the right of Fig. 1.15. For the two-dimensional case we have $n = n_x + n_y$. In the center of mass coordinates we may replace m by the reduced mass μ.

For three dimensions the number of states for a given energy, or the degeneracy, is simply the number of different n_x, n_y, and n_z for a given n or

$$\text{degeneracy } (E_n) = \frac{(n + 1)(n + 2)}{2} \qquad\qquad (1.91)$$

Solving Schrödinger's equation for the quantum mechanical energies one arrives at the same result, Eq. (1.90), with $n_x \to n_x + 1/2$, $n_y \to n_y + 1/2$, and $n_z \to n_z + 1/2$. Thus when $n_x = n_y = n_z = 0$ there is still a *zero-point* energy $(3/2)h\nu$ for the three-dimensional oscillator.

The classical trajectories for the three-dimensional isotropic oscillator are ellipses tangent to a box with sides $\pm a_{n_x}$, $\pm a_{n_y}$, $\pm a_{n_z}$. As in the case of the hydrogen atom, it is worthwhile to quantize the three-dimensional harmonic oscillator in a spherical coordinate system with quantization conditions given by Eq. (1.41). The integrals for p_ϕ and p_θ, as in the case of the hydrogen atom, give rise to the spatial and azimuthal quantum numbers m and ℓ. Again ℓ is the angular momentum of the oscillator, which is perpendicular to the plane of the ellipse and $\ell_z = m\hbar$. The integral for p_r determines the eccentricity of the ellipse such that the semimajor axis a and semiminor axis b of the ellipse are given by

$$a^2 b^2 = \ell^2 \qquad\qquad (1.92a)$$

$$\frac{a^2 + b^2}{2} = n \qquad\qquad (1.92b)$$

where a and b are in units of $a_1 = \sqrt{h/2\pi m\nu}$, where a_1 is the radius of a circular orbit (a = b) for n = 1 as the Bohr radius is defined for the hydrogen atom. The limits of m and ℓ are

$$m = -\ell, -\ell + 1, \ldots, \ell \qquad\qquad (1.93a)$$

$$\ell = n, n - 2, \ldots, 1 \text{ or } 0 \qquad\qquad (1.93b)$$

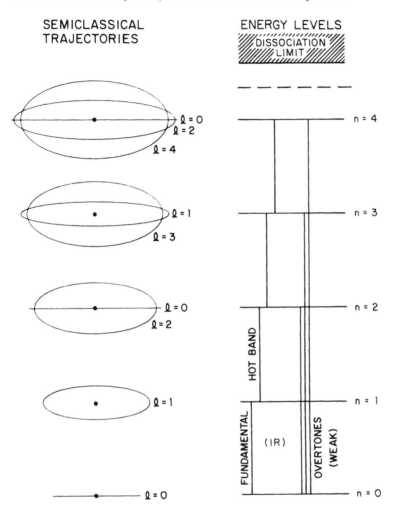

FIG. 1.15 Energy levels and semiclassical trajectories for the three-dimensional, isotropic, harmonic oscillator. For a given quanta n of the oscillator, the eccentricity of the semiclassical elliptical orbits is determined by the angular momentum quantum number ℓ of the vibration. The semimajor axis, a, and semiminor axis, b, have a $\sqrt{n}$ dependence. The strong electric dipole transitions allow sequential steps up the vibrational ladder, whereas the relatively weak overtone transitions can skip rungs of the ladder.

so that ℓ is even (odd) whenever n is even (odd). These limits preserve the degeneracy for a given n in Eq. (1.91). For the two-dimensional oscillator m = $\pm\ell$ and ℓ may have any integer value given in Eq. (1.93b). On the left of Fig. 1.15 we draw the corresponding elliptical trajectories for the first few levels of the three-dimensional harmonic oscillator using n $\rightarrow$ n + 3/2 and $\ell^2 \rightarrow \ell(\ell + 1)$ for the corresponding quantum solutions.

1.4.1B. Wavefunctions

In Cartesian coordinates the wavefunction for the three-dimensional harmonic oscillator is the product of the three wavefunctions for the one-dimensional harmonic oscillator, namely

$$\Psi_n = \psi_{n_x}(x)\psi_{n_y}(y)\psi_{n_z}(z)$$

As mentioned previously, for a spherically symmetric interaction, the angular dependence of the wavefunction in spherical coordinates is given by the spherical harmonics $\psi_{\ell m}(\theta,\phi)$. The only difference in the wavefunctions of the hydrogen atom and the three-dimensional harmonic oscillator will be in their radial functions $R_{n\ell}(r)$. The quantum mechanical probability of finding the three-dimensional harmonic oscillator at distance r is compared with the classical probability in Fig. 1.16. Here r is measured from the center of the ellipse, so that the classical turning points are located at the semimajor and semiminor axes a and b. Note that the classical turning points give roughly the limits of the quantum probabilities. Also, for large n the classical distribution is the average of the quantum mechanical distribution. Again, we see that the most radially localized states occur for ℓ = n since the classical orbits are nearly circular.

1.4.1C. Transitions

We may expand the electric dipole in terms of the displacement x from the equilibrium position of the harmonic oscillator

$$\underline{\mu}(x) = \underline{\mu}(0) + \underline{\mu}'(0)x + \cdots \tag{1.94}$$

The first term on the right is the permanent dipole responsible for the rotational transitions of the molecule. The second term is the dipole derivative responsible for vibrational transitions. For molecules with permanent dipoles, $\underline{\mu}(0)$ can be measured in debye while dipole derivatives are usually tenths of debye. The dipole matrix elements between harmonic oscillator wavefunctions $\psi_i(x)$,

$$\underline{\mu}_{if}(x) = \underline{\mu}(0)\int \psi_i\psi_f \, dx + \underline{\mu}'(0)\int \psi_i \, x \, \psi_f \, dx \tag{1.95}$$

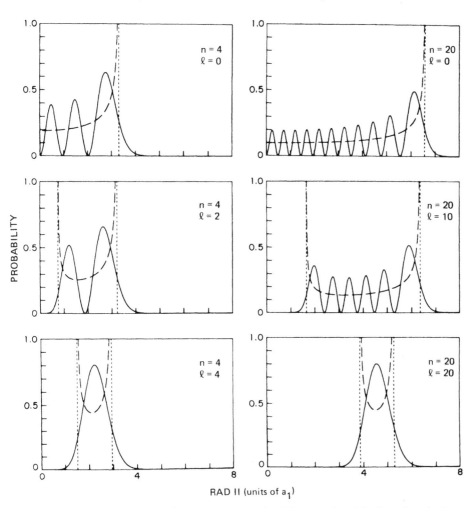

FIG. 1.16 Comparison of the quantum (solid curve) with the classical (dashed curve) probabilities for finding the three-dimensional, iso-tropic, harmonic oscillator displaced distance r. For high n the classical probability distribution is the average of the quantum distribution. The classical turning points at the semiminor axis a and the semimajor axis b of the respective ellipses (see Fig. 1.15) determine the limits of the distribution as shown by the vertical dashed lines.

lead to the following selection rules

$$\text{Permanent dipole } \underline{\mu}(0): \quad \Delta n_x = 0 \tag{1.96a}$$

$$\text{Dipole derivative } \underline{\mu}'(0): \quad \Delta n_x = \pm 1 \tag{1.96b}$$

Similar selection rules result for the three-dimensional harmonic oscillator. For vibrational transitions in terms of Cartesian coordinate wavefunctions we have

$$\Delta n_i = \pm 1 \quad i = (x, \, y, \, \text{or } z) \tag{1.97a}$$

$$\Delta n = \pm 1 \tag{1.97b}$$

whereas in terms of the spherical coordinate wavefunctions,

$$\Delta m = 0, \, \pm 1 \tag{1.98a}$$

$$\Delta \ell = \pm 1 \tag{1.98b}$$

$$\Delta n = \pm 1 \tag{1.98c}$$

Higher-order terms in the dipole expansion, Eq. (1.94), result in higher allowed Δn. We shall see later that anharmonic terms in the potential energy of order x^3, x^4, etc., also enable higher allowed Δn.

Normally, at room temperature most of the molecules are in the ground ($n = 0$) vibrational state. Transition from the ground state to $n = 1$ are called fundamental transitions, while the weaker transitions to $n = 2$ or 3, for example, with $\Delta n = 2$ or 3 are called the first or second overtone transitions, as shown in Fig. 1.15. Also shown are transitions arising from thermally populated vibrational levels with $n > 0$ (hot bands). For equally spaced vibrational levels the hot bands will have the same frequencies as the fundamentals and overtones. The relative population between two vibrational levels is given by the Boltzmann distribution

$$\frac{N_i}{N_j} = \frac{g_i e^{-E_i/kT}}{g_j e^{-E_j/kT}} \tag{1.99}$$

In Fig. 1.15 are shown the possible ladder transitions for the three-dimensional harmonic oscillator. Because of the $\Delta n = \pm 1$ selection rule for strong transitions, a molecule usually absorbs and emits radiation by sequential steps up and down the vibrational ladder without skipping rungs in the ladder.

The classical frequency of oscillation from Eq. (1.53) is $f = \nu$, which is independent of the amplitude of motion when the oscillator is harmonic and the quantum energy levels are equally spaced. For typical vibrational wavelengths, say $\lambda = 10$ μm, the classical period of oscillation is $\tau = 1/2\pi\nu \backsim 5 \times 10^{-15}$ s. In contrast, the period for the $n = 1$ circular orbit of the hydrogen electron is 1×10^{-16} s. Electronic transition frequencies are typically 10 to 100 times greater than vibrational transition frequencies in molecules. As a result the Born-Oppenheimer approximation is valid and we may consider the electronic motion independently of the vibrational motion.

1.4.2 Normal Modes

In general, there are as many independent oscillatory modes in a molecule as there are degrees of freedom. Each such *normal mode* in a molecule can vibrate independently without exciting any other mode for small-amplitude motion. Six of the "modes" have zero frequency and correspond to the six possible overall translations and rotations of the molecule. For a linear molecule, rotation about the axis of symmetry is impossible since it requires infinite energy (for point nuclei) and there are only five such zero-frequency modes. Ignoring these zero-frequency translational and rotational modes, a molecule with N atoms will have 3N degrees of freedom and $3N - 6$ normal-mode frequencies (or $3N - 5$ for linear molecules).

The force on the jth atom of a molecule in the ith direction, F_{ij}, is related to the displacement of the ℓth atom in the kth direction, $x_{k\ell}$, by the molecular force constants and Newton's law:

$$F_{ij} = - \sum_{k=1}^{3} \sum_{\ell=1}^{N} k_{ij;k\ell} x_{k\ell} = m_j \ddot{x}_{ij}$$

or

$$-m^{-1}k\underline{x} = \ddot{\underline{x}} \qquad (1.100)$$

We may diagonalize the acceleration matrix $-m^{-1}k$ to find the normal modes and their frequencies. All zero-frequency modes may be disregarded. For example, let us find the normal modes of a triatomic linear molecule XY_2 shown in Fig. 1.17, such as CO_2. Let the displacements of the particles from equilibrium be x_1, x_2, and x_3 and the masses be m and M, as shown. We ignore the remaining degrees of freedom, which do not correspond to displacements in the x direction. Assuming a force constant k for the X−Y bond, the acceleration matrix is

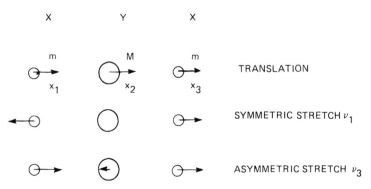

FIG. 1.17 Normal modes in the x direction for a linear XY_2 molecule.

$$-m^{-1}k = \begin{matrix} & x_1 & x_2 & x_3 \\ & \begin{bmatrix} k/m & -k/m & 0 \\ k/M & 2k/M & -k/M \\ 0 & -k/m & k/m \end{bmatrix} \end{matrix} \tag{1.101}$$

Diagonalizing, we find the frequencies corresponding to the normal modes shown in Fig. 1.17:

x translation: $\nu = 0$

$$x_1 = x_2 = x_3 \tag{1.102a}$$

Symmetric stretch: $\nu_1 = \dfrac{1}{2\pi} \sqrt{\dfrac{k}{m}}$

$$x_1 = x_3 \quad x_2 = 0 \tag{1.102b}$$

Asymmetric stretch: $\nu_3 = \dfrac{1}{2\pi} \sqrt{\dfrac{k(2m + M)}{mM}}$

$$x_1 = x_3 \quad x_2 = -\dfrac{2mx_1}{M} \tag{1.102c}$$

The remaining degrees of freedom correspond to the two bending ν_2 modes, the zero frequencies y and z translation, and the two zero-frequency rotations perpendicular to the molecular axis. Since the molecule bends in the y and z directions with the same frequency, the ν_2 mode is a two-dimensional isotropic harmonic oscillator.

The ν_1 frequency of CO_2 has been measured at ~ 1337 cm^{-1}. From the ν_1 frequency we may determine k, giving $\nu_3 = 2114$ cm^{-1}, whereas the actual experimental asymmetric stretch frequency is 2349 cm^{-1} [Herzberg, 1945]. Of course, if one added an additional O—O force constant, one would then have two force constants to determine the two stretching frequencies, which is not very satisfying. This example shows the difficulties in determining spectroscopic parameters from force-field models.

For a large molecule with no symmetry the diagonalization above must be done on a computer. For a molecule with a high degree of symmetry, such as the octahedral molecules SF_6, one may also use the group-theoretical techniques given in standard texts. In Fig. 1.18 we show some of the $3N - 6 = 15$ possible normal modes of SF_6. The modes are labeled according to octahedral symmetry species A_{1i}, A_{2i}, E_i, T_{1i}, and T_{2i}, where i = g or u determine whether the normal mode has even or odd (gerade or ungerade) parity under inversion through the center of symmetry of the octahedron: that is, whether the normal mode remains the same or changes sign after it is inverted through the origin of the octahedron at the S-atom site.

In Fig. 1.18 the A modes are singly degenerate, the E modes are doubly degenerate, and the T modes are triply degenerate. Only one of the degenerate modes is shown for the T species. Note that if one includes the degeneracy the total number of modes is 15, as required.

The modes v_1, v_2, and v_3 are primarily S-F stretching motion, while the modes v_4, v_5, and v_6 are primarily S-F bending motion. Since stretching is stiffer than bending, the stretching modes have higher frequencies than the bending modes. Because of symmetry, the electric dipole active modes must have species T_{1u}, corresponding to the threefold degenerate modes v_3 and v_4. Thus only the v_3 and v_4 modes are infrared active. The higher-frequency v_3 mode in SF_6 and UF_6 has the larger isotope shift, as seen below, and can be pumped with infrared lasers to dissociation in an isotopically selective manner. Before discussing the multiphoton vibrational ladder in these molecules we examine the local-mode character, which differentiates the vibrations of SF_6 and UF_6.

1.4.3 Local Modes

In Table 1.2 we see that the separation of normal modes into two groups of stretching frequencies and bending frequencies is much more distinct in UF_6 than in SF_6. This is primarily because the heavy central uranium atom provides a barrier or filter that prevents the opposite fluorine atoms from interacting. For SF_6 the comparatively light sulfur atom does not provide such an effective barrier. In both molecules the F—F bonds are fairly weak. In the limit of an infinite mass central atom X and no F—F bonds, the v_1, v_2, and v_3 modes would all have the same frequency, corresponding to the six possible degenerate X—F stretches, and the v_4, v_5, and v_6 modes would have

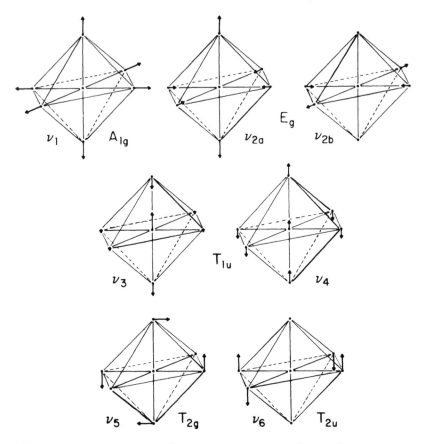

FIG. 1.18 Normal modes of SF$_6$ using a recent force-field model
[McDowell, 1976a]. Only one of each triply degenerate T mode is
shown.

the same frequency, corresponding to the nine possible X−F bends.
Such motion is an example of *local-mode* motion [Halonen, 1983].
Clearly, UF$_6$ is closer to local-mode motion than SF$_6$, as is evident by
comparing the ν_3 and ν_4 modes of SF$_6$ and UF$_6$ in Fig. 1.19.

Also evident in Fig. 1.19 is the fact that the ν_3 motion in UF$_6$ is
primarily an F−U−F asymmetric stretch similar to the ν_3 mode of CO$_2$
in Fig. 1.17. This is true to a lesser extent for the ν_3 mode of SF$_6$
as well. The threefold degeneracy of the ν_3 mode can now be interpreted as belonging to the three asymmetric stretches in the x, y,
and z directions, although only one direction is shown in Figs. 1.18
and 1.19.

TABLE 1.2

	SF$_6$[a] (cm^{-1})	UF$_6$[b] (cm^{-1})
ν_1	775	667
ν_2	643	534
ν_3	948	628
ν_4	615	186
ν_5	524	200
ν_6	348	143

[a]From McDowell et al. [McDowell, 1976a].
[b]From McDowell et al. [McDowell, 1974].

It is also apparent in Fig. 1.19 that of the two infrared active modes, the ν_3 mode will have the largest frequency shift for isotopic substitution of the central atom, since the displacement of the central atom is largest in this mode. Indeed, we may use the formula for the frequency of the asymmetric stretch [Eq. (1.102c)] to estimate the isotope shifts per amu (atomic mass unit) for the ν_3 frequency of SF$_6$ and UF$_6$. Using the known ν_3 frequencies in Table 1.2, we can calculate force constant k and then calculate the new frequency resulting from a new central mass M. For SF$_6$ we find the isotope shift is 7.7 cm^{-1}/amu and for UF$_6$ we find 0.18 cm^{-1}/amu, which can be compared to the published values of 8.7 cm^{-1}/amu for SF$_6$ [McDowell, 1976a] and 0.20 cm^{-1}/amu [Takami, 1984] for UF$_6$.

1.4.4 The Anharmonic Oscillator

Although near equilibrium the molecular potential-energy surface is harmonic, this will no longer be the case far from equilibrium. An empirical "surface" was suggested by Morse in 1929 for a one-dimensional "diatomic" oscillator with reduced mass μ valid for bond lengths up to dissociation. His potential is given by the curve

SF$_6$ UF$_6$

ν_3

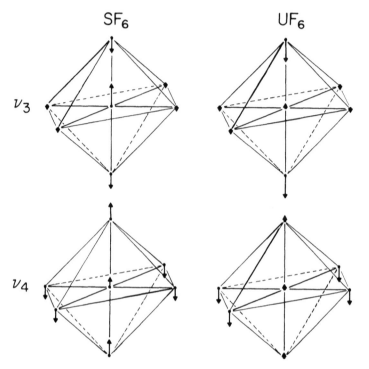

ν_4

FIG. 1.19 Comparison of infrared active normal modes for SF$_6$ and UF$_6$. The vibrations of UF$_6$ are seen to have more of a local-mode character involving pure stretching or bending motion.

$$V(x) = D_e(1 - e^{-\beta x})^2 \qquad (1.103)$$

where D_e is the dissociation energy (in cm^{-1}), which is the limit of the potential for large x. Note that V(x) has a minimum at the equilibrium displacement x = 0.

Morse has shown that the quantized energies for the anharmonic one-dimensional potential above is

$$E_n = \left(n_x + \frac{1}{2}\right)\omega + X\left(n_x + \frac{1}{2}\right)^2 \qquad n_x = 0, 1, 2, \ldots \qquad (1.104)$$

where

$$\nu = E_1 - E_0 = \omega + 2X \qquad (1.105)$$

is the fundamental frequency,

$$\omega = \beta \sqrt{\frac{D_e h}{\pi c \mu}}$$ (1.106a)

is the harmonic frequency, and

$$X = \frac{-\hbar \beta^2}{4\pi c \mu}$$ (1.106b)

is the anharmonicity constant, which is normally negative (all in cm^{-1}).
The second term, on the right of Eq. (1.104), causes an anharmonic
shift from the equally spaced levels of the harmonic oscillator.

As a result of the anharmonicity, the energy levels have a quadratic dependence on n_x and become more closely spaced with increasing energy. Near the dissociation limit the density of states within a
given energy interval can become very large.

An important consequence of the anharmonicity is that transitions
due to the dipole derivative are allowed with $\Delta n > 1$. For a given Δn
such transitions have similar strength to those due to higher dipole
derivatives in the expansion of μ in Eq. (1.94). The anharmonic
shifts depress the overtone frequencies from the simple multiples of
the fundamental frequency. Also, the hot-band frequencies will become anharmonically shifted to the "red" or to lower frequencies than
the ground-state transition.

For a polyatomic molecule the potential surface will have a $3N - 6$
dimensionality. In general, there can be both anharmonic shifts and
splitting of degenerate levels due to anharmonic interactions between
the normal modes. We shall call the vibrational sublevels resulting
from this splitting the vibrational *fine structure*. For large-amplitude
motion on the potential surface, the anharmonic interactions can dominate, so that the normal-mode picture is no longer valid and a local-
mode picture can be more appropriate. We may generalize the anharmonic energy levels given by Eq. (1.104) for a polyatomic molecule.
We find

$$
\begin{aligned}
E_n = {} & \omega_1\left(n_1 + \frac{1}{2}\right) + \omega_2\left(n_2 + \frac{1}{2}\right) + \omega_3\left(n_3 + \frac{1}{2}\right) + \cdots \\
& + X_{11}\left(n_1 + \frac{1}{2}\right)^2 + X_{22}\left(n_2 + \frac{1}{2}\right)^2 + X_{33}\left(n_3 + \frac{1}{2}\right)^2 + \cdots \\
& + X_{12}\left(n_1 + \frac{1}{2}\right)\left(n_2 + \frac{1}{2}\right) + X_{13}\left(n_1 + \frac{1}{2}\right)\left(n_3 + \frac{1}{2}\right) \\
& + X_{23}\left(n_2 + \frac{1}{2}\right)\left(n_3 + \frac{1}{2}\right) + \cdots
\end{aligned}
$$ (1.107)

so that the fundamental frequencies are

$$\nu_1 = \omega_1 + 2X_{11} + \frac{1}{2}X_{12} + \frac{1}{2}X_{13} + \cdots$$

$$\nu_2 = \omega_2 + 2X_{22} + \frac{1}{2}X_{12} + \frac{1}{2}X_{23} + \cdots$$

$$\nu_3 = \omega_3 + 2X_{33} + \frac{1}{2}X_{13} + \frac{1}{2}X_{23} + \cdots$$

Besides *fundamental* transitions ν_i and *overtone* transitions $n_i\nu_i$ from the ground state, it is also possible to observe *combination* bands $n_i\nu_i + n_j\nu_j$ as a result of the anharmonicities. From the experimental observation of these transitions one can determine the anharmonic constants X_{ij} together with the harmonic frequencies ω_i.

1.4.5 Vibrational Fine Structure

The anharmonic constants X_{ij} cause shifting of the levels in polyatomic molecules. It is worthwhile to investigate other anharmonic terms which can cause splitting of vibrational levels for a three-dimensional oscillator, such as the ν_3 or ν_4 vibrations in UF_6 or SF_6, since these terms contribute to the vibrational fine structure.

We have seen in Sec. 1.4.1 that the energy levels of a three-dimensional isotropic oscillator can be labeled by (n_x, n_y, n_z) or by (n, ℓ). In the case of the ν_3 mode of SF_6 or UF_6 the labels (n_x, n_y, n_z) determine the number of quanta of the asymmetric stretch in the x, y, or z directions, respectively. In this case the motion in the x, y, and z directions may be thought of as independent. On the other hand, ℓ is the vibrational angular momentum that results from coupling the independent Cartesian motions so that the vibrational motion is elliptical.

Because of such coupling phenomena, the fine-structure splitting of the degenerate ν_3 vibrational energy levels in Fig. 1.15 can be represented by two terms that account for anharmonicities of motion both along the X–F bonds and perpendicular to them. The ν_3 energy of the nth level is then [Patterson, 1985]

$$E_n = n\omega_3 + X_{33}n(n-1) + (G_{33} + 2T_{33})\underline{\ell}^2 + T_{33}(10\underline{r} - 8n - 6n^2)$$

$$(1.108)$$

where the bar (_) denotes tensor terms which can cause splitting. The first anharmonic term, with the coefficients $(G_{33} + 2T_{33})$, causes splitting of the different (n, ℓ) levels due to precession of the three-dimensional elliptical classical orbits. The last anharmonic term with coefficient $10T_{33}$ causes splitting of the different (n_x, n_y, n_z) levels due to anharmonicities of the independent one-dimensional oscillators in the x, y, and z directions. We have

$$\underline{\ell}^2 |n, \ell\rangle = \ell(\ell + 1) |n, \ell\rangle \tag{1.109}$$

and

$$\underline{r} |n_x, n_y, n_z\rangle = (n_x^2 + n_y^2 + n_z^2) |n_x, n_y, n_z\rangle \tag{1.110}$$

Thus the degree to which $|n, \ell\rangle$ or $|n_x, n_y, n_x\rangle$ form a good vibrational basis depends on the anharmonicity constants T_{33} and G_{33}.

Because of the octahedral symmetry of UF_6 and SF_6 each vibrational level splits into a vibrational sublevel labeled by A_1, A_2, E, T_1, or T_2 (ignoring the u and g labels). The energies and labels above also apply to a tetrahedral molecule such as SiF_4. The directions and ordering of the vibrational-level shifting and splitting are shown in Fig. 1.20 for the three molecules UF_6, SF_6, and SiF_4. There is an opposite limiting behavior for UF_6 and SiF_4 with SF_6 intermediate between the two extremes.

For UF_6 we find that $10T_{33} \gg (G_{33} + 2T_{33})$, so that n_x, n_y, and n_z are nearly good quantum numbers and refer to uncoupled motions in the three cartesian directions. These three quantum numbers describe excitations of three independent *anharmonic* oscillators. Thus in Fig. 1.20 the major energy shifts are labeled by (n_x, n_y, n_z), while the fine-structure splittings due to $\underline{\ell}^2$ are labeled with symmetry species labels.

For SiF_4 we find that $(G_{33} + 2T_{33}) \gg 10T_{33}$, so that the vibrational angular momentum ℓ is a nearly good quantum number. Thus, in Fig. 1.20 the major energy shifts are labeled with ℓ, while the fine-structure splittings due to $\underline{r}$ are labeled with symmetry species labels.

The oscillator motions for UF_6 and SiF_4 labeled by (n_x, n_y, n_z) and ℓ, respectively, may be viewed as examples of "local modes" of vibrational overtones in the sense that the classical trajectories are highly localized in position-momentum phase space in both cases; the classical trajectories evolve on a very limited region of the potential surface [Patterson, 1985]. It can be shown that for intermediate cases, such as SF_6, the classical trajectories for high overtones are chaotic and cover a large part of phase space.

The complex fine-structure sublevel patterns for the overtones of the three molecules above facilitates the multiple photon absorption of radiation, leading to molecular dissociation, as recently observed [Lyman, 1976]. For a one-dimensional anharmonic oscillator, the energy levels given by Eq. (1.104) become anharmonically shifted out of resonance with the laser pump. The level shifts and fine-structure split-

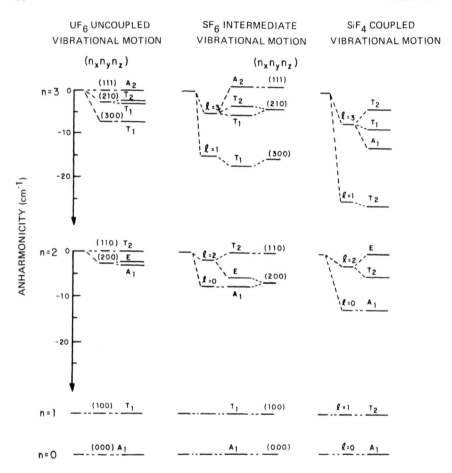

FIG. 1.20 The lower rungs of the ν_3 vibrational ladders of UF_6, SF_6, and SiF_4, showing the vibrational fine-structure shifts and splittings due to anharmonicities. The vibrational sublevels are labeled with their appropriate symmetry species (ignoring μ and g labels).

ting of the three-dimensional oscillator allow for resonant absorption up the vibrational overtone ladder. This is important since in the case of SF_6 some 30 infrared photons are required for dissociation, while even more are required in the case of UF_6 and SiF_4. The drastic difference in the vibrational substructure of energy levels between UF_6 and SiF_4 also leads to different transition selection rules, as given by Eqs. (1.97) and (1.98), respectively. As a result, the multiple photon pathways for these two molecules are quite different [Patterson, 1983]. Since SF_6 is in an intermediate region, it shares some of the selection rules of both UF_6 and SiF_4 and its multiple photon pathways to dissociation are the most complicated. In addition to the vibrational fine structure, the rotational structure of the molecule also enables lower-frequency resonant transitions to occur [Hodgkinson, 1982; Galbraith, 1983]. The rotational structure of polyatomic molecules will be discussed later.

1.4.6 Vibronic Structure

In the Born-Oppenheimer approximation we assume that the internuclear potential energy is only a function of the nuclear coordinates. Thus each electronic energy level gives rise to its own anharmonic potential energy surface from which the vibrational wavefunctions can be derived. Hence we may write the total *vibronic* wavefunction as a product of the electronic and vibrational wavefunction

$$\Psi_{\varepsilon n}(r_e, r_v) = \psi_\varepsilon(r_e, r_v)\psi_n(r_v) \tag{1.111}$$

where r_e and r_v represent the electronic and nuclear coordinates. An electric dipole vibronic transition between two electronic levels with total wavefunctions Ψ and Φ has a transition dipole given by (omitting differentials)

$$\mu_{nn'} = \int \Psi_{\varepsilon n}(r_e, r_v) \mu(r_e, r_v) \Phi_{\varepsilon' n'}(r_e, r_v) \tag{1.112}$$

We may resolve the dipole moment into an electric and nuclear dipole,

$$\mu = \mu(r_e) + \mu(r_v) \tag{1.113}$$

so that using Eqs. (1.111)-(1.113) we have

$$\mu_{nn'} = \int \psi_\varepsilon \mu(r_e) \phi_{\varepsilon'} \int \psi_n \phi_{n'} + \int \psi_\varepsilon \phi_{\varepsilon'} \int \psi_n \mu(r_v) \phi_{n'} \tag{1.114}$$

or since the electronic wavefunctions ψ_ε and $\phi_{\varepsilon'}$ are orthogonal, we have

$$\mu_{nn'} = M \int \psi_n \phi_{n'} \tag{1.115}$$

where M is the electric transition dipole given by

$$M = \int \psi_\varepsilon \mu(r_e) \phi_{\varepsilon'} \tag{1.116}$$

The square of the integral on the right of Eq. (1.115) is called the Franck-Condon factor for the vibronic transition.

We have seen that for vibrational transitions *within* the same potential surface of a given electronic level, strong transitions occur only for $\Delta n = \pm 1$. This selection rule is no longer valid for transitions *between* potential surfaces of two electronic levels, due to the Franck-Condon factor.

In Fig. 1.21 we show a typical vibronic transition between two potential curves for a one-dimensional oscillator. The lower curve would correspond to the lower electronic state ε with vibrational frequencies higher than the upper state ε' because the atoms are more tightly bound. For the excited electronic state we have drawn both a bound and a dissociative potential. The discrete vibrational energies occur only for the bound excited-state potential since the dissociative potential would have a continuum of energy levels. The discreteness and continuum of excited-state energy levels is reflected in the vibronic Franck-Condon spectra shown on the vertical axes of Fig. 1.21.

The overall maximum in the spectra occurs at the energy where the vibrational wavefunction overlap of ψ and ϕ is greatest. For a given excited-state energy the dominant part of the overlap integral occurs at the classical turning point of the upper potential curve x'_A, where the excited-state wavefunction has its greatest amplitude and wavelength. The oscillatory behavior of the excited-state wavefunction far from the turning point results in a negligible contribution to the overlap integral. The frequency spread of the Franck-Condon spectra is determined by the turning points of the lower potential curve x_A and x_B for a given ground-state energy. The projection of these turning-point displacements on the upper-state potential curve delimits the vibronic spectral range. Clearly, the larger the horizontal displacement between the upper ε' and lower ε potential curves, the more extensive is the frequency spread of the vibronic spectrum and the longer is the vibrational progression for discrete transitions. Note that the greater the slope of the upper curve, the larger the spectral range. When the horizontal displacement between potential curves is large, the maximum in the spectra will occur for large Δn.

For polyatomic molecules with many degrees of freedom, the potential surfaces can be extremely complicated, and calculating wavefunctions for highly excited levels becomes computationally impossible. The problem is further complicated by the fact that coupling between the vibrational modes can be large and some modes may be dissociative, while others may not be. In such cases a dynamic time-dependent picture is often better than the previously described static energy picture of Fig. 1.21.

Franck-Condon spectra in the time domain have been extensively developed by Heller and coworkers [Heller, 1981] and are shown in Fig. 1.22. The initial vibrational wavefunction on the lower ε potential surface at $t = 0$ is $\psi(0)$. Let this function be a Gaussian wave-

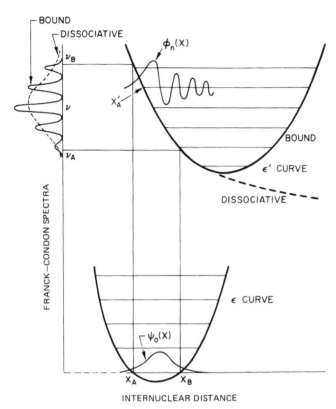

FIG. 1.21 Franck-Condon spectra in the energy picture. The probability of transitions between the lower potential curve ε and the upper potential curve ε' is proportional to the square of the overlap of their wavefunctions. The absorption spectrum arising from state ψ_0 on the lower curve is shown for both a bound and a dissociative upper potential curve. The largest contribution to the overlap occurs at the turning point x_A' of the upper curve. Thus Franck-Condon spectra are delimited by the "reflection" of the lower turning points x_A and x_B on the upper curve at ν_A and ν_B.

packet as it would be for a ground vibrational state. The motion of this wavepacket on the upper ε' potential surface is denoted by $\phi(t)$ and is shown in Fig. 1.22a for a two-dimensional oscillator. This motion is essentially the classical trajectory on this surface.

The time-dependent Franck-Condon overlap $\langle \psi(0)|\phi(t)\rangle$ is shown in Fig. 1.22b. The time t_1 corresponds to the initial decay of the overlap due to the steepness of the potential surface since the classical trajectory will initially follow the maximum slope of the surface.

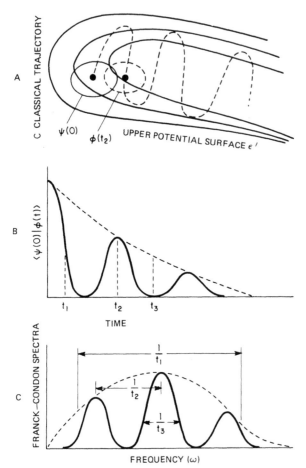

FIG. 1.22 Franck-Condon spectra in the time picture. (A) The classical trajectory of the initial lower surface wavepacket $\psi(0)$ is shown on
the upper surface ϵ' after one vibrational period t_2. (B) The temporal
overlap of the evolved wavepacket $\phi(t)$ with the initial wavepacket
$\psi(0)$ is shown. (C) The Franck-Condon spectrum is simply the fre-
quency-transformed temporal overlap. The line widths shown are approximately the reciprocals of the time constants.

After one oscillation of the wavepacket at time t_2 it returns close to
its initial position. Despite this oscillation the wavepacket moves
steadily down the surface, resulting in an overlap decay time t_3. The
latter decay coordinate can be due to dissociation or intramolecular
vibrational relaxation.

The Franck-Condon spectrum in Fig. 1.22c is then simply the Fourier transform of the time-dependent overlap,

$$F(\omega) = C \int e^{i\omega t} <\psi(0)|\phi(t)> \, dt \qquad (1.117)$$

The three time scales for the overlap result in three line widths for the Franck-Condon spectra. The initial decay t_1 determines the frequency spread $\sim 1/t_1$ of the spectra. As in the energy picture, the larger the slope of the potential surface, the greater the spectral range. From the period of oscillation t_2 we obtain the excited-state frequency $\sim 1/t_2$, which gives rise to a vibrational progression in the vibronic spectra. Finally, the overall "dissociative" decay time t_3 determines the vibronic line widths $\sim 1/t_3$. Thus it is still possible to have vibronic structure in a dissociative band for those modes which are nondissociative [Pack, 1976]. The advantage of the time picture is that one need only calculate trajectories for times comparable to t_3. When there is no overall decay constant t_3, the wavepackets can reflect and interfere, eventually creating the same wavefunctions needed in the energy picture, and all advantages are lost.

1.5 ROTATIONAL STRUCTURE OF POLYATOMIC MOLECULES

Since our emphasis is on high-resolution spectroscopy we shall investigate in detail rotational fine, superfine, and hyperfine structure which has only recently been observed. Experimental techniques currently enable the spectroscopist to resolve the substructure of highly excited rotational levels. Fortunately, these highly excited quantum rotors behave in many respects according to classical principles of mechanics, so that both a qualitative and quantitative understanding of rotational substructures can be attained through more intuitive semiclassical techniques. We first discuss the simple rotational structure of spherical top and linear polyatomic molecules before we treat the more complicated fine and hyperfine structure of molecular rotors. For a review, see the book by Townes [Townes, 1975].

1.5.1 Simple Rotational Structure

Since there are three degrees of rotational freedom for a nonlinear polyatomic molecule, we expect three semiclassical quantization conditions, giving rise to three distinct quantum numbers. One of these quantum numbers corresponds to the total rotational angular momentum $|J|$ of the molecule which must be conserved. We have the rotational angular momentum quantization

$$|\underline{J}|^2 = J(J + 1)\hbar^2, \quad J = 0, 1, 2, \ldots \tag{1.118}$$

As in the case of the hydrogen atom and the three-dimensional harmonic oscillator, there will also be a space quantization of the rotational angular momentum $\underline{J}$ such that the projection of $\underline{J}$ on the laboratory z axis is

$$J_z = M\hbar \tag{1.119}$$

where

$$M = -J, -J + 1, \ldots, J \tag{1.120}$$

Since the energy of a freely rotating molecule cannot depend on the molecule's orientation in the laboratory, one has a spatial $2J + 1$ degeneracy for a given J energy level.

In the case of the hydrogen atom and the three-dimensional harmonic oscillator, the spatial quantum number determines the orientation of the elliptical classical trajectory with respect to the laboratory z axis. However, for a rotating molecule its orientation is not entirely determined by $\underline{J}$ and J_z. A molecule may precess about the laboratory fixed angular momentum vector $\underline{J}$. Conversely, the angular momentum vector $\underline{J}$ is not fixed in the molecular body frame defined by its constituent atoms. Let $\bar{J}_z$ be the projection of $\underline{J}$ on the molecular frame or body z axis. Then $\bar{J}_z$ specifies the orientation of the precessing molecule with respect to $\underline{J}$. The body quantization of the rotational angular momentum provides the third quantum number.

$$\bar{J}_z = K\hbar \tag{1.121}$$

where

$$K = -J, -J + 1, \ldots, J \tag{1.122}$$

Let $\bar{J}_x$, $\bar{J}_y$, and $\bar{J}_z$ be the angular momenta of a freely rotating molecule in the body frame where $\bar{x}$, $\bar{y}$, and $\bar{z}$ are the *principal axes*, in which the moment of inertia matrix I is diagonal with components I_C, I_B, and I_A, respectively:

$$I_C \equiv I_{\bar{x}\bar{x}} = \sum_i m_i(r_i^2 - \bar{x}_i^2)$$

$$I_{\bar{x}\bar{y}} = -\sum_i m_i \bar{x}_i \bar{y}_i = 0 \quad \text{etc.} \tag{1.123}$$

where the sum is over atoms in the molecule. The rotational kinetic energy of the molecule (in cm^{-1}) is

$$E = \frac{\bar{J}_x^2 / 2I_C + \bar{J}_y^2 / 2I_B + \bar{J}_z^2 / 2I_A}{hc}$$

$$= \frac{C\bar{J}_x^2 + B\bar{J}_y^2 + A\bar{J}_z^2}{\hbar^2} \qquad (1.124)$$

with the rotational constants given by

$$C = \frac{\hbar}{4\pi c I_C} \qquad B = \frac{\hbar}{4\pi c I_B} \qquad A = \frac{\hbar}{4\pi c I_A} \qquad (1.125)$$

The principal axes are defined such that $A \geqslant B \geqslant C$.

1.5.1A. Rigid Spherical Top Molecules

For a spherical top molecule such as UF_6, SF_6, or SiF_4, the three rotational constants are equal and

$$E_J = \frac{B(\bar{J}_x^2 + \bar{J}_y^2 + \bar{J}_z^2)}{\hbar^2} = BJ(J + 1) \qquad J = 0, 1, 2, \ldots \qquad (1.126)$$

The rotational energy of a spherical top molecule is independent of M and K, so that each J energy level has a degeneracy of $(2J + 1)^2$. The K degeneracy is a result of the fact that for a spherical top molecule the instantaneous axis of rotation always coincides with the rotational angular moment J vector. Any axes fixed in the molecule may be considered the body z axis. We shall see later that there are additional centrifugal distortion terms in the energy of the nonrigid spherical top which gives rise to rotational fine-structure splitting which breaks the K degeneracy.

At room temperature, the relatively "heavy" spherical top molecules above will have significant population in high-J levels. The population of molecules with angular momentum J relative to the J = 0 rotational ground state is given by the Boltzmann factor,

$$N_J = N_0(2J + 1)^2 e^{-BJ(J+1)hc/kT} \qquad (1.127)$$

where $(2J + 1)^2$ is the degeneracy factor and B is in cm^{-1}.

The maximum of this distribution occurs at

$$J \cong \sqrt{\frac{kT}{B}} \tag{1.128}$$

At room temperature the maximum population occurs at $J \sim 48$ for SF_6 ($B = 0.091$ cm^{-1}) and at $J \sim 61$ for UF_6 ($B = 0.055$ cm^{-1}). For such high rotational levels we show later that the rotational fine structure has simple features which are interpretable semiclassically.

1.5.1B. Linear Molecules (Including Diatomics)

For a linear molecule there are only two degrees of rotational freedom since the molecule cannot rotate about its symmetry $\bar{z}$ axis. From Eq. (1.124) we see that to do so would require very large energy since the moment of inertia I_A about the symmetry axis is nearly zero. For a linear molecule we may then let $\bar{J}_z = 0$, or equivalently, $K = 0$. Since $B = C$, the energy for a linear molecule becomes

$$E_J = BJ(J + 1) \quad J = 0, 1, 2, \ldots \tag{1.129}$$

Since $K = 0$, each J energy level has only a M degeneracy of $2J + 1$ and there can be no rotational fine-structure splitting.

1.5.1C. Pure Rotational Transitions

We have seen that pure rotational transitions occur when the molecule has a permanent electric dipole moment. In this case the rotation of the molecule itself produces an oscillating dipole which can then interact with the electric field of the radiation. For a molecule with an axis of symmetry, the permanent dipole moment must lie along this axis. Thus a spherical top molecule which has several axes of symmetry cannot have a permanent dipole moment and has no pure rotational spectrum.

Linear molecules (including diatomic molecules) without a center of symmetry will always have a permanent dipole moment. As was the case for vibrational and electronic angular momentum, the selection rules for rotational angular momentum are

$$\Delta M = 0, \pm 1 \tag{1.130}$$

$$\Delta J = 0, \pm 1 \quad \text{not } J = 0 \rightarrow J = 0 \tag{1.131}$$

The $\Delta J = -1$ transition would correspond to stimulated emission for which a population inversion is necessary ($N_2 > N_1$), while the $\Delta J = 0$ transition would have zero frequency. For $\Delta J = 1$ transitions the absorption frequency is (in cm^{-1})

$$\nu(\text{rotational}) = \Delta E_J = E_{J+1} - E_J = 2B(J + 1) \tag{1.132}$$

giving rise to a rotational series of equally spaced lines. For large J this is equal to the classical rotational angular frequency,

$$f_J \sim \frac{dE_J}{dJ} = 2BJ \tag{1.133}$$

although the classical J is continuous while the quantum J is discrete. Since the rotational constant varies from tenths of wavenumbers to a few wavenumbers, pure rotational spectra of linear molecules are in the microwave to far-infrared region.

We shall now consider molecules with rotational energy levels which depend on the K quantum number. We show that we may define a rotational energy (RE) surface which leads to the occurrence of rotational fine structure.

1.5.2 Rotational Fine Structure

In analogy with the anharmonic vibrational potential-energy surface (PE) we define here the rotational energy (RE) surface in order to understand rotational fine and superfine structure of virtually all poly-atomic molecules. Using this surface, we show the complicated classical trajectories of the rotational angular momentum J vector in the molecular frame and determine the semiclassical rotational fine-structure energy levels.

The vibrational motion of a polyatomic molecule is generally much faster than the rotational motion. Thus, as a molecule rotates, the nuclei have time to adjust to the vibrational internuclear forces. We may then define a rotational energy surface which depends only on the direction $\hat{J}$ of the rotational angular momentum J vector in the body frame of the molecule with the magnitude $|J|$ held fixed.

We may plot the energy E_J in Eq. (1.124) for arbitrary A, B, and C as a function of $\bar{J}_x$, $\bar{J}_y$, and $\bar{J}_z$ subject to the constraint J = constant. That is, the energy is a function of only the orientation of the J vector in the body frame. Let this orientation be specified using standard Euler angles (α, β, γ) [Edmonds, 1960], which may be related to the *body* spherical angles $(-\beta, -\gamma)$ [Harter, 1978]. The body angular momenta may then be written in terms of these body spherical coordinates.

$$\bar{J}_x = -\sqrt{J(J + 1)} \sin\beta \cos\gamma \tag{1.134a}$$

$$\bar{J}_y = \sqrt{J(J + 1)} \sin\beta \sin\gamma \tag{1.134b}$$

$$\bar{J}_z = \sqrt{J(J + 1)} \cos\beta \tag{1.134c}$$

Then the *rotational energy surface* $E_J(\beta, \gamma)$ may be found by substituting the $\bar{J}_x$, $\bar{J}_y$, $\bar{J}_z$ of Eq. (1.134) into Eq. (1.124). We shall now use the RE surface $\bar{E}_J(\beta, \gamma)$ to determine the rotational fine structure for various rotational constants A, B, and C.

1.5.2A. Symmetric Top Molecules

For a symmetric top molecule two of the moments of inertia are equal and the molecule is nonlinear. For a *prolate* symmetric top molecule defined by A > B = C, the RE surface for J = 10 is shown in Fig. 1.23a. The constant E curves shown on the surface must correspond to the classical rotational trajectories, where both J and E are necessarily conserved. The curves clearly show the precession of the J vector about the molecular $\bar{z}$ axis and are equivalent to precessional motion of the molecule about the lab fixed J vector. Note that for a given classical trajectory $\bar{J}_z$ is a constant of the motion. Indeed, to determine the body quantum number we use the quantization condition

$$\int \bar{J}_z \, d\gamma_z = K_A h$$

or

$$\bar{J}_z = K_A \hbar, \quad K_A = -J, \, -J + 1, \, \ldots, \, J \tag{1.135}$$

Note that γ_z is the angle *conjugate* to $\bar{J}_z$ since it defines rotations about the $\bar{z}$ axis.

The energy for a prolate symmetric top is then

$$E_{J,K} = \frac{B(\bar{J}_x^2 + \bar{J}_y^2 + \bar{J}_z^2)}{\hbar^2} + \frac{(A - B)\bar{J}_z^2}{\hbar^2}$$

$$= BJ(J + 1) + (A - B)K_A^2 \tag{1.136}$$

The curves drawn in Fig. 1.23a are indeed the semiclassical trajectories corresponding to energies for *integer* K_A with J = 10. There are 2J + 1 = 21 different trajectories for each of the different possible K_A's. It is obvious from the RE surface that trajectories with $\pm K_A$ have the same energy which is reflected by the K_A^2 term for $E_{J,K}$ in Eq. (1.136). The energy is also independent of the quantum number M. Equation (1.136) shows that the energy of the rotating molecule is dependent on the orientation of the angular momentum J vector with respect to the molecular axes but not with respect to the laboratory axis. The degeneracy of any given energy level is 2(2J + 1) for $K_A \neq 0$ and (2J + 1) for $K_A = 0$.

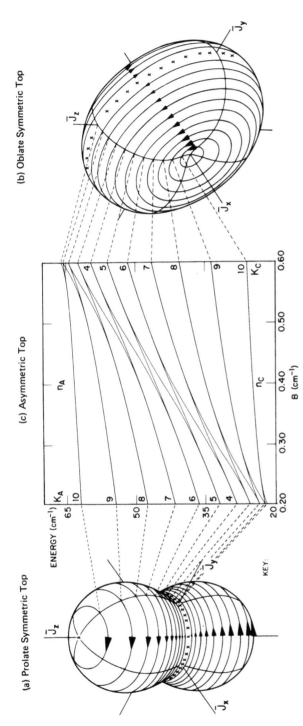

FIG. 1.23 Rotational energy surface for (a) a prolate symmetric top molecule and (b) an oblate symmetric top molecule. In both cases the curves on the surface are the $J = 10$ semiclassical trajectories for which the angular momentum about the molecular symmetry axis z or x is quantized by $\bar{J}_z = K_A\hbar$ or $\bar{J}_x = K_C\hbar$, respectively. The arrows indicate the speed and direction of the circular J-vector's precession about the respective molecular symmetry axes.   (c) Correlation diagram for the $J = 10$ rotational sublevels of a prolate ($B = C$) and oblate ($B = A$) symmetric top molecule.   When the rotational constant B is in between $C = 0.2$ cm^{-1} and $A = 0.6$ cm^{-1}, the fine-structure sublevels are those of an asymmetric top molecule and are labeled by quantum numbers n_A and n_C as described in the text.   Note that for any B, the sublevels occur in pairs or "clusters" except along the diagonal of the correlation diagram.

67

The RE surface for an oblate symmetric top, where $A = B > C$ is shown in Fig. 1.23b for $J = 10$. Again the constant E curves on the surface are the classical rotational trajectories and correspond to the precession of the J vector about the molecular $\bar{x}$ axis. For a given classical trajectory $\bar{J}_x$ is now a constant of the motion and we find that

$$\int \bar{J}_x \, d\gamma_x = K_C h$$

or

$$\bar{J}_x = K_C \hbar, \quad K_C = -J, \ -J + 1, \ \ldots . \ J \tag{1.137}$$

where γ_x is the angle about the $\bar{x}$ axis conjugate to $\bar{J}_x$. The energy for an oblate symmetric top is then

$$E_{J,K} = BJ(J + 1) - (B - C)K_C^2 \tag{1.138}$$

The curves drawn in Fig. 1.23b are again the semiclassical trajectories corresponding to energies for integer K_C with $J = 10$. There are $2J + 1 = 21$ different trajectories for each of the different possible K_C, and trajectories with $\pm K_C$ have the save energy.

The splitting of the J levels for a symmetric top is a result of the precessional motion of the J vector. This motion does not occur for a rigid spherical top or a linear molecule. The classical precessional frequency corresponds to the spacing between K levels. For a prolate top this precessional frequency (in cm^{-1}) about the $\bar{z}$ axis is

$$\nu(\text{precessional}) = \Delta E_K = E_{J,K+1} - E_{J,K} = (A - B)(2K + 1)$$

$$\tag{1.139}$$

or

$$f_K = \frac{\partial E_{J,K}}{\partial K} = (A - B)2K \quad (K \equiv K_A) \tag{1.140}$$

For an oblate top the precession is about the $\bar{x}$ axis and in the opposite direction. The arrows in Fig. 1.23 indicate the speed of precession. For trajectories with K_A or K_C zero, denoted by (xxx), there is no precession and the J vector is fixed in the molecular frame.

Because the permanent dipole of a symmetric top molecule is along the molecular symmetry axes, rotation about the symmetry axes will not cause the dipole moment to oscillate. For symmetric top molecules we have the selection rule, in addition to Eqs. (1.130) and (1.131) for pure rotational transitions:

Prolate: $\Delta K_A = 0$ (1.141a)

Oblate: $\Delta K_C = 0$ (1.141b)

Thus, if we ignore higher-order centrifugal distortion terms in the energy, the pure rotational spectra of a symmetric top molecule is the same as for a linear molecule with equally spaced levels.

1.5.2B. Asymmetric Top Molecules

An asymmetric top molecule is defined by $C < B < A$, so that the rotational constant B is somewhere between A and C. In Fig. 1.24 the RE surface for $J = 10$ is shown for a most-asymmetric top, where B is halfway between A and C or $B = (A + C)/2$. Interestingly, the most-asymmetric top shares the features of both the prolate and oblate symmetric top. In Fig. 1.24 we see an equal number of semiclassical trajectories about the z axis and x axis of the molecule. Note, however, that $\bar{J}_z$ and $\bar{J}_x$ are not constants of the motion since the J vector not only precesses about the z and x axes but *nutates* as well. That is, the classical trajectories are *not* perpendicular to the $\bar{J}_z$ or $\bar{J}_x$ axes.

The dotted line that divides the x trajectories from the z trajectories in Fig. 1.24 is called the separatrix. Motion for which the J vector is on the separatrix is unstable. Thus if a molecule is set spinning near its $\bar{y}$ axis with intermediate moment of inertia I_B, subsequent motion will cause the molecule to flip wildly to the $-y$ axis and back again along the separatrix.

We may find the semiclassical rotational fine-structure energies from the quantization conditions about the respective symmetry axes:

$\bar{z}$ axes: $\displaystyle\int \bar{J}_z \, d\gamma_z = n_A h \quad n_A = J, J - 1, \ldots$ (1.142a)

$\bar{x}$ axes: $\displaystyle\int \bar{J}_x \, d\gamma_x = n_C h \quad n_C = J, J - 1, \ldots$ (1.142b)

which must be integrated numerically since $\bar{J}_z$ and $\bar{J}_x$ are not constants. Quantization occurs for integers n_A equal to J, J − 1, ... until the J vector crosses the separatrix on the RE surface and is no longer a trajectory about its axis of quantization. The position of the separatrix depends on the rotational constant B relative to A and C. In Fig. 1.24 are drawn the $J = 10$ trajectories at the quantized energies according to the equations above. We have chosen $C = 0.2$ cm^{-1}, $B = 0.4$ cm^{-1}, and $A = 0.6$ cm^{-1} for a "heavy" asymmetric top so that

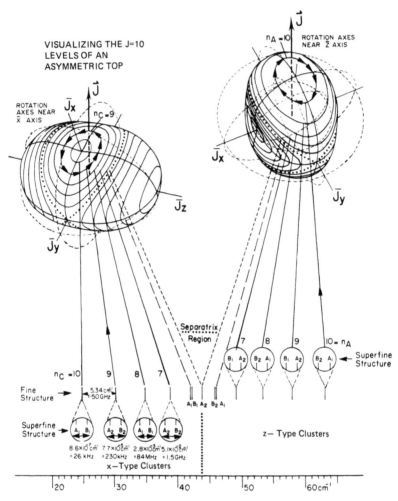

FIG. 1.24 The rotational energy surface for a most-asymmetric top with B = (A + C)/2. The curves on the surface are the 2J + 1 = 21 possible semiclassical trajectories for J = 10. Semiclassical trajectories about the molecular symmetry axes $\bar{z}$ and $\bar{x}$ are labeled by n_A and n_C, respectively. Note that the J vector is nutating as well as precessing so that $\bar{J}_z$ and $\bar{J}_x$ are no longer constants of the motion and K_A and K_C are not good quantum numbers. The dotted curve or separatrix denotes the unstable region which separates the two types of trajectories. The trajectories are correlated with their rotational fine-structure sublevels below, corresponding to the B = 0.4 cm^{-1} sublevels in Fig. 1.23c. Most sublevels are doubly degenerate because there are two semiclassical trajectories with the same energy about a given symmetry axis. The clustered sublevels are magnified, revealing their superfine structure splitting and are labeled according to symmetry species of a bent XY$_2$-type molecule.

at room temperature the J = 10 rotational levels would be appreciably populated. For the $\bar{z}$ trajectories we have 10 curves with $n_A = \pm 10$, $\pm 9, \ldots, \pm 6$ and for the $\bar{x}$ trajectories we have 10 curves with $n_C = \pm 10$, $\pm 9, \ldots, \pm 6$. Along with the separatrix curve with $n_A = n_C = \pm 5$ we have 2J + 1 = 21 semiclassical trajectories. Note that n_A about the $\bar{z}$ axis is $-n_A$ about the opposite $-\bar{z}$ axis. Each energy is twofold degenerate except at the separatrix, since the trajectories with $\pm n_A$ or $\pm n_C$ have the same energy. The degenerate energy levels are called fine-structure clusters. As for the symmetric top, the spacing of the fine-structure levels is related to the classical frequency of precession and nutation for a complete orbit of a trajectory.

The exact quantum fine-structure energy levels are shown below the RE surface in Fig. 1.24 and are correlated to the semiclassical trajectories on the RE surface. The individual rotational level species are labeled A_1, B_1, A_2, and B_2 according to the symmetry of a molecule like H_2O. Note that the fine-structure clusters show splitting of the semiclassically degenerate energy levels, which are referred to as superfine structure. The superfine structure is due to quantum tunneling between equivalent trajectories, which will be discussed later.

The fine-structure clusters are not a result of our particular choice of rotational constants. In Fig. 1.23c we show the J = 10 energy levels of a molecule where we vary B between C and A such that C ⩽ B ⩽ A. On the left of Fig. 1.23c where B = C the energy levels are labeled with K_A for a prolate symmetric top, while on the right where B = A the energy levels are labeled with K_C for an oblate symmetric top. In between it is common practice to label the levels with (K_A, K_C) according to how the asymmetric top levels correlate with the symmetric top levels on the extreme left and right. Thus the $n_A = \pm 10$ cluster pair would be labeled (10,0) and (10,1), while the $n_C = \pm 10$ cluster pair would be labeled (0,10) and (1,10).

Interestingly, the fine-structure levels occur in clustered pairs irrespective of the rotational constant B. Only on the extreme left and right are K_A and K_C actually good quantum numbers. In between, clusters should be labeled with n_A and n_C. Note that for energies corresponding to the separatrix region of the RE surface of the asymmetric top, the fine-structure clusters abruptly switch their partners in Fig. 1.23c and momentarily become nondegenerate. However, for any rotational constant B, the separatrix region constitutes only a small fraction of the spectrum at J = 10 corresponding to the diagonal of Fig. 1.23c. This fraction becomes even smaller for higher J.

Pure rotational transitions for asymmetric top molecules are not as simple as those for symmetric top molecules, for two reasons. First, asymmetric top molecules do not necessarily have an axis of symmetry, so that the permanent dipole moment may be arbitrarily oriented. Thus, in general, we have the selection rules

$$\Delta K_A = 0, \pm 1 \tag{1.143a}$$

$$\Delta K_C = 0, \pm 1 \tag{1.143b}$$

when the dipole is not aligned with the $\bar{z}$ axis or the $\bar{x}$ axis, respectively. Second, we have seen that even though it is standard practice to label the asymmetric top energy levels by (K_A, K_C), these are *not good quantum numbers*. Each energy level is actually a linear combination of K_A or K_C which can be determined by the extent to which the trajectories in Fig. 1.24 nutate out of the plane perpendicular to $\bar{J}_z$ or $\bar{J}_x$, respectively. In general, the selection rules above allow for transition between almost all asymmetric top levels. Only for the extremal energy levels for a given J, where the trajectories are nearly circular such that K_A or K_C are nearly good quantum numbers, do the selection rules above restrict the possible transitions.

1.5.2C. Spherical Top Molecules

We have already seen that a rigid spherical top molecule has a simple rotational structure. However, the centrifugal distortion of the nonrigid molecule will result in a fine-structure splitting of the rotational J levels which exhibit properties similar to the asymmetric top molecule. A spherical top molecule like SF_6 will distort more when rotating about a trigonal symmetry axis which promotes S$-$F bending motion than when rotating about a tetragonal symmetry axis which promotes S$-$F stretching motion. As a result, the RE surface is anisotropic and the energy is given by

$$E_J = BJ(J+1) + \frac{10t_{044}(\bar{J}_x^4 + \bar{J}_y^4 + \bar{J}_z^4)}{\hbar^4} - 6t_{044}J(J+1) + DJ^2(J+1)^2$$

$$\tag{1.144}$$

In Fig. 1.25 we plot the RE surface for J = 30 of an octahedral molecule like SF_6, where B = 0.091 cm^{-1}, t_{044} = 5.44 Hz, and D = 0 for the vibrational and electronic ground states. The RE surface for a tetrahedral molecule like SiF_4 would look "inverted" since t_{044} would be negative. In Fig. 1.25 we see about twice as many semiclassical trajectories about the high-energy tetragonal-axis "hills" than about the low-energy trigonal-axis "valleys." Let us denote the angular momentum about the hills and valleys by $\bar{J}_4$ and $\bar{J}_3$, respectively. Clearly, near the top of the hill and the bottom of the valley the trajectories are nearly circular and $\bar{J}_4$ and $\bar{J}_3$ will be nearly constants of the motion, respectively. Thus the fine structure of the hills will be similar to a prolate symmetric top with K_4 a good quantum number,

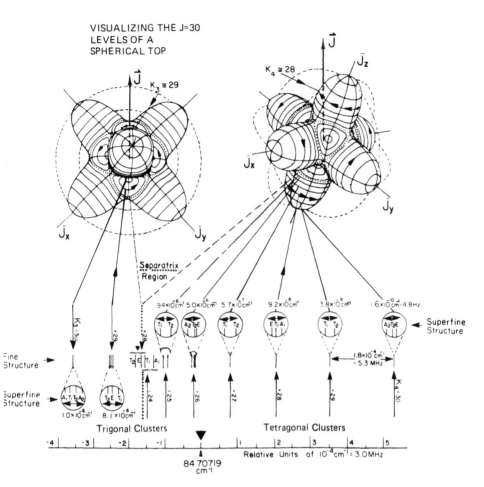

FIG. 1.25 The rotational energy surface a nonrigid spherical top XY_6 molecule. The curves on the surface are the $2J + 1 = 61$ possible semi-classical trajectories for $J = 30$. Semiclassical trajectories are nearly perpendicular to the tetragonal and trigonal symmetry axes corresponding to simple precession of the J vector and are labeled K_4 and K_3, respectively. The dotted curve or separatrix denotes the unstable region which separates the two types of trajectories. The trajectories are correlated with their rotational fine-structure sublevels below for a SF_6 molecule. The fine-structure splitting is relative to $BJ(J + 1)$. Fine-structure sublevels are sixfold or eightfold degenerate because of the number of semiclassical trajectories with the same energy about tetragonal or trigonal axes, respectively. The clustered sublevels are magnified, revealing their superfine structure splitting and are labeled according to octahedral symmetry species (ignoring u and g labels).

while the fine structure of the valleys will be similar to an oblate symmetric top with K_3 a good quantum number for a given J level. The
separatrix in Fig. 1.25, drawn with a dotted line, separates the hill
and valley regions and delineates the two types of classical trajectories. Again motion for which the J vector is on the separatrix is unstable.

We find the semiclassical rotational fine-structure energies from
the quantization conditions about the respective symmetry axes:

tetragonal axis: $\quad \displaystyle\int \bar{J}_4 \, d\gamma_4 = K_4 h$

or

$$\bar{J}_4 \sim K_4 \hbar \quad K_4 = J, \, J + 1, \, \ldots \qquad\qquad (1.145a)$$

and

triagonal axes: $\quad \displaystyle\int \bar{J}_3 \, d\gamma_3 = K_3 h$

or

$$\bar{J}_3 \sim K_3 \hbar \quad K_3 = J, \, J + 1, \, \ldots \qquad\qquad (1.145b)$$

Again, the quantization occurs for values of K_4 or K_3 until the J vector crosses the separatrix. We have drawn in Fig. 1.25 the $2J + 1 =$
61 semiclassical trajectories for which the energy is quantized using
the equations above. For each energy above the separatrix, there
are equivalent trajectories for each of the six hills, while for each energy below the separatrix there are equivalent trajectories for each of
the eight valleys. The fine structure for the deformable spherical top
resulting from the RE surface quantization consists of sixfold and
eightfold degenerate clusters of levels.

In Fig. 1.25 are shown the exact quantum fine-structure energy
levels below the RE surface. The semiclassical trajectories corresponding to each fine-structure cluster is also shown in this figure. For an
octahedral molecule like SF_6 the individual fine-structure levels are
labeled by singly degenerate species A_1 and A_2, doubly degenerate
species E, and triply degenerate species T_1 and T_2 (ignoring the u
and g labels). The high-energy sixfold-degenerate fine-structure
levels on the right of Fig. 1.25 are comprised of (A_2,T_2,E), (T_1,T_2),
and (E,T_1,A_1) clusters of species, while the low-energy eightfold degenerate levels on the left are comprised of (A_1,T_1,T_2,A_2) and
(T_2,E,T_1) clusters of species. These clusters can also be labeled according to their semiclassical quantum numbers K_4 and K_3, as seen in
the figure.

The rotational fine-structure clusters were first seen in high-resolution IR diode spectra of SF_6 [McDowell, 1976b] and were later simulated with matrix diagonalizations of the energy [Fox, 1977]. The general semiclassical explanation of clusters given here using RE surfaces for asymmetric and spherical top molecules has only recently been developed [Harter, 1984].

As in the case of the asymmetric top rotor, the fine-structure clusters of the spherical top rotor exhibit superfine splitting into individual octahedral species as seen in Fig. 1.25.

1.5.3 Rotational Superfine Structure

We have seen that the total degeneracy of each rotational fine-structure cluster corresponds to the number of distinct classical trajectories with the same energy in equivalent regions of the RE surface. The topology of the RE surface determines the equivalent body axes of quantization and the separatrix, which delineates motion about different axes.

In actuality, it is possible for the J vector to tunnel between *equivalent* classical trajectories across the separatrix. As a result of this tunneling, fine-structure clusters are no longer degenerate but are split into individual rotational symmetry species. The splitting frequency is simply the frequency at which the tunneling occurs. In Fig. 1.24 the superfine splitting of the $n_A = \pm 10$ cluster of the J = 10 level of the most-asymmetric top is 26 kHz, and in Fig. 1.25 the superfine splitting of the $K_4 = 30$ cluster of the J = 30 level of SF_6 is 4.8 Hz. In the latter case this means that the SF_6 molecule will precess about a tetragonal axis for about 0.2 s before tunneling to a nearby equivalent tetragonal axes. For physical processes that occur in times shorter than this, the molecule can be viewed as effectively "stuck" on one of its axes. For these physical processes we say that the molecule undergoes *dynamical symmetry breaking* since SF_6 is effectively behaving like a prolate symmetric top about one of its tetragonal axes. Dynamical symmetry breaking can occur for equivalent axes of quantization on the RE surface of any polyatomic molecule. We shall see that dynamical symmetry breaking has important consequences for the hyperfine structure of polyatomic molecules.

Tunneling is largest near the separatrix, where the tunneling barrier between the equivalent trajectories is smallest. Generally, the larger the angle between equivalent trajectories, the harder it is to tunnel and the smaller the superfine cluster splitting. The superfine splitting is smaller for K_4 clusters than for K_3 clusters since the angle between tetragonal axes (90°) is greater than the angle between trigonal axes (55°). Also, for a given J, a most-asymmetric top, with 180° between symmetry axes, will have smaller cluster splittings than a spherical top. For example, the ratio of superfine splitting to fine-structure splitting in the most-asymmetric top for J = 10 is less than

that in the spherical top for J = 30 when comparing extremal cluster
energies as seen in Figs. 1.24 and 1.25. Also, for an extremal cluster
the superfine splitting will decrease exponentially with increasing J.
For example, the $J = K_4 = 88$ cluster of SF_6 [Kim, 1979] has a superfine splitting of less than 6×10^{-19} Hz, corresponding to tunneling
once in over 50,000 years!

1.5.4 Hyperfine Structure

The interaction of the nuclear quadrupole moment with the electric
field gradient and the interaction of the nuclear spin with the rotational magnetic field predominates in the hyperfine splitting of rotational
fine and superfine structure levels. We shall first find the degeneracy
of the nuclear spin levels and then consider the hyperfine splitting
for polyatomic molecules.

1.5.4A. Nuclear Spin and Statistical Weights

For atomic spectra we introduced a tableau notation to describe
the permutational characteristics of the electronic spatial and nuclear
spin wavefunctions. We may now use these same tableaus for the rotational wavefunction in the body frame and for the nuclear spin.

The permutation of identical nuclei in a molecule, according to the
Pauli principle, must be antisymmetric for nuclei with half-integer
spin (such as H or F) and symmetric for nuclei with integer spin (such
as D or O). The rotational species of the molecule determines its spatial permutational character, which can be described by one or more
tableaus. The nuclear spin states must then correspond to *conjugate*
tableau for half-integer spin nuclei (fermions) or to the *same* tableau
for integer spin nuclei (bosons) such that the product of spatial and
spin wavefunctions is antisymmetric or symmetric, respectively.

The only difficulty in the foregoing procedure is correlating the
rotational symmetry species with the permutational tableaus. This is
done in Table 1.3 for various molecular symmetries [Harter, 1978].
We note that for a XY_6 molecule the u and g labels are often omitted
since such a molecule has never been seen to invert. In actuality a
species such as T_1 in Fig. 1.25 refers to an inversion doublet
$T_{1u} + T_{1g}$ for a XY_6 molecule.

For a given nuclear spin it is a simple matter to count the possible tableaus associated with the rotational symmetry species in Table
1.3 in order to determine the nuclear spin degeneracy. This degeneracy is often called the statistical weight which multiplies the Boltzmann factor [Eq. (1.127)] when calculating a given rotational species
population. Since K state degeneracies are now split, the degeneracy factor in Eq. (1.127) is $2J + 1$. For spin-1/2 nuclei, as for
spin-1/2 electrons, the number of unpaired nuclei in a tableau, as
shown in Fig. 1.10, determines the total nuclear spin I.

Thus for an XY_2 molecule such as water we have the following
total nuclear spin I and statistical weight $g = 2I + 1$:

TABLE 1.3

NUCLEAR SPIN TABLEAU		ROTATIONAL SPECIES			
FERMIONS	BOSONS				

A_1	A_2	B_1	B_2	XY_2 (BENT)
1	1	•	•	
•	•	1	1	

A_1	A_2	E	T_1	T_2	XY_4 (TETRAHEDRAL)
1	1	•	•	•	
1	1	•	•	•	
•	•	2	•	•	
•	•	•	1	1	
•	•	•	1	1	

XY_6 (OCTAHEDRAL)

A_{1g}	A_{2g}	E_g	T_{1g}	T_{2g}	A_{2u}	A_{1u}	E_u	T_{2u}	T_{1u}
1	•	•	•	•	•	•	•	•	•
•	•	1	•	•	•	•	•	•	1
1	•	1	•	1	•	•	•	1	•
•	1	•	1	•	•	•	•	1	1
•	1	•	•	•	1	•	•	•	1
•	•	1	1	1	•	•	1	1	1
1	•	•	•	1	•	1	•	•	•
•	•	•	1	1	•	1	•	1	•
•	•	•	1	•	1	•	1	•	1
•	•	•	•	1	•	•	1	•	•
•	•	•	•	•	1	•	•	•	•

$$A_1 \quad A_2 \quad B_1 \quad B_2$$

I: 0   0   1   1

g: 1   1   3   3

The $I = 0$ is often called the para modification and $I = 1$ the ortho modification. In Fig. 1.24 we see that both the n_A and n_C clusters occur in singlet-triplet pairs with a total statistical weight of $g = 4$. For a XY_4 molecule such as CH_4 (methane) or SiF_4, we have

$$A_1 \quad A_2 \quad E \quad T_1 \quad T_2$$

I: 2   2   0   1   1

g: 5   5   2   3   3

The E species has a statistical weight $g = 2$ since it is repeated twice in Table 1.3. For a XY_6 molecules such as SF_6 or UF_6, we have

	A_{1g}	A_{2g}	E_g	T_{1g}	T_{2g}	A_{1u}	A_{2u}	E_u	T_{1u}	T_{2u}
I:	0	—	—	1	0,2	0	1,3	1,2	1	—
g:	1	—	—	3	6	1	10	8	3	—

so that the inversion doublets A_1, A_2, E, T_1, T_2 have statistical weights 2, 10, 8, 6, 6, respectively, where we add u and g species.

For a molecule like CD_4, where D has a nuclear spin of 1, we may use the Boson tableaus in Table 1.3 and count the possible tableaus as we did for p $\ell = 1$ orbitals in Sec. 1.3.6 (see Fig. 1.11). The tableaus work for *any* nuclear spin.

1.5.4B. Hyperfine Splitting

We shall treat molecules with no quadupole moment so that the main contribution to hyperfine splitting is due to the interaction of the total nuclear spin with the magnetic field created by the rotating molecule. This interaction leads to a magnetic hyperfine energy of the form

$$E(\text{hyperfine}) = \frac{M_{aa} I_x \bar{J}_x + M_{bb} I_y \bar{J}_y + M_{cc} I_z \bar{J}_z}{\hbar^2} \tag{1.146}$$

where the angular momenta are defined along the principal axes of the molecule. Physically, the nuclear spin I vector will precess about the J vector due to the rotational magnetic field. The precession of the I vector about the J vector will be much slower than the precession of

the J vector about a body axis. In other words, the hyperfine splitting is much smaller than the fine-structure splitting. To find a trajectory-averaged body component of the I vector, we may then simply
resolve $\underline{J}$ about $\underline{I}$ and find the trajectory-averaged body component of
the J vector. The magnetic hyperfine energy then becomes

$$E(\text{hyperfine}) = \frac{\underline{J} \cdot \underline{I}}{J(J + 1)} \; \frac{M_{aa} \bar{J}_x^2 + M_{bb} \bar{J}_y^2 + M_{cc} \bar{J}_z^2}{\hbar^2} \tag{1.147}$$

where $\bar{J}_x^2$, $\bar{J}_y^2$, and $\bar{J}_z^2$ are averaged over the fine-structure trajectory.
If we define the total angular momentum by $\underline{F} = \underline{J} + \underline{I}$ as we did for the
hyperfine levels in atoms, then

$$E(\text{hyperfine}) = \frac{F(F + 1) - J(J + 1) - I(I + 1)}{2J(J + 1)} \; \frac{M_{aa} \bar{J}_x^2 + M_{bb} \bar{J}_y^2 + M_{cc} \bar{J}_z^2}{\hbar^2}$$

$$\tag{1.148}$$

and the hyperfine levels are split into $2I + 1$ components for $J \geqslant I$.
Thus the nuclear spin degeneracy g given previously is split by the
magnetic hyperfine interaction into separate hyperfine levels. For
prolate or oblate symmetric top molecules we have $M_{cc} = M_{bb}$ or $M_{bb} = M_{aa}$, respectively, with $\bar{J}_z = K_A \hbar$ or $\bar{J}_x = K_C \hbar$. For a spherical top
molecule we have $M_{aa} = M_{bb} = M_{cc} = M$ and the hyperfine splitting is
simply

$$E(\text{hyperfine}) = \frac{M}{2} [F(F + 1) - J(J + 1) - I(I + 1)] \tag{1.149}$$

Magnetic hyperfine splittings are usually on the order of tens of
kilohertz or less. For example, for H_2O the hyperfine constants are
$M_{aa} = 35.3$ kHz, $M_{bb} = -30.8$ kHz, and $M_{cc} = -33.1$ kHz [Bluyssen,
1967], while for SF_6 the hyperfine constant is $M = -5.27$ kHz [Ozier,
1977]. The A_2 line of SF_6 would split into an overlapping triplet and
septet with $I = 1$ and $I = 3$ according to the prescription above. We
shall see this splitting in the next section, where we discuss the hyperfine structure in the rotation-vibration infrared spectra of SF_6.

As in the atomic case, transitions occur only for $\Delta F = 0, \pm 1$. Hyperfine structure for pure rotational transitions is again in the far infrared and microwave region and cannot occur for spherical top molecules. We should mention that when the superfine splitting is less
than the hyperfine splitting, it is then possible to have a mixing of
rotational symmetry species and nuclear spin modifications, which has

been called superhyperfine structure. Such species mixing was analyzed by J. Bordé (Bordé, 1982) in saturation spectra of SF_6 taken by Ch. Bordé (Bordé, 1979).

1.5.5 Ro-vibrational Structure

With sufficient infrared resolution, the rotational structures of polyatomic molecules may be seen superimposed on the vibrational spectra in an analogous fashion to vibronic spectra. We shall not treat the even more complicated ro-vibronic structure, although many of the principles described here apply.

We have seen that for a two- or three-dimensional isotropic oscillator we may define a vibrational angular momentum ℓ which is quantized according to Eqs. (1.90) and (1.93). We may have doubly degenerate normal modes in both linear and symmetric top molecules, while triply degenerate normal modes are possible in spherical top molecules. The same applies to degenerate electronic states which carry electronic angular momentum when considering vibronic spectra.

The total ro-vibrational angular momentum for normal mode ν_i is then

$$\underline{J} = \underline{R} + \zeta_i \underline{\ell} \tag{1.150}$$

where $\underline{R}$ is the angular momentum of the molecular rotor itself and ζ_i is the *Coriolis constant* for effective vibrational angular momenta in the rotating frame of the molecule [Louck, 1976] acting on the coupled basis $|J, R, \ell\rangle$, where it is understood that

$$
\begin{aligned}
J &= |R + \ell|, \ |R + \ell - 1|, \ \ldots, \ |R - \ell| \\
K_R &= -R, \ -R + 1, \ \ldots, \ R
\end{aligned}
\tag{1.151}
$$

For normal modes of distinct vibrational symmetry species it can be shown that $\zeta_i = 1$. We shall see that the rotor angular momentum R and K_R plays the same role in ro-vibrational spectra as does J and K in pure rotational spectra. The kinetic energy (in cm^{-1}) of the rotor in the body frame is

$$E_{\ell J} = \frac{CR_x^2 + BR_y^2 + AR_z^2}{\hbar^2} \tag{1.152}$$

For a prolate symmetric top, the vibrational angular momentum must be in the z direction ($m = \pm\ell$) and

$$E_{\ell J} = BJ(J + 1) + (A - B)K_A^2 - \frac{2A\zeta_i J_z \ell_z}{\hbar^2}$$

where we have ignored the constant energy term $A\zeta_i^2 \ell_z^2$ since it may be considered as part of the vibrational energy. For a fundamental with $\ell = 1$ and $\ell_z = \pm\hbar$, we have the energy expression

$$E_{\ell J} = BJ(J + 1) + (A - B)K_A^2 \begin{cases} + 2A\zeta_i K_A & \quad R_z = K_A + 1 \quad (+) \text{ level} \\ \\ - 2A\zeta_i K_A & \quad R_z = K_A - 1 \quad (-) \text{ level} \end{cases}$$

We may obtain the expression for an oblate symmetric top by exchanging A and C.

For a spherical top the ro-vibrational energy expression for a triply degenerate mode ν_i is

$$E_{\ell J} = BJ(J + 1) - \frac{2B \zeta_i J \cdot \ell}{\hbar^2}$$

$$= BJ(J + 1) + B\zeta_i(R(R + 1) - J(J + 1) - \ell(\ell + 1))$$

For a fundamental with $\ell = 1$ we have

$$E = BJ(J + 1) \begin{cases} + 2B\zeta_i(J) & R = J + 1 \quad (+) \text{ level} \\ - 2B\zeta_i & R = J \quad\quad\ (0) \text{ level} \\ - 2B\zeta_i(J + 1) & R = J - 1 \quad (-) \text{ level} \end{cases} \quad (1.154)$$

Note that the vibrational degeneracy is split by the Coriolis term into Coriolus levels labeled by (+), (0), and (−) for both symmetric and spherical top molecules.

The usual selection rules for total rotational angular momentum J are still applicable for ro-vibrational transitions and give rise to three *branches* for $\Delta J = -1$, 0, and 1, which are labeled P(J), Q(J), and R(J), where J = R is the angular momentum of the ground vibrational state from which the transition is assumed to arise. These branches give rotational progressions resulting from the transition energy differences given by

$-2B(J + 1)$ P branch

0 Q branch   (1.155)

$2BJ$ R branch

Thus the Q branch is the narrowest, with a width due to the rotational fine structure.

The ro-vibrational transition selection rules depend on the induced dipole moment in the direction of vibrational displacement. For a doubly degenerate mode in a symmetric top molecule this displacement is perpendicular to the molecular symmetry axes and the resultant spectrum is called a *perpendicular band*. Since the symmetric top rotor has no dipole moment perpendicular to its symmetry axis, we have the selection rule $\Delta K_R = 0$. Thus, for transitions to the (+) and (−) Coriolis levels in Eq. (1.153) from the vibrational ground state where $K_R = K_A \hbar$, we have the selection rules

p branch: $\Delta K_A = -1$ to (+) level   (1.156a)

r branch: $\Delta K_A = +1$ to (−) level   (1.156b)

In all, we have six branches for perpendicular bands labeled $^rP_k(J)$, $^pP_k(J)$, $^rQ_k(J)$, $^pQ_k(J)$, $^rR_k(J)$, and $^pR_k(J)$, where $k = K_A$ for the vibrational ground state.

A singly degenerate mode of a symmetric top molecule has no vibrational angular momentum. In this case the vibrational dipole is in the direction of the symmetry axes and the resultant spectrum is called a *parallel band*. The rule $\Delta K_R = 0$ then leads to the same selection rules as for pure rotational microwave transitions:

q branch: $\Delta K_A = 0$ (1.157)

We have three branches for parallel bands labeled $^qP_k(J)$, $^qQ_k(J)$, and $^qR_k(J)$.

For the selection rules for an oblate symmetric top we simply change A to C in the equations above. Vibronic transitions will have the same selection rules for doubly degenerate electronic states that have electronic angular momentum.

For a spherical top molecule the rotor has no dipole moment, so that $\Delta R = 0$. Thus, for transitions to the (+), (0), and (−) Coriolis levels in Eq. (1.154) from the vibrational ground state where $R = J$, we have the selection rules

P branch: $\Delta J = -1$ to (+) level

Q branch: $\Delta J = 0$ to (0) level   (1.158)

R branch: $\Delta J = 1$ to (−) level

Because of the Coriolis splitting, the rotational transition frequencies given in Eq. (1.155) will have an effective rotational constant $B_{eff} = B - \zeta_i$.

Superimposed on the ro-vibrational spectra due to Coriolis splitting will be structure due to fine, superfine, and hyperfine splitting in the ground and vibrationally excited states. Figure 1.26 summarize the various regimes of molecular structure discussed in the preceding two sections corresponding to increasing spectral resolution. In Fig. 1.26a we show the infrared spectrum of the triply degenerate ν_3 band of SF_6 under low (0.05 cm^{-1}) resolution. The three branches of unresolved rotational structure are easily seen and we have indicated the contribution from the individual rotational transitions by labels in J. In Fig. 1.26b is shown a diode spectrum of the fine structure of the R(27)-R(31) manifold taken at 0.001-cm^{-1} Doppler-limited resolution. The fine-structure levels are labeled according to their K_3 and K_4 quantum numbers as discussed in Sec. 1.5.2. The fine-structure splitting of J levels results in the overlap of J manifolds, causing a very complicated spectra. Note that the fine structure of R(30) in Fig. 1.26b, although dominated by splitting in the ν_3 excited state, is similar to the ground-state splitting shown in Fig. 1.25. The saturation spectra in Fig. 1.26c shows the superfine splitting of the $K_3 = 27$ cluster of the R(28) manifold into A_1, T_1, T_2, and A_2 rotational species. The position and splitting of the fine-structure clusters can be accurately calculated using semiclassical methods described above. Finally, the hyperfine splitting of the A_2 rotational species is shown in the saturation spectrum in Fig. 1.26d. This spectrum is consistent with a triplet of levels superimposed on a septet of levels arising from the $I = 0$ and $I = 3$ total nuclear spin, as discussed previously.

1.6 SUMMARY

We have seen how the classical motion of electrons in atoms and vibration-rotations in molecules can be viewed in terms of elliptical trajectories. These trajectories have degenerate energies labeled by "principal" quantum numbers n for electrons, n for vibrations, and J for rotations. In all three cases this degeneracy is lifted by the precession of the ellipse, leading to energy levels with well-defined angular momentum ℓ of an electron, ℓ of vibration or K of rotation, respectively. These angular momentum states are the bases of electronic, vibrational, and rotational fine structure in atoms and molecules as seen in high-resolution spectra.

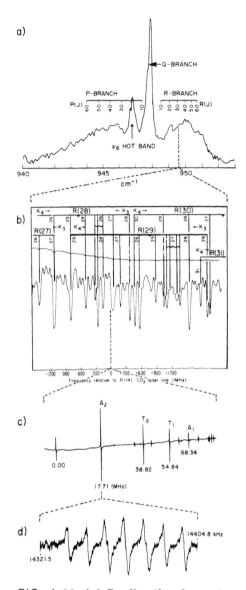

FIG. 1.26 (a) Rovibrational spectra of the ν_3 fundamental of SF_6 taken at 0.05-cm^{-1} resolution and $T = 200$ K. (b) Doppler-limited spectra of the R(27)-R(31) region showing the rotational fine structure [McDowell, 1976b]. (c) Saturation spectra with $\sim 30\text{-kHz}$ resolution of the R(28) $K_3 = 27$ cluster showing the superfine (A_2, T_2, T_1, A_1) splitting [Bordé, 1982]. (d) Saturation spectra with $\sim 1\text{-kHz}$ resolution of the A_2 species above, revealing the overlapping hyperfine splitting of the $I = 1$ and $I = 3$ sublevels [Bordé, 1979, 1982].

In all three cases the classical motion is nearly circular for highly excited states, with the angular momentum quantum number nearly equal to the principal quantum number, and the corresponding quantum wavefunction is very localized in phase space. Such classical "local-mode" states play an important role in highly excited atoms and molecules. In the hydrogen atom classical local-mode states can be shown to lead to the recently discovered level anticrossing anomalies in Rydberg levels split by static electric and magnetic fields [Kleppner, 1981].

The presence of local-mode vibrations in spherical top molecules such as SiF_4, SF_6, and UF_6 is important in understanding their multiphoton excitation and dissociation [Patterson, 1985]. Vibrational-level anticrossing anomalies as a function of anharmonicity have been demonstrated in the reference above for the highly excited fine-structure states of spherical top molecules and can be explained adequately by classical local-mode trajectories. Local-mode behavior is evident in the vibrations of other molecules that have been similarly modeled ([Clodius, 1984] and [Skodje, 1985]). It is possible that the classical description of these level anticrossings will lead to a framework for understanding chaotic vibrational motion for highly excited states and its quantum analog. As in the case of atoms, these level anticrossing anomalies are a result of dynamical symmetry breaking due to local-mode motion. Finally, the recent wavepacket techniques to describe vibronic Franck-Condon spectra depend largely on a semiclassical description of molecular vibrations ([Heller, 1981] and [DeLeon, 1984]).

Rotational local modes or "clusters" have been observed in high-resolution spectra of asymmetric and spherical top molecules [McDowell, 1976b]. The dynamical symmetry breaking of these localized rotations leads to mixing of nuclear spin modifications which has recently been observed in saturation spectra of SF_6 (Bordé, 1982).

High-resolution spectroscopy of highly excited levels of atoms and molecules is revealing unexpected structural details, which in many cases can most easily be understood, both qualitatively and quantitatively, using a semiclassical description. We are now seeing a reemergence of semiclassical techniques after some 50 years of obscurity.

ACKNOWLEDGMENTS

Work performed at the University of California, Los Alamos National Laboratory, Los Alamos, N.M., under the auspices of the U.S. Department of Energy.

The author would like to thank Drs. Robin McDowell, William Harter, and Eric Heller for their interest in this work and for their helpful suggestions.

BIBLIOGRAPHY

Cowan, R. D. (1981). *The Theory of Atomic Structure and Spectra*, University of California Press, Berkeley, Calif. Thorough quantum treatment of complex spectra using angular momentum algebra.

Edmonds, A. R. (1960). *Angular Momentum in Quantum Mechanics*, Princeton University Press, Princeton, N.J. Useful presentation of angular momentum coupling techniques in atomic and molecular spectroscopy.

Harter, W. G., and Patterson, C. W. (1976). A unitary calculus for electronic orbitals, in *Lecture Notes in Physics*, Vol. 49, Springer-Verlag, Berlin. Application of tableau techniques to complex atomic structure.

Heitler, W. (1954). *The Quantum Theory of Radiation*, Clarendon Press, Oxford. Thorough quantum mechanical treatment of radiative transitions.

Heller, E. J. (1981). *Potential Surface Properties and Dynamics from Molecular Spectra: A Time-Dependent Picture*, in *Potential Energy Surfaces and Dynamics Calculations* (D. G. Truhlar, ed.), Plenum Press, New York. Survey of semiclassical treatment of vibronic spectra.

Herzberg, G. (1945). *Molecular Spectra and Molecular Structure*, Vol. 2: *Infrared and Raman Spectra of Polyatomic Molecules*, Van Nostrand Reinhold, New York. Mostly classical and semiclassical treatments of vibrational and rotational molecular structure with many intuitive arguments.

Kuhn, H. G. (1969). *Atomic Spectra*, 2nd ed., Longmans, Green, London. Intermediate text emphasizing the vector model and nuclear effects in atomic spectra.

Loudon, R. (1983). *The Quantum Theory of Light*, 2nd ed., Clarendon Press, Oxford. Semiclassical and quantum treatments of absorption and emission of radiation.

Pauling, L. and Wilson, E. B. (1935). *Introduction to Quantum Mechanics*, McGraw-Hill, New York. Good semiclassical treatment of the hydrogen atom.

Shore, B. W., and Menzel, D. H. (1968). *Principles of Atomic Spectra*, Wiley, New York. Presentation of the quantum mechanics of simple and complex atoms.

Sobelman, I. I. (1979). *Atomic Spectra and Radiative Transitions*, Springer-Verlag, New York. Systematic treatment of coupling schemes in complex atomic spectra and transition multipole selection rules.

Steinfeld, J. I. (1985). *Molecules and Radiation: An Introduction to Modern Molecular Spectroscopy*, 2nd ed., MIT Press, Cambridge, Mass. Simple quantum treatment of electronic, vibrational, and rotational molecular structure, including quantum optical effects.

Townes, C. H., and Schawlow, A. L. (1975). *Microwave Spectroscopy*, Dover, New York. Discusses microwave transitions in atoms and molecules. Standard treatment of atomic and molecular fine and hyperfine structure.

White, H. E. (1934). *Introduction to Atomic Spectra*, McGraw-Hill, New York. Semiclassical treatment of atomic structure.

REFERENCES

[Allen, 1975] L. Allen and J. H. Eberly. *Optical Resonance and Two-Level Atoms*, Wiley-Interscience, New York, 1975.

[Bluyssen, 1967] H. Bluyssen, A. Dymanus, J. Reuss, and J. Verhoeven. Spin-rotation constants in H_2O, HDO, and D_2O, *Phys. Lett.*, A 25, 584 (1967).

[Bordé, 1979] C. J. Bordé, M. Ouhayoun, A. van Lerberghe, C. Saloman, S. Avrillier, C. D. Cantrell, and J. Borde. High resolution saturation spectroscopy with CO_2 lasers. Application to the ν_3 bands of SF_6 and OsO_4, in *Laser Spectroscopy*, 4th ed. (H. Walther and K. W. Rothe, eds.), Springer-Verlag, 1979, New York.

[Bordé, 1982] J. Bordé and Ch. J. Bordé, Superfine and hyperfine structures in the ν_3 band of SF_6, *Chem. Phys.* 71, 417-441 (1982).

[Clodius, 1984] W. B. Clodius and R. B. Shirts. Enhancement of Intramolecular Vibrational Energy Transfer Through Rotational Coupling, *J. Chem. Phys.* 81, 6224 (1984).

[Cowan, 1981] R. D. Cowan. *The Theory of Atomic Structure and Spectra*, University of California Press, Berkeley, Calif., 1981.

[De Leon, 1984] N. De Leon and E. J. Heller. Semiclassical Spectral Quantization: Application to Two and Four Coupled Molecular Degrees of Freedom, *J. Chem. Phys.* 81, 5957 (1984).

[Edmonds, 1960] A. R. Edmonds. *Angular Momentum in Quantum Mechanics*, Princeton University Press, Princeton, N.J., 1960.

[Fox, 1977] K. Fox, H. W. Galbraith, B. J. Krohn, and J. D. Louck. Theory of level splitting: spectrum of the octahedrally invariant fourth-rank tensor operator, *Phys. Rev.*, *A15*, 1363 (1977).

[Galbraith, 1983] H. W. Galbraith, J. R. Ackerhalt, and P. W. Milonni, Chaos in the multiple photon excitation of molecules due to vibration-rotation coupling at lowest order, *J. Chem. Phys. 79*, 5345-5350 (1983).

[Halonen, 1983] L. Halonen and M. S. Child, Model stretching overtone eigenvalues for SF_6, WF_6, and UF_6, *J. Chem. Phys. 79*, 559-570 (1983).

[Hänsch, 1971] T. W. Hänsch, I. S. Shahin, and A. L. Schawlow. High-resolution saturation spectroscopy of the sodium D lines with a pulsed dye laser, *Phys. Lett.*, *27*, 707 (1971).

[Harter, 1976] W. G. Harter and C. W. Patterson. A unitary calculus for electronic orbitals, in *Lecture Notes in Physics*, Vol. 49, Springer-Verlag, Berlin, 1976.

[Harter, 1978] W. G. Harter, C. W. Patterson, and F. J. da Paixao. Frame transformation relations and multipole transitions in symmetric polyatomic molecules, *Rev. Mod. Phys.*, *50*, 37 (1978).

[Harter, 1984] W. G. Harter and C. W. Patterson. Rotational energy surfaces and high-J eigenvalue structure of polyatomic molecules, *J. Chem. Phys.*, *80*, 4241 (1984).

[Heller, 1981] E. J. Heller. Potential surface properties and dynamics from molecular spectra: a time-dependent picture, in *Potential Energy Surfaces and Dynamics Calculations* (D. G. Truhlar, ed.), Plenum Press, New York, 1981.

[Herzberg, 1945] G. Herzberg. *Molecular Spectra and Molecular Structure: II. Infrared and Raman Spectra of Polyatomic Molecules*, Van Nostrand Reinhold, New York, 1945.

[Hodgkinson, 1982] D. P. Hodgkinson, A. J. Taylor, D. W. Wright, and A. G. Robiette, Non-linear infrared spectroscopy of SF_6 at 30K, *Chem. Phys. Lett. 90*, 230-234 (1982).

[Kim, 1979] K. C. Kim, W. B. Person, D. Seitz, and B. J. Krohn. Analysis of the ν_4 ··· spectra of SF_6, *J. Mol. Spectrosc.*, *76*, 322 (1979).

[Kleppner, 1981] D. Kleppner, M. G. Littman, and M. L. Zimmerman. Highly excited atoms, *Sci. Am.*, 130 May, 1981.

[Louck, 1976]   J. D. Louck and H. W. Galbraith.   Eckart vectors, Eckart frames, and polyatomic molecules, *Rev. Mod. Phys.*, *48*, 69 (1976).

[Loudon, 1983]   R. Loudon, *The Quantum Theory of Light*, 2nd ed., Clarendon Press, Oxford, 1983.

[Lyman, 1976]   J. L. Lyman and S. D. Rockwood.   Enrichment of boron, carbon, and silicone isotopes by multiple-photon absorption of $10.6 \mu m$ laser radiation, *J. Appl. Phys.*, *47*, 595 (1976).

[McDowell, 1974]   R. S. McDowell, L. B. Asprey, and R. T. Paine. Vibrational spectrum and force field of uranium hexafluoride, *J. Chem. Phys.*, *61*, 3571 (1974).

[McDowell, 1976a]   R. S. McDowell, J. P. Aldridge, and R. F. Holland.   Vibrational constants and force field of SF_6, *J. Phys. Chem.*, *80*, 1203 (1976).

[McDowell, 1976b]   R. S. McDowell, H. W. Galbraith, B. J. Krohn, C. D. Cantrell, and E. D. Hinkley.   Identification of the SF_6 transitions pumped by a CO_2 laser, *Opt. Commun.*, *17*, 178 (1976).

[McDowell, 1982]   R. S. McDowell, C. W. Patterson, and W. G. Harter, The modern revolution in infrared spectroscopy, in *Los Alamos Science*, Los Alamos National Laboratory, Winter/Spring, (1982).

[Ozier, 1977]   I. Ozier, P. N. Yi, and N. F. Ramsey.   Rotational magnetic momentum spectrum of SF_6, *J. Chem. Phys.*, *66*, 143 (1977).

[Pack, 1976]   R. T. Pack.   Simple theory of diffuse vibrational structure in continuous UV spectra of polyatomic molecules, *J. Chem. Phys.*, *65*, 4765 (1976).

[Patterson, 1983]   C. W. Patterson and A. S. Pine, Predictions of multiphoton resonances in SF_6 and SiF_4, *Opt. Commun.* *44*, 170-174 (1983).

[Patterson, 1985]   C. W. Patterson.   Quantum and semiclassical description of a triply degenerate anharmonic oscillator, *J. Chem. Phys.*, *83*, 4618 (1985).

[Pauling, 1935]   L. Pauling and E. B. Wilson, *Introduction to Quantum Mechanics*, McGraw Hill, New York, 1935.

[Skodje, 1985]   R. T. Skodje, F. Borondo, and W. P. Reinhardt. The semiclassical quantization of nonseparable systems using the method of adiabatic switching, *J. Chem. Phys.* *82*, 4611 (1985).

[Snider, 1983] D. R. Snider, Elliptical orbits in quantum mechanics, *Am. J. Phys. 51*, 801-803 (1983).

[Sobelman, 1979] I. I. Sobelman. *Atomic Spectra and Radiative Transitions*, Springer-Verlag, New York, 1979.

[Takami, 1984] M. Takami, T. Oyama, T. Watanabe, S. Namba, and R. Nakane. Cold jet infrared absorption spectroscopy: the ν_3 band of UF_6, *Jpn. J. Appl. Phys. Lett.*, *23*, L88 (1984).

[Townes, 1975] C. H. Townes and A. L. Schawlow. *Microwave Spectroscopy*, Dover, New York, 1975.

[White, 1934] H. E. White. *Introduction to Atomic Spectra*, McGraw-Hill, New York, 1934.

2

Lasers for Spectroscopy

JOHN R. MURRAY   *Lawrence Livermore National Laboratory,
Livermore, California*

2.1 INTRODUCTION

This chapter reviews the underlying physics and the technology of
the most common spectroscopic laser sources. It assumes that the
reader is reasonably familiar with the basic principles of optical
physics but may not be familiar with the issues governing which laser
sources and technologies one might choose to implement a particular
spectroscopic experiment. The purpose of the chapter is to aid in
that choice. The basic principles of optics and laser physics are
treated in a number of standard texts, such as Yariv (1971), Sieg-
n.an (1971), Svelto (1982), and Tarasov (1983).

Section 2.2 describes the properties of the most important spectroscopic lasers. Tunable dye lasers are emphasized since they are
the most common sources for laser spectroscopy, and tunability is also
emphasized in the other lasers discussed there. Section 2.3 discusses
the physical principles and engineering implementation of laser resonator design, wavelength selection, and tuning. It also briefly discusses the special problems of nanosecond-pulse systems and laser
oscillator-amplifier chains. Dye lasers are again emphasized, but the
same principles apply to any laser medium. Section 2.4 discusses the
physics and technology of nonlinear frequency conversion as used to
generate tunable sources at wavelengths or output powers not accessible with the lasers described in Section 2.2.

There are standard laser handbooks that contain extensive lists
of laser line frequencies, important parameters of nonlinear crystals,
and the like. Certain articles in these references will be called out by
name in the references, but there are also many other articles in
these handbooks which discuss topics of importance to spectroscopic
lasers. The two most comprehensive handbooks are the Laser Hand-

book, Vols. 1 and 2 (Arecchi and Schulz-Dubois, ed., 1972), Vol. 3
(Stitch, ed., 1979); and the CRC Handbook of Laser Science and
Technology (Weber, ed., 1982), which is an updated version of the
CRC Handbook of Lasers (Pressley, ed., 1971). The earlier edition
is still useful since it contains some data on incoherent sources, propagation, and materials that are not included in the 1982 edition. Tables
of Laser Lines in Gases and Vapors, 3rd ed. (Beck, 1980) is also useful. The term "standard references" in this chapter means handbooks
such as these.

There are no references to specific manufacturers or product data
in this chapter since there are numerous spectroscopic laser systems
available commercially and the model designations and performance
specifications of these change frequently. The standard references
for a quick survey of systems currently available from manufacturers
are the Laser Focus Buyer's Guide published annually by Laser Focus
magazine, Box 1111, Littleton, MA 01460 and the Lasers and Applications Designer's Handbook and Product Directory published annually
by Lasers and Applications magazine, 3220 W. Sepulveda, Torrance,
CA 90505.

The journal publications in the list of references are a representative sample of recent publications dealing with topics important for
spectroscopic lasers, but this list is not a comprehensive review of all
such publications. Papers discussing the design or physics of spectroscopic lasers appear most frequently in the journals listed in Table
2.1, and many further references can be found there.

It is often necessary to measure output parameters of spectroscopic
lasers. The Laser Parameter Measurements Handbook (Heard, 1968) is
a good introduction to basic techniques for such measurements.

2.2 PROPERTIES OF THE PRINCIPAL SPECTRO-
   SCOPIC LASER SOURCES

2.2.1 Dye Lasers

Optically pumped liquid solutions of organic dyes are the most flexible
and the most common lasers for spectroscopy from about 340 nm in the
ultraviolet through the visible to about 1200 nm in the infrared. In
this context a "dye" is an organic molecule containing a chain of con-
jugated double bonds, which is an electronic structure leading to a
broad absorption band at relatively low energy [Schäfer 1972, 1973;
Nair, 1982]. Many of these molecules reemit absorbed photons with
high efficiency and a subset of these efficient fluorescers have energy
levels structured in such a way that population inversion and laser oscillation are possible. There are more than 100 dyes which will oscillate in dye lasers, and each of these can be tuned continuously over a
limited range of wavelengths. This gives overlapping coverage of the

TABLE 2.1 Journals That Frequently Publish Papers Describing
Spectroscopic Laser Systems

Letter journals

Applied Physics Letters, American Institute of Physics, New York

Optics Communications, North-Holland, Amsterdam

Optics Letters, Optical Society of America, Washington, D.C.

Regular journals

Applied Optics, Optical Society of America, Washington, D.C.

Applied Physics B: Photophysics and Laser Chemistry,
Springer-Verlag, Berlin

IEEE Journal of Quantum Electronics, Institute of Electrical and
Electronic Engineers, New York

Kvantovaya Elektronika, Radio i Svyaz', Moscow; English translation, *Soviet Journal of Quantum Electronics*, American Institute of
Physics, New York

Optics and Quantum Electronics, Chapman & Hall, London

Review journal

Progress in Quantum Electronics, Pergamon Press, Oxford

---

spectrum so that any chosen wavelength between these limits can be
generated in a dye laser, usually in several different dyes. The many
commercial suppliers of dye laser apparatus and dyes can provide detailed tuning ranges for specific dyes in those systems, and the information is also available in many reviews and reference articles [Drex-
hage, 1973; Nair, 1982; Maeda, 1984; Wallenstein, 1979]. This chapter
will focus on general physical properties.

Figure 2.1 shows a schematic energy-level diagram of a laser dye
molecule. An electronic state of the molecule has a minimum energy
for some particular equilibrium position of the nuclei composing the
molecule, so if we draw the energy as a function of a generalized "normal coordinate" of the nuclear configuration, each electronic state will
have a potential well with its minimum at the equilibrium position of
the nuclei. Vibrational excitation raises the internal energy of the
molecule above the bottom of the well and the normal coordinate oscillates between the classical turning points where the internal energy
of the molecule intersects the sides of the potential well. There are of
course very many normal vibrational coordinates for a molecule con-

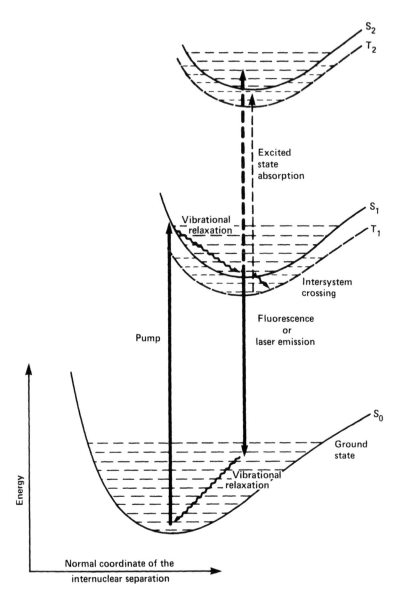

FIG. 2.1 Schematic representation of the energy levels of a laser dye molecule. Absorption of a pump photon takes the molecule to a high vibrational level of the S_1 excited state. Vibrational relaxation takes the molecule to the bottom of the S_1 state, where there is an inversion and gain for transitions to the empty upper vibrational levels of the S_0 ground state. Intersystem crossing which takes the molecule from S_1 to T_1 and excited-state absorption from the S_1 and T_1 states can impede laser oscillation.

taining a large number of nuclei, as well as three sets of rotational
levels superposed on these, but the highly simplified picture of Fig.
2.1 is adequate for most discussions of the laser properties of dyes.
These spectroscopic features are discussed in more detail by Schäfer
[Schäfer, 1972, 1973] with particular emphasis on laser dyes. Complex dye molecules have closely spaced vibrational and rotational levels,
and relaxation between these levels is very fast. This broadens the
individual levels so that they overlap to form a quasi-continuous spectrum rather than showing discrete rotational and vibrational structure.

A molecule in an excited electronic state frequently has a different
equilibrium position of its constituent nuclei than in the ground state,
so that the minimum of the excited-state potential well is shifted with
respect to the ground-state well, as shown in Fig. 2.1. The Franck-
Condon principle requires that an electronic transition in a molecule
must occur without a change in the position of the nuclei, so a molecule in a low vibrational level of the ground state which absorbs a photon must go to a high vibrational level of the excited electronic state
S_1. The high vibrational levels of the excited state relax to low vibrational levels in a picosecond or less in a typical dye molecule in solution, which is much faster than the S_1 radiative lifetime of a few nanoseconds, so the fluorescence from S_1 to the ground state comes from
low vibrational levels. The shift in the potential wells requires that
the fluorescence terminate on a high vibrational level of the ground
state, and this vibration also quenches in less than a picosecond.
This arrangement of pump, fluorescence, and fast relaxation processes is exactly that required for the four-level laser oscillator discussed
in standard texts [Yariv, 1971; Siegman, 1971; Svelto, 1982; Tarasov,
1983]. The fluorescent emission is at longer wavelength or smaller
photon energy than the pump. The shift between the peaks of the absorption and fluorescence spectra has been named the Stokes shift in
honor of an early investigator of fluorescence.

Figure 2.2 shows the S_1 absorption band and the fluorescence
band of one of the most important laser dyes, rhodamine 6G. Note
that there is still a significant overlap between the fluorescence and
absorption even though the fluorescence is shifted to longer wavelength. Net gain and laser oscillation are possible only over the region of the spectrum where gain in greater than absorption, so a typical rhodamine 6G dye laser has a short-wavelength tuning limit of perhaps 555 nm or longer even though the fluorescence extends to much
shorter wavelengths and the gross gain is proportional to fluorescence intensity. The long-wavelength limit of laser oscillation is set
by other losses in the system and is typically about 640 nm: peak
laser output is at about 590 to 600 nm since the net gain (but not the
fluorescence) is highest there.

The widest tuning ranges are found in dyes with a Stokes shift
large enough to give very small overlap of fluorescence and absorption spectra. Figure 2.3 shows the spectra of one such dye, 4-dicy-

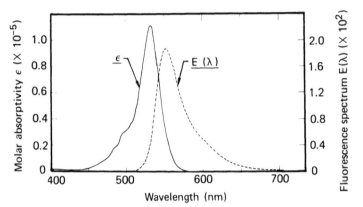

FIG. 2.2 Absorption and fluorescence spectra of the laser dye
rhodamine 6G. The absorption spectrum is the solid line and the
fluorescence spectrum the dashed line. (Courtesy of P. R. Hammond,
reproduced by permission.)

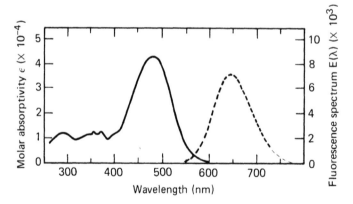

FIG. 2.3 Absorption and fluorescence spectra of the laser dye DCM.
The dye DCM has a large Stokes shift in its fluorescence, which gives
a very wide tuning range, and broad absorption bands which couple
very well to broadband radiators such as flash lamps (compare Fig.
2.2). (Courtesy of P. R. Hammond, reproduced by permission.)

anomethylene-2-methyl-6-p-dimethylaminostyryl-4H-pyran (DCM),
which was developed specifically for use in dye lasers and is becoming
one of the more important high-efficiency laser dyes [Hammond, 1978].
A DCM dye laser can be tuned from about 600 to a little longer than
720 nm.

The S_1 state of Fig. 2.1 may undergo nonradiative deactivation or
"internal conversion" to the S_0 ground state in which the electronic
excitation is lost without emission of fluorescence. This does not prevent laser oscillation but does decrease laser efficiency. The best
laser dyes have low internal conversion rates, or, to state the same
point differently, high fluorescence efficiency.

There are other excited electronic states in dye molecules which
can influence laser properties. There are allowed electronic absorptions from the ground state to higher-lying electronic states as well
as to the lowest singlet state S_1. The absorption from 250 to 400 nm
in the DCM spectrum of Fig. 2.3, for example, takes the molecule to
these higher-lying states. In all but a few cases (which do not include laser dyes) these excited singlet states are quenched to the
first excited singlet state much faster than they can radiate so absorption at wavelengths shorter than the S_1 absorption band can also be
used to pump the laser. These absorptions are broad and intense in
DCM dye, which makes it a good candidate for pumping by short-
wavelength sources or broadband radiators such as xenon flash lamps
[Hammond, 1978; Weber, 1983]. As Fig. 2.2 shows, rhodamine 6G has
very little absorption from about 360 to 450 nm, so sources in this
wavelength range do not pump this dye well. There is significant absorption from 360 nm to shorter wavelength (see Drexhage, 1973, Fig.
4.4) and rhodamine 6G can be used with pump wavelengths in that region.

For each excited singlet state of a dye molecule there is an excited
triplet state which lies at a slightly lower energy [Schäfer, 1973; Nair,
1982]. Optical transitions between singlet and triplet states are forbidden to lowest order and are typically about 1000 times weaker than
allowed singlet-singlet or triplet-triplet transitions. The first excited
triplet state T_1 can therefore have a lifetime of microseconds in solution. There is a small intersystem crossing rate for conversion of S_1
to T_1 in solution, so the T_1 state population can grow for microseconds
before it reaches a steady state. Molecules in the T_1 state have allowed absorption bands to higher triplet states and these may overlap
the S_1-S_0 fluorescence spectrum and impede laser oscillation. Intersystem crossing rates and triplet absorptions have not been systemati-
cally studied for all dyes, but are known for a few. The triplet absorption of rhodamine 6G, for example, extends from about 560 nm to
the near infrared [Dempster, 1974] and can suppress laser oscillation
over the entire tuning range of that dye if the triplet concentration
rises high enough. Since the triplet population accumulates on a mi-

crosecond time scale, triplet absorption is most important in dye lasers with pump pulses of a microsecond or longer. If molecules such as cyclooctatetraene (COT), which have low-energy triplet states, are added to the solution, dye triplets will transfer energy to them in collisions, and this can be a useful way to keep the dye triplet population low if the absorptions of the triplet quencher do not interfere with the laser gain or pump bands. Molecular oxygen dissolved in the dye solution quenches triplets very well but can also increase the rate of S_1-T_1 intersystem crossing. Dissolved O_2 improves the performance of rhodamine 6G dye lasers at moderate concentrations, but it can degrade the laser performance of some other dyes.

There is an absorption from S_1 to higher singlet states similar to the absorption from T_1 to higher triplets, and this excited-state absorption can also limit the performance of dye lasers. It is particularly difficult to study such absorptions because of the short S_1 radiative lifetime, but there have been a few measurements [Dolan, 1976; Sabar, 1977; Hammond, 1979; Magde, 1981]. In the absence of data we may infer that those dyes which perform well in lasers have fairly low S_1 absorption at the laser and pump wavelengths, while other efficient fluorescers with appropriate Stokes shifts which lase poorly or not at all probably have high excited-state absorption. The latter category includes a number of well-known efficient fluorescers such as perylene, which was one of the first compounds tested as a potential laser dye in 1964. The first successful dye-laser experiments were published two years later (reviewed by Schäfer [Schäfer, 1973]).

Some dyes, such as the rhodamine 6G of Fig. 2.2, do not absorb well over a considerable part of the spectrum, and this reduces their performance with broadband pumping. There may also be convenient laser pump sources in these regions which cannot be used with the dye. Several researchers have attempted to correct this situation using a mixture of two dyes, one of which is a converter that absorbs pump radiation and reradiates it into the absorption band of a laser dye. This can sometimes improve performance but has never given such a dramatic improvement that it has become standard practice. One difficulty is that the loss processes and absorptions in the converter dye must now be considered in addition to those in the laser dye. Pavlopoulos [Pavlopoulos, 1981] shows an increase in output from the dye mixture perylene-rhodamine 110 of about a factor of 2 compared to flash-lamp pumping of pure rhodamine 110, which has an absorption spectrum similar to the rhodamine 6G spectrum of Figure 2.2 but shifted to slightly shorter wavelengths. Marason [Marason, 1982] discusses a mixture of DCM and the dye LD700 for operation in the infrared out to 805 nm using a 515-nm argon ion laser pump which is poorly absorbed by LD700 alone. These papers also have references to a selection of earlier papers on dye mixtures.

The energy of the pump photon absorbed by a dye is adequate to break chemical bonds, and some small fraction of the excited dye molecules may dissociate or react with other molecules in the solution rather than fluoresce back to the ground state. This causes a gradual loss of dye molecules and may also generate other species which absorb at the laser or pump wavelengths and reduce laser output. In general, the most stable dyes are those emitting in the near-ultraviolet and red-orange, while the least stable are those in the blue-green and infrared. Dye degradation is most severe with ultraviolet pump photons since these carry the most energy. The relative lifetime of a number of laser dyes pumped by a 308-nm XeCl laser has recently been studied in some detail [Antonov, 1983].

The pump intensity must be rather large to reach oscillation threshold in a laser dye, ranging from a low of a few kilowatts per square centimeter to a high of perhaps a hundred kilowatts per square centimeter, depending on the dye, cavity losses, and gain length [Snavely, 1973; Schäfer, 1973; Nair, 1982]. It is easy to reach these power densities over small areas using laser pump sources, and most dye lasers used for spectroscopy are pumped by other lasers, but intense flash lamps filled with xenon or other gases can also be used.

The high pump intensity in the typical dye laser causes a rapid increase in the temperature of the dye solution during the pulse which can change the index of refraction, causing optical distortion. High temperature in the pumped volume can also decompose the dye. The typical continuous-wave (CW) dye laser controls thermal effects by using a CW laser pump of several watts focused to a spot a few micrometers in diameter on a jet of dye solution moving at about 10 m/s. This gives a residence time of only about 1 µs for any particular volume of dye in the laser-active region and reduces thermal effects. The high-speed flow also "mechanically quenches" the buildup of triplet states by sweeping them out of the active volume. The most successful CW dye lasers use a free, windowless jet of dye solution in air since this eliminates any problems with thermal loading and boundary layers at window surfaces. The solvent used in a free jet must have a viscosity high enough that the jet does not break up into droplets for at least a few millimeters after exiting the nozzle, so CW dye lasers use viscous solvents such as ethylene glycol rather than the light alcohols and similar solvents used in pulsed systems.

Laser-pumped pulsed dye lasers usually use pump pulses between about 5 and 50 ns, and there are only minor thermal effects on this time scale. There must be hydrodynamic motion in the solution to convert a temperature gradient to a density gradient and hence a change in the index of refraction before the laser-beam quality can be affected by thermal gradients. The characteristic hydrodynamic velocity is the speed of sound, which is about 1 µm/ns in liquids, so density disturbances can develop over distances of at most tens of micrometers

during these short pump pulses, and the refractive index change due
to thermal effects is negligible. It is still desirable to flow the dye
solution to avoid building up an index gradient between pulses or
overheating a small volume of dye in the laser-active volume. Flash-
lamp-pumped pulsed dye lasers, on the other hand, have output pulse
lengths in the microsecond-to-millisecond range, and thermally induced
index gradients have a severe effect on their operation and beam quality.

2.2.2 Semiconductor Lasers

The radiative recombination of carriers in direct-gap semiconductors
is an efficient process which leads to an important class of lasers. The
physics of these devices will be sketched briefly here with many omis-
sions: more specialized reviews should be consulted for further details [Stern, 1972; Kressel, 1977; Thompson, 1980].

Figure 2.4 is a representation of the allowed energy of states in
the valence and conduction bands of a direct-gap semiconductor as a
function of the carrier momentum in the crystal. The momentum carried by a photon is small compared to the momentum of an electron or
hole, so optical transitions between the valence and conduction bands
are nearly vertical lines on this diagram. Absorption of a photon takes
an electron from the valence band to the conduction band, leaving a
hole in the valence band. The electron and hole momenta relax very
rapidly toward thermal equilibrium and fill a range of states near $k = 0$
up to a Fermi energy E_{fe} for electrons in the conduction band and
down to an energy E_{fh} for holes in the valence band. The relaxed
distributions of electrons and holes can recombine, with the emission
of a photon having an energy slightly greater than the band-gap energy, and there is gain for this emission for frequencies $h\nu < E_{fe} - E_{fh}$,
since there are no electrons in the valence band which can reabsorb at
these frequencies. There can be no emission or gain at energies less
than the band-gap energy since that energy is the minimum separation
between the valence and conduction bands.

A semiconductor laser can be pumped by any mechanism which is
capable of rapidly filling the minimum of the conduction band and simultaneously emptying the top of the valence band. Optical pumping
at a wavelength shorter than the laser emission obviously does just
that. The valence-band electrons might also be excited to the conduction band by collisions with energetic electrons injected into the semiconductor (electron-beam pumping), or accelerated within the semiconductor by very high applied electric fields (impact ionization pumping).

The most common semiconductor lasers use a somewhat different
mechanism to generate an inversion near a pn junction in the material,
although there are a few semiconductors in which pn junctions cannot

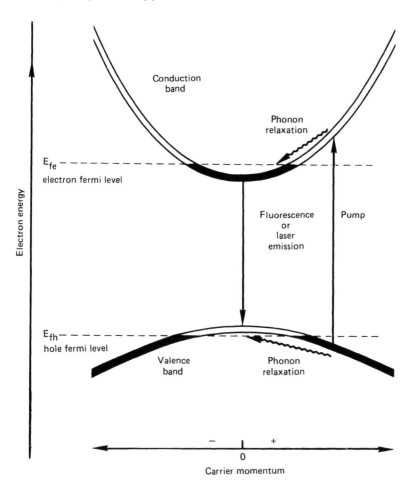

FIG. 2.4 Schematic representation of the allowed energy levels of carriers in a semiconductor. Absorption of a pump photon, or another excitation mechanism, excites an electron from the valence band to the conduction band. The electron in the conduction band and the "hole" in the valence band relax quickly to near zero momentum. There is then gain on the recombination transition between electrons at the bottom of the conduction band and holes at the top of the valence band.

be fabricated using present technology and these materials must use optical or electron-beam pumping. Suppose that there is a current flowing through a pn junction between a block of material doped to have an excess of electrons in the conduction band (type n) and a block doped to have an excess of holes in the valence band (type p).

Electrons enter the type-n material from the external circuit and diffuse in the direction of the electric field to the pn junction. At the
same time electrons are leaving the type-p material into the external
circuit, generating holes that diffuse back to the junction in the opposite direction along the electric field. Near the junction the carriers
diffuse into the oppositely doped material for a short distance, perhaps
a few micrometers. In this very narrow region, then, there is a population of electrons in the conduction band, a population of holes in the
valence band, and gain for a recombination laser. The radiative recombination at the junction completes the electrical circuit through the
diode.

The best modern semiconductor diode lasers take advantage of
Junction diode lasers can have active regions comparable to or
even smaller than the wavelength of their radiation. Optical absorption coefficients and electron stopping power are large in these materials, so laser or electron-beam-pumped semiconductor lasers also typically have gain regions with dimensions measured in micrometers
[Roxlo, 1982], although some devices have been built using two-photon
rather than single-photon absorption as a pumping technique, which
gives larger gain volumes [Vaucher, 1982; Basov, 1966]. The semiconductor crystals have a high index of refraction and that index may
vary significantly with doping. Diffraction, waveguiding, and other
electromagnetic propagation effects are therefore much more important
in these lasers than in other classes of lasers, where they can frequently be neglected.

The best modern semiconductor diode lasers take advantage of
waveguiding effects to give lower laser threshold current and better
beam quality and have a structure composed of several layers rather
than a single simple pn junction. Figure 2.5 shows the more complex
multilayered structure of a particularly popular design called a "double heterojunction" as executed in a mixed lead-tin telluride [Groves,
1974; Tomasetta, 1974]. The design principles of this and other laser
structures are discussed in the reviews listed in the first paragraph
of this section.

The most common semiconductor laser cavity dispenses with external optics and depends on the reflection at the index discontinuity between the semiconductor crystal and air for optical feedback, but precise tuning is easier in a controlled, external cavity with wavelength-
selective elements and antireflection coatings on the crystal.

It is clear from Fig. 2.4 that semiconductor lasers can be tuned
over a finite bandwidth. As an example, a typical GaAs pn-junction
laser operated in an external cavity will tune from about 815 to 826 nm,
or about 1.5% of its center wavelength [Fleming, 1981]. For spectroscopic applications it is more common to tune these lasers by varying
the diode composition and temperature as discussed below.

In binary semiconductors such as GaAs, PbSe, and CdTe, it is
possible to make mixed semiconducting solid solutions in which other
elements from the same column of the periodic table substitute for

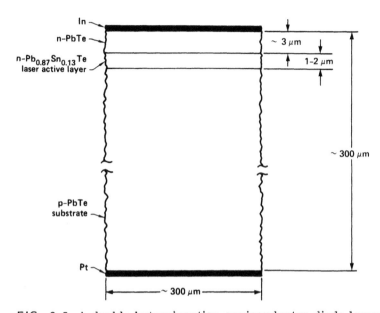

FIG. 2.5 A double heterojunction semiconductor diode laser executed in lead-tin telluride. The figure shows a cross section perpendicular to the laser axis. The length along the laser axis is typically 500 to 600 μm. The external circuit is connected to the indium and platinum metal layers. The end faces, which serve as laser mirrors, are polished or cleaved flat. The side faces are roughened to reduce feedback transverse to the desired laser axis, or the upper metal layer near the junction may be a narrow "stripe" which confines the current to a narrow region transverse to the laser axis. Some designs use a mixed $PbSe_yTe_{1-y}$ alloy in place of the PbTe layers shown in the figure since this mixed alloy gives a better match between the lattice spacing of the PbSnTe laser-active layer and the other layers in the structure [Kasemset, 1980]. (Adapted with permission from Tomasetta [Tomasetta, 1974] and Kasemset [Kasemset, 1980].)

either of the two components. This substitution can shift the band gap of the material over wide ranges. The band gap of $Pb_xSn_{1-x}Se$, for example, can shift from 0.17 eV in pure PbSe smoothly to zero at a Sn fraction of 40%. Laser diodes of mixed lead-tin selenide can be designed to have center wavelengths between about 8 and 34 μm in the infrared. Similar lead salt diodes can be designed with sulfur or tellurium replacing selenium. The band gap of these materials has a strong temperature dependence, so a change in temperature can give a rather large tuning range in some advanced designs. Groves [Groves, 1974] describes an example in $Pb_{0.88}Sn_{0.12}Te$ which tunes from 8.2 μm at 80 K to 10.5 μm at 12 K, for a tuning range of 25% of center wavelength. The lead salt diode lasers are particularly useful for vibrational spectroscopy in the infrared. They are limited to CW, single-mode output power of less than a few milliwatts, so they are useful only for experiments that do not require an intense source.

It is possible to build optically pumped lead salt lasers using, for example, a GaAs semiconductor diode laser or Nd:YAG laser as a pump [Mooradian, 1973], but junction diode lasers are much more common. Lead salt diode lasers are available from several commercial suppliers.

2.2.3 Color-Center Lasers

Color centers are optically active lattice defects in the crystal structure of solids [Mollenauer, 1975, 1982; Arkhangel'skaya, 1980]. The typical color center is a negative-ion vacancy which leads to an excess positive charge in a small region of the crystal. A free electron can become trapped in this potential well. Optically allowed transitions between bound states of the electron in the potential well then add new absorption bands to the spectrum of the crystal. An electron in an excited state of the color center has a different charge distribution than in its ground state, so the equilibrium positions of nearby ions will be slightly different for the various electronic states, and absorption from the ground state will take the color center to a "vibrationally excited" level of the excited state. Phonon (vibrational) relaxations of crystals are rapid, so the excited state will quickly relax to its equilibrium ion configuration and may then fluoresce to a "vibrationally excited" level of the ground state. This sequence is identical to the energy transfer in organic dye lasers which we have just discussed, and consequently many of these color centers can be used to construct broadly tunable lasers which have properties very similar to dye lasers. Color-center lasers are particularly useful in the near infrared from 0.8 to 4 μm, where they extend the range of broadly tunable lasers beyond the 1-μm practical cutoff of dye lasers. Like the dyes, color-center lasers can operate either pulsed or CW.

The limiting processes discussed above for dye lasers also apply to color centers: the Stokes shift must be adequate to separate the absorption and fluorescence, excited-state absorption must be absent, and nonradiative deactivation must be slow. These processes eliminate some types of color centers from consideration as lasers as they do for the dyes.

The proper type of color center must be deliberately induced in a crystal to prepare it for use in a laser, which requires very careful processing and handling [Mollenauer, 1978b; Litfin, 1978]. Lattice vacancies can migrate through the crystal and this may convert the laser-active centers into other types not useful for lasers. Migration can be reduced by cooling the crystal, so these lasers are often operated or even stored at liquid-nitrogen temperature (77 K). Color centers can be stabilized by association with an impurity ion which traps the center and reduces migration significantly. These impurity-associated centers can have a reasonable storage lifetime at room temperature. Some of the impurity-associated centers have significant non-radiative deactivation at room temperature, however, so they are usually cooled during operation to improve the laser performance.

Table 2.2 summarizes the properties of a number of color-center lasers. The F_2^+ centers are notable for high CW output powers of order 1 W: the other lasers typically have an output in the tens of milliwatts. Most color-center lasers have been laser pumped, but flashlamp pumping is also possible [Litfin, 1983]. Prepared color-center-laser crystals and apparatus are available from at least one commercial supplier.

2.2.4 Impurity-Ion Lasers in Solids

2.2.4A. General

Transitions between electronic levels of optically pumped impurity ions in crystals or glasses are both the oldest and one of the most important classes of lasers. The two leading examples of such lasers are the first laser ever constructed, a 3d-3d electronic transition in Cr^{3+}-doped Al_2O_3 (ruby) at 694 nm, and a 4f-4f electronic transition in Nd^{3+}-doped $Y_3Al_5O_{12}$ (yttrium aluminum garnet, commonly known as YAG) at 1064 nm. There are numerous other crystals and glasses in which transitions in these and other impurity ions have been made to lase. Extensive tables of these transitions and their properties can be found in standard references. A few of these lasers use transitions which are electric-dipole allowed in the free ion, but most, such as the two just mentioned, use transitions between states of a single electron configuration which are forbidden in the free ion. An ion in a crystal lattice is distorted by the crystal field and the states are no longer those of a pure single configuration but have a small admixture of other configurations to which electric dipole transitions are allowed, although the transitions are still much weaker than fully allowed transitions.

Figure 2.6 shows the energy levels of Nd^{3+} in a solid host as a representative of this class of laser transitions. The neodymium ion absorbs light at wavelengths shorter than 900 nm, which raises the ion to the $^4F_{3/2}$ upper laser level or to one of the higher levels shown in the figure. These higher levels are rather closely spaced, so the energy gaps between them are comparable to the energy of lattice vibrations and the excitation cascades rapidly down to the $^4F_{3/2}$ upper laser level, exciting lattice phonons to dispose of the excess energy. The $^4F_{3/2}$-$^4I_{11/2}$ laser transition has an energy large compared to a lattice vibration, so relaxation across that gap requires a multiphonon process which has a very low probability, and most of the excited ions

TABLE 2.2 Color-Center Lasers

Crystal	Center type[a]	Tuning range (m)	Reference
LiF	F_2^+	0.82-1.05	[Mollenauer, 1978a]
LiF	F_2^-	1.09-1.23	[Basiev, 1982]
NaF:Mn^{2+}	$(F_2^+)^*$	0.99-1.22	[Mollenauer, 1980, 1981]
NaF:OH$^-$	$(F_2^+)^{**}$	1.08-1.38	[Mollenauer, 1981]
KF	F_2^+	1.22-1.50	[Mollenauer, 1979]
NaCl	F_2^+	1.40-1.75	[Mollenauer, 1982]
KCl:Li	$(F_2^+)_A$	2.00-2.5	[Schneider, 1980]
KCl:Li	$F_A(II)$	2.5-2.9	[Koch, 1979]
KCl:Na	$(F_2^+)_A$	1.62-1.91	[Schneider, 1979]
KCl:Na	$F_B(II)$	2.25-2.65	[Koch, 1979]
KCl:Na,Li	$(F_2^+)_A$	1.67-2.46	[Schneider, 1983]
KI:Li,Pd	$(F_2^+)_A, F_2^+$	1.98-3.85	[Schneider, 1982]
CaO	F^+	0.36-0.40	[Henderson, 1981]

[a]Asterisks denote various types of impurity-stabilized center.

accumulate in the upper laser level and generate gain on the laser transition. The $^4I_{11/2}$ lower laser level has a small energy gap to the sublevels of the $^4I_{9/2}$ ground state of the ion, so ions in the lower laser level also relax very rapidly to lower levels, emptying the laser lower level and maintaining the inversion. This mechanism is typical of the impurity ion lasers.

Most impurity ion lasers in crystals have rather narrow gain bandwidths of order 1 part in 10^4 of the laser wavelength. The center of the band can be shifted slightly by changing the crystal temperature, but usually not to any useful degree. Transitions such as the 1064-nm Nd^{3+} transition, which oscillate well in disordered solids (glass) rather than crystals, have somewhat broader tuning ranges of order 1 to 3% of the center wavelength [Booth, 1979]. Even this tuning range is small compared to broadly tunable systems such as dye or color-center lasers, which might tune over a range between 5 and 20% of the center wavelength.

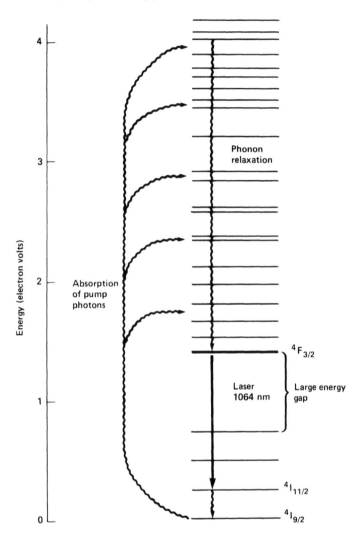

FIG. 2.6 Electronic energy levels of the Nd^{3+} ion used in the 1064-nm solid-state neodymium laser.

2.2.4B. Phonon-Terminated Lasers

There is a class of impurity ion lasers which are broadly tunable in the near infrared and which are likely to have important practical applications in the future although they have not been widely used in the past. Broadly tunable lasers have a wide tuning range because they can make transitions between many sublevels of the electronic states of the system. Some ions of the transition metals have 3d electronic states which interact sufficiently with the crystal lattice that the lattice assumes a slightly different equilibrium configuration when the ion is in an excited electronic state. The fluorescence transition down to the ground state then terminates on a vibrationally excited level of the lattice in the ground electronic state of the ion. There are many possible levels of this sort, and a broadly tunable "phonon-terminated" laser is the result. The electronic states of 4f and 5f impurity-ion lasers (rare earths and actinides) are well shielded from the crystal lattice, and vibronic or photon-terminated transitions are usually so weak as to be negligible, although one such phonon-terminated laser has been demonstrated in $Ho^{3+}:BaY_2F_8$ near 3050 nm [Johnson, 1974]. Vibronic effects are also negligible in some 3d transitions, such as the 694-nm Cr^{3+} R line of the ruby laser.

Table 2.3 shows the properties of a number of phonon-terminated lasers. Alexandrite lasers have been built with CW output powers in the tens of watts and short pulse energies of several joules and are being considered for industrial applications. The $Ni^{2+}:MgF_2$ laser pumped by a 1064-nm $Nd^{3+}:YAG$ laser has demonstrated 37% conversion efficiency with a CW output power of about 1 W [Moulton, 1979]. Several of these lasers have excited-state absorption, which restricts their tuning range and efficiency. This absorption has been demonstrated in alexandrite [Shand, 1983] and in Ni^{2+} in MgF_2 [Moulton, 1982] and $KMnF_3$ [Johnson, 1983]. The phonon-terminated lasers are in a phase of rapid development at present, and several more tunable sources in the red and infrared should be available in the near future.

2.2.4C. Allowed-Transition Lasers

Transitions between 4f and higher-lying 5d states in the rare earths are allowed by dipole selection rules. These transitions are often vibronically broadened and are appealing candidates for tunable lasers. Since most of these transitions lie at rather high energy, they are particularly susceptible to excited-state absorption, and it is not clear that they will ever be of much practical value, although they might make useful spectroscopic sources. The only two such transitions that have been demonstrated are an ultraviolet Ce^{3+} transition near 300 nm and an infrared Sm^{2+} transition near 730 nm. The Ce^{3+} laser tunes from 306 to 315 and from 323 to 328 nm in a $LiYF_4$ (YLF) crystal [Ehrlich, 1978]: it has also oscillated at the peak of the fluorescence band at 286 nm in LaF_3, although the performance was worse than in YLF [Ehrlich, 1980]. It has been demonstrated that excited-state absorption prevents oscillation on this transition in YAG. The Sm^{2+} transition tunes from about 708 to 745 nm in CaF_2, but there are some gaps within this range [Vagin, 1969].

TABLE 2.3 Phonon-Terminated Lasers

Laser material	Tuning range (μm)	Reference
Co^{2+}:MgF_2	1.6-2.3	[Moulton, 1979, 1982]
V^{2+}:MgF_2	1.05-1.3	[Moulton, 1979, 1982]
Ni^{2+}:MgF_2	1.6-1.75	[Moulton, 1979, 1982]
Ni^{2+}:$KMnF_3$	1.59	[Johnson, 1983]
Ni^{2+}:MgO	1.32-1.41	[Moulton, unpublished]
Cr^{3+}:$BeAl_2O_4$ (alexandrite)	0.72-0.82	[Walling, 1980]
Cr^{3+}:$Be_3Al_2(SiO_3)_6$ (emerald)	0.72-0.83	[Shand, 1982]
Cr^{3+}:$Gd_3Sc_2Al_3O_{12}$	0.72-0.82	[Drube, 1984]
Cr^{3+}:$Gd_3(Ga,Sc)_2Ga_3O_{12}$	0.74-0.84	[Struve, 1983]
Cr^{3+}:$ZnWO_4$	0.98-1.10	[Kolbe, 1984]
Ti^{3+}:Al_2O_3	0.66-0.99	[Moulton, 1984]

2.2.5 Excimer Lasers

There are molecular systems which have bound excited states but no bound ground states, as shown in the energy-level diagram of Fig. 2.7. The leading examples of these systems are rare-gas halide molecules such as KrF, which have excited ion-pair states correlating to Kr^+F^- at large internuclear separation and an unbound ground state correlating to neutral atoms. The ground state of these molecules dissociates very rapidly, so there is automatically a population inversion on the transition between the excited and ground states. The typical excimer laser uses a fast (20 to 50 ns) electrical discharge pulse to excite the upper laser level, as reviewed elsewhere [Ewing, 1979; Shaw, 1979; Hutchinson, 1980; Rhodes, 1984]. Electron-beam pumping can be used to construct very large excimer lasers, but these are too large to be of any interest for spectroscopy.

The properties of the most important rare-gas halide and halogen excimer lasers are summarized later (Table 2.4). There are other excimer laser transitions in rare-gas dimers, rare-gas halides, and oxides, dihalogens, and other species which are less useful for practical applications: these are cataloged in standard references and the reviews listed above. Some of these molecules, specifically XeCl, XeF, and F_2,

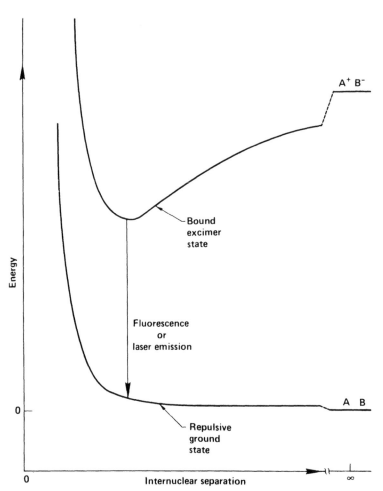

FIG. 2.7 Schematic representation of the energy levels of an excimer molecule such as a rare-gas halide. The rare-gas atom A forms a positive ion A^+, and a halogen atom B easily attaches an electron to form a negative ion B^-. These ions attract each other on a Coulomb potential. Collisions between core electrons cause a repulsion between the atoms which grows exponentially as the internuclear separation decreases, so there will be a minimum in the potential at an internuclear separation where Coulomb attraction and repulsion just balance. The resulting bound state is called an "ion-pair" state. If the electron on the B^- ion now radiatively recombines with the A^+ ion, the molecule will fall to the neutral AB ground-state potential. There is no Coulomb attraction here, so the core electron repulsion causes a rapid dissociation of the molecule.

have slightly bound lower levels and therefore are not, strictly speaking, "excimers," but they share so many properties and kinetic mechanisms with the true excimers that they are commonly called such. These bound-bound transitions have the usual structured emission bands with resolved vibrational and rotational lines rather than the true continuum emission characteristic of the bound-free transition of the true excimer. The true excimers, such as ArF, KrF, and KrCl, are continuously tunable over a small region, perhaps half a percent of center frequency, near the peak of their emission bands; quasi-excimers such as XeF and XeCl can be tuned over their rather densely packed line structure but not continuously. The excimer lasers are particularly useful tunable sources in the ultraviolet, where they give higher output power and energy than most other options. The F_2 laser is the most convenient reasonably energetic discharge laser source for the vacuum ultraviolet [Pummer, 1979].

Reliable discharge-pumped rare-gas halide excimer lasers are available from a number of commercial sources. Typical characteristics of these lasers are summarized in Table 2.4.

2.2.6 Line-Tunable Lasers

There are numerous vibrational and rotational laser transitions in simple molecular gases [Willett, 1974; Beck, 1980]. These transitions extend throughout the infrared and have often been used for spectroscopic experiments. These lasers oscillate on discrete rotation-vibration lines and are not continuously tunable, but for some experiments a line-tunable source is acceptable. Extensive tables of these lines covering the range from about 2.6 to 500 μm can be found in standard references. These laser transitions are most frequently pumped using pulsed electric discharges in gas mixtures, but some can operate with continuous discharges, and a few are optically pumped by other lasers, most frequently by the 10.6-μm, CO_2 laser [Grasyuk, 1980].

The most important of the vibrational infrared lasers is the electric-discharge-excited CO_2 laser. The most important laser lines in this molecule fall between about 9 and 11 μm, although there are other useful transitions outside this range. The two vibrational transitions responsible for the 9- to 11-μm emission and the pumping mechanism that generates the laser inversion are shown in Fig. 2.8.

The CO_2 laser mixture typically contains nitrogen, carbon dioxide, and a light gas such as helium. When the nitrogen molecule collides with a discharge electron at an energy of a few volts, there is a high probability that it will form a short-lived N_2^- negative ion. The internuclear separation in N_2^- is larger than in neutral N_2, so when the electron leaves, the N_2 molecule finds itself in a vibrationally excited state. The symmetric stretch vibrational mode of CO_2 is very close in energy to the N_2 vibration, so in a collision with vibrationally excited

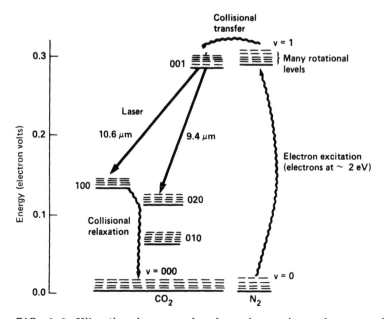

FIG. 2.8 Vibrational energy levels and pumping scheme used in the
9 to 11-μm CO_2 infrared gas laser. The CO_2 vibrational energy levels
are described by the notation abc where a, b, and c are the number
of quanta in the asymmetric stretch, bending, and symmetric stretch
modes, respectively.

N_2 there is a high probability that the excitation will be transferred to
the CO_2, populating the laser upper level, which is designated as 001
since it has one quantum of excitation in the symmetric stretch and no
quanta of excitation in any other mode. This gives gain on the two
transitions down to the 100 and 020 levels, and these transitions are
the 9- to 11-μm laser lines. Collisions between CO_2 molecules rapidly
convert these laser lower levels to the 010 level, and collisions with
the light molecules in the mixture convert the 010-level excitations to
translational motion (heat). Relatively small and simple CO_2 lasers
can generate several tens of watts of continuous power or several
megawatts pulsed, and larger laser systems are available for those applications that require them. The CO_2 laser is also among the easiest
lasers to build and operate. It has been very widely used for linear
and nonlinear spectroscopy as well as for materials processing, ranging, and many other applications.

   If a line-tunable gas laser has closely spaced lines and the laser
kinetics will permit it to operate at high pressure, it may be possible
to collision broaden the lines until they overlap and give truly continuous tuning. This is just a less extreme case of the processes that

give continuous tuning in the laser dyes. The 10.6-μm carbon dioxide laser, for example, begins to have a small continuous tuning range at a pressure of several atmospheres. At a pressure of 15 to 20 atm the continuous tuning range of a CO_2 laser extends over four bands (P and R branches of the 9.4- and 10.6-μm laser bands), which cover wavelength ranges of about 9.16-9.30, 9.38-9.47, 10.14-10.33, and 10.44-10.62 μm, for a total continuous tuning range of about 6% of center wavelength. This particular system has been studied extensively.

It is possible, although not at all trivial, to design a stable pulsed electrical discharge to excite a CO_2 laser mixture at a pressure of several atmospheres [Bagratshvili, 1971; Miller, 1975; Taylor, 1979; Yi, 1983]. Optical pumping can also be used to build continuously tunable spectroscopic sources, and several optical pumping techniques are available for CO_2. The CO_2 upper laser level can be pumped directly using a 4.3-μm HBr laser [Chang, 1977]. A DF laser can be used to resonantly pump DF in a DF-CO_2-He mixture: the vibrationally excited DF then transfers its energy to the CO_2 upper laser level in collisions [Stenersen, 1983]. The HBr laser can be used to pump CO_2, which then transfers its excitation in collisions to N_2O: the N_2O has laser transitions in the same general wavelength range as CO_2 but it is not symmetric and therefore has rotational lines with both odd and even angular momenta in the laser band, whereas only even lines are allowed in CO_2. The more closely spaced lines in N_2O give better line overlap at reasonable pressures [Chang, 1977]. The CO_2 in any of these lasers can have a carefully chosen mixture of isotopes enriched in normally rare carbon and oxygen isotopes so that there are many additional rotation-vibration lines displaced significantly from the normal $^{12}C^{16}O_2$ lines. The greater density of lines gives line overlap and continuous tuning at a pressure of perhaps 4 to 5 atm rather than the 15 to 20 atm required with normal CO_2 [Gibson, 1979].

Molecular bands with a high density of lines are also found in electronic transitions in the visible and ultraviolet, and many of these transitions are between states having energy levels similar to Fig. 2.1, which lead to population inversion and laser oscillation. The history and performance of optically pumped dimer lasers on transitions in I_2, Br_2, S_2, Te_2, Na_2, K_2, and Li_2 are reviewed by Wellegehausen [Wellegehausen, 1977]. These lasers offer a multitude of lines between about 365 and 1300 nm, many of which will oscillate CW as well as pulsed. Most of them require a high temperature in order to have a useful dimer vapor density. The Se_2 laser, as an example, will oscillate CW on a number of lines between 387 and 709 nm and might be a useful spectroscopic reference source over that wavelength region [Wellegehausen, 1982].

There are a number of mercury, cadmium, and zinc halide species which oscillate in line-tunable bands from the near infrared to the near ultraviolet using discharge or photodissociation pumping. These transitions include HgCl (552 to 559 nm), HgBr (499 to 505 nm), HgI (443 to 445 nm), CdBr (811 nm), CdI (656 to 658 nm), and ZnI (600 to 604 nm) [Burnham, 1978; McCown, 1981; Ediger, 1982; Greene, 1983].

The line-tunable dimer and metal halide lasers could be useful in spectroscopy, although they compete with highly developed dye laser technology and have not had much practical application in the past. The HgBr laser at 502 nm has been a candidate for a source for optical communication in the transmission band of seawater [White, 1977].

Molecular hydrogen and deuterium have vacuum-ultraviolet laser transitions which oscillate with pulsed discharge [Knyazev, 1975] or electron-beam [Dreyfus, 1974] excitation and are line tunable over the wavelength range from 110 to 160 nm. A very fast pulsed discharge is the simplest excitation technique for spectroscopic applications. These transitions are useful spectroscopic sources in the vacuum ultraviolet, although it is easier to get high pulse energy from the 158-nm F_2 discharge laser if energetic pulses are required.

2.2.7 The Principal Pump Lasers

Many of the tunable lasers discussed here are optically pumped systems requiring an intense pump source, which is most often another laser. In this section we describe the most common choices for these pump lasers. Dye lasers will be emphasized again, although these pumps are also used with semiconductor, color-center, and impurity-ion lasers. Table 2.4 summarizes the properties of the pump lasers to be described here. Dye lasers are also used as pump lasers when no better options are available.

The 337-nm molecular nitrogen second-positive-band ultraviolet laser is one of the most popular pump sources for spectroscopic dye lasers. Figure 2.9 shows the electronic energy levels of nitrogen which are important for this laser transition. A very fast, high-voltage electrical pulse accelerates electrons to a voltage high enough to excite ground-state $X^1\Sigma_g$ nitrogen molecules to the $C^3\Pi_u$ state, which has a larger cross section for electron excitation than the lower-lying A or B states shown in the figure. This gives an inversion and laser oscillation on various rotational-vibrational transitions of the $C^3\Pi_u$-$B^3\Pi_g$ transition near 337 nm. Laser oscillation terminates when the B state population rises to equal that of the C state. It is possible under some circumstances to have the laser oscillate simultaneously on the B-A transition in the near infrared.

Nitrogen lasers are inexpensive, very reliable, and have low operating costs. The 337-nm wavelength allows pumping of most laser dyes. A quite adequate nitrogen laser can be built for less than $100 using hand tools and common materials [Small, 1976], although of course serious spectroscopy will be easier if the system has somewhat

TABLE 2.4 Principal Pump Lasers for Laser-Pumped Spectroscopic Sources[a]

Laser type	Wavelength (nm)	Pulse length (ns)	Pulse energy (mJ)	Pulse rate (Hz)
Nitrogen	337	1-10	1-10	1-200
Excimer		10-30		1-100
F_2	158		2-15	
ArF	193		2-400	
KrCl	222		4-60	
KrF	249		10-1000	
XeCl	308		5-1000	
XeF	351 + 353		5-500	
Copper	510 + 578	30	1-10	5000
Gold	628	30	0.4	5000
Nd^{3+}:YAG	1064	5-10	200-1000	1-10
Doubled	532		100-200	
Tripled	355		30-100	
Quadrupled	266		20-60	

			Average power (W)	
Nd^{3+}:YAG	1064	CW	1-100	
Argon	All lines	CW	1-20	
ion	351 + 364		0.04-3	
	488		0.7-6	
	515		1-8	
Krypton	All lines	CW	1-10	
ion	337 + 356 + 351		0.07-1	
	521 + 568 + 531		0.6-4	
	647 + 676		0.6-5	
	753 + 799		0.25-2	

[a]Standard commercial products could be purchased with output characteristics roughly within the ranges shown in 1984. Some very specialized systems have output characteristics outside these ranges.

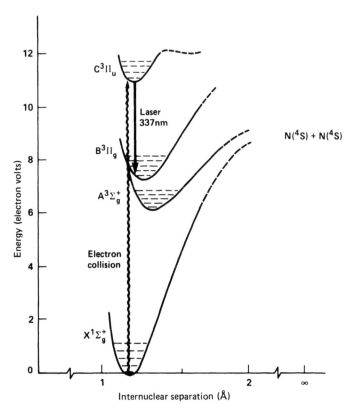

FIG. 2.9 Electronic energy levels of the nitrogen molecule used in the 337-nm N_2 second-positive-band laser.

more sophistication. The nitrogen laser output power is typically a few hundred kilowatts for a few nanoseconds and is high enough to pump dyes easily using inexpensive optics and dye cells constructed from common materials such as microscope slides and cover glasses [Capelle, 1970]. The output from the dye laser is usually well above 1 kW, which is adequate for many experiments investigating saturation and nonlinear effects as well as simple linear spectroscopy. The 428-nm N_2^+ first-negative-band laser uses similar discharge techniques and could also be a useful pump laser [Collins, 1984], although it has not been widely used in the past.

Excimer lasers such as the 308-nm XeCl laser give much higher pulse energy and longer pulses than the nitrogen laser. Operating costs are higher since expensive rare gases such as xenon are required, although the rare gases can be cleaned and recycled after the laser gas mixture degrades during operation. The electrical and mechanical design of the laser discharge chamber and pulsed power system are more complex than those of a nitrogen laser. Highly corrosive halogen compounds such as HCl and F_2 are required for these lasers, which is a hazard and also increases the cost of components. Excimer lasers offer a range of pump wavelengths on different excimer transitions with a change in gas fill (see Table 2.4) and limited tunability in the excimer transition itself.

Copper vapor lasers are useful nanosecond-pulse pump sources in the visible when a high pulse repetition frequency and high average power are required [Isaev, 1977; Bokhan, 1978; Smilanski, 1978, 1979; Hargrove, 1979, 1980; Broyer, 1984; Duarte, 1984]. Figure 2.10 shows a level diagram of the copper laser transitions. The s-p transition from the $3d^{10}4s$ ^2S ground state to the $3d^{10}4p$ ^2P laser upper level has a very large cross section for excitation by discharge electrons, much larger than the cross section to excite the d-s transition to the $3d^94s^2$ ^2D lower laser level. The transition probability from ^2P to ^2D is large enough to give a reasonably large laser gain on those transitions. Radiative s-d relaxation from ^2D to the ^2S ground state is also very slow, so laser oscillation terminates when the ^2D-level population rises to equal the population of the upper laser level. Between laser pulses the lower level empties slowly to the ground state by collision

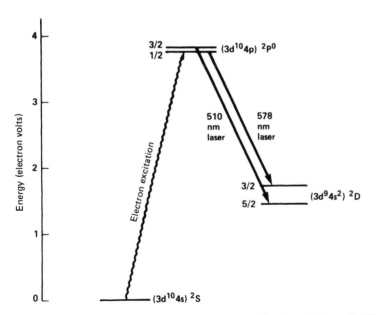

FIG. 2.10 Electronic energy levels used in the 510- and 578-nm copper vapor laser.

with the walls of the discharge tube and by a rather small cross section for deactivation to ^2S by slow discharge electrons. Some similar
transitions such as the gold vapor line at 628 nm and the lead vapor
line at 723 nm could be useful pump sources for infrared dye and color
center lasers but have not been exploited for that in the past. Metal
vapor lasers operate at temperatures between perhaps 800 and 1800 K,
and these high temperatures place serious restrictions on construction
materials and engineering design. The metal vapor lasers operate with
only a few torr of gas in the active volume, so thermally induced
gradients in the index of refraction are negligible and these lasers can
run at high average power without the elaborate gas flow systems
which would be required in lasers that operate at higher pressures,
such as CO_2 or excimer lasers.

The Nd:YAG laser has reached a high level of engineering development and reliability [Koechner, 1976] and is an excellent choice for
a pump source. CW Nd:YAG lasers are frequently used to pump phonon-terminated and color-center lasers in the near infrared. Pulsed
Nd:YAG lasers can be doubled, tripled, and quadrupled quite reliably
and efficiently to give pump beams at 532, 355, and 266 nm with pulse
lengths typically 10 to 20 ns and average power up to several watts
[Liu, 1984]. These harmonics are frequently used to pump dye lasers,
and numerous commercial systems of this sort are available. CW Nd:
YAG lasers can be doubled to 532 nm [Geusic, 1968] but advances in
nonlinear materials will be required before CW doubling becomes a
standard laboratory technique.

Argon and krypton ion lasers are the most common of the CW pump
lasers at wavelengths shorter than the near infrared. Figure 2.11
shows a schematic level diagram of the laser excitation in the argon ion
laser; krypton has similar transitions. The upper laser levels (of
which there are several) belong to the $3p^44p$ electron configuration of
the ion and are populated by several processes in a vigorous discharge, including direct excitation by discharge electrons, cascade
from higher levels, and sequential excitation from lower levels. The
lower laser levels of the $3p^44s$ configuration are also populated by
these processes, but they radiate to the $3p^5$ ionic ground state so
rapidly that there is a population inversion between the $3p^44p$ and
$3p^44s$ configurations. A typical 10 to 15-W argon laser requires about
20 to 30 kW of electrical power and 15 to 30 liters/min of cooling water.
The active diameter of the discharge tube is about 2mm, so the thermal
loading is severe and the discharge-tube lifetime is limited to a few
thousand hours. Krypton ion lasers give a wider range of output
wavelengths than do argon lasers, at the cost of lower output power
and lower gain. This makes them more sensitive to misalignment or
other deviations from optimum operating conditions.

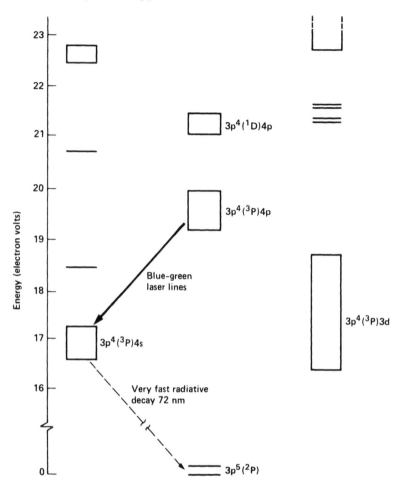

FIG. 2.11 Electronic energy levels used in the argon ion laser.
The blue-green laser lines are from several levels of the $3p^44p$ configuration of Ar^+ down to levels of the $3p^44s$ configuration. The lower levels are emptied by very rapid radiation to the $3p^5Ar^+$ ground
configuration. The UV laser lines are from similar $4p \rightarrow 4s$ transitions
in Ar^{2+}. The krypton laser lines are from similar $5p \rightarrow 5s$ transitions.

2.3 DESIGN OF SPECTROSCOPIC LASER SYSTEMS

2.3.1 Pump Geometries

Lasers can be set up with one of two generic pump geometries, as
shown in Fig. 2.12. Transverse pumping avoids mechanical interferences between the pump beam and the laser resonator but gives non-

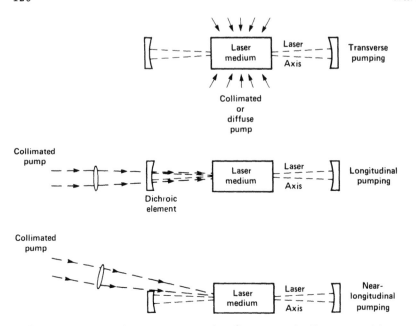

FIG. 2.12 General pump geometries for an optically pumped laser.

uniform excitation across the aperture of the optically pumped volume.
The nonuniformity can be reduced and beam quality improved by
pumping from two opposing sides [Hargrove, 1980]. Most simple systems pump from one side and sacrifice either uniformity or the efficiency with which they use pump photons to avoid the additional complexity. Frequently, a cylindrical lens is used to form a line focus on
the optically pumped laser medium to increase the pump power per unit
area. A typical nitrogen-laser transverse-pumped dye laser, for example, will have a pump area with a height of a millimeter or less and
a length of 0.5 to 2 cm with the dye concentration adjusted to give a
pump penetration depth roughly equal to the height of the pumped region. Diffuse sources such as flash lamps also require transverse
pump geometries.

Longitudinal pumping allows better control of the transverse gain
profile of the laser but interferes with the mechanical layout of the
laser resonator. If the wavelengths of the pump and tunable lasers
are sufficiently far apart, a dichroic multilayer dielectric mirror can
be designed to transmit the pump and reflect the tunable laser. The
two beams can also be made collinear by combining them in a dispersive
element such as a prism. Laser designers often avoid this problem by
resorting to "near-longitudinal" pumping, in which the pump and

laser beams cross at an angle large enough that the pump beam misses the tunable laser cavity mirror as shown in Fig. 2.12. This near-longitudinal geometry is almost universal in CW dye lasers.

2.3.2 Resonators

Many lasers will emit an intense beam of amplified spontaneous emission even in the absence of feedback from an optical resonator, but the laser must be constrained to oscillate in the modes of an optical resonator or "cavity" if high beam quality and narrow line width are desired. There are two general types of resonator used in spectroscopic lasers, as shown in Fig. 2.13 and described in more detail in standard texts [Yariv, 1971; Siegman, 1971; Svelto, 1982; Tarasov, 1983]. The simple linear Fabry-Perot resonator has two mirrors aligned parallel to each other on opposite sides of the gain volume. There are resonances which give a high-intensity standing wave between the mirrors for any wavelength such that the length of the resonator is an integral number of half-wavelengths of the radiation, from which it follows that the resonances are separated in frequency by $c/2\ell$, where ℓ is the length of the resonator and c the velocity of light in the medium. The high-intensity standing waves at the resonance frequencies will cause enhanced stimulated emission from the laser into these modes.

Since there is a standing wave in a simple linear resonator, there will be regions in the laser gain medium which are near nodes of zero intensity in the intensity distribution of a particular mode. The rate of stimulated emission into the mode is low in these regions if we try to operate the laser on a single mode. The high gain remaining in these regions is not only wasted energy but makes single-mode operation more difficult since it gives very high gain for other modes which have antinodes near the nodes of the desired mode [Pike, 1974]. This effect can be eliminated by using a ring resonator with a traveling wave circulating in only one direction, as shown in Fig. 2.13. This resonator has no standing wave and the standing-wave mode-competition effects do not occur. The ring resonant modes are separated in frequency by c/d, where d is the total optical path around the ring. It is sometimes necessary to introduce losses for one propagation direction around the ring to ensure that there will be a traveling wave in only one direction, although the loss asymmetry need not be very large to force unidirectional oscillation [Green, 1973]. The most generally useful technique for this is to insert an "optical diode" in the laser beam path, consisting of a Faraday rotator which rotates the polarization through a small angle, a crystal quartz rotator rotating through the same angle, and a polarizer. The polarizer might be only the differential loss associated with components placed at Brewster's angle in the beam path. The Faraday rotation changes sign with the direction of propagation of the laser, but the crystal rotation does not.

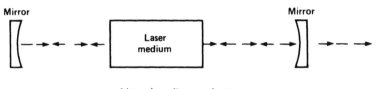

Linear (standing wave) resonator

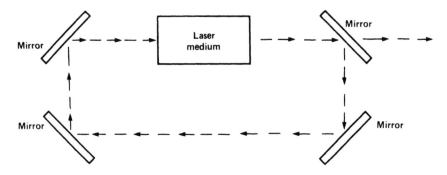

Ring (traveling wave) resonator

FIG. 2.13 Simple linear (standing wave) and ring (traveling wave) laser resonators.

In one direction, then, the rotations cancel and there is no change in the beam polarization. In the other direction the two rotations add and the change in polarization introduces additional loss at polarizing elements [Jarrett, 1979; Johnson, 1980]. Most narrow-line-width ring lasers in the laser literature are CW dye lasers [Marowsky, 1975; Schröder, 1977], but the techniques are general [Beigang, 1977].

Many of the spectroscopic lasers discussed here, in particular the CW dye lasers, have gain volumes only a few micrometers in diameter, yet must have resonators many centimeters long to give space for other components. A long resonator designed to have a very small mode diameter has severe alignment problems [Snavely, 1973]. Two modified resonator designs that give better performance under these conditions are shown in Fig. 2.14. Two identical concave mirrors of short focal length are used to match the small spot size in the laser gain volume to a long resonator. The mirrors are used off-axis, which introduces astigmatism. In the three-mirror design it is possible to

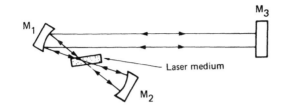

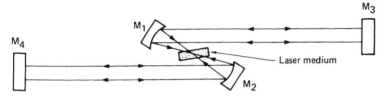

FIG. 2.14 CW dye-laser resonators using two identical short-focal-length mirrors M_1 and M_2 to match a small gain region to a long laser resonator. The four-mirror cavity has better correction of the aberrations induced by the off-axis reflections from the short-focal-length mirrors M_1 and M_2.

compensate for some of the astigmatism with the astigmatism introduced by the cell or jet of dye at Brewster's angle [Kogelnik, 1972]. The symmetric arrangement of two identical mirrors in the four-mirror design gives somewhat better cancellation of the aberrations [Johnson, 1972; Marowsky, 1975].

2.3.3 Line-Width Control

There are many resonator modes within the gain bandwidth of broadly tunable lasers; consequently, it is difficult to get narrow line widths in these devices unless additional optical elements are added to the resonator to restrict the wavelength range over which the laser can oscillate. Figure 2.15 shows three common examples of a single simple wavelength-selecting element added to a Fabry-Perot laser resonator. The grating, which in practice is most often used with some modifications to be discussed later, can give higher resolution than the other two schemes but has higher loss and is less suitable for lasers having low gain and unable to tolerate such high losses. The prism has very low loss if designed so that the beam intercepts the faces near Brewster's angle. It is frequently used to select lines in CW argon and krypton ion lasers and has also been used with many other systems, including phonon-terminated and color-center lasers. Marowsky [Marowsky, 1973, 1975] describes a mirrorless ring resonator using four

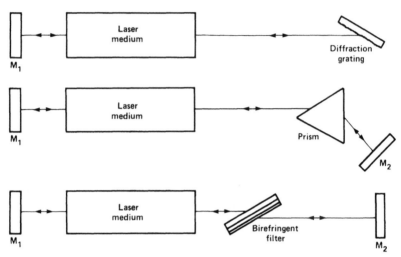

FIG. 2.15 Three common wavelength-selecting elements as they are used in laser resonators. The grating- and prism-tuned resonators disperse wavelengths in angle so that the mirrors M_1 and M_2, or M_1 and the grating reflection, are aligned parallel over only a small range of wavelengths. The birefringent filter rotates the polarization of unwanted frequency components so that they have high losses at polarization-selecting elements such as Brewster angle windows.

90° constant-deviation prisms. Other prism-tuned ring geometries have also been demonstrated [Schäfer, 1973].

The birefringent filter uses the polarization-rotating properties of a birefringent crystal to give a loss that depends on laser wavelength. A linearly polarized beam incident on a birefringent crystal is resolved into two components propagating at different speeds along the "fast" and "slow" axes of the crystal. The retardation of one component with respect to the other converts the original linear polarization to a mixed or elliptical state of polarization. If there are an integral number of wavelengths of retardation in a transit through the crystal, the polarization of the output beam is the same linear polarization as the input beam, while other retardations give a polarization having a smaller component along the original axis of polarization. When the crystal is placed in a resonator that contains polarization-selecting elements, such as Brewster angle windows, there will be low loss if the crystal has an integral number of wavelengths of retardation but higher loss if it does not, since the orthogonal linear polarization component has higher losses. The retardation in waves is proportional to wavelength and consequently the birefringent element acts as a wavelength-selecting element. The optical axis of the crystal can be oriented so that rotation of the crystal plate about an axis perpendicular to the plate

changes the direction of propagation of the laser beam with respect to
the optical axis, so this rotation then changes the resonant wavelength
at which a beam passes through the crystal with its polarization unchanged. Rotation of the plate in its own plane has no effect on the
alignment of other optical elements in the resonator, which is very convenient for the mechanical design of the laser. In practice the crystal
plate is placed at Brewster's angle so that it acts as its own polarizer.
There are many different wavelengths that have an integral number of
waves of retardation in a single plate, but a set of plates of different
thickness can be chosen so that the resonances of the entire assembly
are widely spaced [Bloom, 1974; Holtom, 1974]. Birefringent filters
have very low loss at peak transmission, so they are most useful in
low-gain CW systems, such as CW dye lasers. They do not have particularly high discrimination against secondary wavelength resonances
of the filter stack unless the stack has many precise elements and a
complex design, so they are less useful with high-gain, short-pulse
lasers.

The simple grating of Fig. 2.15 is a common wavelength-selecting
element and has better resolution than simple prism and birefringent
filters, but its performance is still not adequate for very high resolution spectroscopy. The resolution of a grating instrument is proportional to the number of grooves illuminated on the grating, and this
number is rather small in the millimeter-or-less beam diameters common
in tunable lasers. The beam can be expanded to illuminate a wider
section of the grating using several techniques, and most modern
spectroscopic lasers use one or a combination of these. All of these
beam expansion techniques increase the losses in the laser resonator,
and this must be kept in mind.

Figure 2.16 shows a number of beam-expanded laser resonators
which have been used with pulsed dye lasers and some other systems.
Typical properties of the first three of these configurations are shown
in Table 2.5, which is based on a review by Trebino et al. [Trebino,
1982] which also discusses some other properties of this and other
beam-expanded, grating-tuned laser cavities.

The Gallilean telescope beam expander was introduced by Hänsch
[Hänsch, 1972]. The grating is mounted at the Littrow angle at which
the most intense first-order blazed grating reflection feeds directly
back along the input beam. This gives the highest possible reflectivity from the grating. The line width in a well-designed dye oscillator
of this sort is typically about 30 GHz, although there is considerable
variation in the line widths quoted in the literature, depending on details of the design. A 30-GHz line width still includes about 50 resonator modes in a typical resonator with a length of 25 cm, so an additional wavelength-selecting element must be added if a narrower line width
or true single-mode operation is desired. The usual choice is a Fabry-
Perot etalon consisting of two multilayer, dielectric-coated, partially

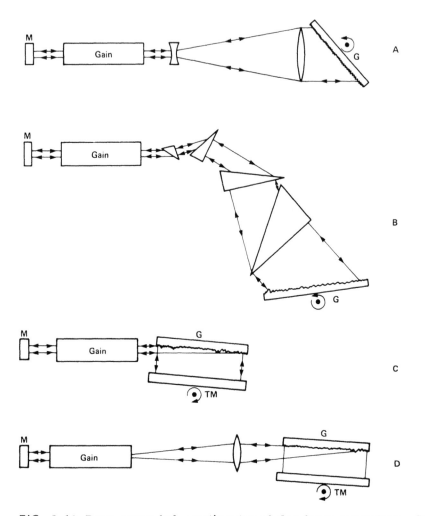

FIG. 2.16 Beam-expanded, grating-tuned dye-laser resonators. In the Gallilean telescope (A) and prism-expander (B) resonators, the wavelength is adjusted by rotating the grating about an axis as shown. The resonance wavelength of the grazing-incidence-grating (C) or relayed grazing-incidence-grating (D) resonators is changed by rotating the tuning mirror TM. The resonator mirror M is often a partial reflector so that the light transmitted through it becomes the output beam from the laser oscillator.

TABLE 2.5 Comparison of Typical Nanosecond-Pulse, Grating-Tuned
Dye Laser Cavities

	Gallilean telescope	Prism expander (four prisms)	Grazing-incidence grating
Alignment	Difficult	Easy	Very easy
Magnification	25	40-100	50
Line width (no etalon)	30 GHz	6 GHz	3 GHz
Principal losses	Grating 15%	Grating 15% prism surfaces	Grating 99%
Efficiency[a] (no etalon)	25%	35%	3%
Efficiency[b] (single mode)	3%	5%	2%
Cavity length	25-40 cm	15 cm	12 cm

[a]Typical efficiency (dye laser out/pump laser in) for a 10-ns, 532-nm doubled Nd:YAG-pumped, rhodamine 6G dye laser in the simplest cavity configuration.
[b]Single-mode efficiencies vary considerably in the literature. These values should be considered order-of-magnitude estimates.

transmitting mirrors aligned parallel to each other and inserted in the beam path at a slight angle to the optical axis so that transmission resonances of the etalon affect the beam circulating in the resonator but reflections from the etalon do not feed back into the resonator. The angle between the etalon and resonator axes causes the multiply reflected beams within the etalon to "walk off" the resonator axis, and this reduces the overlap and interference of multiple reflections which give the wavelength selectivity. The etalon is usually placed in the expanded part of the beam between the telescope and grating to minimize walk-off effects. Etalons are also used in prism- and birefringent-filter-tuned systems, although these will not be discussed in detail here. Cassegrainian and off-axis reflecting telescopes have been used in place of the Gallilean telescope, but it is not clear that they have any general advantages [Trebino, 1982].

The prism beam expander of Fig. 2.16 uses the prisms only for beam expansion: all wavelength selectivity is provided by the Littrow grating. The prisms expand the beam in only one dimension, which is an advantage since beam expansion in a direction parallel to the grooves on the grating has no effect on wavelength selectivity but makes the system much more sensitive to mechanical flaws and angular drifts in the grating mount. The expander of Fig. 2.16 is a four-prism achromatic expander design introduced by Novikov and Terty-shnik [Novikov, 1975] and Klauminzer [Klauminzer, 1977], but any number of prisms can be used. The earliest designs have only one [Myers, 1971; Stokes, 1972; Hanna, 1975]. A single-prism expander is not achromatic, so prism dispersion has some effect on tuning [Zhang, 1981]. The multiprism expanders have the advantage that the beam can have a lower angle of incidence on the prism faces at high magnifications than it could in a single-prism expander. This makes it easier to design the expander for low loss since it is very difficult to design good antireflection coatings for high angles of incidence. Loss-es can also be reduced by choosing an angle of incidence near Brew-ster's angle for one face of the prism [Klauminzer, 1977]. Barr [Barr, 1984] discusses design principles in detail.

A multiprism beam expander can give higher magnification in a very short resonator than can a Gallilean telescope designed using lenses which are reasonably easy to fabricate, so the multiprism ex-pander gives a somewhat narrower line width. An etalon can be in-serted between the expander and grating as it was in the Gallilean de-sign.

The grazing-incidence grating resonator uses the grating itself as a one-dimensional beam expander in a very simple configuration [Sho-shan, 1977, 1978; Littman, 1978a,b]. The grating is used at a high an-gle of incidence (typically about 89°) and in a very high grating order, where the grating efficiency is very low, so the feedback into the amplifying medium is typically 1% or less of the intensity incident on the grating. Holographic gratings with very high spatial frequency (1800 to 2400 lines per millimeter) give somewhat better performance than do standard ruled gratings, although the losses are still very large [Wilson, 1979; Mory, 1981]. Most of the loss from the grating is in the zero-order specular reflection from the grating, and this beam is often used as the laser output beam. The grazing-incidence grating resonator can give about an order-of-magnitude narrower line width than the Gallilean beam expander, at the cost of about an order-of-magnitude lower output power because of the high losses.

Normally, one would propose to put an etalon in the expanded beam to get single-mode operation, but the losses are so high in graz-ing-incidence grating resonators that other techniques are often used, although an etalon has been used successfully [Mory, 1981]. Etalons have been used as passive filters on the output beam and as resonant reflectors to replace the resonator mirror opposing the grating [Saik-an, 1978; Chang, 1980]. The tuning mirror can be replaced by a Littrow grating to give additional wavelength discrimination [Littman, 1978; Iles, 1980]. Some authors mention jitter in the output frequency of single-mode lasers using grazing-incidence gratings. This has been attributed to thermal effects in the dye cell [Littman, 1978]. Lisboa et al. [Lisboa, 1983] note that diffractive spreading of the beam reflected from the grating and tuning mirror reduces both the feedback efficiency and selectivity in grazing-incidence grating resonators and suggests that a relay lens should be added to image the dye cell back onto itself since diffractive effects are suppressed when an image is relayed through an optical system. He observes a line-width reduction of a factor of 3, to 1.1 GHz, and a significantly higher output power in a relayed cavity. Yodh et al. [Yodh, 1984] discuss further features of this configuration. Armandillo et al. [Armandillo, 1984] show that if the tuning mirror and grating of a grazing-indicdence grating resonator are close together and aligned parallel to each other, multiple reflections between them can serve as a tuning etalon to reduce the laser line width. He describes an XeF excimer laser using this technique which has a bandwidth of about 40 MHz. Littman [Littman, 1984] describes a carefully engineered, miniaturized grazing-incidence dye laser oscillator using near-longitudinal pumping, which gives a line width of about 150 MHz continuously tunable over 450 GHz.

It should be obvious that many other combinations of tuning elements are possible. The combination of prism beam expanders and the grazing-incidence grating is an interesting example [Racz, 1981; Duarte, 1984]. A ring resonator using an unstable resonator as a beam expanding element has also been demonstrated [Teschke, 1980].

There is a problem common to all of these lossy resonators containing very high gain amplifying regions which should be mentioned here. The laser gain is so high that "resonators" using feedback from very weak stray reflections from various components or even single-pass amplified spontaneous emission can add highly undesirable broadband background signals to the laser output beam. Stray reflections from the antireflection-coated telescope lenses of a Gallilean beam expander frequently add a few percent of broadband background in that geometry, and there is obviously a potential for similar problems in the relayed grazing-incidence resonator. A few percent broadband background is sometimes seen in simple grazing-incidence grating resonators also, possibly because the gain must be so high to overcome the exceptionally high resonator losses that single-pass amplified spontaneous emission becomes important, although careful design can reduce the background to about 0.01% [Littman, 1984]. There is considerable variation in the background signals quoted in the literature, and a number of authors have ignored them. The prism beam expander should be able to maintain somewhat lower broadband background than the other designs since it has relatively low losses and can be designed to direct all stray reflections out of the laser cavity before they intersect the gain region. Bor [Bor, 1981] points out that the distributed feedback dye laser gives excellent control of amplified spontaneous emission because of its very short effective resonator length. This laser design has not been used much for spectroscopy but may deserve wider application. Bor has references to earlier work on distributed feedback systems, which will not be discussed here.

2.3.4 Continuous Tuning

Figure 2.17 shows the mode resonances of a laser resonator superposed on the wider passband of a wavelength-selecting element such as an etalon, grating, or laser gain profile. The laser will naturally seek to oscillate at the peak of the mode resonance and as near as possible to the peak of the wider passband. Suppose that the laser is oscillating on a single mode, as shown in the figure, because only that mode has a loss low enough to be above oscillation threshold. If we attempt to tune the laser by moving the wider passband with respect to the modes, the actual laser frequency will not move at all, apart from small effects which we will ignore here, but at some point the original mode will have such high loss that oscillation on that mode will cease and the laser will begin oscillating on an adjacent mode which now has less loss. If we change the length of the laser resonator rather than moving the gain profile, the resonator modes will move and the laser will tune continuously but only over a range somewhat less than the resonator mode spacing. It is clear that truly continuous tuning requires that we tune the length of the resonator and the passband of the wavelength-selecting elements simultaneously, although if a frequency resolution somewhat greater than the separation between resonator modes is adequate for the particular experiment under consideration, we can allow discontinuous mode hops between resonator modes and control only the wavelength-selecting elements.

The most straightforward way to tune these elements simultaneously is to design an electromechanical apparatus to do just that. CW dye-laser systems of this sort usually tune continuously over a range of about 10 to 30 GHz in a laser which has a resonator mode spacing of about 300 MHz [Johnson, 1982]. With sophisticated computer control of the various tuning elements and of a precise reference wavemeter the continuous tuning range can be extended to about 10 THz, which is about 2% of the dye center frequency. The computer synthesizes this range from many individual scans of about 10 GHz, however [Williams, 1983; Raymond, 1984].

Cleverness substitutes for sophistication in one continuous tuning design. If all of the wavelength-selecting elements are based on interference of laser beams propagating in free space, such as gratings, etalons, or the large etalon we have named the laser resonator, the entire laser can be embedded in a medium with a variable index of refraction and a change in the index will tune all of these elements si-

A. Tune wavelength selector

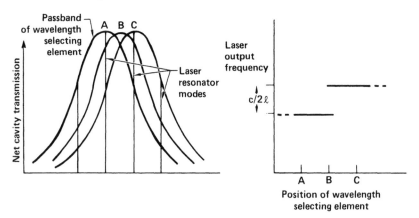

B. Tune resonator modes

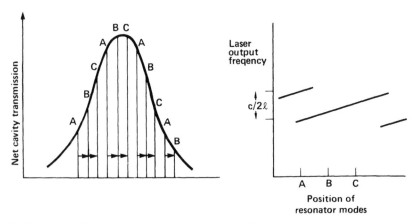

FIG. 2.17 The response of the output frequency of a laser when the passband of a wavelength-selecting element (A) or the laser resonator modes (B) are tuned independently. In (A), the position of the laser resonator mode is fixed and the laser oscillates on the mode (or group of modes) nearest the peak of the passband of the wavelength-selecting element where the resonator has the lowest loss. When the passband moves, the laser will cease oscillating on the original mode and the output frequency will jump discontinuously by the mode spacing of $c/2\ell$ as the second mode begins to oscillate. In (B), the position of the laser resonator modes moves while the passband of the wavelength-selecting element remains fixed. Now the output frequency tunes continuously but the oscillating mode moves off the center of the passband and the output frequency shifts discontinuously back to the adjacent mode, which now has the lowest loss.

multaneously over precisely the same frequency range [Wallenstein, 1974]. A change in pressure of 1 atm of air gives a tuning range of 150 GHz, while an atmosphere of a gas with a higher index, such as propane or CF_2Cl_2 [McGee, 1981], tunes about four times faster. There is a practical limit of a few atmospheres pressure change in a laser which must maintain precise mechanical alignment, and this limits the tuning range. Even without this practical limit the continuous-tuning range will be limited because some of the tuning elements, in particular the laser resonator itself, contain components such as the laser amplifier cell whose optical path length is not a function of the index of the surrounding medium. These components usually occupy a small fraction of the length of the resonator, but over large tuning ranges they will prevent the tuning elements from tracking precisely together as the index changes in the rest of the resonator.

If the tuning mirror in a grazing-incidence grating resonator pivots about exactly the right point, the change in resonator length and the change in the passband of the grating can be made to track precisely together and give continuous tuning [Liu, 1981]. The pivot point must be located with an accuracy of a few micrometers if the continuous-tuning range is to be very large. Littman [Littman, 1984] describes a design which tunes continuously over 450 GHz with a line width of 150 MHz.

2.3.5 Special Problems of Nanosecond-Pulse Systems

The uncertainty principle of quantum mechanics sets a minimum line width for a laser with a finite pulse length and this limit approaches the line widths of interest for high-resolution spectroscopy in nanosecond-pulse systems. A pulse that has the minimum possible line width has a spread in frequency which is the Fourier transform of its shape in time and is therefore called a transform-limited pulse. Pulses in real systems will always have a frequency spread of this width or greater. A transform-limited pulse with a Gaussian shape (i.e., $\exp[-(t/t_0)^2]$) in time has a Gaussian shape in frequency, and the pulse length and bandwidth are related by $\Delta \nu \, \Delta t = 0.44$, where the widths are defined as the full width at half-maximum intensity and frequency is measured in hertz. Pulses of other shapes have somewhat different time-bandwidth products, but the product will be within a factor of 2 or so of this Gaussian case unless the particular shape puts a large fraction of its energy into long, low-intensity "wings" or other odd features. A typical 5-ns, more-or-less-Gaussian laser pulse, then, cannot have a bandwidth of less than about 80 to 100 MHz, and this width is too large for some very sensitive spectroscopic measurements.

Long-pulse or CW lasers also have line width limits imposed by quantum mechanics, but the practical line width limit in these systems is usually set by more mundane effects, such as mechanical vibration and thermal expansion. A typical short-term line width is between a few kilohertz and a few megahertz, with slow thermal drifts giving rather worse long-term stability unless the laser is actively locked to a narrow spectroscopic feature. Helmcke et al. [Helmcke, 1982] and Hough [Hough, 1984] describe representative CW dye lasers with active control designed to give a short-term line width of about 1 kHz.

Nanosecond-pulse lasers are also affected by transient effects in optical resonators and wavelength-selecting elements. The very narrow mode resonances in a laser resonator are caused by the interference between beams which have made multiple passes through the resonator: indeed, it is not hard to show that the ratio of the resonator mode width to the resonator mode spacing is the number of round trips which the typical photon makes in the resonator (an exercise for the reader using information found in Section 4.6 of Yariv, 1971). As an example, the 5-ns pulse mentioned previously can make only about 2.5 round trips through a 30-cm-long laser resonator during the pulse, so the mode cannot appear to be narrower than about a half to a third of the mode spacing, even though the resonator finesse (ratio of mode width to mode spacing for CW operation) can easily be 10 or 20 times larger. In a 1-m resonator the 5-ns pulse makes less than one round trip, no interference can occur, and the resonator has no effect at all on laser performance. Wavelength-selecting elements such as gratings or prisms are also affected by the number of round trips through those elements during the laser pulse since an element with a transmission function $T(\nu)$ has an effect proportional to $[T(\nu)]^n$, where n is the number of passes through the element.

It will be obvious that a short resonator is essential in a nanosecond-pulse system. The grazing-incidence grating and prism beam-expander resonators have a significant advantage for this application, as shown by the typical resonator lengths listed in Table 2.5.

Nanosecond-pulse lasers that operate with the narrowest possible line width frequently have rather unstable output pulse shape, line width, and pulse energy, particularly when part of the frequency selectivity comes from filtering through an external passive etalon [Curry, 1973], although the problem is not unknown in other laser resonators where the gain pulse length is too short to reach steady-state conditions [Kuizenga, 1973]. The fluctuation in pulse shape in frequency and time can be attributed to the quantum statistics of energy buildup in the resonator [Curry, 1973] or, for somewhat longer pulses, to small contributions from nearby resonator modes which vary randomly from pulse to pulse. Interference between these weak modes and the desired single mode modulates the output at the mode spacing frequency. These weak modes will eventually die away if the pulse is long enough, but they can be important until gain saturation by the desired mode and the $[T(\nu)]^n$ wavelength discrimination suppress them, which may take longer than the time available in a nanosecond-pulse system.

If the ultimate in frequency and amplitude stability is required in a nanosecond-pulse system, the proper choice is a CW or very long pulse oscillator followed by pulsed nanosecond amplifiers [Drell, 1979; Eesley, 1980; Egger, 1981].

2.3.6 Mode-Locking and Subnanosecond Pulses

The simple Fabry-Perot laser resonator has mode resonances spaced by $\nu = c/2\ell$ over the entire bandwidth reflected by the resonator mirrors. If n of these modes oscillate simultaneously and they are properly locked in phase, the Fourier transform of the n-mode spectrum is a train of pulses coming at a rate of ν per second with a pulse width of order $1/n\nu$. Broadly tunable lasers have such wide gain bandwidths that many hundreds of modes can oscillate simultaneously. If these modes are properly locked in phase, a train of picosecond or even sub-picosecond pulses can be generated [Shank, 1973]. The most sophisticated designs, using mode-locked rhodamine dye oscillators, can generate pulses as short as 30 to 70 fs [Fork, 1981; Mourou, 1982; Shank, 1982].

The short-pulse mode-locked laser can be thought of as a single short pulse circulating in a laser resonator as shown in Fig. 2.18. Each time the pulse strikes the output mirror a small fraction is transmitted, so the interval between pulses is the round-trip time in the resonator $2\ell/c$. If the cavity has a loss (or gain) which is modulated at the mode spacing frequency of $c/2\ell$, a short, circulating pulse can strike the region of low loss or high gain on each trip through the resonator, and its growth will be favored over some other combination of oscillating modes which cannot be represented as a short circulating pulse. If the gain or loss is modulated externally, the laser is said to be actively mode locked. A saturable absorber which has a high transmission at high intensity and low transmission at low intensity can also be used to modulate the resonator loss: a laser containing such an absorber is said to be passively mode locked. Combinations of active and passive mode locking are used to generate the shortest and most stable pulses. A frequency chirp combined with a dispersive element can also be used to compress pulses either inside a resonator [Corkum, 1983; Mollenauer, 1984] or as a separate element [Treacy, 1969; Shank, 1982; Nikolaus, 1983; Tomlinson, 1984].

Picosecond pulses have enabled revolutionary advances in the study of fast kinetic processes and transient species [Shank, 1973; Shapiro, 1977], including the spectroscopy of short-lived species such as excited states [Magde, 1981] and transient reaction intermediates. Self phase modulation can be used to broaden the spectrum of high-intensity picosecond pulses to give a "picosecond white light continuum," which is a very useful source for studying the absorptions of transient species [Alfano, 1971; Tashiro, 1974; Knox, 1984]. Water and deuterated water are particularly useful media for generating such white light continua [Busch, 1973; Varma, 1973].

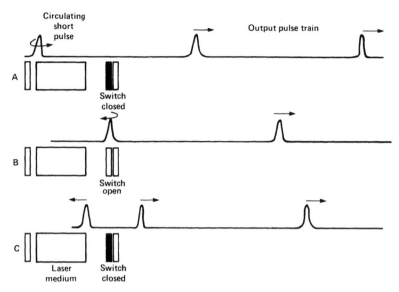

FIG. 2.18 The mode-locked oscillator viewed as a short pulse circulating in a laser resonator. At time A the pulse reflects from a resonator mirror and passes through the laser medium, where it is amplified. A switch (active or passive) near the other mirror is in the closed position to prevent any other pulses from circulating in the cavity. At time B the short circulating pulse strikes the switch, which is now open. The pulse reflects from a partially transmitting mirror and resumes circulating in the cavity. That part of the pulse which is transmitted through the mirror becomes the next pulse in the output pulse train from the oscillator, with pulses separated in time (or space) by one round trip through the resonator. At time C the switch has again closed to suppress other pulses which might begin to grow from noise.

2.3.7 Oscillator-Amplifier Chains

An oscillator with a very narrow line width unavoidably has a resonator with high losses as a consequence of the large number of optical components that it must contain to achieve that line width. If mode-limiting apertures are used to improve the spatial beam quality of the oscillator, the losses will be even higher. These apertures often restrict the oscillator active volume to a size much smaller than the pumped volume, which further reduces the oscillator output energy. One usually finds, then, that a high-quality, narrow-line-width laser oscillator has a rather low output power. If an application requires more output than such oscillators can provide, it is easier to add amplifier stages to a low-power oscillator than to design an oscillator which can deliver high beam quality and narrow line width at high energy. These master oscillator-power amplifier (MOPA) systems are common in high peak- and average-power tunable laser systems. Figure 2.19 shows a typical, although highly schematic, layout of a MOPA chain such as one might find in a pulsed dye-laser system. The high-quality output beam from the oscillator is amplified in several amplifier stages of increasing size until the output is adequate for the application at hand.

A MOPA chain must be designed to suppress amplified spontaneous emission and parasitic oscillations from stray reflections in the amplifiers so that the output contains only the high-quality, narrow-linewidth signal injected by the oscillator [Haag, 1983]. Stray reflections back into the oscillator must also be avoided since they may upset its stability. The most popular techniques for controlling propagation in MOPA chains will be described here very briefly. There are many MOPA designs in the literature using various combinations of these techniques, and occasionally one finds a simple amplifier with none of them.

Transit-time isolation. A nanosecond or subnanosecond pulse system has such short gain pulses that the amplifiers can be pumped in sequence so that the gain region moves down the amplifier chain at the same speed as the oscillator pulse. If the transit time between ampli-

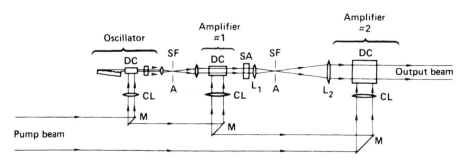

FIG. 2.19 A typical dye-laser master oscillator-power amplifier (MOPA) system. The pump beam enters from the left and a small fraction is split off by mirrors M to pump the oscillator and the first amplifier (number 1). Most of the beam pumps the final amplifier, which is number 2 in this case. Cylindrical lenses CL adjust the pump beam size to coincide with the size of the dye-laser beam in each dye cell DC. Between stages are spatial filters SF consisting of two lenses spaced at the sum of their focal lengths with an aperture A at the common focal point. A saturable absorber cell SA may be included to give further isolation against amplified spontaneous emission and parasitic oscillation. The amplifiers could also use longitudinal pumping (see Fig. 2.12).

fiers is greater than the pulse length, any radiation scattered backward down the chain will reach earlier amplifiers or the oscillator long after the output pulse has left them and cannot possibly affect their behavior. The MOPA of Fig. 2.19 shows how the pump and signal beams usually propagate in such a system.

Spatial filtering. A pair of lenses spaced a distance equal to the sum of their focal lengths will transform a parallel beam exiting one amplifier, such as amplifier 1 in Fig. 2.19, into a parallel beam for input into a second amplifier. An aperture on the optical axis at the common focus allows only a limited range of input angles to be transmitted through the filter. This reduces the amount of amplified spontaneous emission or stray parasitic oscillation that can enter amplifier 2 and also suppresses high-frequency spatial noise on the amplified beam. Note that any stray "ghost" reflections from lenses and apertures can be a source of feedback for parasitic oscillations.

Image relaying. With a little extra thought, a chain of spatial filters can perform another function. If the positions of the lenses and amplifiers are chosen properly, lens L_2 will form an image of amplifier 1 superposed on amplifier 2. The net "optical propagation distance" from the original object is zero when an image forms in an optical system, so sharp edges and other features in the object reproduce themselves in the image without distortion due to diffraction. This makes it much easier to design a laser chain in which the amplifiers are filled with beams of uniform intensity. It also limits the growth of diffraction fringes from sharp edges or other obscurations [Simmons, 1981]. These properties are useful but not essential in small spectroscopic laser systems.

Saturable absorption. In some absorbers the ground state is repopulated so slowly that a high-intensity short pulse can depopulate the ground state and "saturate" the ground-state absorption. Saturable absorbers have a much higher transmission for short, high-intensity pulses than for low-intensity amplified spontaneous emission or other weak parasitics, so they can be used to improve the contrast between desired and undesired output in a MOPA chain. Some organic or organometallic dyes and some color centers have population kinetics which allow them to function as saturable absorbers. Vibrational absorptions in molecules have also been used as saturable absorbers in infrared MOPA systems.

Intensity stabilization by saturation. Highly saturated amplifiers have an output intensity which is rather insensitive to the input intensity and can therefore reduce the intensity noise which is prevalent in narrow-line-width oscillators [Curry, 1973].

Electro-optic, acousto-optic, and magneto-optic isolation. All types of beam-deflecting and polarization-rotating switches have been

used at one time or another to isolate amplifiers in MOPA systems.
Magneto-optic Faraday rotators are frequently used to prevent back-
reflection into early amplifiers and the oscillator.

There are many MOPA system designs in the literature and only a
few of these will be listed here. Curry et al. [Curry, 1973] describe
a typical nitrogen-laser-pumped dye-laser MOPA. Salour [Salour,
1977], Drell and Chu [Drell, 1979], Eesley et al. [Eesley, 1980], Wallenstein and Zacharias [Wallenstein, 1980], and Egger et al. [Egger,
1981] describe similar but later versions. MOPA systems for picosecond pulses have been a very active area of development recently.
Sizer et al. [Sizer, 1981], Koch et al. [Koch, 1982], Wokaun et al.
[Wokaun, 1982], and Knox [Knox, 1984] describe such systems.

2.3.8 Injection-Locked Slave Oscillators

The MOPA chain is straightforward but has a rather large number of
components. It is sometimes possible to simplify a system by using a
large amplifier as a "slave oscillator" or "regenerative amplifier" controlled by a weak injected signal from an oscillator, as shown in Fig.
2.20. The injected "seed" signal from the master oscillator must be
strong enough to compete with spontaneous emission into all of the
modes which are allowed to oscillate in the resonator. It is therefore
useful to reduce the number of transverse modes which are allowed to
oscillate, either by inserting a spatial filter with a mode-limiting aperture (as shown in Fig. 2.20) or by using an unstable resonator on the
slave oscillator.

The most popular unstable resonator is the positive branch confocal unstable resonator shown in Fig. 2.21. The resonator consists
of concave and convex mirrors M_1 and M_2 which have a common focal
point. A plane wave that strikes M_2 becomes a spherical wave that reflects from M_1 to give a plane wave output beam, part of which strikes
M_2 again and continues circulating in the resonator. The master oscillator beam enters the resonator through a small hole or partially
transmitting coating on one of the mirrors.

The length of the slave resonator must be adjusted so that the injected signal falls at the peak of one of the slave resonator modes for
optimum performance, although this adjustment is not particularly important if the slave resonator losses are very large and the modes
therefore have a large line width, which is usually the case in unstable resonators.

Bigio and Slatkine [Bigio, 1983] and Ganiel et al. [Ganiel, 1976]
review the literature on injection locking and discuss design issues in
greater detail. Blit et al. [Blit, 1977], Okada et al. [Okada, 1979],
Goldhar et al. [Goldhar, 1980], Bhattacharyya [Bhattacharyya, 1981],
and Park [Park, 1984] present some specific examples. Note that if an
injection-locked oscillator is tuned off the peak of the laser gain curve,
the saturated laser gain will just equal the resonator loss at the oper-

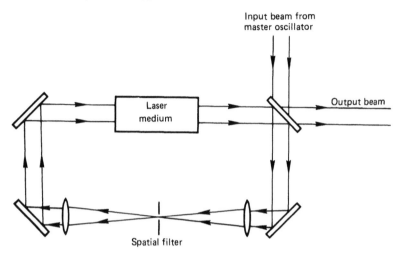

FIG. 2.20 An injection-locked, ring resonator slave oscillator. Ring resonators such as this are often used for slave oscillators. The input and output beams can go through the same partially transmitting mirror as shown. The resonator length of the slave oscillator must be adjusted so that the injected signal hits a mode resonance of the slave cavity if the best performance is required, although it is often possible to operate slave oscillators without exact mode matching in resonators with very high losses or short excitation pulses.

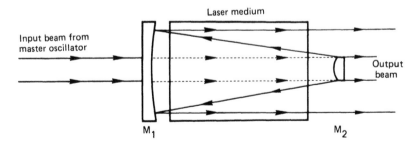

FIG. 2.21 An injection-locked, confocal unstable resonator slave oscillator. Mirror M_1 has a small transmission so that the master oscillator beam can enter the cavity. The injected beam could also enter through a small hole in the mirror, or could be injected at M_2.

ating frequency, but some small net gain will remain at the peak of the gain curve. Amplified spontaneous emission at the peak of the curve may then grow and will eventually dominate the output, although of course its growth rate will be much slower than it would have been in the absence of an injected signal.

2.4 WAVELENGTH CONVERSION

2.4.1 General

The tunable lasers just described cover a wide range of wavelengths but do not satisfy all needs for spectroscopic sources. The lead salt diodes make excellent tunable sources in the mid-infrared but have very low CW power output, which is inadequate for many experiments in nonlinear spectroscopy, remote sensing, and other fields. At wavelengths shorter than 340 nm there are few tunable lasers and those do not have particularly wide tuning ranges, so there are wide gaps in spectral coverage. Useful pump sources such as the Nd:YAG laser have too long a wavelength to pump dye lasers directly. Nonlinear optical effects such as harmonic generation, parametric oscillation, frequency mixing, and stimulated Raman scattering make it possible to convert intense, high-quality sources to other spectral regions, which corrects some of these deficiencies.

2.4.2 Basic Physics of Harmonic Conversion and Mixing

When an electromagnetic wave passes through a material, the electric field induces a polarization in the bound electrons. It is customary to assume that the induced polarization is proportional to the electric field, in which case the induced polarization oscillates sinusoidally, following the applied field, and manifests itself as the dielectric constant ε or index of refraction n = $\sqrt{\varepsilon}$. It can be shown that this interaction can be modeled as bound electrons moving in purely parabolic potential energy wells, which have a linear restoring force.

Any real energy well cannot be parabolic forever: for large enough displacements the restoring force must have nonlinear components. If the driving electric field becomes large enough, the induced polarization will not be linear in the applied field and will be distorted from its sinusoidal shape. Distortion of a sine wave corresponds to the introduction of harmonics which are integral multiples of the driving frequency, so there are now weak polarization components oscillating at these harmonics which can radiate new electromagnetic waves at those frequencies.

The input wave propagates through the material with a phase velocity c/n and the induced polarization varies in space in exactly the same way. If the harmonic wave radiated by the induced polarization propagates at the same velocity, it will stay in phase with the polarization and continue to grow as it propagates. If the index of refraction is different for the harmonic wave, the wave will slip out of phase with the polarization as it propagates and there will be destructive interference between harmonic waves radiated from different parts of the material, which reduces the harmonic intensity. The phase-matched condition where the two waves remain in phase is necessary if there is to be a high efficiency for conversion to the harmonic wave.

Note that if the potential well is symmetric along the direction of the electric field, the induced polarization will also be symmetric. It is easy to show that a symmetric waveform, however distorted, contains only odd harmonics. Even harmonics are generated only when there is an asymmetry in the potential well along the direction of the applied electric field, in which case the material is said to lack inversion symmetry in that direction.

We have assumed that there is only one input electromagnetic wave here, but it is easy to generalize the problem to include several waves of different frequency, in which case sum and difference frequencies appear in the polarization and the output. This brief introduction to the underlying physics of nonlinear conversion processes omits many important points, but a complete treatment is beyond the scope of this chapter. Bloembergen [Bloembergen, 1965], Zernicke and Midwinter [Zernicke, 1973], Hanna et al. [Hanna, 1979], Shen [Shen, 1984], and many other authors treat nonlinear processes in greater detail.

2.4.3 Frequency Doubling in Birefringent Crystals

The most important nonlinear harmonic conversion process is frequency doubling in uniaxial birefringent nonlinear crystals where the birefringence is exploited to give a phase-matched interaction. Biaxial crystals can be used, but the analysis is more complex through a minor extension of the treatment of uniaxial materials, so biaxial crystals will not be explicitly treated here. Of course, the crystals must also lack inversion symmetry, as discussed previously.

A uniaxial birefringent crystal allows two modes of propagation of an electromagnetic wave. A wave polarized perpendicular to the optical axis of the crystal sees an index of refraction n_0 independent of its direction of propagation in the crystal, and this is called an "ordinary" wave. A wave propagating in the same direction but polarized perpendicular to the polarization of the ordinary wave is called the "extraordinary" wave: it sees an index of refraction that varies elliptically between n_0 when the vector normal to the wavefront is parallel to the optical axis and a different index n_e when it propagates perpendicular to the optical axis. The Poynting vector $\underline{E} \times \underline{H}$, which

gives the direction of energy flow in the wave, does not coincide with
the wave normal in birefringent crystals, and this introduces some
minor complications that we shall ignore here.

Phase-matched second harmonic generation requires that the second harmonic wave propagate at the same velocity as the nonlinear
polarization that tracks the fundamental wave, yet the index of refraction, which is inversely related to propagation velocity, increases at
short wavelengths. If we propagate one of these waves as an ordinary
wave and the other as an extraordinary wave in a birefringent crystal,
it may be possible to adjust the index seen by the extraordinary wave
so that the fundamental and second harmonic see the same index and
propagate at the same velocity. Assume, for example, that we have
$n_e < n_o$, or a negative uniaxial crystal. We might make the fundamental an ordinary wave, which has a high index, and the second harmonic an extraordinary wave. The extraordinary index can then be adjusted by rotating the beam with respect to the optical axis until the
ordinary index at the fundamental and the extraordinary index at the
second harmonic are equal, provided that the ordinary index at ω is
larger than the extraordinary index at 2ω. This arrangement is called
type I phase matching. We might also make the fundamental a mixture
of ordinary and extraordinary waves and generate a purely ordinary
or purely extraordinary wave in the second harmonic: this is called
type II phase matching. Both of these cases have fundamental and
second harmonic beams propagating in exactly the same direction.
There are numerous other phase-matching possibilities with noncollin-
ear beams, but most applications use one of these two collinear cases.

A real laser beam is not a perfect plane wave but has some angular spread. Angle-phase-matched systems such as those just presented are not the most convenient choice for use with real beams since
they are phase matched over only a very narrow angular range which
is frequently smaller than the angular divergence of the laser beam.
If the crystal can be oriented so that the interaction is phase-matched
when the waves propagate at 90° to the optical axis, n_e is at its extreme value and its variation with angle is second order, which gives
a much larger angular range over which there is an adequate phase
match. The Poynting vector walk-off, which we are ignoring here, is
also minimized under these conditions, which can be significant for
beams with very small diameter. Many crystals have indices of refraction (usually the extraordinary) which vary with temperature, and the
wavelength at which 90° or "noncritical" phase matching occurs can be
varied over a small range by changing the temperature. As an example, the common nonlinear crystal ammonium dihydrogen phosphate
(ADP) has 90° phase matching for type I doubling with a fundamental
wavelength of about 497 nm at 180 K and 542 nm at 360 K.

Sometimes special crystal compositions can be designed to have
noncritical phase matching at a desired wavelength. Returning to the

example of ADP, the ammonium ion might be replaced by an alkali metal, the hydrogen by deuterium, and the phosphorus by arsenic. This particular class of KDP isomorphs (KDP for potassium dihydrogen phosphate, KH_2PO_4) includes many of the most common nonlinear crystals. The KDP isomorphs do not have particularly large nonlinearities but are relatively easy to grow in large sizes and have excellent optical quality and resistance to damage by intense laser pulses when compared to most other nonlinear crystals. It may also be possible to grow mixed-crystal species within this class, giving an even wider range of noncritical phase-matching materials, which is a subject of active study at present. Other important nonlinear materials include lithium niobate ($LiNbO_3$), lithium iodate ($LiIO_3$), barium sodium niobate ($Ba_2NaNb_5O_{15}$) [Geusic, 1968], and potassium titanyl phosphate ($KTiOPO_4$) [Liu, 1984], which are all used to double neodymium lasers. The last two of these are biaxial crystals. Proustite (Ag_3AsS_3) and cadmium selenide (CdSe) are interesting phase-matchable materials in the infrared. There are many more materials listed in standard references [Pressley, 1971; Zernicke, 1973].

So far we have discussed phase matching but have ignored the magnitude of the nonlinear coefficients of these crystals. Since the electric field of the incident wave drives the bound electrons parallel to its polarization, the direction of motion of the electrons in their potential wells depends on the propagation direction of the wave through the crystal, and the nonlinear coefficient is a function of that direction. It is often possible to choose propagation directions in which the potential well is symmetric and the second harmonic generation is zero. In urea [$CO(NH_2)_2$], as an example, the nonlinear coefficient for doubling vanishes at exactly the angle that gives noncritical type I phase matching [Halbout, 1979]. In the KDP isomorphs the doubling coefficient vanishes for noncritical type II phase matching.

The nonlinearity of useful phase-matchable materials varies over several orders of magnitude, so the laser electric field required to get a useful conversion efficiency in a reasonable length varies considerably depending on the particular crystal, wavelength, laser beam quality, and experimental arrangement. A high-efficiency doubler using a pump source in the visible or near infrared and a nonlinear crystal such as one of the KDP isomorphs would typically require a laser intensity of 10^8 to 10^9 W/cm^2 to reach a conversion efficiency of 50 to 70% in a crystal about 1 cm long [Zernicke, 1973; Hon, 1979]. It is not too difficult to run such a crystal at this intensity for nanosecond pulses, but optical damage can become a serious problem with long pulses under these conditions. A material with a higher nonlinearity, such as lithium niobate, might require one or two orders of magnitude less intensity to reach a second harmonic conversion of a few tens of percent, but is somewhat more vulnerable to damage than KDP isomorphs even at that lower intensity. A carefully optimized KDP doubler using a 4.5-cm crystal in a noncritically phase-matched geometry located inside the cavity of a several-watt argon ion laser which has an intercavity intensity of about 3×10^5 W/cm^2 can reach about 50% conversion efficiency, but thermal effects prevent true CW operation at this level [Dowley, 1968a,b]. Doublers in CW [Geusic, 1968] or CW mode-locked [Liu, 1984] Nd:YAG lasers can be more than 50% efficient. A more typical result with a few watts of argon, dye, or other laser power focused on an external noncritically phase-matched doubler would be a conversion efficiency on the order of 1% [Jain, 1973].

2.4.4 Other Nonlinear Conversion Processes in Crystals

There are many possibilities for nonlinear conversion processes other than simple frequency doubling. Sum and difference frequencies of all types can be generated by mixing various intense beams: common processes include mixing the fundamental and a second harmonic beam to generate the third harmonic [Craxton, 1980; Seka, 1980], mixing a fundamental with a beam that has been doubled twice to produce the fifth harmonic [Kato, 1980], and mixing a tunable dye laser with various harmonics of an intense pump laser to generate tunable short-wavelength sources [Massey, 1976]. The conversion efficiency is higher if the pump intensities are raised by operating the nonlinear crystal in a resonator [Couillaud, 1982; Hemmati, 1983]. These processes all also require phase matching and laser intensities of the same order of magnitude as frequency doubling if the conversion is to be efficient.

A single strong input wave at ω_p can generate two waves of lesser frequency ω_s and ω_i such that $\omega_s + \omega_i = \omega_p$. This is a special case of difference frequency generation in which a pump wave ω_p and idler wave ω_i generate a signal wave ω_s. There is gain for both the signal and idler waves in the presence of the pump, so those waves can grow from amplified spontaneous emission to high intensity, just as in any other laser [Harris, 1969; Smith, 1972; Baumgartner, 1979]. This device is called a parametric amplifier or oscillator, by analogy with similar devices used at radio frequencies. (The "parameter" being varied here is the index of refraction; at radio frequencies it is commonly a nonlinear capacitance.) An ADP parametric oscillator pumped by the fourth harmonic of a neodymium laser will tune across the visible from about 410 to 720 nm [Yarborough, 1971] but is rather more difficult to operate than the dye lasers with which it competes. Lithium niobate oscillators that tune across a wide range from about 700 to 2000 nm in the near infrared when pumped with ruby or doubled neodymium lasers are more useful [Smith, 1972; Zernicke, 1973]. A cadmium selenide parametric oscillator pumped by a hydrogen fluoride laser or other laser near 2.7 μm can be tuned from about 14.1 to 16.4 μm in the mid-infrared [Davydov, 1973; Wenzel, 1976].

Frequency-mixing processes need not be restricted to optical fields. Electrooptic materials have an index of refraction which varies significantly with applied voltage in the radio-frequency (RF) or microwave regions of the electromagnetic spectrum, and electrooptic modulators driven by tunable RF sources can modulate a nontunable laser source to generate very precisely tunable sidebands over a small frequency range near the laser frequency. Magerl et al. [Magerl, 1982] and Cheo [Cheo, 1984] discuss the design of electro-optic modulators which allow tuning over a range of 15 to 20 GHz on either side of any of the many lines of a CO_2 laser in the region 9 to 11 μm. Similar modulators have been used with other nontunable laser sources.

2.4.5 Odd-Harmonic Generation and Mixing in Gases

Gases are isotropic, so the generation of even harmonics is forbidden unless the applied electric fields are so large that they destroy the symmetry [Kiyashko, 1985]. The third and higher odd harmonics can be generated with weaker fields. Gas harmonic generators are particularly useful for generating coherent sources in the vacuum ultraviolet at wavelengths shorter than 200 nm since nonlinear crystals are not available for this region. Hanna et al. [Hanna, 1979] review the nonlinear optics of gases in detail.

Efficient harmonic generation in gases, as in crystals, requires a phase-matched interaction. The anomalous dispersion on the short-wavelength side of an absorption line can be exploited to give phase matching as shown in Fig. 2.22 [Miles, 1973]. An absorption line causes a decrease in the index of refraction (anomalous dispersion) in a small wavelength region just to the blue side of the line. In an ideal case where this was the only absorption, the index would go through unity at the center of the absorption line and fall below unity to the blue side, but in the real world the dispersion due to any particular absorption line modulates a background index due to all the other absorptions in the system, as shown in Fig. 2.22. If the third harmonic frequency lies just to the blue side of an absorption line, it will see an index lower than the background index and this dip may be deep enough that the index is less than the index at the fundamental frequency. The background index and its wavelength dependence can be changed by adding other gases to the cell if the anomalous dispersion is too large for phase matching. The near resonance with an absorption may also enhance the nonlinear index (an advantage) or multiphoton absorption (a disadvantage).

Phase-matched tripling from 1064 nm to 355 nm has been demonstrated in mixtures of alkali metal vapor and xenon [Bloom, 1975]. The conversion efficiency to third harmonic was as high as 10% at a fundamental input power of order 10^{10} to 10^{11} W/cm^2. A second tripling step from 355 nm to 118 nm can be phase matched using a xenon-argon mixture, but other nonlinear effects limit the conversion efficiency to a few tenths of a percent [Zych, 1978]. Resonance with

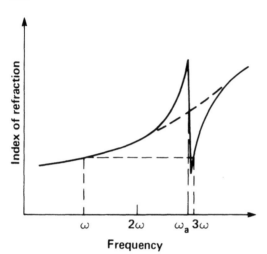

FIG. 2.22 Phase matching using anomalous dispersion near an absorption. The index of refraction falls below its background value on the short-wavelength side of an absorption line. If 3ω falls just on the blue side of an absorption, it may be possible to have an equal index of refraction at ω and 3ω over a small range of frequencies.

an intermediate state can enhance the conversion efficiency even if the interaction is not phase matched [Kiyashko, 1979]. Hillig and Wallenstein [Hillig, 1983] describe a typical, tunable, nonresonant VUV source.

Useful intensities of even higher harmonics can be generated in short interaction regions at high intensity, although the interaction cannot be phase matched and the conversion efficiency is very low. The fifth and seventh harmonics of a 266-nm quadrupled neodymium laser [Reintjes, 1978] and 248-nm KrF laser [Bokor, 1983] have been seen from beams focused to extreme intensities in a low density of helium.

Nonlinear mixing processes other than harmonic generation can also be observed in gases. Various mixing processes using two-photon resonances have been used to generate tunable radiation over small ranges in the vacuum ultraviolet using strontium [Hodgson, 1974], xenon [Yiu, 1982], zinc [Jamroz, 1982], and other atoms. A phase-matched mixture of krypton and magnesium vapor has been used to generate CW vacuum ultraviolet radiation [Timmerman, 1983]. Molecular gases can also be used: Valee and Lukasik [Valee, 1982] describe an example using CO gas.

Harmonic generation and nonlinear mixing in gases is of course not restricted to the ultraviolet, although there are fewer competing sources in the ultraviolet than in other spectral regions. Kildal [Kildal, 1977], for example, describes a phase-matched gas tripler for 10.6-μm CO_2 laser radiation. A number of other infrared experiments are reviewed in a book edited by Shen [Shen, 1977].

2.4.6 Stimulated Raman Scattering

There are processes that generate terms in the nonlinear index of refraction which oscillate at frequencies determined by resonances in the material itself rather than the input frequency, and these processes can also be used to generate new frequencies from laser sources. The most important of these processes is Raman scattering from index perturbations produced by molecular vibrations and rotations, and these will be used as an example in this discussion. There are similar processes (which can be treated in much the same way) using atomic or molecular electronic transitions, plasma waves, spin states of carriers in semiconductors in a magnetic field (spin-flip Raman scattering), coherent acoustic waves (Brillouin scattering), and other nonlinear phenomena [Kaiser, 1972; Penzkofer, 1979; Shen, 1984].

The electronic structure of a molecule changes slightly as it vibrates, which modulates the polarizability of the molecule (and hence the index of refraction of the medium) at the vibrational frequency. The modulation of the index phase-modulates any electromagnetic wave propagating through the medium. In frequency space the phase modulation corresponds to the generation of Raman sidebands, which are new electromagnetic waves having frequencies shifted up and down by the vibrational frequency from the original input frequency of the wave. The sideband shifted to lower frequency is called the Stokes wave by analogy with fluorescence, and the sideband shifted to higher frequency has become known as the "anti-Stokes" wave.

Figure 2.23 shows the energy-level diagram of a vibrational Raman scatterer in a quantum mechanical picture. A molecule in its ground vibrational level can absorb a photon from the pump wave propagating through the medium and simultaneously reradiate a Stokes-shifted photon which leaves the molecule in its first vibrational level. Similarly, a molecule already in the first vibrational level could absorb a pump photon and simultaneously reradiate an anti-Stokes photon which leaves the molecule in its ground vibrational level. One can think of the "virtual state" through which the molecule passes in this process as the far wing of an allowed electronic transition of the molecule. This discussion will consider only Raman processes which are far from resonance with the allowed electronic transition. In these processes the Raman scattering cross section is small and has a slow dependence on the frequency of the pump wave. If the pump wave is tuned near resonance with an allowed electronic transition, the Raman scattering can become much more intense and also varies rapidly with pump frequency.

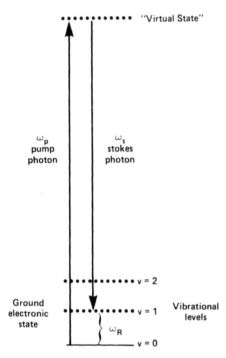

FIG. 2.23 Energy levels used in stimulated Raman scattering. In the presence of an intense pump wave a molecule can simultaneously absorb a photon of frequency ω_p and emit a Stokes photon of frequency ω_s. The molecule is left in an excited vibrational state with an energy corresponding to the vibrational frequency ω_R.

Suppose now that most of the molecules are in the ground vibrational level. In the presence of a pump wave there is a population inversion between the virtual state of Fig. 2.23 and the first excited vibrational level of the molecule. An inverted population implies laser gain, so spontaneous emission at the Stokes frequency is amplified and can grow to an intensity comparable to the pump intensity. This condition is known as "stimulated Raman scattering." The Raman gain is proportional to the "population" of the virtual state, which is the product of the density of molecules in the vibrational ground state and the pump photon density or intensity. The gain is then

$\exp[\gamma \rho \int I_p \, d\ell]$, where the integral of I with respect to length ℓ is along the Stokes propagation direction. In this equation γ is a Raman gain parameter, ρ is the density of the ground vibrational level (or, more accurately, the difference in population between the ground and first excited levels), and I_p is the pump intensity.

The vibrational or rotational frequencies of the most useful Raman scatterers are listed in Table 2.6, and a more complete list will be found in standard references. The many organic liquids which have been used for stimulated Raman scattering also show serious beam distortions due to self-focusing and other nonlinear index effects at the high laser intensities in stimulated Raman conversion cells, so they are less useful for spectroscopic sources than the materials listed in Table 2.6.

When the Stokes waves grows so intense that it has an amplitude comparable to that of the pump, the Stokes wave can itself serve as a pump wave and generate higher-order Stokes waves by successive Raman scattering steps. The Stokes wave at a frequency $\omega_s = \omega_p - \omega_R$ produces gain at a frequency $\omega_{s2} = \omega_s - \omega_R$, and the second Stokes wave ω_{s2} shifted $2\omega_R$ from the pump can grow to a high intensity. Higher-order Stokes waves can follow in succession by the same process.

In principle the Stokes wave can propagate in any direction with respect to the pump, but for a variety of reasons [Akhamanov, 1974; Murray, 1979; Penzkofer, 1979] that will not be discussed here, the gain for a simple Raman process is always highest for forward scattering in which the Stokes wave propagates in exactly the same direction as the pump, so forward scattering is the first process to go over threshold when a Stokes wave grows from amplified spontaneous emission.

There is another type of interaction between waves in a Raman-active medium which is very important for explaining their behavior. Strong pump and Stokes waves propagating in a Raman-active medium prepare a phased array of vibrating molecules (in quantum mechanical terms, a coherent superposition of the ground and first vibrational levels), which has a very interesting property. Consider a pump wave that propagates as $\exp[i\underline{k}_p \cdot \underline{r}]$ and a Stokes wave that propagates as $\exp[i\underline{k}_s \cdot \underline{r}]$. The phase of these waves drifts with respect to each other as $\exp[i(\underline{k}_p - \underline{k}_s) \cdot \underline{r}]$ as they propagate, so the phase of molecules driven by these waves and vibrating at position $\underline{r}$ must also vary as $(\underline{k}_p - \underline{k}_s) \cdot \underline{r}$. This phase-matching condition is a restatement of the conservation of momentum in the scattering process and can be represented as the $\underline{k}$-vector diagram of Fig. 2.24. Now consider what happens if the two waves are propagating in the exact forward direction, as shown in Fig. 2.25, and assume for the moment that the index of refraction of the Raman medium is not a function of frequency. Since $\underline{k} = n\omega/c$, the condition that $\underline{k}_p - \underline{k}_s = \underline{k}_R$ becomes

TABLE 2.6 Raman Frequency Shifts of the Most Often Used Raman
Spectral Converters

		Frequency shift, w_R (cm^{-1})
H_2 gas	Vibration	4155
	Rotation	587, 354
D_2 gas	Vibration	2991
	Rotation	414, 179
CH_4 gas	Vibration	2916
SF_6 gas	Vibration	775
N_2 liquid	Vibration	2331

identical to the energy conservation condition $\omega_p - \omega_s = \omega_R$, which is automatically satisfied for any arbitrary input wave ω_p. The result is that the phased array of vibrating molecules prepared by a stimulated Raman process using a pump ω_p and Stokes ω_s can shift any other input wave ω_p' to either a higher or a lower Raman sideband, and this interaction will be phase matched so that the new waves can grow to high intensity. The interaction is particularly useful for producing very short wavelength sources by successive anti-Stokes scattering from the original pump and Stokes waves, as shown in Fig. 2.26. In real systems the change in the slope of the index of refraction as a function of frequency (index dispersion) introduces enough phase mismatch in the mixing process to limit the efficiency of conversion to the higher anti-Stokes orders. The anti-Stokes conversion efficiency is very critically dependent on the intensity of the pump in space and

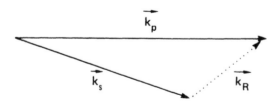

FIG. 2.24 Phase-matching conditions between the $\underline{k}$ vectors of the pump wave $\underline{k}_p$, Stokes wave $\underline{k}_s$, and the phased array of vibrating molecules $\underline{k}_R$.

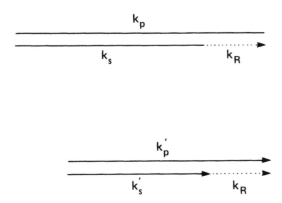

FIG. 2.25 Phase-matching condition between the $\underline{k}$ vectors in the exact forward direction in a material where $|\underline{k}|$ is directly proportional to ω (no index dispersion). The vector condition for momentum conservation $\underline{k}_p = \underline{k}_s + \underline{k}_R$ is now identical to the scalar condition for energy conservation $\omega_p = \omega_s + \omega_R$, and consequently the same phased array $\underline{k}_R$ can couple any arbitrary pump and Stokes waves such as $\underline{k}'_p$ and $\underline{k}'_s$.

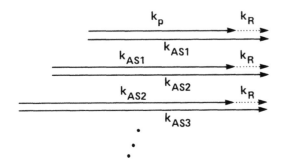

FIG. 2.26 Generation of high-order anti-Stokes waves by successive scattering from the phased array $\underline{k}_R$. The pump wave $\underline{k}_p$ generates a first anti-Stokes wave $\underline{k}_{AS1}$, which then generates a second anti-Stokes wave $\underline{k}_{AS2}$, and so on to higher orders until the index dispersion makes the phase-matched interaction distance so short that the conversion to the higher order is negligible.

time, so the higher anti-Stokes orders have much larger fluctuations
in amplitude than does the pump. This "four-wave mixing" or "parametric generation" process and the simple Raman gain term proportional to pump intensity are both important in Raman systems whenever
there are two strong copropagating waves, and the interaction between
them can be very complex. It is also important to note that this anti-
Stokes generation process is quite different from another process in
which the population of the ground and excited states of the Raman
scatterer are inverted so that there is simple Raman gain for the anti-
Stokes wave rather than for the Stokes wave.

The phase-matched four-wave mixing process in forward-stimulated
Raman scattering has several interesting consequences. It can be
shown that forward four-wave mixing causes the stimulated Raman gain
produced by a broadband pump to be identical to that produced by a
single-frequency pump but that this happens only when the bandwidth—and, in fact, the detailed phase fluctuations—of the Stokes
wave exactly reproduce those of the pump [Akhmanov, 1974; Trutna,
1979]. If one attempts to amplify a narrowband Stokes wave in a forward Raman amplifier driven by a broadband pump, the Stokes wave
will see a low gain at first, but other Stokes frequency components
will be generated by four-wave mixing until the Stokes bandwidth
equals that of the pump and then the gain will rise [Zubarev, 1976].
A Raman system operated under conditions where forward four-wave
mixing is important will therefore generate anti-Stokes and Stokes
waves with a bandwidth equal to that of the pump wave.

The relative importance of forward four-wave mixing can be controlled to some degree by altering the operating conditions of the
Raman converter. The phase mismatch in the four-wave mixing process will be proportional to $d\ell$, where d is an index mismatch parameter proportional to the number of molecules per unit volume and ℓ is
the length over which the interaction takes place. To maximize four-
wave mixing, then, we must use short interaction lengths and a low
particle density, which implies a very high pump wave intensity to get
enough simple Raman gain that the first Stokes wave can grow from
noise to a high intensity. There are other important nonlinearities,
such as self-focusing and two-photon absorption, in these materials,
so arbitrarily high pump intensities are not feasible in practice. A
Raman converter that optimizes the production of high anti-Stokes orders by four-wave mixing might then use a short-focal-length lens to
generate a very high intensity over a short interaction region in a low
density of a gas such as hydrogen, which has a low index and low index dispersion. A converter designed to defeat four-wave mixing and
optimize the sequential conversion of energy to higher Stokes orders
by the simple Raman gain process would use just the opposite: a low
pump intensity over a long interaction distance in a dense and highly
dispersive medium.

Wilke and Schmidt [Wilke, 1979] present results of a study of a Raman converter designed to provide a tunable ultraviolet spectroscopic source using anti-Stokes generation. A typical converter uses a lens with a 30-cm focal length to focus a 40-mJ, 10-ns, 558-nm dye-laser pulse into a cell containing about 5 atm of hydrogen. About 10% of the pump energy is converted to the first anti-Stokes at 458 nm and successively smaller energies to higher anti-Stokes orders out to the eighth anti-Stokes order at 195 nm, which contains about 10^{-4} of the pump energy. Brunk and Proch [Brunk, 1982] and Schomburg et al. [Schomburg, 1983] describe similar experiments.

Paisner [Paisner, 1979] generates tunable vacuum ultraviolet radiation using a 40-mJ, 20-ns argon fluoride excimer laser pump pulse focused with a 50-cm lens into 2 to 5 atm of hydrogen. He detects anti-Stokes orders out to the seventh order at 124 nm. The first anti-Stokes order contains about 10% of the pump energy, and each successive anti-Stokes order contains about 10% of the energy in the next lowest order. Döbele et al. [Döbele, 1984a, 1984b] have published an extensive study of vacuum ultraviolet sources of this sort.

There are many older experiments in the literature, of which only one will be mentioned here. Johnson et al. [Johnson, 1967] generate simultaneous rotational and vibrational Raman scattering in hydrogen with a neodymium laser and shows that complex multiple intercombination orders of vibrational and rotational scattering can be generated. This gives a closely spaced array of Stokes and anti-Stokes lines over a wide spectral range.

Several experiments have demonstrated that four-wave mixing can be enhanced by driving an intense phased molecular array with a frequency widely separated from that which is being converted. Brosnan et al. [Brosnan, 1977] use stimulated vibrational Raman scattering in hydrogen with a Nd:YAG or doubled Nd:YAG pump to prepare a phased array of molecules which then shifts the frequency of a $LiNbO_3$ parametric oscillator to give a tunable source from 3.5 to 13 μm. Byer and Trutna [Byer, 1978] use stimulated rotational Raman scattering in hydrogen driven by a neodymium laser to prepare a molecular array which generates a 16.9-μm signal by four-wave mixing from a 10.6-μm input. The 16.9-μm signal is then amplified by simple Raman gain driven by the 10.6-μm pump. This technique allows conversion to 16.9 μm at a far lower 10.6-μm intensity than would be required if the 16.9-μm signal had to grow by amplified spontaneous emission. White et al. [White, 1981] show that the efficiency of conversion of a doubled dye laser to the fourth anti-Stokes order in hydrogen is enhanced by a factor of 5 when 40 mJ of undoubled dye radiation is focused into the Raman converter together with the 15 mJ of doubled dye radiation.

Rabinowitz et al. [Rabinowitz, 1979] and Carlsten and Wenzel [Carlsten, 1983] discuss Raman systems that convert 50 to 85% of the input 10.6-μm pump photons to 16.9 μm in hydrogen using focusing multipass cells. Carlsten uses a CdSe parametric oscillator as a 16.9-μm source, while Rabinowitz lets the 16.9-μm Stokes wave grow by amplified spontaneous emission. DeMartino et al. [deMartino, 1980] describe a source tunable over a much wider range in the infrared. Hartig and Schmidt [Hartig, 1979], Chraplyvy and Bridges [Chraplyvy, 1981], and Berry et al. [Berry, 1982] discuss Stokes converters in which the pump and Stokes waves are confined to a waveguide. Jain et al. [Jain, 1977] discuss a similar converter in a glass fiber. Frey and Pradere [Frey, 1980] discuss the conditions under which a Raman amplifier can avoid four-wave mixing and amplify a Stokes wave with a bandwidth narrower than that of the pump. Komine [Komine, 1982] demonstrates efficient conversion of an excimer laser to the visible region. Grasyuk and Zubarev [Grasyuk, 1978] review the physics of Raman lasers and discuss many earlier Raman conversion experiments.

Near-resonant electronic stimulated Raman scattering has been used to generate Joule-level pulses in the blue-green [Brosnan, 1982] and to tune over a small range near the lines of a dimer laser [Luhs, 1983], although for spectroscopic purposes these devices could be replaced by dye lasers. Simple Raman gain in an atom in which an electronic-level population is inverted with respect to the ground state can be used to generate short-wavelength anti-Stokes Raman lasers [White, 1984]. Electronic Raman scattering of a dye laser in cesium vapor gives an infrared source tunable from about 1.7 to 11 μm [Harris, 1984].

Several commercial dye-laser manufacturers offer a Raman conversion accessory to expand the tuning range of their products. In the future, Raman converters will be important sources of high-energy tunable infrared pulses and of tunable ultraviolet radiation at wavelengths shorter than those conveniently accessible by other techniques.

2.5 SUMMARY

Lasers have been available for a quarter of a century now, and in that time have revolutionized spectroscopy. Powerful new techniques such as those described in this book have opened up entirely new subfields, most of which were impractical with thermal sources. In the early days, the construction and fine adjustment of the laser itself was often the major task in the experiment, but laser engineering has now progressed to the point that for many experiments the laser is a minor part of the apparatus. The more sophisticated tunable lasers are not quite yet sealed "black boxes," but every new model the manufacturers introduce takes them further in that direction. Spectroscopic sources are still in a developmental stage for wavelengths less than about 340 nm, subnanosecond pulses, and to some degree at very long wavelength in the far infrared, but almost any issue of the journals listed in Table 2.1 contains a further advance in these areas. We can be confident that in the future it will be even more common than it is now to ask, of a spectroscopic experiment, not "did you consider a laser technique" but "did you consider anything else?"

BIBLIOGRAPHY

Laser Handbook, Vols. 1 and 2 (F. T. Arecchi and E. O. Schultz-Dubois, eds.), 1972; Vol. 3 (M. Stitch, ed.), 1979; North-Holland, Amsterdam. Contains authoritative review articles on most topics in laser physics and technology, and is an excellent starting point for investigating a particular subject in the field.

Demtröder, W. (1982). *Laser Spectroscopy: Basic Concepts and Instrumentation*, Springer-Verlag, New York. Gives an excellent introductory survey of laser spectroscopy and instrumentation. Covers many topics discussed in the present book but with some differences in emphasis.

Heard, H. G. (1968). *Laser Parameter Measurements Handbook*, Wiley, New York. Reviews basic techniques for measuring such parameters as beam intensity distribution and divergence, gain, line width and frequency stability, and pulse shape. A person wishing to measure accurate intensity profiles with a minimum of apparatus should also consider variations on the self-calibrating photographic technique described by Winer (1966).

Koechner, W. (1976). *Solid State Laser Engineering*, Springer-Verlag, New York. Reviews the engineering problems of lasers based on impurity ions in solids, including harmonic conversion.

Kressel, H., and Butler, J. K. (1977). *Semiconductor Lasers and Heterojunction LEDs*, Academic Press, New York. Reviews the major issues in the design, fabrication, and performance of semiconductor lasers.

Pressley, R. J., ed., (1971). *CRC Handbook of Lasers*, CRC Press, Boca Raton, Fla. Contains extensive tables of the properties of various laser transitions with short review articles and copious references to the literature.

Schäfer, F. P., ed. (1977). *Dye Lasers*, 2nd rev. ed., Springer-Verlag, New York. An excellent review of the history, physics, and technology of dye lasers.

Weber, M. J., ed. (1982). *CRC Handbook of Laser Science and Technology*, Vols. 1 and 2, CRC Press, Boca Raton, Fla. An updated version of Pressley, above.

Zernicke, F., and Midwinter, J. E. (1973). *Applied Nonlinear Optics*,
Wiley, New York. Reviews nonlinear optics in crystals with a
strong emphasis on practical devices using second harmonic generation, frequency mixing, and parametric amplification in
birefringent crystals.

REFERENCES

[Akhmanov, 1974] S. A. Akhmanov, Y. E. D'yakov, and L. I. Pavlov.
Statistical phenomena in Raman scattering stimulated by a broadband pump, *Sov. Phys. JETP*, *39*, 249 (1974).

[Alfano, 1971] R. R. Alfano and S. L. Shapiro. Picosecond spectroscopy using the inverse Raman effect, *Chem. Phys. Lett.*, *8*,
631 (1971).

[Antonov, 1983] V. S. Antonov and K. L. Hohla. Dye stability under excimer laser pumping: I. Method and modelling for infrared
dyes, *Appl. Phys. B*, *30*, 109; II. visible and UV dyes, *Appl.
Phys.*, *B 32*, 9 (1983).

[Arecchi, 1972] F. T. Arecchi and E. O. Schulz-Dubois, eds. *Laser
Handbook*, Vols. 1 and 2, North-Holland, Amsterdam, 1972.

[Arkhangel'skaya, 1980] V. A. Arkhangel'skaya and P. P. Feofilov.
Tunable lasers using color centers in ionic crystals (review),
Sov. J. Quantum Electron., *10*, 657 (1980).

[Armandillo, 1984] E. Armandillo, P. V. M. Lopatriello, and G. Giuliani. Single-mode, tunable operation of a XeF excimer laser employing an original interferometer, *Opt. Lett.*, *9*, 327 (1984).

[Bagratshvili, 1971] V. N. Bagratshvili, I. N. Knyazev, and V. S.
Letokhov. On the tunable infrared gas lasers, *Opt. Commun.*,
4, 154 (1971).

[Barr, 1984] J. R. M. Barr. Achromatic prism beam expanders,
Opt. Commun., *51*, 41 (1984).

[Basiev, 1982] T. T. Basiev, Y. K. Voron'ko, S. B. Mirov, V. V.
Osiko, A. M. Prokhorov, M. S. Soskin, and V. B. Taranenko.
Efficient tunable LiF:F_2^- crystal lasers, *Sov. J. Quantum Electron.*,
12, 1125 (1982).

[Basov, 1966] N. G. Basov, A. Z. Grasyuk, I. G. Zubarev, V. A.
Katulin, and O. N. Krokhin. Semiconductor quantum generator
with two-photon optical excitation, *Sov. Phys. JETP*, *23*, 366
(1966).

[Baumgartner, 1979] R. Baumgartner and R. L. Byer. Optical Parametric amplification, *IEEE J. Quantum Electron.*, QE-16, 432 (1979).

[Beck, 1980] R. Beck, W. Englisch, and K. Gürs. *Tables of Laser Lines in Gases and Vapors*, 3rd ed., Springer-Verlag, Berlin, 1980.

[Beigang, 1977] R. Beigang, G. Liftin, and H. Welling. Frequency behaviour and linewidth of CW single mode color center lasers, *Opt. Commun.*, 22, 269 (1977).

[Berry, 1982] A. J. Berry, D. C. Hanna, and D. B. Hearn. Low threshold operation of a waveguide H_2 Raman laser, *Opt. Commun.*, 43, 229 (1982).

[Bhattacharyya, 1981] S. K. Bhattacharyya, P. Eggett, and L. Thomas. Locking of a flashlamp-pumped dye laser at small injection powers, *Opt. Commun.*, 37, 387 (1981).

[Bigio, 1983] I. J. Bigio and M. Slatkine. Injection-locking unstable resonator excimer lasers, *IEEE J. Quantum Electron.*, QE-19, 1426 (1983).

[Blit, 1977] S. Blit, U. Ganiel, and D. Treves. A tunable injection-locked, flashlamp-pumped dye laser, *Appl. Phys.*, 12, 69 (1977).

[Bloembergen, 1965] N. Bloembergen. *Nonlinear Optics*, Benjamin, New York, 1965.

[Bloom, 1974] A. L. Bloom. Modes of a laser resonator containing tilted birefringent plates, *J. Opt. Soc. Am.*, 64, 447 (1974).

[Bloom, 1975] D. M. Bloom, G. W. Beckers, J. F. Young, and S. E. Harris. Third harmonic generation in phase-matched alkali metal vapors, *Appl. Phys. Lett.*, 26, 687 (1975).

[Bokhan, 1978] P. A. Bokhan, V. A. Gerasimov, V. I. Solomonov, and V. B. Shcheglov. Stimulated emission mechanism of a copper vapor laser, *Sov. J. Quantum Electron.*, 8, 1220 (1978).

[Bokor, 1983] J. Bokor, P. H. Bucksbaum, and R. R. Freeman. Generation of 35.5 nm coherent radiation, *Opt. Lett.*, 8, 217 (1983).

[Booth, 1979] D. J. Booth, T. Kobayashi, and H. Inaba. A widely tunable narrow linewidth Nd-glass laser, *Opt. Quantum Electron.*, 11, 370 (1979).

[Bor, 1981] Z. Bor. Amplified spontaneous emission from N_2-pumped dye lasers, *Opt. Commun.*, 37, 383 (1981).

[Brosnan, 1977] S. J. Brosnan, R. N. Fleming, R. L. Herbst, and R. L. Byer. Tunable infrared generation by coherent Raman mixing in H_2, *Appl. Phys. Lett.*, *30*, 330 (1977).

[Brosnan, 1982] S. J. Brosnan, H. Komine, E. A. Stappaerts, M. J. Plummer, and J. B. West. High efficiency Joule-level Raman generation in Pb vapor, *Opt. Lett.*, *7*, 154 (1982).

[Broyer, 1984] M. Broyer, J. Chevaleyre, G. Delacretaz, and L. Wöste. CVL-pumped dye laser for spectroscopic application, *Appl. Phys.*, *B35*, 31 (1984).

[Brunk, 1982] D. J. Brunk and D. Proch. Efficient tunable ultraviolet source based on stimulated Raman scattering of an excimer-pumped dye laser, *Opt. Lett.*, *7*, 494 (1982).

[Burnham, 1978] R. Burnham. Discharge pumped mercuric halide dissociation lasers, *Appl. Phys. Lett.*, *33*, 156 (1978).

[Busch, 1973] G. E. Busch, R. P. Jones, and P. M. Rentzepis. Picosecond spectroscopy using a picosecond continuum, *Chem. Phys. Lett.*, *18*, 178 (1973).

[Byer, 1978] R. L. Byer and W. R. Trutna. 16 μm generation by CO_2-pumped rotational Raman scattering in H_2, *Opt. Lett.*, *3*, 144 (1978).

[Capelle, 1970] G. Capelle and D. Phillips. Tuned nitrogen-laser-pumped dye laser, *Appl. Opt.*, *9*, 2742 (1970).

[Carlsten, 1983] J. L. Carlsten and R. G. Wenzel. Stimulated rotational Raman scattering in CO_2-pumped *para*-H_2, *IEEE J. Quantum Electron.*, *QE-19*, 1407 (1983).

[Chang, 1977] T. I. Chang and O. R. Wood II. Optically pumped continuously tunable high pressure molecular lasers, *IEEE J. Quantum Electron.*, *QE-13*, 907 (1977).

[Chang, 1980] T. Chang and F. Y. Li. Pulsed dye laser with grating and etalon in a symmetric arrangement, *Appl. Opt.*, *19*, 3651 (1980).

[Cheo, 1984] P. K. Cheo. Frequency synthesized and continuously tunable IR laser sources in 9-11 μm, *IEEE J. Quantum Electron.*, *QE-20*, 700 (1984).

[Chraplyvy, 1981] A. R. Chraplyvy and T. J. Bridges. Infrared generation by means of multiple-order stimulated Raman scattering in CCl_4- and $CBrCl_3$-filled hollow silica fibers, *Opt. Lett.*, *6*, 632 (1981).

[Collins, 1984] C. B. Collins. The nitrogen ion laser pumped by charge transfer, *IEEE J. Quantum Electron.*, *QE-20*, 47 (1984).

[Corkum, 1983] P. B. Corkum. High-power, subpicosecond 10-μm pulse generation, *Opt. Lett.*, *8*, 514 (1983).

[Couillaud, 1982] B. Couillaud, P. Dabkiewicz, L. A. Bloomfield, and T. W. Hänsch. Generation of continuous-wave ultraviolet radiation by sum-frequency mixing in an external ring cavity, *Opt. Lett.*, *7*, 265 (1982).

[Craxton, 1980] R. S. Craxton. Theory of high efficiency third harmonic generation of high power Nd-glass laser radiation, *Opt. Commun.*, *34*, 474 (1980).

[Curry, 1973] S. M. Curry, R. Cubeddu, and T. W. Hänsch. Intensity stabilization of dye laser radiation by saturated amplification, *Appl. Phys.*, *1*, 153 (1973).

[Davydov, 1973] A. A. Davydov, L. A. Kulevskii, A. M. Prokhorov, A. D. Savelev, V. V. Smirnov, and A. V. Shirkov. A tunable infrared parametric oscillator in a CdSe crystal, *Opt. Commun.*, *9*, 234 (1973).

[deMartino, 1980] A. deMartino, R. Frey, and F. Pradere. Near to far infrared tunable Raman laser, *IEEE J. Quantum Electron.*, *QE-16*, 1184 (1980).

[Dempster, 1974] D. N. Dempster, T. Morrow, and M. Quinn. Photochemical characteristics of rhodamine 6G-ethanol solutions, *J. Photochem.*, *2*, 343 (1974).

[Demtröder, 1982] W. Demtröder. *Laser Spectroscopy: Basic Concepts and Instrumentation*, Springer-Verlag, New York, 1982.

[Döbele, 1984a] H. F. Döbele and B. Rückle. Application of an argon fluoride laser system to the generation of VUV radiation by stimulated Raman scattering, *Appl. Opt.*, *23*, 1040 (1984).

[Döbele, 1984b] H. F. Döbele, M. Röwenkamp, and B. Rückle. Amplification of 193 nm radiation in argon fluoride and generation of tunable VUV radiation by high order anti-Stokes Raman scattering, *IEEE J. Quantum Electron.*, *QE-20*, 1284 (1984).

[Dolan, 1976] G. Dolan and C. R. Goldschmidt. A new method for absolute absorption cross-section measurements: rhodamine 6G excited singlet-singlet absorption spectrum, *Chem. Phys. Lett.*, *39*, 320 (1976).

[Dowley, 1968a] M. W. Dowley. Efficient CW second harmonic generation to 2573 Å, *Appl. Phys. Lett.*, *13*, 395 (1968).

[Dowley, 1968b] M. W. Dowley and E. G. Hodges, Studies of high-power CW and quasi-CW parametric UV generation by ADP and KDP in an argon-ion laser cavity, *IEEE J. Quantum Electron*, *QE-4*, 552 (1968).

[Drell, 1979] P. Drell and S. Chu. A megawatt dye laser oscillator-amplifier system for high resolution spectroscopy, *Opt. Commun.*, *28*, 343 (1979).

[Drexhage, 1973] K. H. Drexhage. Structure and properties of laser dyes, in *Dye Lasers* (F. P. Schäfer, ed.), Springer-Verlag, New York, 1973, Chap. 4.

[Dreyfus, 1974] R. W. Dreyfus and R. T. Hodgson. Molecular hydrogen laser: 1098–1613 Å, *Phys. Rev.*, *A 9*, 2635 (1974).

[Drube, 1984] J. Drube, B. Struve, and G. Huber. Tunable room temperature CW laser action in Cr^{3+}:GdScAl garnet, *Opt. Commun.*, *50*, 45 (1984).

[Duarte, 1984] F. J. Duarte and J. A. Piper. Narrow-linewidth, high PRF copper-laser-pumped dye laser oscillators, *Appl. Opt.*, *23*, 1391 (1984).

[Ediger, 1982] M. N. Ediger, A. W. McCown, and J. G. Eden. CdI and CdBr photodissociation lasers at 655 and 811 nm, *Appl. Phys. Lett.*, *40*, 99 (1982).

[Eesley, 1980] G. L. Eesley, M. D. Levinson, D. E. Nitz, and A. V. Smith. Narrow band pulsed dye laser system for precision nonlinear spectroscopy, *IEEE J. Quantum Electron.*, *QE-16*, 113 (1980).

[Egger, 1981] H. Egger, T. Srinivasan, K. Hohla, H. Scheingraber, C. R. Vidal, H. Pummer, and C. K. Rhodes. A tunable, ultrahigh spectral brightness ArF excimer laser source, *Appl. Phys. Lett.*, *39*, 37 (1981).

[Ehrlich, 1978] D. J. Ehrlich, P. F. Moulton, and R. M. Osgood. Ultraviolet solid state Ce:YLF laser at 325 nm, *Opt. Lett.*, *4*, 184 (1978).

[Ehrlich, 1980] D. J. Ehrlich, P. F. Moulton, and R. M. Osgood. Optically-pumped Ce:LaF_3 laser at 286 nm, *Opt. Lett.*, *5*, 339 (1980).

[Ewing, 1979] J. J. Ewing. Excimer lasers, in *Laser Handbook*, Vol. 3 (M. Stitch, ed.), North-Holland, Amsterdam, 1979, Chap. A4.

[Fleming, 1981] M. W. Fleming and A. Mooradian. Spectral characteristics of external cavity controlled semiconductor lasers, *IEEE J. Quantum Electron.*, *QE-17*, 44 (1981).

[Fork, 1981] R. L. Fork, B. I. Greene, and C. V. Shank. Generation of optical pulses shorter than 0.1 picosecond by colliding-pulse mode-locking, *Appl. Phys. Lett.*, *38*, 671 (1981).

[Frey, 1980] R. Frey and F. Pradere. High efficiency narrow line-width Raman amplification and spectral compression, *Opt. Lett.*, 5, 374 (1980).

[Ganiel, 1976] U. Ganiel, A. Hardy, and D. Treves. Analysis of injection locking in pulsed dye laser systems, *IEEE J. Quantum Electron.*, QE-12, 704 (1976).

[Geusic, 1968] J. Geusic, H. Levinstein, S. Singh, R. Smith, and L. Van Uitert. Continuous 0.532 μm solid state source using $Ba_2NaNb_5O_{15}$, *Appl. Phys. Lett.*, 12, 306 (1968).

[Gibson, 1979] R. B. Gibson, K. Boyer, and A. Javan. Mixed isotope multiatmosphere CO_2 laser, *IEEE J. Quantum Electron.*, QE-15, 1224 (1979).

[Goldhar, 1980] J. Goldhar, W. R. Rapoport, and J. R. Murray. An injection-locked, unstable resonator rare-gas-halide discharge laser of narrow linewidth and high spatial quality, *IEEE J. Quantum Electron.*, QE-16, 235 (1980).

[Grasyuk, 1978] A. Z. Grasyuk and I. G. Zubarev. High power tunable IR Raman lasers, *Appl. Phys.*, 17, 211 (1978).

[Grasyuk, 1980] A. Z. Grasyuk, V. S. Letokhov, and V. V. Lobko. Molecular infrared lasers using resonant laser pumping, *Prog. Quantum Electron.*, 6, 245 (1980).

[Green, 1973] J. M. Green, J. P. Hoheimer, and F. K. Tittel. Travelling wave operation of a tunable CW dye laser, *Opt. Commun.*, 7, 349 (1973).

[Greene, 1983] D. P. Greene and J. G. Eden. Lasing on the B-X band of cadmium monoiodide in a UV preionized transverse discharge, *Appl. Phys. Lett.*, 43, 418 (1983).

[Groves, 1974] S. H. Groves, K. W. Nill, and A. J. Strauss. Double heterostructure $Pb_xSn_{1-x}Te$-PbTe lasers with CW operation at 77°K, *Appl. Phys. Lett.*, 25, 331 (1974).

[Haag, 1983] G. Haag, M. Munz, and G. Marowsky. Amplified spontaneous emission (ASE) in laser oscillators and amplifiers, *IEEE J. Quantum Electron.*, QE-19, 1149 (1983).

[Halbout, 1979] J.-M. Halbout, S. Blit, W. Donaldson, and C. L. Tang. Efficient phase-matched second harmonic generation and sum-frequency mixing in urea, *IEEE J. Quantum Electron.*, QE-15, 1176 (1979).

[Hammond, 1978] P. R. Hammond. Laser dye DCM, its spectral properties, synthesis, and comparison with other dyes in the red, *Opt. Commun.*, 29, 331 (1978).

[Hammond, 1979] P. R. Hammond. Spectra of the lowest excited singlet states of rhodamine 6G and rhodamine B, *IEEE J. Quantum Electron.*, *QE-15*, 624 (1979).

[Hanna, 1979] D. C. Hanna, M. A. Yuratich, and D. Cotter. *Nonlinear Optics of Free Atoms and Molecules*, Springer-Verlag, New York, 1979.

[Hanna, 1975] D. C. Hanna, P. A. Karkkainen, and R. Wyatt. A simple beam expander for frequency narrowing of dye lasers, *Opt. Quantum Electron.*, *7*, 115 (1975).

[Hänsch, 1972] T. W. Hänsch. Repetitively pulsed tunable dye laser for high resolution spectroscopy, *Appl. Opt.*, *11*, 895 (1972).

[Hargrove, 1980] R. S. Hargrove and T. Kan. High power, efficient dye amplifier pumped by copper vapor lasers, *IEEE J. Quantum Electron*, *QE-16*, 1108 (1980).

[Hargrove, 1979] R. S. Hargrove, R. Grove, and T. Kan. Copper vapor laser unstable resonator oscillator and oscillator-amplifier characteristics, *IEEE J. Quantum Electron.*, *QE-15*, 1228 (1979).

[Harris, 1984] A. L. Harris, J. K. Brown, M. Berg, and C. B. Harris. Generation of widely tunable nanosecond pulses in the vibrational infrared by stimulated Raman scattering from the cesium 6s-5d transition, *Opt. Lett.*, *9*, 47 (1984).

[Harris, 1969] S. E. Harris. Optical parametric oscillators, *Proc. IEEE*, *57*, 2096 (1969).

[Hartig, 1979] W. Hartig and W. Schmidt. A broadly tunable IR waveguide Raman laser pumped by a dye laser, *Appl. Phys.*, *18*, 235 (1979).

[Heard, 1968] H. G. Heard. *Laser Parameters Measurements Handbook*, Wiley, New York, 1968.

[Helmcke, 1982] J. Helmcke, S. A. Lee, and J. L. Hall. Dye laser spectrometer for ultrahigh spectral resolution, *Appl. Opt.*, *21*, 1686 (1982).

[Hemmati, 1983] H. Hemmati, J. C. Bergquist, and W. M. Itano. Generation of continuous-wave 194 nm radiation by sum frequency mixing in an external ring cavity, *Opt. Lett.*, *8*, 73 (1983).

[Henderson, 1981] B. Henderson. Tunable visible lasers using F^+ centers in oxides, *Opt. Lett.*, *6*, 437 (1981).

[Hillig, 1983] R. Hillig and R. Wallenstein. Tunable XUV radiation generated by nonresonant frequency tripling in argon, *Opt. Commun.*, *44*, 283 (1983).

[Hodgson, 1974] R. T. Hodgson, P. P. Sorokin, and J. J. Wynne. Tunable coherent VUV generation in atomic vapors, *Phys. Rev. Lett.*, *32*, 343 (1974).

[Holtom, 1974] G. Holtom and O. Teschke. Design of a birefringent filter for high-power dye lasers, *IEEE J. Quantum Electron.*, *QE-10*, 577 (1974).

[Hon, 1979] D. T. Hon. High average power efficient second harmonic generation, in *Laser Handbook*, Vol. 3 (M. Stitch, ed.), North-Holland, Amsterdam, 1979, Chap. B2.

[Hough, 1984] J. Hough, D. Hils, M. D. Rayman, L.-S. Ma, L. Hollberg, and J. L. Hall. *Appl. Phys.*, *B 33*, 179 (1984).

[Hutchinson, 1980] M. H. R. Hutchinson. Excimers and excimer lasers, *Appl. Phys.*, *21*, 95 (1980).

[Iles, 1980] M. K. Iles, A. P. D'Silva, and V. A. Fassel. Single mode nitrogen pumped dye laser, *Opt. Commun.*, *35*, 133 (1980).

[Isaev, 1977] A. A. Isaev and M. A. Kazaryan. Investigation of pulsed copper vapor laser, *Sov. J. Quantum Electron.*, *7*, 253 (1977).

[Jain, 1973] R. K. Jain and T. K. Gustavson. Second harmonic generation of several argon ion laser lines, *IEEE J. Quantum Electron.*, *QE-9*, 859 (1973).

[Jain, 1977] R. K. Jain, C. L. Lin, R. H. Stohlen, and A. Ashkin. A tunable, multiple Stokes, CW fiber Raman oscillator, *Appl. Phys. Lett.*, *31*, 89 (1977).

[Jamroz, 1982] W. Jamroz, P. E. LaRocque, and B. P. Stoicheff. Generation of continuously tunable coherent vacuum ultraviolet radiation (140 to 106 nm) in zinc vapor, *Opt. Lett.*, *7*, 617 (1982).

[Jarrett, 1979] S. M. Jarrett and J. F. Young. High efficiency single frequency CW ring dye laser, *Opt. Lett.*, *4*, 176 (1979).

[Johnson, 1967] F. M. Johnson, J. A. Duardo, and G. L. Clark. Complex stimulated Raman rotational-vibrational spectra in hydrogen, *Appl. Phys. Lett.*, *10*, 157 (1967).

[Johnson, 1972] W. D. Johnson and P. K. Runge. An improved astigmatically compensated resonator for CW dye lasers, *IEEE J. Quantum Electron.*, *QE-8*, 724 (1972).

[Johnson, 1974] L. F. Johnson and H. J. Guggenheim. Electronic and phonon-terminated laser emission from Ho^{3+} in BaY_2F_8, *IEEE J. Quantum Electron.*, *QE-10*, 442 (1974).

[Johnson, 1980] T. F. Johnson, Jr. and W. Proffitt. Design and performance of a broad-band optical diode to enforce one-direction travelling wave operation of a ring laser, *IEEE J. Quantum Electron, QE-16*, 483 (1980).

[Johnson, 1982] T. F. Johnson, Jr., R. H. Brady, and W. Proffitt. Powerful single-frequency ring dye laser spanning the visible spectrum, *Appl. Opt.*, *21*, 2307 (1982).

[Johnson, 1983] L. F. Johnson, H. J. Guggenheim, D. Bahnck, and A. M. Johnson. Phonon-terminated laser emission from Ni^{2+} ions in $KMnF_3$, *Opt. Lett.*, *8*, 371 (1983).

[Kaiser, 1972] W. Kaiser and M. Maier. Stimulated Rayleigh, Brillouin, and Raman spectroscopy, in *Laser Handbook*, Vol. 2 (F. T. Arecchi and E. O. Schulz-Dubois, eds.), North-Holland, Amsterdam, 1972, Chap. E2.

[Kasemset, 1980] D. Kasemset, S. Rotter, and C. G. Fonstad. $Pb_{1-x}Sn_x$ $Te/PbTe_{1-y}Se_y$ lattice-matched buried heterostructure lasers with CW single mode output, *IEEE Electron Device Lett.*, *EDL-1*, 75 (1980).

[Kato, 1980] K. Kato. High efficiency high power UV generation at 2128 Å in urea, *IEEE J. Quantum Electron.*, *QE-16*, 810 (1980).

[Kildal, 1977] H. Kildal. Infrared third harmonic generation in phase-matched CO gas, *IEEE J. Quantum Electron.*, *QE-13*, 109 (1977).

[Kiyashko, 1979] V. A. Kiyashko and V. P. Timofeev. Ruby laser third harmonic generation in thallium vapor, *Sov. J. Quantum Electron.*, *9*, 1064 (1979).

[Kiyashko, 1985] V. A. Kiyashko, A. K. Popov, V. P. Timofeev, N. P. Makarov, and V. S. Epstein. Resonant generation of even-order harmonics in metal vapors, *Appl. Phys.*, *B 36*, 53 (1985).

[Klauminzer, 1977] G. K. Klauminzer. New high-performance short-cavity dye laser design, *IEEE J. Quantum Electron.*, *QE-13(9)*, 92D (1977).

[Knox, 1984] W. H. Knox, M. V. Downer, R. L. Fork, and C. V. Shank. *Opt. Lett.*, *9*, 552 (1984).

[Knyazev, 1975] I. N. Knyazev, V. S. Letokhov, and V. G. Movshev. Efficient and practical hydrogen vacuum ultraviolet laser, *IEEE J. Quantum Electron.*, *QE-11*, 805 (1975).

[Koch, 1979] K. P. Koch, G. Litfin, and H. Welling. Continuous wave laser oscillation with extended tuning range in F_A (II)-F_B (II) color center crystals, *Opt. Lett.*, *4*, 387 (1979).

[Koch, 1982] T. L. Koch, L. C. Chiu, and A. Yariv. Gain saturation of a picosecond dye laser amplifier chain, *Opt. Commun.*, *40*, 364 (1982).

[Koechner, 1976] W. Koechner. *Solid State Laser Engineering*, Springer-Verlag, New York, 1976.

[Kogelnik, 1972] H. W. Kogelnik, E. P. Ippen, A. Dienes, and C. V. Shank. Astigmatically compensated cavities for CW dye lasers, *IEEE J. Quantum Electron.*, QE-8, 373 (1972).

[Kolbe, 1984] W. Kolbe, K. Petermann, and G. Huber. Tunable laser action in Cr:ZnWO$_4$ near 1 µm wavelength, paper WI1, *Conf. Lasers Electro-Opt.*, CLEO-84, Anaheim, Calif, June 1984.

[Komine, 1982] H. Komine and E. A. Stappaerts. Higher Stokes order Raman conversion of XeCl laser in hydrogen, *Opt. Lett.*, 7, 157 (1982).

[Kressel, 1977] H. Kressel and J. K. Butler. *Semiconductor Lasers and Heterojunction LEDs*, Academic Press, New York, 1977.

[Kuizenga, 1973] D. J. Kuizenga, D. W. Phillion, T. Lund, and A. E. Siegman. Simultaneous Q-switching and mode-locking in the CW Nd:YAG laser, *Opt. Commun.*, 9, 221 (1973).

[Lisboa, 1983] J. A. Lisboa, S. Ribeiro-Teixeira, S. L. S. Gunha, R. E. Francke, and H. P. Grieneisen. A grazing-incidence dye laser with an intracavity lens, *Opt. Commun.*, 44, 393 (1983).

[Litfin, 1978] G. Litfin and R. Beigang. Design of tunable colour centre lasers, *J. Phys. E11*, 984 (1978).

[Litfin, 1983] G. Litfin, D. Wandt, and D. Huhn. Flashlamp pumped color center lasers, *Opt. Commun.*, 48, 270 (1983).

[Littman, 1978a] M. G. Littman and H. J. Metcalf. Spectrally narrow pulsed dye laser without beam expander, *Appl. Opt.* 17, 2224 (1978).

[Littman, 1978b] M. G. Littman. Single mode operation of grazing-incidence pulsed dye laser, *Opt. Lett.*, 3, 138 (1978).

[Littman, 1984] M. G. Littman. Single mode pulsed tunable dye laser, *Appl. Opt.*, 23, 4465 (1984).

[Liu, 1981] K. Liu and M. G. Littman. Novel geometry for single-mode scanning of tunable lasers, *Opt. Lett.*, 6, 117 (1981).

[Liu, 1984] Y. S. Liu, D. Dentz, and R. Belt. High average power intracavity second harmonic generation using KTiOPO$_4$ in an acousto-optically Q-switched Nd:YAG oscillator at 5 kHz, *Opt. Lett.*, 9, 76 (1984).

[Luhs, 1983] W. Luhs and B. Wellegehausen. Raman tuning of optically-pumped continuous dimer lasers, *Opt. Commun.*, 46, 121 (1983).

[Maeda, 1984] M. Maeda. *Laser Dyes*, Academic Press, Orlando, Fla., 1984.

[Magde, 1981] D. Magde, S. T. Gaffney, and B. F. Campbell. Excited singlet absorption in blue laser dyes: measurement by picosecond flash photolysis, *IEEE J. Quantum Electron.*, QE-17, 489 (1981).

[Magerl, 1982] G. Magerl, W. Schupita, and E. Bonek. A tunable CO_2 laser sideband spectrometer, *IEEE J. Quantum Electron.*, QE-18, 1214 (1982).

[Marason, 1982] E. G. Marason. Energy transfer dye mixture for argon-pumped dye laser operation in the 700-800 nm region, *Opt. Commun.*, 40, 212 (1982).

[Marowsky, 1973] G. Marowsky. A tunable, flashlamp pumped dye ring laser of extremely narrow bandwidth, *IEEE J. Quantum Electron.*, QE-9, 245 (1973).

[Marowsky, 1975] G. Marowsky. Astigmatism and coma-free prism ring dye laser, *Appl. Phys. Lett.*, 26, 647 (1975).

[Massey, 1976] G. A. Massey and J. C. Johnson. Wavelength-tunable optical mixing experiments between 208 nm and 259 nm, *IEEE J. Quantum Electron.*, QE-12, 721 (1976).

[McCown, 1981] A. W. McCown and J. G. Eden. ZnI B-X laser at 600-604 nm, *Appl. Phys. Lett.*, 39, 371 (1981).

[McGee, 1981] T. J. McGee and J. Burris, Jr. CF_2Cl_2 as a dye laser tuning gas: refractive index measurements, *Appl. Opt.*, 20, 3483 (1981).

[Miles, 1973] R. B. Miles and S. E. Harris. Optical third harmonic generation in alkali metal vapors, *IEEE J. Quantum Electron.*, QE-9, 470 (1973).

[Miller, 1975] J. L. Miller, A. H. M. Ross, and E. V. George. Gain characteristics of a multiatmosphere CO_2 laser, *Appl. Phys. Lett.*, 26, 523 (1975).

[Mollenauer, 1975] L. F. Mollenauer and D. H. Olsen. Broadly tunable lasers using color centers, *J. Appl. Phys.*, 46, 3109 (1975).

[Mollenauer, 1977] L. F. Mollenauer. Dyelike lasers for the 0.9-2-μm region using F_2^+ centers in alkali halides, *Opt. Lett.*, 1, 164 (1977).

[Mollenauer, 1978a] L. F. Mollenauer, D. M. Bloom, and A. M. DelGaudio. Broadly tunable CW lasers using F_2^+ centers for the 1.26-1.48 and 0.82-1.07 μm bands, *Opt. Lett.*, 3, 48 (1978).

[Mollenauer, 1978b] L. F. Mollenauer. Apparatus for the coloration of laser quality alkali halide crystals, *Rev. Sci. Instrum.*, 49, 809 (1978).

[Mollenauer, 1979] L. F. Mollenauer and D. M. Bloom. Color center laser generates picosecond pulses and several watts CW over the 1.24-1.45 μm range, *Opt. Lett.*, *4*, 247 (1979).

[Mollenauer, 1980] L. F. Mollenauer. Room temperature stable, F_2^+-like center yields CW laser tunable over the 0.99-1.22 μm range, *Opt. Lett.*, *5*, 188 (1980).

[Mollenauer, 1981] L. F. Mollenauer. Laser-active, defect-stabilized F_2^+ center in $NaF:OH^-$ and dynamics of defect-stabilized center formation, *Opt. Lett.*, *6*, 342 (1981).

[Mollenauer, 1982] L. F. Mollenauer. Color center lasers, in *CRC Handbook of Laser Science and Technology*, Vols. 1 and 2 (M. J. Weber, ed.), CRC Press, Boca Raton, Fla., 1982, Chap. 2.1.3.

[Mollenauer, 1983] L. F. Mollenauer, R. H. Stolen, J. P. Gordon, and W. J. Tomlinson. Extreme picosecond pulse narrowing by means of soliton effect in single-mode optical fibers, *Opt. Lett.*, *8*, 289 (1983).

[Mollenauer, 1984] L. F. Mollenauer and R. H. Stolen. The soliton laser, *Opt. Lett.*, *9*, 13, 105 (1984).

[Mooradian, 1973] A. Mooradian, A. J. Strauss, and A. Rossi. Broadband laser emission from optically-pumped $Pb\ S_{1-x}\ Se_x$, *IEEE J. Quantum Electron.*, *QE-9*, 347 (1973).

[Mory, 1981] S. Mory, A. Rosenfeld, S. Polze, and G. Korn. Nanosecond dye laser with a high-efficiency holographic grating for grazing incidence, *Opt. Commun.*, *36*, 342 (1981).

[Moulton, 1979] P. F. Moulton and A. Mooradian. Broadly tunable CW operation of $Ni:MgF_2$ and $Co:MgF_2$ lasers, *Appl. Phys. Lett.*, *35*, 838 (1979).

[Moulton, 1982] P. F. Moulton. Pulse-pumped operation of divalent transition metal lasers, *IEEE J. Quantum Electron.*, *QE-18*, 1185 (1982).

[Moulton, 1984] P. F. Moulton. Recent advances in solid state lasers, paper WA2, *Conf. Lasers Electro-Opt.*, *CLEO-84*, Anaheim, Calif., June 1984.

[Moulton, unpublished]

[Mourou, 1982] G. A. Mourou and T. Sizer, II. Generation of pulses shorter than 70 fs with a synchronously-pumped dye laser, *Opt. Commun.*, *41*, 47 (1982).

[Murray, 1979] J. R. Murray, J. Goldhar, D. Eimerl, and A. Szöke. Raman pulse compression of excimer lasers for application to laser fusion. *IEEE J. Quantum Electron.*, *QE-15*, 342 (1979).

[Myers, 1971] S. A. Myers. An improved line narrowing technique for a dye laser excited by a nitrogen laser, *Opt. Commun.*, *4*, 187 (1971).

[Nair, 1982] L. G. Nair. Dye lasers, *Prog. Quantum Electron.*, *7*, 153 (1982).

[Nikolaus, 1983] B. Nikolaus and D. Grischkowsky. 90-fs tunable optical pulses obtained by two-stage pulse compression, *Appl. Phys. Lett.*, *43*, 228 (1983).

[Novikov, 1975] M. A. Novikov and A. D. Tertyshnik. Tunable dye laser with a narrow emission spectrum, *Sov. J. Quantum Electron.*, *5*, 848 (1975).

[Okada, 1979] T. Okada, M. Maeda, and Y. Miyazoe. Spectral narrowing of a flashlamp-pumped high energy dye laser by two stage injection locking, *IEEE J. Quantum Electron.*, *QE-15*, 616 (1979).

[Pavlopoulos, 1981] T. G. Pavlopoulos. The dye mixture perylene-rhodamine 110, *Opt. Commun.*, *38*, 299 (1981).

[Paisner, 1979] J. A. Paisner and R. S. Hargrove, *Laser Program Annual Report-1979, Volume 3*, pp. 9.26-9.35, Lawrence Livermore National Laboratory publication UCRL 50021-79.

[Park, 1984] Y. K. Park, G. Giuliani, and R. L. Byer. *IEEE J. Quantum Electron.*, *QE-20*, 117 (1984).

[Penzkofer, 1979] A. Penzkofer, L. Laubereau, and W. Kaiser. High intensity Raman interactions, *Prog. Quantum Electron.*, *6*, 56 (1979).

[Pike, 1974] C. T. Pike. Spatial hole burning in CW dye lasers, *Opt. Commun.*, *10*, 14 (1974).

[Pressley, 1971] R. J. Pressley, ed. *CRC Handbook of Lasers*, CRC Press, Boca Raton, Fla., 1971.

[Pummer, 1979] H. Pummer, K. Hohla, M. Diegelmann, and J. P. Reilly. Discharge pumped F_2 laser at 1580 Å, *Opt. Commun.*, *28*, 104 (1979).

[Rabinowitz, 1979] P. Rabinowitz, A. Stein, R. O. Brickman, and A. Kaldor. Efficient tunable H_2 Raman laser, *Appl. Phys. Lett.*, *35*, 739 (1979).

[Racz, 1981] B. Racz, Z. Bor, S. Szatmari, and G. Szabo. Comparative study of beam expanders used in nitrogen laser pumped dye lasers, *Opt. Commun.*, *36*, 399 (1981).

[Raymond, 1984] T. D. Raymond, S. T. Walsh, and J. W. Keto. Narrow band dye laser with a large scan range, *Appl. Opt.*, *23*, 2062 (1984).

[Reintjes, 1978] J. Reintjes, C. Y. She, and R. C. Eckhart. Generation of coherent radiation in the XUV by fifth and seventh order frequency conversion in rare gases, *IEEE J. Quantum Electron.*, QE-14, 581 (1978).

[Rhodes, 1984] C. K. Rhodes, ed. *Excimer Lasers*, 2nd ed., Springer-Verlag, New York, 1984.

[Roxlo, 1982] C. B. Roxlo, R. S. Putnam, and M. M. Salour. Optically pumped semiconductor platelet lasers, *IEEE J. Quantum Electron.*, QE-18, 338 (1982).

[Sabar, 1977] E. Sabar and D. Treves. Excited singlet state absorption in dyes and their effect on dye lasers, *IEEE J. Quantum Electron.*, QE-13, 962 (1977).

[Saikan, 1978] S. Saikan. Nitrogen laser pumped single mode dye laser, *Appl. Phys.*, 17, 41 (1978).

[Salour, 1977] M. M. Salour. Powerful dye laser oscillator-amplifier system for high resolution and coherent pulse spectroscopy, *Opt. Comm.*, 22, 202 (1977).

[Schäfer, 1972] F. P. Schäfer. Liquid lasers, in *Laser Handbook*, Vols. 1 and 2 (F. T. Arecchi and E. O. Schulz-Dubois, eds.), North-Holland, Amsterdam, 1972, Chap. B3.

[Schäfer, 1973] F. P. Schäfer, ed. *Dye Lasers*, 2nd rev. ed., Springer-Verlag, New York, 1977.

[Schneider, 1979] I. Schneider and M. J. Marrone. Continuous wave laser action of $(F_2^+)_A$ centers in sodium-doped KCl crystals, *Opt. Lett.*, 4, 390 (1979).

[Schneider, 1980] I. Schneider and C. Marquardt. Tunable CW laser action using $(F_2^+)_A$ centers in Li-doped KCl, *Opt. Lett.*, 5, 214 (1980).

[Schneider, 1981] I. Schneider and C. L. Marquardt. Broadly tunable laser action beyond 3 μm from $(F_2^+)_A$ centers in lithium-doped KI, *Opt. Lett.*, 6, 627 (1981).

[Schneider, 1982] I. Schneider. Continuous tuning of a color center laser between 2 and 4 μm, *Opt. Lett.*, 7, 271 (1982).

[Schneider, 1983] I. Schneider and S. C. Moss. Color center laser continuously tunable from 1.67 to 2.46 μm, *Opt. Lett.*, 8, 7 (1983).

[Schomburg, 1983] H. Schomburg, H. F. Döbele, and B. Rückle. Generation of tunable narrow bandwidth VUV radiation by anti-Stokes SRS in H_2, *Appl. Phys.*, B30, 131 (1983).

[Schröder, 1977] H. W. Schröder, L. Stein, D. Frölich, and H. Welling. A high power, single-mode CW ring dye laser, *Appl. Phys.*, 14, 377 (1977).

[Seka, 1980] W. Seka, S. D. Jacobs, J. E. Rizzo, R. Boni, and R. S. Craxton. Demonstration of high efficiency third harmonic conversion of high power Nd:glass laser radiation, *Opt. Commun.*, *34*, 469 (1980).

[Shand, 1982] M. L. Shand and J. C. Walling. A tunable emerald laser, *IEEE J. Quantum Electron.*, *QE-18*, 1829 (1982).

[Shand, 1983] M. L. Shand. Temperature dependence from 28-290 °C of excited state absorption in the gain region of alexandrite, *IEEE J. Quantum Electron.*, *QE-19*, 480 (1983).

[Shank, 1973] C. V. Shank and E. P. Ippen. Mode locking of dye lasers, in *Dye Lasers* (F. P. Schäfer, ed.), Springer-Verlag, New York, 1973, Chap. 3.

[Shank, 1982] C. V. Shank, R. L. Fork, R. Yen, R. H. Stolen, and W. J. Tomlinson. Compression of femtosecond optical pulses, *Appl. Phys. Lett.*, *40*, 761 (1982).

[Shapiro, 1977] S. L. Shapiro, ed. *Ultrashort Light Pulses: Picosecond Techniques and Applications*, Springer-Verlag, Berlin, 1977.

[Shaw, 1979] M. T. Shaw. Excimer lasers, *Prog. Quantum Electron.*, *6*, 3 (1979).

[Shen, 1977] Y. R. Shen, ed. *Nonlinear Infrared Generation*, Topics in Applied Physics, Vol. 16, Springer-Verlag, Berlin, 1977.

[Shen, 1984] Y. R. Shen. *The Principles of Nonlinear Optics*, Wiley, New York, 1984.

[Shoshan, 1977] I. Shoshan, N. Danon, and U. P. Oppenheim. Narrowband operation of a pulsed dye laser without intracavity beam expansion, *J. Appl. Phys.*, *48*, 4495 (1977).

[Shoshan, 1978] I. Shoshan and U. P. Oppenheim. The use of a diffraction grating as a beam expander in a dye laser cavity, *Opt. Commun.*, *25*, 375 (1978).

[Siegman, 1971] A. E. Siegman. *Introduction to Lasers and Masers*, McGraw-Hill, New York, 1971.

[Simmons, 1981] W. W. Simmons, J. T. Hunt, and W. E. Warren. Light propagation through large laser systems, *IEEE J. Quantum Electron.*, *QE-17*, 1727 (1981).

[Sizer, 1981] T. Sizer, II, J. D. Kafka, A. Krisloff, and G. Mourou. Generation and amplification of sub-picosecond pulses using a frequency-doubled Nd:YAG pumping source, *Opt. Commun.*, *39*, 259 (1981).

[Small, 1976] J. G. Small. The $30 dye laser, in *Laser Photochemistry, Tunable Lasers, and Other Topics; Physics of Quantum Electronics*, Vol. 4 (S. F. Jacobs et al., eds.) Addison-Wesley, Reading, Mass; see also "The Amateur Scientist," conducted by C. L. Strong, *Sci. Am.*, June 1974, 122; J. G. Small and R. Ashari, A simple pulsed nitrogen 3371 Å laser with a modified Blumlein excitation method, *Rev. Sci. Instrum.*, *43*, 1205 (1972).

[Smilanski, 1978] I. Smilanski, A. Herman, L. A. Levin, and G. Erez. Scaling of the discharge heated copper vapor laser, *Opt. Commun.*, *25*, 79 (1978).

[Smilanski, 1979] I. Smilanski, G. Erez, A. Kerman, and L. A. Levin. High power, high pressure, discharge heated copper vapor laser, *Opt. Commun.*, *30*, 70 (1979).

[Smith, 1972] R. G. Smith. Optical Parametric oscillators, in *Laser Handbook*, Vols. 1 and 2 (F. T. Arecchi and E. O. Schulz-Dubois, eds.), North-Holland, Amsterdam, 1972, Chap. C8.

[Snavely, 1973] B. B. Snavely. Continuous wave dye lasers, in *Dye Lasers* (F. P. Schäfer, ed.), Springer-Verlag, New York, 1973, Chap. 2.

[Stenersen, 1983] K. Stenersen and G. Wang. Continuously tunable optically-pumped high pressure $DF-CO_2$ transfer laser, *IEEE J. Quantum Electron.*, *QE-19*, 1414 (1983).

[Stern, 1972] F. Stern. Semiconductor lasers, in *Laser Handbook*. Vols. 1 and 2 (F. T. Arecchi and E. O. Schulz-Dubois, eds.), North-Holland, Amsterdam, 1972, Chap. B4.

[Stitch, 1979] M. Stitch, ed. *Laser Handbook*, Vol. 3, North-Holland, Amsterdam, 1979.

[Stokes, 1972] E. D. Stokes, F. B. Dunning, R. F. Stebbings, G. K. Walters, and R. D. Rundel. A high efficiency dye laser tunable from the UV to the IR, *Opt. Commun.*, *5*, 267 (1972).

[Struve, 1983] B. Struve, G. Huber, V. V. Laptev, I. A. Shcherbakov, and E. V. Zharikov. Tunable room temperature CW laser action in Cr^{3+}:GdScGa garnet, *Appl. Phys.*, *B 30*, 117 (1983).

[Svelto, 1982] O. Svelto. *Principles of Lasers*, 2nd ed. (D. Hanna, transl.) Plenum Press, New York, 1982.

[Tarasov, 1983] L. V. Tarasov. *Laser Physics*, English edition (R. S. Wadhwa, transl.) Mir, Moscow; Russian edition, Radio i Svyaz', Moscow, 1981.

[Tashiro, 1974] H. Tashiro and T. Yajima. Picosecond absorption spectroscopy of excited states of dye molecules, *Chem. Phys. Lett.*, *25*, 582 (1974).

[Taylor, 1979] R. S. Taylor, A. J. Alcock, W. J. Sarjeant, and K. E. Leopold. Electrical and gain characteristics of a multiatmosphere UV-preionized CO_2 laser, *IEEE J. Quantum Electron.*, QE-15, 1131 (1979).

[Teschke, 1980] O. Teschke and S. Ribiero-Teixeira. Unstable ring resonator nitrogen-pumped dye laser, *Opt. Commun.*, 32, 287 (1980).

[Thompson, 1980] G. H. B. Thompson. *Physics of Semiconductor Laser Devices*, Wiley, New York, 1980.

[Timmermann, 1983] A. Timmermann and R. Wallenstein. Generation of tunable single frequency coherent VUV radiation, *Opt. Lett.*, 8, 517 (1983).

[Tomasetta, 1974] L. R. Tomasetta and C. G. Fonstad. Threshold reduction in PbSnTe laser diodes through the use of double heterojunction geometries, *Appl. Phys. Lett.*, 25, 440 (1974).

[Tomlinson, 1984] W. J. Tomlinson, R. H. Stolen, and C. V. Shank. Compression of optical pulses chirped by self-phase-modulation in fibers, *J. Opt. Soc. Am. B1*, 139 (1984).

[Treacy, 1969] E. B. Treacy. Optical pulse compression with diffraction gratings, *IEEE J. Quantum Electron.*, QE-5, 454 (1969).

[Trebino, 1982] R. Trebino, J. P. Roller, and A. E. Siegman. A comparison of the Cassegrain and other beam expanders in high-power pulsed dye lasers, *IEEE J. Quantum Electron.*, QE-18, 1208 (1982).

[Trutna, 1979] W. R. Trutna, Jr., Y. K. Park, and R. L. Byer. The dependence of Raman gain on pump laser bandwidth, *IEEE J. Quantum Electron.*, QE-15, 648 (1979).

[Vagin, 1969] Y. S. Vagin, V. M. Marchenko, and A. M. Prokhorov. Spectrum of a laser based on electron-vibrational transitions in a $CaF_2:Sm^{2+}$ crystal, *Sov. Phys. JETP*, 28, 904 (1969).

[Valee, 1982] F. Valee and J. Lukasik. Vacuum ultraviolet generation in phase-matched carbon monoxide, *Opt. Commun.*, 43, 287 (1982).

[Varma, 1973] C. A. G. O. Varma and P. M. Rentzepis. Time resolution and characteristics of a broadband picosecond continuum and light gate, *J. Chem. Phys.*, 58, 5237 (1973).

[Vaucher, 1982] A. M. Vaucher, W. L. Cau, J. D. Ling, and C. H. Lee. Generation of tunable picosecond pulses from a bulk GaAs laser, *IEEE J. Quantum Electron.*, QE-18, 187 (1982).

[Wallenstein, 1974] R. Wallenstein and T. W. Hänsch. Linear pressure tuning of a multielement dye laser spectrometer, *Appl. Opt.*, *13*, 1625 (1974).

[Wallenstein, 1980] R. Wallenstein and H. Zacharias. High power narrowband pulsed dye laser oscillator-amplifier system, *Opt. Commun.*, *32*, 429 (1980).

[Wallenstein, 1979] R. Wallenstein. Pulsed dye lasers, in *Laser Handbook*, Vol. 3 (M. Stitch, ed.), North-Holland, Amsterdam, 1979, Chap. A6.

[Walling, 1980] J. C. Walling, O. G. Peterson, H. P. Jenssen, R. C. Morris, and E. W. O'Dell. Tunable alexandrite lasers, *IEEE J. Quantum Electron.*, *QE-16*, 1302 (1980).

[Weber, 1982] M. J. Weber, ed. *CRC Handbook of Laser Science and Technology*, Vols. 1 and 2, CRC Press, Boca Raton, Fla., 1982.

[Weber, 1983] P. G. Weber. Long pulse DCM dye laser, *IEEE J. Quantum Electron.*, *QE-19*, 1200 (1983).

[Wellegehausen, 1977] B. Wellegehausen. Optically pumped CW dimer lasers, *IEEE J. Quantum Electron.*, *QE-13*, 1108 (1977).

[Wellegehausen, 1982] B. Wellegehausen, A. Topouzkhanian, C. Effantin, and J. d'Incan. Optically-pumped continous multiline Se_2 laser, *Opt. Commun.*, *41*, 437 (1982).

[Wenzel, 1976] R. G. Wenzel and G. P. Arnold. HF oscillator-amplifier-pumped CdSe parametric oscillator tunable from 14.1 µm to 16.4 µm, *Appl. Opt.*, *15*, 1322 (1976).

[White, 1977] M. B. White. Blue-green lasers for ocean optics, *Opt. Eng.*, *16*, 145 (1977).

[White, 1981] J. C. White, J. Bokor, R. R. Freeman, and D. Henderson. Tunable ArF excimer laser source, *Opt. Lett.*, *6*, 293 (1981).

[White, 1984] J. C. White. Upconversion of excimer lasers via stimulated anti-Stokes Raman scattering, *IEEE J. Quantum Electron.*, *QE-20*, 185 (1984).

[Wilke, 1979] V. Wilke and W. Schmidt. Tunable coherent radiation source covering a spectral range from 185 to 880 nm, *Appl. Phys.*, *18*, 177 (1979).

[Willett, 1974] C. S. Willett. *Introduction to Gas Lasers: Population Inversion Mechanisms*, Pergamon Press, Oxford, 1974.

[Williams, 1983] G. H. Williams, J. L. Hobart, and T. F. Johnson, Jr. 10 THz scan range dye laser with 1 MHz resolution and integral wavelength readout, paper WS1, *Conf. Lasers Electro-Opt.*, *CLEO-83*, Baltimore, Md., May 1983.

[Wilson, 1979] I. J. Wilson, B. Brown, and E. G. Loewen. Grazing incidence grating efficiencies, *Appl. Opt.*, *18*, 426 (1979).

[Winer, 1966] I. M. Winer. A self-calibrating technique for measuring laser beam intensity distributions, *Appl. Opt.*, *5*, 1437 (1966).

[Wokaun, 1982] A. Wokaun, P. F. Liao, R. R. Freeman, and R. H. Storz. High energy picosecond pulses: design of a dye laser amplifier system, *Opt. Lett.*, *7*, 13 (1982).

[Yarborough, 1971] J. M. Yarborough and G. A. Massey. Efficient high-gain parametric generation in ADP continously tunable across the visible spectrum, *Appl. Phys. Lett.*, *18*, 438 (1971).

[Yariv, 1971] A. Yariv. *Introduction to Optical Electronics*, Holt, Rinehart and Winston, New York, 1971.

[Yi, 1983] W. C. Yi, C. Schwab, W. Fuss, and K. L. Kompa. A self-sustained discharge, multiatmosphere CO_2 laser with electron beam preionization, *Opt. Commun.*, *46*, 311 (1983).

[Yiu, 1982] Y. M. Yiu, K. D. Bonin, and T. J. McIlrath. Two-photon-resonant upconversion in xenon, *Opt. Lett.*, *7*, 268 (1982).

[Yodh, 1984] A. G. Yodh, Y. Bai, J. E. Golub, and T. W. Mossberg. Grazing incidence dye lasers with and without intracavity lenses: a comparative study, *Appl. Opt.*, *23*, 2040 (1984).

[Zernicke, 1973] F. Zernicke and J. E. Midwinter. *Applied Nonlinear Optics*, Wiley, New York, 1973.

[Zhang, 1981] G. W. Zhang, H. Grethen, H.-D. Kronfeldt, and R. Winkler. The influence of the beam expanding prism in a dye laser resonator on the linewidth and its dependence on the expansion ratio, *Opt. Commun.*, *40*, 49 (1981).

[Zubarev, 1976] I. G. Zubarev, A. G. Mironov, and S. I. Mikhailov. Influence of nonmonochromaticity of the pump on the gain of monochromatic Stokes radiation, *JETP Lett.*, *23*, 642 (1976).

[Zych, 1978] L. Y. Zych and J. F. Young. Limitation of 3547 to 1182 Å conversion efficiency in xenon, *IEEE J. Quantum Electron.*, *QE-14*, 147 (1978).

3

Resonance Photoionization Spectroscopy

Jeffrey A. Paisner and Richard W. Solarz *Lawrence Livermore National Laboratory, Livermore, California*

3.1 INTRODUCTION

3.1.1 Overview

The development of high-power tunable lasers has increased the popularity and range of application of many branches of spectroscopy. Perhaps the greatest expansion in effort has occurred in the field of resonance photoionization spectroscopy. Before the invention of the laser, photoionization spectroscopy was only of limited application because it relied on the use of electrical discharge and arc lamps. These conventional light sources were of limited utility because of the short wavelengths and large spectral intensities required to induce atomic or molecular photoionization. In contrast, lasers now commercially available can be used to generate the wavelengths and intensities necessary for efficient photoionization via resonant stepwise and nonresonant multiphoton pathways. Consequently, resonance photoionization spectroscopy has been the subject of several review articles (see Table 3.1) and is being used in a vast assortment of fundamental physics and chemistry studies. It also forms the basis for the potentially important commercial application of laser separation of uranium and other high-valued isotopes.

The goal of this chapter is to give the reader a functional understanding of the principles and applications of resonance photoionization spectroscopy. The growth of this branch of spectroscopy is briefly reviewed in Sec. 3.1.2. A theoretical description of the relevant excitation dynamics is given in Secs. 3.2.1 to 3.2.7. These sections should familiarize the reader with theoretical formalisms found in the literature and provide a basis for understanding the limit of validity of each approximation. An overall comprehension of the underlying physics is stressed above computational precision. The multiplicity of

TABLE 3.1 Resonance Photoionization—Some Review Articles and
Conference Proceedings

Relevant section	Reference	Topic
3.1	[Weissler, 1972]	Conventional methods used in photoionization spectroscopy
3.1-3.4	[Eberly 1977, 1979b, 1979c, 1980, 1981, 1984]	Index of multiphoton excitation literature
3.2	[Shore, 1980]	Theory of resonant multistep photoionization
3.2	[Morellec, 1982]	Nonresonant multiphoton ionization
3.2	[Camus, 1983]	Optogalvanic detection and its applications—conference proceedings
3.3	[Feneuille, 1977]	Spectroscopy and applications of highly excited states in atoms and molecules—conference proceedings
3.3	[Letokhov, 1977b]	Selective photoionization of atoms
3.3	[Hurst, 1979]	Resonance ionization spectroscopy and one-atom detection
3.3	[Johnson, 1981]	Multiphoton ionization of molecules
3.3	[Parker, 1983]	Multiphoton ionization of molecules and mass spectrometry
3.4	[Wynne, 1984]	Current trends in atomic spectroscopy—report of National Research Council

applications discussed in Sec. 3.3 have associated with them an array
of experimental configurations and ionization methods. These are discussed in Sec. 3.2.8, together with arrangements employing conventional photon or electron-impact sources commonly used before the advent of the laser. Section 3.2.9 describes the ionization pathways used
to develop the applications discussed in Sec. 3.3. Section 3.4 contains
a summary of the status of resonance photoionization spectroscopy.

3.1.2 Evolution of Resonance Photoionization Spectroscopy

The photoelectric ionization of atomic cesium vapor was first observed over 50 years ago by Foote and Mohler [Foote, 1925]. Their investigators were soon followed by the classic work of Penning [Penning, 1928]. These studies were made possible long before the invention of the laser due to the previous development of high-gain detectors for ions and electrons, in particular: the proportional counter of Rutherford and Geiger [Rutherford, 1908], the thermionic diode of Kingdon and Hertz [Kingdon, 1923; Hertz, 1923], the pulse ionization chamber of Greinacher [Greinacher, 1926], and the Geiger-Müller counter [Geiger, 1928]. These detectors were of simple design, usually consisting of suitably biased plates inserted in a vacuum-tight envelope with windows for optical access. Furthermore, these devices were easily shielded against unwanted background ions and could collect signal ions with approximately unit quantum efficiency. These features were important because ion yields were small due to two experimental limitations: weak light sources and small ground-state photoionization cross sections.

Until recently, signal photons in the optical region (ultraviolet to infrared) could not be detected with such high quantum efficiencies. Coupled with the weak spectral brightness of available light sources and the presence of ambient background light, insensitive detection often limited the signal-to-noise ratio achievable in experiments. Nevertheless, important atomic ionization phenomena were studied in this way. A good example is the process of autoionization. Autoionization was first suggested for optical spectra by Wentzel in 1927 [Wentzel, 1927]. The effect was studied soon after using spectra recorded on photographic film by Shenstone [Shenstone, 1931].

The use of conventional light sources and other nonlaser sources (such as synchrotron radiation) in spectroscopy continues even today. This situation is due primarily to the limited tuning range of available laser sources in the vacuum ultraviolet (VUV). For example, Parr and Elder [Parr, 1968] measured the ionization potential of heavy atoms using broadband ultraviolet radiation from arc lamps filtered by monochromators. Using similar light sources, Berkowitz and Chupka [Berkowitz, 1966], in a beautiful series of experiments, studied the autoionization spectra of simple molecules and the associative ionization of atoms. Similarly, photoionization cross sections in carbon were measured by Hofmann and Weissler [Hofmann, 1971]. The ionization behavior of molecular hydrogen was predicted by Berry [Berry, 1966] and later studied experimentally by Chupka and Berkowitz [Chupka, 1968]. Additional studies of molecular photoionization were published by Murad and Inghram [Murad, 1964], and Wainfan et al. [Wainfan, 1955]. An excellent review of photoionization measurements using conventional techniques is given by Weissler [Weissler, 1962, 1972]. During the last 20 years these methods have become less attractive because of the advent of more sensitive and widely applicable laser techniques that avoid the need for short-wavelength lasers.

The ionization potential (IP) of most atoms and molecules is in the range 5 to 20 eV or 40,000 to 160,000 cm^{-1} [see Fig. 3.1 and Tables 3.2(a) and 3.2(b)]. The earliest lasers were of fixed wavelength and could only be used to ionize atoms and molecules by the process of nonresonant multiphoton ionization. The 1960s and early 1970s saw theoretical treatments [Keldysh, 1965; Geltman, 1963] and experimental observations [Voronov, 1965; Fox, 1971; Delone, 1972; Held, 1973; Lecompte, 1974; Bayfield, 1974] aimed at understanding the phenomena of breakdown in gases induced by high laser intensity. In spite of progress in this area, the efficient and selective ionization of atoms and molecules cannot usually be realized by nonresonant multiphoton ionization. An important exception is discussed in Sec. 3.3.8.

With the development of tunable dye lasers in the visible spectrum, a number of workers recognized that efficient photoionization of atoms and molecules could be achieved through resonant stepwise excitation. The atomic or molecular system serves as a summing device, which stores the energy of each successive resonant photon until the system reaches an energy state where it either spontaneously ionizes or ionizes in a collision. To achieve high efficiency for this process, the up-pumping action to each successive level must compete successfully with the radiative or collisional decay of the lower level in each resonance step. In other words, each step of the process must be strongly pumped or saturated. For an atomic transition, this frequently requires a fluence (energy/area) per lifetime (typically hundreds of nanoseconds) of only a few microjoules per square centimeter. Correspondingly, molecular transitions in the visible spectrum usually require less than a joule per square centimeter per lifetime. Lasers currently available can usually meet these requirements. Most atoms and molecules have resonant transitions between 1 and 5 eV (8000 to 40,000 cm^{-1}), which are entirely accessible using the fundamental and harmonic tuning range of dye lasers. Consequently, resonant multistep laser excitation is a very general ionization technique.

V. S. Letokhov, at the Institute of Spectroscopy in the USSR, was one of the first to outline the primary requirements for the stepwise excitation and ionization of atoms and molecules. Two-step selective photoionization of rubidium atoms was reported in 1971 by Ambartzumian et al. [Ambartzumian, 1971]. In 1973, Levy and Janes [Levy, 1973] were issued a patent for the use of resonant multistep laser ionization in the isotope separation of uranium. Photoionization of cesium dimers was reported in 1974 by Popescu et al. [Popescu, 1974]. Rydberg-state photoionization cross sections in atomic helium were measured by Dunning and Stebbings in 1974 using laser techniques [Dunning, 1974]. One of the first applications of this method to heteronu-

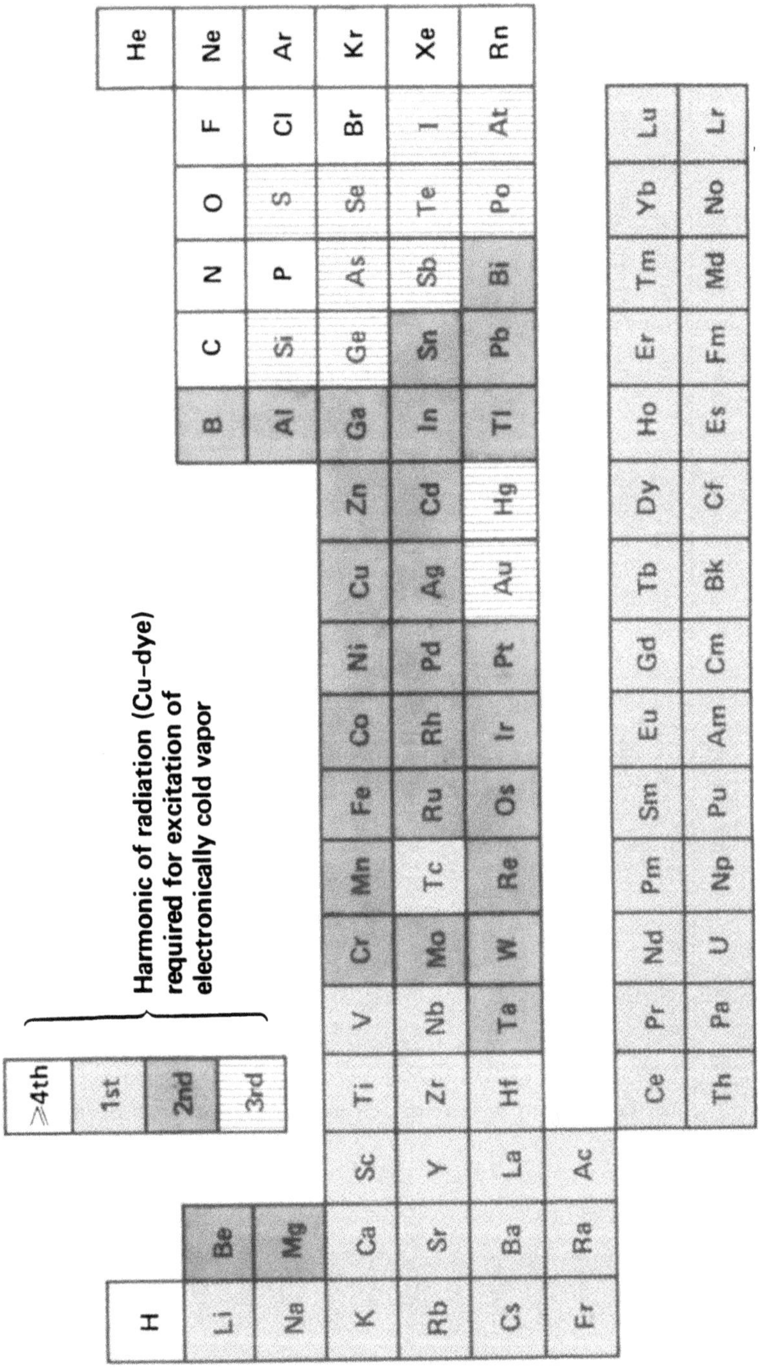

FIG. 3.1 Harmonics (first through fourth) of copper-vapor-pumped dye lasers need to resonantly photoionize the elements. This technology is under development in the Atomic Vapor Laser Isotope Separation Program at the Lawrence Livermore National Laboratory.

TABLE 3.2(a) First Ionization Potentials of the Elements[a]

Atomic number	Element	Symbol	Ionization potential (eV)	Atomic number	Element	Symbol	Ionization potential (eV)	Atomic number	Element	Symbol	Ionization potential (eV)
1	Hydrogen	H	13.599	35	Bromine	Br	11.814	69	Thulium	Tm	6.184
2	Helium	He	24.588	36	Krypton	Kr	14.000	70	Ytterbium	Yb	6.254
3	Lithium	Li	5.392	37	Rubidium	Rb	4.177	71	Lutetium	Lu	5.426
4	Beryllium	Be	9.323	38	Strontium	Sr	5.695	72	Hafnium	Hf	6.65
5	Boron	B	8.298	39	Yttrium	Y	6.38	73	Tantalum	Ta	7.89
6	Carbon	C	11.260	40	Zirconium	Zr	6.84	74	Tungsten	W	7.98
7	Nitrogen	N	14.534	41	Niobium	Nb	6.88	75	Rhenium	Re	7.88
8	Oxygen	O	13.618	42	Molybdenum	Mo	7.099	76	Osmium	Os	8.7
9	Fluorine	F	17.423	43	Technetium	Tc	7.28	77	Iridium	Ir	9.1
10	Neon	Ne	21.565	44	Ruthenium	Ru	7.37	78	Platinum	Pt	9.0
11	Sodium	Na	5.139	45	Rhodium	Rh	7.46	79	Gold	Au	9.226
12	Magnesium	Mg	7.646	46	Palladium	Pd	8.34	80	Mercury	Hg	10.438
13	Aluminum	Al	5.986	47	Silver	Ag	7.576	81	Thallium	Tl	6.108
14	Silicon	Si	8.152	48	Cadmium	Cd	8.994	82	Lead	Pb	7.417
15	Phosphorus	P	10.487	49	Indium	In	5.786	83	Bismuth	Bi	7.289
16	Sulfur	S	10.360	50	Tin	Sn	7.344	84	Polonium	Po	8.42

No.	Element	Symbol	Value	No.	Element	Symbol	Value	No.	Element	Symbol	Value
17	Chlorine	Cl	12.968	51	Antimony	Sb	8.642	85	Astatine	At	
18	Argon	Ar	15.760	52	Tellurium	Te	9.010	86	Radon	Rn	10.749
19	Potassium	K	4.341	53	Iodine	I	10.451	87	Francium	Fr	
20	Calcium	Ca	6.113	54	Xenon	Xe	12.130	88	Radium	Ra	5.279
21	Scandium	Sc	6.54	55	Cesium	Cs	3.894	89	Actinium	Ac	5.17
22	Titanium	Ti	6.82	56	Barium	Ba	5.212	90	Thorium	Th	6.08
23	Vanadium	V	6.74	57	Lanthanum	La	5.577	91	Protactinium	Pa	5.89
24	Chromium	Cr	6.766	58	Cerium	Ce	5.539	92	Uranium	U	6.194
25	Manganese	Mn	7.437	59	Praseodymium	Pr	5.473	93	Neptunium	Np	6.266
26	Iron	Fe	7.902	60	Neodymium	Nd	5.525	94	Plutonium	Pu	6.06
27	Cobalt	Co	7.86	61	Promethium	Pm	5.582	95	Americium	Am	5.99
28	Nickel	Ni	7.635	62	Samarium	Sm	5.644	96	Curium	Cm	6.02
29	Copper	Cu	7.726	63	Europium	Eu	5.670	97	Berkelium	Bk	6.23
30	Zinc	Zn	9.394	64	Gadolinium	Gd	6.150	98	Californium	Cf	6.30
31	Gallium	Ga	5.999	65	Terbium	Tb	5.864	99	Einsteinium	Es	6.42
32	Germanium	Ge	7.900	66	Dysprosium	Dy	5.939	100	Fermium	Fm	6.50
33	Arsenic	As	9.81	67	Holmium	Ho	6.022	101	Mendelevium	Md	6.58
34	Selenium	Se	9.752	68	Erbium	Er	6.108	102	Nobelium	No	6.65

[a]Conversion factor: 1 eV = 8065.479 cm^{-1}, where 1 eV $\equiv$ 1.6022 $\times$ 10^{-19} J and 1 cm^{-1} = 3 $\times$ 10^{10} Hz $\equiv$ 30 GHz.
Source: Adapted from Cowan, [Cowan, 1981].

TABLE 3.2(b) Ionization Potentials of some Molecules and Radicals

Diatomics			Triatomics			Polyatomics		
Mass (amu)	Molecule	Ionization potential (eV)	Mass (amu)	Molecule	Ionization potential (eV)	Mass (amu)	Molecule	Ionization potential (eV)
							4 atoms	
2	H_2	15.427	14	CH_2	10.396	15	CH_3	9.83
4	D_2	15.46	16	NH_2	11.4	17	NH_3	10.20
12	BH	9.77	18	H_2O	12.60	26	C_2H_2	11.40
13	CH	11.13	20	D_2O	12.60	30	CH_2O	10.88
15	NH	13.10	27	HCN	13.80	34	H_2O_2	11.00
17	OH	13.17	29	CHO	9.80	34	PH_3	9.98
20	HF	15.77	30	Li_2O	6.80		5 atoms	
24	C_2	12.00	33	HO_2	11.53	16	CH_4	12.60
26	CN	14.30	34	H_2S	10.40	27	C_2H_3	9.40
28	CO	14.013	36	C_3	12.60	34	CH_3F	12.85
28	N_2	15.576	44	CO_2	13.769	50	CH_3Cl	11.30
30	BF	11.30	44	N_2O	12.894	95	CH_3Br	10.53

Mass	Species	IE	Mass	Species	IE	Mass	Species	IE
30	NO	9.25	46	NO_2	9.79	143	CH_3I	9.54
32	O	12.063	48	O_3	12.30	152	CCl_4	11.47
33	HS	10.50	50	CF_2	11.80		6 atoms	
36	HCl	12.74	52	NF_2	11.90	26	C_2H_4	10.50
38	F_2	15.70	54	OF_2	13.60	32	CH_3OH	10.84
42	LiCl	10.10	60	COS	11.17		7 atoms	
46	Na_2	4.90	64	SO_2	12.34	29	C_2H_5	8.40
46	AlF	9.80	76	CS_2	10.08	146	SF_6	19.30
59	CaF	5.80	169	XeF_2	11.50		8 atoms or more	
70	Cl_2	11.48	188	$MgBr_2$	10.65	30	C_2H_6	11.50
81	HBr	11.62				78	C_6H_6	9.24
115	BrCl	11.10				93	C_6H_7N	7.70
128	HI	10.39				170	n-C_3H_7I	9.26
160	Br	10.54				170	iso-C_3H_7I	9.27
162	ICl	10.31						
207	IBr	9.98						
254	I_2	9.28						

Source: Adapted from Weast [Weast, 1982].

(a) Photoionization

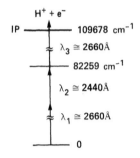

(b) Autoionization

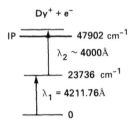

(c) Infrared ionization

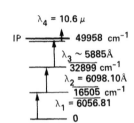

(d) Microwave ionization

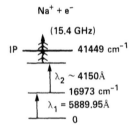

(e) Field ionization

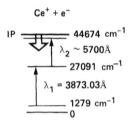

(f) Laser driven ionization

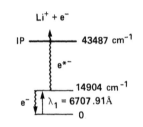

(g) Collisional ionization
     by electrons

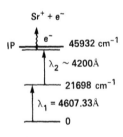

(h) Collisional ionization
     by neutrals

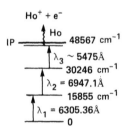

(i) Associative ionization

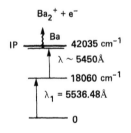

FIG. 3.2 Examples of resonance photoionization pathways: (a) photoionization [Bjorklund, 1978]; (b) autoionization [Worden, 1978a];
(c) infrared ionization [Solarz, 1976]; (d) microwave ionization [Pillet,
1984]; (e) field ionization [Worden, 1978a]; (f) laser-driven ionization
[McIlrath, 1977]; (g) collisional ionization by electrons [Worden,
1978b]; (h) collisional ionization by neutrals [Worden, 1978a];
(i) associative ionization [Solarz, 1980].

clear molecules was reported for nitric oxide by Johnson in 1976 [Johnson, 1976]. The use of resonant multistep ionization as a spectroscopic tool to study heavy atoms was first reported in 1975 by Janes et al. [Janes, 1975] and in 1976 by Carlson et al. [Carlson, 1976]. The capability of this technique to detect single atoms was reported in 1977 by Hurst et al. [Hurst, 1977] and Bekov et al. [Bekov, 1978a, 1979]. Laser resonance photoionization continues to be popular in many research fields because of the number of ionization pathways (see Fig. 3.2), the ease of applying this new method, and the diversity of applications. In fact, the technique has given rise to areas of study that were previously impossible to investigate. These new applications are discussed in Sec. 3.3. First, we briefly review in the next section the fundamentals of resonance photoionization.

3.2 THEORETICAL AND EXPERIMENTAL METHODS

3.2.1 Introduction

Under many conditions, a completely valid description of the response of a collection of oscillators (atomic or molecular) can be obtained by the use of population rate equations as discussed in Chapter 1. Rate equations predict the simple and smooth transfer of oscillator population from one set of energy levels to another. Although mathematically simple, rate equations ignore the coherent aspects of the process. Under the conditions of intense narrowband excitation, the atomic or molecular population will exhibit coherent Rabi oscillations. This is the trademark of a coherent process and can only be described by applying a Schrödinger equation or density matrix formalism. Normally, however, modest-to-large bandwidth lasers coupled with modest ionization time scales allow for a reasonably valid description by the use of simple rate equations. Recent work by Ackerhalt and Shore [Ackerhalt, 1977] and by Eberly and O'Neil [Eberly, 1979a] indicates that the total ionization produced in a system can be predicted from rate equations even when there are bottlenecks in the population transfer arising from the coherent character of the excitation process. Rate equations are thus valid under a wide variety of conditions. Following the approach of Shore [Shore, 1980], we present a more comprehensive discussion of excitation dynamics and show how the coherent description evolves into the rate-equation limit under certain experimental conditions.

3.2.2 Coherent Dynamics

3.2.2A. Schrödinger's Equations

As a starting point, consider the N-level photoionization scheme depicted in Fig. 3.3, in which each laser is resonant with a specific

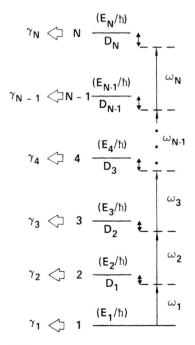

FIG. 3.3 N-level photoionization schemes. $\{D_k\}$ are the cumulative de-tunings (angular frequencies) of the light fields from the atomic states [see Eq. (3.6)]. $\{\omega_k\}$ are the laser angular frequencies, $\{\gamma_k\}$ are the population decay rates of the levels in the excitation sequence, and γ_N is the ionization rate of the Nth level.

transition in the ladder. Under the experimental conditions that the pulse lengths of the respective laser fields are shorter than the radiative emission time T_1 or collisional time T_2 associated with each state, the atomic dynamics,

$$\left(i\hbar \frac{\partial}{\partial t} - H\right)\psi(t) = 0 \quad H = H_0 - \underline{d} \cdot \underline{E} \tag{3.1}$$

can be described by the time-dependent solution to Schrödinger's equation:

$$\psi(t) = \sum_{n} C_n(t) \, e^{(-i\alpha_n t)} \psi_n \tag{3.2}$$

Here $\{\psi_n\}$ are the wavefunctions of H_0 having energy $\{E_n\}$. The phases $\{\alpha_n\}$ can be set arbitrarily. Using classical fields and making the appropriate rotating-wave approximation yields

$$\underset{\sim}{E} = \sum_\lambda Re(E_{\lambda q} \underset{\sim}{\epsilon}_q) \tag{3.3a}$$

$$E_{\lambda q} = \epsilon_{\lambda q} e^{-i\omega_\lambda t} \tag{3.3b}$$

where

λ = step or wavelength
q = polarization at any time, fixed for each λ
ω_λ = instantaneous frequency for each λ
$\{\epsilon_{\lambda q}\}$ = real field amplitudes

and

$$\hbar\alpha_1 = E_1$$

$$\hbar\alpha_2 = \hbar\alpha_1 + \omega_1$$

$$\cdot$$
$$\cdot \tag{3.4}$$
$$\cdot$$

$$\hbar\alpha_{n+1} = \hbar\alpha_n + \omega_n$$

The Schrödinger equation reduces to

$$\frac{dC_n(t)}{dt} = -i \sum_m W_{nm} C_m(t) \tag{3.5a}$$

or in matrix form,

$$\frac{d\underset{\sim}{C}}{dt} = -i\underset{\sim}{W}\underset{\sim}{C} \tag{3.5b}$$

where the ionization probability P_I is given by

$$\frac{dP_I}{dt} = \gamma_N |C_N|^2 \tag{3.5c}$$

Here $\{C_n\}$ are the state amplitudes (i.e., $\{|C_n(t)|^2\}$ are the state populations) and the interaction matrix is given by

$$
W_{nm} = \begin{cases} D_n \dfrac{-i\gamma_n}{2} = \dfrac{1}{\hbar}(E_n - E_1 - \hbar\omega_1 - \cdots - \hbar\omega_n) - \dfrac{i\gamma_n}{2} & n = m \\[4mm] \dfrac{1}{2}\Omega_{nm} = -\dfrac{1}{2}\dfrac{1}{\hbar}\langle n|\underset{\sim}{d} \cdot \underset{\sim}{E}|m\rangle & |n - m| = 1 \end{cases}
$$

$$(3.6)$$

Nonzero cumulative detunings $\{D_n\}$ of the light fields from the atomic states can result from nearly resonant excitation or from inhomogeneous broadening mechanisms, such as Doppler broadening. In the latter case, Eq. (3.5) should be solved for all velocity classes. We have included phenomenological population decay rates $\{\gamma_n\}$ to states outside the excitation ladder. γ_N is the ionization rate. The Rabi (angular) frequencies $\{\Omega_{nm}\}$ contain the interaction strengths through the products of the transition dipole moments and the light-field amplitudes. States in the ladder are only pairwise coupled because each laser is resonant with a specific transition. Each of these transitions is allowed by parity selection rules. Consequently, Ω_{nm} has a nonzero value only when $|n - m| = 1$. For a lossless ($\gamma_1 = \gamma_2 = 0$) resonant ($D_1 = D_2 = 0$) two-level system, Ω_{12} is the instantaneous (angular) frequency at which the population "flops" between the ground and excited states. In fact, the atomic dynamics contained in Eqs. (3.5) and (3.6) for an N-level system will usually exhibit these coherent "Rabi oscillations" in each of the state amplitudes as long as the population decay rates do not damp them out. Under resonant and lossless conditions, the atomic evolution is that of N-coupled oscillators whose normal-mode frequencies are given by functions of the Rabi frequencies. Table 3.3 relates the Rabi (angular) frequency to some more familiar spectroscopic parameters.

As an example, consider the $5s5p$ $^1P_1 \leftarrow 5s^2$ $^1S_0$ (21,698 cm$^{-1} \leftarrow 0$ cm^{-1}) transition at 4607.33 Å in atomic strontium. The transition has an absorption oscillator strength $f_{u\ell} = 1.80$, corresponding to an excited-state radiative lifetime $\tau_u = 5.29$ ns and an emission branching ratio $\beta_{\ell u} = 1$ [Dickie, 1973]. At a Rabi frequency $\Omega/2\pi = 200$ MHz, the entire population of ground-state strontium atoms will evolve through a complete excitation-emission cycle ($^1S_0 \leftarrow {}^1P_1 \leftarrow {}^1S_0$) within a radiative lifetime. This can be realized with a laser pulse of duration $\tau_p = 5$ ns and fluence $I\tau_p = 18$ nJ/cm^2 (i.e., $I = 3.6$ W/cm^2). As a general rule, to achieve "efficient" ionization, the Rabi (angular) frequencies at each step should be well matched $\Omega_{12} < \Omega_{23} < \cdots < \Omega_{N,N-1} < \gamma_N$. Under these conditions, nearly 100% photoionization takes place in a time of the order of $\sim 2\pi/\Omega_{12}$. The origin of this rule resides in the ac Stark broadening, which is contained implicitly in the formulation. As any individual intensity or Rabi frequency increases, the associated levels in that transition spectrally broaden. Consequently, the cross section for the other transition coupled to these levels decreases. Increasing the Rabi (angular) frequencies above γ_N does not produce more efficient ionization. Instead, the population will tend to flop rather than become critically damped.

3.2.2B. Density Matrix

In order to treat collisions and spontaneous emission properly, one must use a full density matrix description. The state density matrix is given by

$$\underset{\sim}{\rho} = \underset{\sim}{C}\underset{\sim}{C}^{+} \tag{3.7}$$

The evolution of ρ is given by the Liouville equation with phenomenological decay terms:

$$i\frac{d}{dt}\underset{\sim}{\rho} = \underbrace{\underset{\sim}{W}\underset{\sim}{\rho} - \underset{\sim}{\rho}\underset{\sim}{W}^{+}}_{} - \underbrace{i\bar{R}\underset{\sim}{\rho}}_{} \tag{3.8}$$

Liouville equation   Decay terms

Here

$$(\bar{R}\rho)_{nm} = \sum_{k,l} R_{nk,ml}\rho_{kl} \tag{3.9}$$

where

$$R_{nk,ml} = \delta_{nk}\delta_{ml}R_{nm}(1 - \delta_{mn}) - \delta_{nm}\delta_{kl}\Gamma_{mk}$$

R_{nm} = off-diagonal dephasing rate

Γ_{nk} = $k \to n$ population decay rate arising from collisions and radiative emission

Γ_{nn} = $-\Sigma_{k \neq n}\Gamma_{kn}$ = population decay rate out of state n but into the excitation ladder

One obtains the following master equations for the populations $P_n = \rho_{nn} = |C_n|^2$ and the off-diagonal elements $Q_{nm} = \rho_{nm} = C_nC_m^*$:

TABLE 3.3 Commonly Used Spectroscopic Parameters[a]

Spectroscopic parameters	Definition	Value[b]
Dipole moment transition $u \leftarrow \ell$	$\underline{d} = e\underline{r}$; $\quad |\underline{d}_{u\ell}|^2 = e^2|r_{u\ell}|^2$	$|r_{u\ell}|^2 = \dfrac{1}{3}\dfrac{1}{g_\ell}\displaystyle\sum_{q,m_\ell}|\langle u|\underline{r}\cdot\underline{\epsilon}_q|\ell\rangle|^2$
Einstein A coefficient transition $\ell \leftarrow u$	$A_{\ell u} = \dfrac{32}{3}\pi^3\alpha c\,\dfrac{|r_{\ell u}|^2}{\lambda^3}$	$7.235\times10^{10}\ s^{-1}\,\dfrac{|r_{\ell u}(cm)|^2}{[\lambda(cm)]^3}$
Radiative branching ratio transition $\ell \leftarrow u$	$\beta_{\ell u} = \dfrac{A_{\ell u}}{\sum_{\ell'}A_{\ell'u}}$	$5.744\times10^5\,\dfrac{g_\ell}{g_u}\,\dfrac{\sigma_D(cm^2)}{[\lambda(cm)]^3}\,\tau_u[s]\left[\dfrac{T(K)}{M(amu)}\right]^{1/2}$
Radiative lifetime transition $\ell' \leftarrow u$	$\tau_u = \dfrac{1}{\sum_{\ell'}A_{\ell'u}}$	$1.499\ s\,\dfrac{g_u}{g_\ell}\,[\lambda(cm)]^2\,\dfrac{\beta_{\ell u}}{f_{u\ell}}$
Oscillator strength transition $u \leftarrow \ell$	$f_{u\ell} = \dfrac{1}{8\pi^2}\dfrac{1}{\alpha\,\bar{\lambda}_c\,c}\,\lambda^2\,\dfrac{g_u}{g_\ell}A_{\ell u}$	$1.499\ s\,\dfrac{g_u}{g_\ell}\,[\lambda(cm)]^2]A_{\ell u}(s^{-1})$
Line width homogeneous—radiative + collisional	$\Delta\nu_L = \dfrac{\Gamma}{2\pi} = \dfrac{1}{2\pi}\left(\dfrac{1}{\tau_u}+\dfrac{1}{\tau_\ell}\right)+\dfrac{\gamma_c}{2\pi}$	$0.15915\ Hz\left\{\left[\dfrac{1}{\tau_u(s)}+\dfrac{1}{\tau_\ell(s)}\right]+\gamma_c\left(\dfrac{rad}{s}\right)\right\}$

homogeneous
—Doppler

$$\Delta \nu_D = (8 \ln 2)^{1/2} \left(\frac{kT}{Mc^2}\right)^{1/2} \nu$$

$$7.162 \times 10^{-7} \text{ Hz } \nu(\text{Hz}) \left[\frac{T(K)}{M(\text{amu})}\right]^{1/2}$$

Peak cross section
u ← ℓ Lorentzian

$$\sigma_L = 2\alpha \bar{\lambda} c \frac{f_{u\ell}}{\Delta \nu_L}$$

$$1.690 \times 10^{-2} \text{ cm}^2 \frac{f_{u\ell}}{\Delta \nu_L(\text{Hz})}$$

Doppler

$$\sigma_D = (4\pi \ln 2)^{1/2} \alpha \bar{\lambda} c \frac{f_{u\ell}}{\Delta \nu_D}$$

$$2.493 \times 10^{-2} \text{ cm}^2 \frac{f_{u\ell}}{\Delta \nu_D(\text{Hz})}$$

Rabi (angular) frequency
transition u ← ℓ

$$\Omega_{u\ell}^{\text{Rabi}} = \left(2 \frac{\alpha \bar{\lambda} c}{\hbar} \lambda f_{u\ell} I\right)^{1/2}$$

$$7.310 \times 10^{10} \frac{\text{rad}}{s} \left[\lambda(\text{cm}) f_{u\ell} I\left(\frac{W}{cm^2}\right)\right]^{1/2}$$

[a]*Fundamental constants*

Speed of light: $c = 2.99793 \times 10^{10}$ cm/s   Planck's constant $\hbar = 1.05459 \times 10^{-34}$ J-s

Fine structure: $\alpha = \dfrac{1}{137.036}$ Boltzmann's constant $k = 1.38066 \times 10^{-23}$ J/K

Compton wavelength: $\bar{\lambda}_c = 3.86159 \times 10^{-11}$ cm   Atomic mass unit $\times c^2$ 1 amu $\times c^2 = 1.49244 \times 10^{-10}$ J

[b]Value units are given after the constant. Units in parentheses are to be used in the expression to obtain the value.

$$\left\{\frac{d}{dt} + \gamma_n\right\} P_n = -\frac{i}{2} \sum_{k \neq n} (\Omega_{nk} Q_{kn} - \Omega_{nk}^* Q_{nk}) + \sum_k \Gamma_{nk} P_k \qquad (3.10)$$

$$\left\{\frac{d}{dt} + i(D_n - D_m) + \left[\frac{1}{2}(\gamma_n + \gamma_m) + R_{nm}\right]\right\} Q_{nm}$$

$$= -\frac{1}{2}\Omega_{nm}(P_m - P_n) - \frac{i}{2} \sum_{k \neq m, n} (\Omega_{nk} Q_{km} - \Omega_{mk}^* Q_{nk}) \qquad (3.11)$$

Equations (3.10) and (3.11) constitute N^2 equations that must be solved. With a straightforward redefinition of terms, these equations can also be applied to experimental conditions in which velocity-changing collisions are important. This formulation is very general and mathematically complex. Consequently, it is difficult to obtain physical insight from the equations. In fact, to date they have not been used to describe or design resonant photoionization experiments. Instead, the density matrix approach can be used to derive rate equations appropriate for describing certain photoionization experiments. This is discussed in the next section.

3.2.3 Incoherent Versus Coherent Dynamics: Rate-Equation Limit

We expect to find a rate-equation solution when damping effects become important. The population dynamics in this limit are often called "incoherent" because the system can be described by populations $\{P_n\}$ alone (i.e., one need not consider off-diagonal components $\{Q_{nm}\}$). Consequently, there is no Rabi flopping. Suppose that

$$\frac{d}{dt} Q_{kl} \cong 0 \qquad (3.12)$$

Then Eq. (3.11) constitutes a set of homogeneous simultaneous equations whose solutions are in the form

$$Q_{kn} = \sum_{lm} M_{kn,lm} \Omega_{ml} (P_m - P_l) \qquad (3.13)$$

The populations $\{P_n\}$ are then given by

$$\left\{\frac{d}{dt} + \gamma_n\right\} P_n = \sum_k \Gamma_{nk} P_k + \text{imag}\left\{\sum_{klm} M_{kn,lm} \Omega_{nk} \Omega_{ml} (P_m - P_l)\right\}$$

$$(3.14)$$

For a two-level system, it is easy to show from Eqs. (3.10)-(3.12) that

$$\left\{\frac{d}{dt} + \gamma_{1C} + \gamma_{1R}\right\} P_1 = -\frac{|\Omega|^2}{\Gamma} \frac{(\Gamma/2)^2}{\Delta^2 + (\Gamma/2)^2} (P_1 - P_2) + \Gamma_{12} P_2$$

$$-\Gamma_{21} P_1 + \beta \gamma_{2R} P_2 \qquad (3.15)$$

$$\left\{\frac{d}{dt} + \gamma_{2C} + \gamma_{2R} + \gamma_{2I}\right\} P_2 = \frac{|\Omega|^2}{\Gamma} \frac{(\Gamma/2)^2}{\Delta^2 + (\Gamma/2)^2} (P_1 - P_2) - \Gamma_{12} P_2 + \Gamma_{21} P_1$$

where

Ω = Rabi (angular) frequency
Δ = $D_2 - D_1$ = laser detuning
$\gamma_{1,2C}$ = collisional rate out of ladder
$\gamma_{1,2R}$ = spontaneous emission rate
γ_{2I} = ionization rate
R_{12} = collisional dephasing rate
Γ = $\gamma_{1C} + \gamma_{1R} + \gamma_{2C} + \gamma_{2R} + \gamma_{2I} + 2R_{12}$
$\Gamma_{12,21}$ = collisionally induced $1 \leftarrow 2$, $2 \leftarrow 1$ transition rate
β = spontaneous emission branching ratio

Clearly, the on-resonant ($\Delta = 0$) pumping rate is given by

$$R (\Delta = 0) = \frac{|\Omega|^2}{\Gamma} = \frac{\sigma_L I}{\hbar \omega} \qquad (3.16)$$

which is the familiar rate equation pumping rate. This equality can be easily shown using the entries in Table 3.3. Here σ_L is the peak cross section of the homogeneously broadened transition and I is the intensity of the laser (usually expressed in W/cm^2) at (angular) frequency ω. From Eq. (3.15) the ionization probability for a laser of pulse duration τ_p is given by

$$P_I = 1 - e^{-R(\Delta=0)\tau_p} \qquad (3.17)$$

when the following two conditions are satisfied:

$$\frac{2\pi}{\tau_p} \gg \Gamma - \gamma_{2I} \quad \text{(short pulse)}$$

and

$$\gamma_{2I} \gg \Gamma - \gamma_{2I} \quad \text{(collisionless)}$$

These conditions are often satisfied in photoionization experiments. The laser fluence $I\tau_p$ necessary to "saturate" the homogeneously broadened transition is just $\hbar\omega/\sigma_L$. Saturation fluences for excited-state autoionizing transitions encountered in atoms can be quite modest: for $\sigma_L = 10^{-18}$ to 10^{-15} cm^2 and $\hbar\omega = 2$ eV (~ 6000 Å), saturation fluences range from 300 mJ/cm^2 to 300 μJ/cm^2. The Lorentzian line shape in Eq. (3.15) is a consequence of characterizing the ionization decay of the last level in terms of a simple decay constant. This corresponds to the $q = \infty$ limit for the Beutler-Fano autoionization line shape. The reader is referred to the literature [Fano, 1968; Mizushima, 1970; Cowan, 1981] for a discussion of this interesting topic.

We note that power broadening is implicitly contained in Eq. (3.14). That is, power broadening on one step influences the excitation rate on subsequent steps. This is not the case in the "heuristic" rate-equation limit obtained by setting

$$Q_{rs} = 0 \quad \text{if} \quad |r - s| = 1 \tag{3.18}$$

$$\left\{\frac{d}{dt} + \gamma_n\right\}P_n = -R_{n+1,n}(P_n - P_{n+1}) + R_{n,n-1}(P_{n-1} - P_n) + \sum_k \Gamma_{nk}P_k$$

$$\tag{3.19}$$

where

$$R_{k,n}(\Delta_{kn} = D_k - D_n) = \frac{|\Omega_{nk}|^2}{\gamma_k + \gamma_n + 2R_{kn}}$$

$$\cdot \frac{[(\gamma_k + \gamma_n + 2R_{kn})/2]^2}{(D_k - D_n)^2 + [(\gamma_k + \gamma_n + 2R_{kn})/2]^2}$$

Except for population saturation among adjacent levels, no power-broadening effects are contained by these commonly used heuristic rate equations. However, when operating near optimum conditions for coherent dynamics, as discussed earlier, these equations yield reasonable predictions, as shown in Fig. 3.4.

Obviously, the equation set to use will be dictated by the experimental conditions. Other descriptions of the rate-equation approach have been given in the literature. The reader is referred to the works of Panock and Temkin, Zakheim and Johnson, and Rothberg and others [Panock, 1977; Zakheim, 1980; Rothberg, 1981].

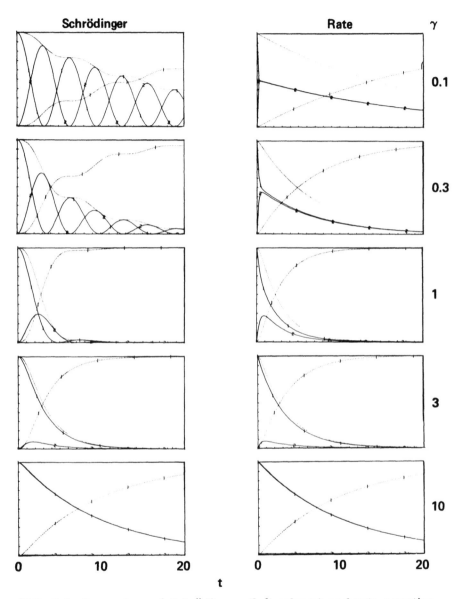

FIG. 3.4 Comparison of Schrödinger (left column) and rate-equation (right column) level populations for resonantly excited two-level atom, Rabi frequency $\Omega = 1$. Successive frames, from top to bottom, are for ionization loss rates $\gamma = 0.1$, 0.3, 1, 3, and 10. Solid lines represent level populations (level 1 starts from 1 at $t = 0$; other levels rise successively from 0 at $t = 0$); dotted lines represent ions (starting from 0 at $t = 0$) and nonionized probability (starting from 1 at $t = 0$) [Shore, 1978].

3.2.4 Doppler Widths and Laser Spectra

As indicated earlier, Doppler broadening is inhomogeneous. It must be handled by solving the excitation dynamics of the state populations $P_n(t,v)$ for each velocity class v, using

$$D_{n+1} - D_n = \omega_n - \frac{E_{n+1} - E_n}{\hbar} - \underset{\sim}{k}_n \cdot \underset{\sim}{v} \qquad (3.20)$$

and then averaging over the entire Gaussian velocity distribution $G(v)$. Here $[\underset{\sim}{k}_n]$ is the wave vector of the exciting lasers. For an atomic vapor of mass M at temperature T, the full width at half maximum (FWHM) Doppler width $\Delta \nu_D$ is proportional to the excitation frequency and is given by

$$\Delta \nu_D = 7.162 \times 10^{-7} \nu \left[\frac{T[K]}{M(amu)} \right]^{1/2} \qquad (3.21)$$

The 4607-Å resonance line of the even isotope of strontium (M = 88) has a room-temperature (T = 293 K) $\Delta \nu_D$ = 0.85 Ghz. This is a typical Doppler width for transitions in the visible spectrum.

To obtain efficient photoionization, all velocity classes must be accessed by the exciting laser fields. Assuming that the homogeneous line width of the transition, which arises from radiative and collisional processes, is small compared to the Doppler width, full access can be achieved in any of four ways: (1) making the transition Rabi frequencies greater than the corresponding Doppler widths, (2) using an "opposed beam" technique, (3) using short-pulse τ_p excitation so that $1/\tau_p > \Delta \nu_D$ (amplitude modulation), or (4) spreading the laser spectra to include the entire atomic distribution on each step (phase modulation).

The first approach, sometimes called power broadening, is simple but involves larger laser intensities than required in the other methods. Recall that the Rabi frequency is proportional to $(\sigma I)^{1/2}$ (see Table 3.3). The second method, opposed beams, will be described in the next subsection. It usually requires narrowband lasers. The third approach involves using tunable picosecond lasers and is not very useful for resonant multistep excitation schemes. The fourth, controlling the laser spectra using phase-modulation techniques, is the most general approach and usually involves the smallest expenditure of laser intensity.

The detunings introduced in Eqs. (3.1)-(3.11) are "instantaneous detunings" since $\omega_i = \omega_i(t)$. Consequently, a judicious choice of $\omega_i(t)$ can ensure that all velocity classes are accessed. This can be accomplished either by spending a sufficient time resonant with any one velocity class or by sweeping across the resonance to induce "adiaba-

tic" rapid passage [Allen, 1975]. Although frequency control during
the laser pulse is within the state of the art using solid-state phase-
modulation devices, this technique has not been widely used in resonant photoionization experiments. Often the lasers used in these studies have well-defined time-averaged (over many pulses) line widths
but have unspecified time-dependent frequency spectra $\{\omega_i(t)\}$. Without adequate control of the laser oscillator (see Chapter 2), $\omega_i(t)$ is
uncharacterized and often varies from shot to shot or within any short
time window. The photoionization fraction will vary similarly. It is
only proper to statistically model the laser bandwidth by a time-independent spectral intensity distribution $S_n(\omega_n)$ if the laser achieves
full bandwidth excursion over a time that is "short" compared to the
Rabi frequency and that is consistent with time bandwidth constraints.
Under these conditions, it is appropriate in the rate equations, although not rigorous, to use average values sampled over the laser
spectral distributions for $\{R_{k,n}\}$:

$$\langle R \rangle_{n+1,n} = \int R_{n+1,n}(\omega_n - \omega')G\left(\omega' - \frac{E_{n+1} - E_n}{\hbar}\right)S_n(\omega_n)\, d\omega_n\, d\omega'$$

$$(3.22)$$

Essentially, one replaces in the rate equations the Lorentzian widths
due to lifetime effects with the convolution of the Doppler and laser
widths.

3.2.5 Two-Photon Excitation

Often excitation pathways are inaccessible using a completely resonant
multistep pathway because of the lack of a good spectral match between
available laser sources and resonance transitions. For example, as indicated in Fig. 3.1, hydrogen and the noble gases have resonance
lines in the VUV spectrum for which tunable laser sources are difficult
to construct. In other circumstances, one wants to minimize the number of laser sources and the number of excitation steps required to
photoionize. For these situations, researchers in the field exploit two-
photon excitation techniques, as illustrated in Fig. 3.5. Here one excites the levels at energy $E_1 + E_2$ by two photons ω_1 and ω_2. Frequency ω_1 may be equal to ω_2. Neither ω_1 nor ω_2 are exactly resonant
with either transition E_1 or E_2. Assuming that $\omega_1(\omega_2)$ is close to resonance with transition $1 \to 2$ ($2 \to 3$), then by adiabatic elimination in
Schrödinger's equation, the net transition $1 \to 3$ can be described by
an effective two-photon Rabi (angular) frequency Ω_{13} given by

$$\Omega_{13} \cong -\frac{\Omega_{12}\Omega_{23}}{2D_2}$$

$$(3.23)$$

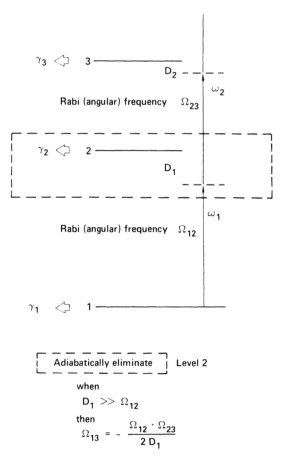

$$\Omega_{13} = -\frac{\Omega_{12} \cdot \Omega_{23}}{2 D_1}$$

FIG. 3.5 For the conditions shown, level 2 may be "adiabatically eliminated"—the interaction of lasers at ω_1 and ω_2 can be expressed as a two-photon Rabi frequency Ω_{13} directly coupling levels 1 and 3.

One replaces (in all previous equations) steps $1 \rightarrow 2$ and $2 \rightarrow 3$ and their respective Rabi (angular) frequencies by the one step $1 \rightarrow 3$, which has Rabi (angular) frequency Ω_{13}.

Two-photon excitation can also be used to eliminate or reduce the effects of Doppler broadening by exploiting the opposed-beam techniques first proposed by Vasilenko et al. in 1970 [Vasilenko, 1970]. In essence, the positive (negative) Doppler shift on one step is canceled by the negative (positive) shift on a subsequent step, as shown in Fig. 3.6. Many workers have used this approach to perform high-resolution spectroscopy in an atomic vapor [Bloembergen, 1976; Letokhov, 1977a]. The method has been suggested for simultaneously ac-

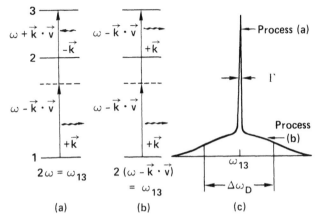

FIG. 3.6 Level diagram of two-photon excitation using: (a) opposed or counter-propagating beams; (b) co-propagating beams; (c) absorption spectra, illustrating Doppler-free narrowed line shape that results from counter-propagating geometry. Γ is the homogeneous line width associated with the transition.

cessing all velocity components and maintaining high selectivity for isotope separation [Hodgkinson, 1981], and for high-sensitivity resonance excitation mass spectrometry [Lucatorto, 1984].

3.2.6 Selection Rules

Until now we have assumed nondegenerate levels in the excitation pathway. Selection rules arising from degenerate levels, as discussed in Chapter 1, can be very important in a multistep resonant photoionization process and can impose polarization constraints on the excitation lasers. Consider using parallel linear polarizations, in the absence of an external magnetic field and collisions, to excite the sequence $J \to J - 1 \to J - 2 \to \cdots \to 0$. This results in a maximum fraction ionized of $1/(2J + 1)$, which is independent of laser intensity since only the $m = 0$ magnetic sublevel in the ground state can find an ionization pathway. As another example, consider the photoionization scheme $J = 0 \to J = 1 \to J = 0 \to J = 1$, where the final state is a $J = 1$ autoionizing level. It is easy to see that if the polarization of λ_1 is perpendicular to that of λ_2, no ground-state atoms or molecules can be excited to the $J = 0$ excited state. Consequently, no photoionization will be observed. The presence of a nuclear spin I can break these selection rules, as shown for another example in Fig. 3.7. Balling and Wright [Balling, 1976] proposed using polarization selection rules to facilitate atomic vapor laser isotope separation of odd ($I \neq 0$) iso-

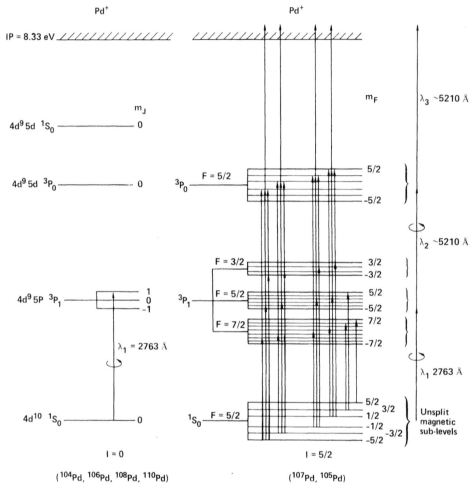

FIG. 3.7 Selective photoionization of palladium by circularly polarized light [Chen, 1981].

topes from even (I = 0) ones when the atomic transition isotope shifts are small compared to their Doppler widths.

A full treatment of resonant multistep photoionization must include a detailed accounting of all magnetic sublevels of each level participating in the excitation. Each magnetic sublevel participates in a transition that has its own dipole moment and consequently its own Rabi frequency. It is this accounting that gives rise to the polarization selection rules. This accounting can be accomplished in a straightforward way using the formalism introduced in this chapter. For most experimental situations, estimating performance using the average Rabi frequency given in Table 3.3 (i.e., using an average dipole moment or cross section) is sufficient as long as the selection rules are kept in mind.

3.2.7 Nonresonant Multiphoton Ionization

So far we have concentrated on the effects of radiation fields on atoms and molecules when the laser is tuned near resonances in the absorbing gas. An important class of experiments is based on the nonresonant photoionization of matter. Some of the earliest experiments in laser photoionization reported the breakdown of various gases in the vicinity of the focal point of Q-switched fixed-frequency lasers [Meyerand, 1963]. The principles of this phenomenon form the basis for plasma shutters and can be used to describe a number of frequency-conversion methods. The classic early work of Bebb and Gold [Bebb, 1966], later extended by Lambropoulos [Lambropoulos, 1974, 1980] and others, used perturbation theory to calculate the rate of ionization of atoms by nonresonant photons at high intensity [Morellec, 1982; Mainfray, 1980].

When the frequency of the laser is far from resonance and in the so-called weak or moderate field limit, the ionization probability W_N for a nonresonant N-photon process [$N\hbar\omega > IP > (N - 1)\hbar\omega$] is

$$W_N = \sigma_N F^N \tag{3.24}$$

where F is the photon flux density (photons/cm^2-s). σ_N, the so-called N-photon ionization cross section (cm^{2N}-s^{N-1}), is given by

$$\sigma_N = \text{constant} \cdot [(2\pi\alpha)^N] m Q \omega_K^N \int |M_{fg}^{(N)}|^2 \, d\Omega_Q \tag{3.25}$$

where α is the fine-structure constant, m the electron mass, and Q the wavenumber of the outgoing electron. The polarization dependence of the process is contained in the integration of the square of the multiphoton dipole moment over angular phase space Ω_Q. M describes the dependence of the cross section on atomic structure and frequency and is written as

$$M_{fg}^{(N)} = \sum_a \frac{<f|r^{(\lambda)}|a_{n-1}> \cdots <a_1|r^{(\lambda)}|g>}{[\omega_{a_{N-1}} - \omega_g - (N - 1)\omega_k] \cdots (\omega_{a_1} - \omega_g - \omega_k)} \tag{3.26}$$

Here a are pure atomic states and $r^{(\lambda)} = \underset{\sim}{r} \cdot \underset{\sim}{\varepsilon}_\lambda$ depends on laser polarization. The wavefunction of the atomic states and the detuning

of each multiple-photon virtual state from each real resonance is then
required to obtain the overall ionization probability. For atomic hydrogen, the Bebb and Gold derived multiphoton ionization rates for
1-W/cm^2 excitation at 10.6 µm (12-photon ionization) is 3.0×10^{-149}/s
and at 5300-Å radiation (6-photon ionization) is 3.3×10^{-70}/s [Morel-
lec, 1982]. Even at a laser intensity of 1 GW/cm^2, the rates are exceedingly small. The ability to tune lasers to resonances on each step
of a multiphoton ladder is clearly of great importance.

3.2.8 Experimental Devices

3.2.8A. Light Sources

The greatest impetus to photoionization spectroscopy has been the
emergence of the laser. The types of lasers available and their principles of operation have already been covered in Chapter 2. Currently available continuous-wave dye lasers can deliver several watts of
radiation in line widths less than 1 MHz. Moderate repetition-rate (10
to 100 Hz) pulsed (10 to 1000 ns) dye lasers can deliver greater than
0.1 J per pulse in line widths to approximately 100 MHz. Trade-offs
can usually be made in pulse energy, line width, and even pulse-repe-
tition rate around these values. Attention has been given to many
new pulsed laser sources, such as copper-vapor-laser-pumped dye
lasers, in which repetition rates up to several tens of kilohertz have
been obtained. A copper-vapor-laser-pumped dye-laser system composed of master oscillators and power amplifiers has been operational
at the Lawrence Livermore National Laboratory since 1976. It delivers
approximately 75 W of radiation in several different colors. The device operates at repetition rates over 10 kHz and at line widths down
to 100 MHz. A system with similar oscillator specifications that will
deliver several thousand watts of average power is currently under
construction [Emmett, 1984].

The most common source of ultraviolet and visible radiation prior
to the development of the laser was an arc lamp or a hot plasma that
operated on principles similar to that of an arc lamp. These devices
emit up to 20 eV radiation and can deliver a total power of some tens
of watts in the VUV. Ten watts of 2000-Å radiation corresponds to
10^{19} photons/s. However, this radiation is usually distributed non-
uniformly over a spectral bandwidth of hundreds to thousands of
angstroms. In addition, the radiation source itself has physical dimensions on the order of centimeters. Consequently, its light output
cannot be optically delivered to an experimental spot much less than
1 cm^2. Single-step photoionization cross sections vary from 10^{-22} to
10^{-15} cm^2, with molecular cross sections for ionization often being the
weakest, hydrogen-like cross sections being in the middle range, and
excited-state cross sections of multielectron atomic systems generally
being the strongest. In order to saturate an optical transition with a

pulsed source, a photon fluence of $1/\sigma_L$ must be delivered to the experiment, as discussed in Sec. 3.2.3. The photon flux density from a continuous source must be greater than $1/\sigma_L\tau$, where τ is a characteristic time in the experiment. The atom or molecule will respond as though the light source is continuous if the excitation to the experiment lasts longer than a characteristic time: (1) the lifetime of the intermediate states if the atom is stepwise excited, or (2) the flow time of an atomic beam through the focal volume of the light.

Typical atomic ground-state photoionization cross sections are 10^{-18} cm^2. Consequently, under the assumption of a 10-μs transit time, photon flux densities of about 10^{23} photons/cm^2-s are required to achieve saturation. A continuous-arc lamp can deliver about 10^{19} photons/s, or about 10^{16} photons/s per angstrom line width. If it is necessary to have narrow-bandwidth radiation in the experiment, this photon flux may be further reduced by several orders of magnitude by the throughput efficiency of a monochromator. As a result, conventional lamps cannot saturate ionization transitions under any conditions. For studies requiring a resolution of 1 Å or less, a typical 10-Hz pulsed dye laser system providing 10-ns pulses can deliver, on the average, greater than 1000× more photons and 1×10^{10} times higher photon fluxes. Furthermore, because the laser is spatially coherent and can be focused to a spot whose area is less than 10^{-4} cm^2, lasers can provide photon flux densities greater than 10^{14} times that obtained using conventional lamps. The advantages of spectral resolution, time resolution, and volume resolution are all overwhelmingly in favor of the laser. The ability of the laser to saturate bound-bound transitions and perform excited-state photoionization studies is exceedingly important. The only advantage of the arc lamp is its ability to generate very short wavelength radiation (i.e., high-energy photons) at low cost.

Although capable of achieving at best only one hundred millionth (10^{-8}) of saturation on each transition, light from conventional lamps can be used for all single-photon ionization studies that do not require high resolution. This is not the case for multiphoton processes. As an example, consider nonresonant multiphoton ionization. Nonlaser light sources are capable of generating photon flux densities of 10^{15} photons/cm^2-s within a useful line width. Tunable laser sources can produce flux densities of 10^{29} photons/cm^2-s. From Eqs. (3.24)-(3.26) it can be shown that typical nonresonant two- and three-step photoionization cross sections for the alkalis are of the order of 10^{-50} cm^4-s and 10^{-80} cm^6-s^2, respectively. The ionization rates from the two-photon, three-photon conventional, and two-photon, three-photon laser-driven cases become 10^{-20}/s, 10^{-35}/s, and 10/s, 1/s, respectively. Lasers are clearly required for the observation of multiphoton processes. Even in the case of resonant multiphoton processes, the conventional source is handicapped by its lack of photons at the required wavelength for the resonant absorption process. A final observation is that above approximately 10^{35} photons/cm^2-s (or ~10^{16} W/cm^2), perturbation theory is no longer valid since the electric field amplitude:

$$E = \left(\frac{8\pi I}{c}\right)^{1/2} = 27.5 \ \frac{V}{cm} \ \left[I\left(\frac{W}{cm^2}\right)\right]^{1/2} \tag{3.27}$$

exceeds the typical binding energy (~2×10^9 V/cm) of the electron to the nucleus.

3.2.8B. Ion and Electron Detectors

We have already noted that sensitive detection for ions and electrons has been in existence for over 50 years. During this period, new types of detectors have been developed that offer faster time response and higher sensitivity than the earlier devices. Detector types can be chosen for reasons of cost, sensitivity, time resolution, ease of operation, or adaptability to a given experimental configuration.

The ionization chamber of Greinacher [Greinacher, 1926] can detect as few as about 10^5 ions. Although not particularly sensitive, it operates at fairly high pressures so that only a simple vacuum system is required. This device is normally operated in a regime in which all ions are collected but no gain is achieved. Operation at much higher voltage and with the addition of a gain medium such as P-10 gas (90% argon plus 10% methane) results in a new mode of operation and can result in an overall gain of 10^4. In this mode of operation, the detector is essentially equivalent to the proportional counter of Rutherford and Geiger [Rutherford, 1908]. In the ideal case, the signal pulse height is linearly proportional to the number of ions produced by the light source. Under other conditions, the device can be made to operate in such a way that the signal height is independent of the number of charge pairs initially produced. Operating at very low initial ion densities, it is possible ot obtain an output pulse of several volts from a single photoion. The *Encyclopedia of Physics* contains a good review of these devices [Korff, 1958].

The Kingdon-Hertz diode usually consists of a long cylindrical anode surrounding a thermionically emitting wire cathode. Ionization occurs by resonant excitation to a high-lying electronic state followed by collisional ionization or by photoionization. The resultant ions are trapped electrostatically in a region surrounding the cathode. Their presence in a space-charge-limited diode affects the flow of current. The trapping of the ions greatly enhances the sensitivity of the device, so that almost 100% collection efficiency and a gain of 10^5 can be achieved. Typical diode currents of nanoamperes to microamperes are possible [Esherick, 1977; Popescu, 1974].

The more recently developed solid-state devices, such as electron multipliers, channeltrons, and microchannel plates, are useful when the system pressures are much lower than for the foregoing devices (10^{-5} torr or lower). Channeltron electron multipliers and microchannel plates are nonmagnetic devices fabricated from a special formulation of glass that is heavily doped with lead. This glass exhibits secondary emissive and resistive properties. A particle of sufficient energy or a photon (< 2000 Å), when incident on the active region of the device, causes the emission of at least one secondary electron. This secondary electron is accelerated by the electrostatic field within the device, causing additional emission and gain. These devices can be operated with gains as high as 10^8. Channeltrons are generally preferred over electron multipliers since they usually operate in a gain-saturated mode and therefore have much more stable or reproducible output signals [Kurz, 1979]. Microchannel plate arrays having 10^4 to 10^7 miniature electron multipliers can be operated with instrument response times of less than 1 ns and with spatial resolution of less than 20 μm [Wiza, 1979].

Optogalvanic detection has been widely used in ionization spectroscopy since the observation of laser-induced changes in the voltage drop across a hollow cathode lamp [Green, 1976]. Essentially, optogalvanic signals are generated whenever atomic or molecular populations are altered in a discharge by resonant excitation. Excitation to higher-lying states produces (1) an increase in cross section for electron-impact ionization, or (2) a perturbation of the equilibrium between the electron temperature and the atomic-excitation temperature, which tends to increase the electron temperature via superelastic collisions. Both mechanisms result in increased conductivity, which can be detected by a voltage change across the discharge. Typical signals range from several millivolts to several volts.

The optogalvanic method of detection has recently received considerable attention due to its simplicity and wide range of application. Optogalvanic spectroscopy has been used in resonant excitation and ionization studies of atomic, ionic, and molecular species in flames as well as in discharges. Perhaps the most practical application of optogalvanic detection is analytical analysis of trace elements and impurities. For example, laser-enhanced ionization (LEI) in flames is being tested at the National Bureau of Standards as a technique for certification of standard reference materials. The current status of this work has recently been reviewed elsewhere. The reader is referred to the articles contained in the proceedings of the International Colloquium on Optogalvanic Spectroscopy and Its Applications [Camus, 1983].

3.2.8C. Vapor Sources

Two recent developments in vapor-source technology have contributed substantially to many fields of spectroscopy, including resonant laser photoionization. By using a collimated beam or a supersonic nozzle, it is possible to further increase the experimental resolution above that attainable with narrowband lasers. While a variety of spectroscopic methods reviewed elsewhere in this text have been developed to obtained spectra not limited by the oscillator's line width (such as Doppler-free two-photon spectroscopy, Lamb-dip spectroscopy, etc.), nozzled beam sources also cool to very low temperatures most of the internal degrees of freedom of the atoms or molecules of interest. This reduction in temperature greatly reduces the number of initial states of the sample in the experiment. The resulting multiphoton ionization spectrum is therefore the convolution of as small a number of initially absorbing states as possible.

Another noteworthy development is that of laser vaporization methods for the generation of large concentrations of refractory metals. Laser vaporization is achieved by focusing an intense (usually a joule-per-pulse Nd:YAG laser) beam onto a refractory metal surface. The fast (of order nanoseconds) addition of the energy to the surface and the limited ability of the metal to conduct the energy away from the locally heated region brings the surface of the material to a very high temperature. The amount of material vaporized from the metal sample is so large that association of the vaporized material just above the surface usually occurs and results in the formation of metal clusters. The "condensation" phenomenon can be accelerated by either bathing the refractory gas-phase material in a gentle stream of cold gas, or by entraining the material in a supersonic nozzled expansion. Some recent results from novel sources of this type will be reported later in this chapter.

3.2.8D. Mass Spectrometers and Post-Ionization Analysis

Mass spectrometers are commonly used in conjunction with photoionization experiments. The mass analysis of laser-produced ions can be accomplished in a number of ways. It is natural that time-of-flight mass spectrometry has received considerable attention since pulsed lasers typically ionize the target samples within a few nanoseconds. Clearly, the major advantage of time-of-flight spectrometers is their ability to obtain a complete mass spectrum on each laser pulse. This is a significant advantage when using low-duty-cycle devices. Once the sample is ionized, the arrival of individual ion pulses at the detector is recorded as a function of time. For ultrahigh resolution, it is necessary to reduce the initial dispersion of velocities during ion fragment formation. By analyzing individual peak widths, one can determine the kinetic energy of ion fragments along the time-of-flight axis [Carney, 1982]. Quadrupole mass filters or magnetic sectors with a resolution of better than 1 amu can be used to measure, at a fixed mass or mass range, the dependence of ion formation on the laser frequency and intensity. A variation on these themes is that of laser-

produced plasma chromatography [Lubman, 1982]. Laser photoions
are produced in a drift tube containing nitrogen carrier gas. The
photoions are then repelled by an electric field and become mass-separ-
ated due to their different mobilities in the carrier gas. Although the
mass resolution is poor, the apparatus itself is quite simple.

3.2.9 Methods of Ionization

3.2.9A. Photoionization

Many of the applications discussed in this chapter exploit direct
photoexcitation to the ionization continuum or to an autoionizing level
above the ionization limit. Photoionization into the continuum corresponds to promoting a single electron to an unbound orbit while the
core electrons are left in a low-lying state, often the ground state of
the ion. Typical continuum photoionization cross sections are 10^{-18}
cm^2, but they vary as a function of detuning from the ionization
threshold. Usually, the cross section starts out high near threshold
and decreases quickly above threshold. In Fig. 3.8 are shown photoionization cross sections ns to ϵp (n = 1 to 100) for atomic hydrogen
as a function of photon energy ϵ.

Double excited states or two-electron excitations above the ionization limit that are quasi-bound or unstable against ionization are called
autoionizing levels. They are observed in all multielectron atoms and
ions and they also occur in molecules. Oscillator strengths of transi-

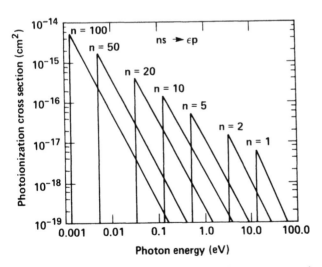

FIG. 3.8 Photoionization cross sections ns → ϵp in atomic hydrogen as
a function of photon energy in electron volts.

tions to these levels are characteristic of bound-bound transitions. However, the peak cross sections are reduced by the usually wide line widths associated with these transitions. For example, the $3s3p$ $^1P_1^0 \rightarrow 3p^2$ $^1S_0$ autoionizing transition in atomic magnesium is found to have a cross section of 8×10^{-16} cm^2 and a line width of 300 cm^{-1} (9×10^{12} Hz) [Bradley, 1976]. The corresponding transition oscillator strength is 0.43. The transition width derives from an autoionizing decay lifetime of 18 fs. The lifetime of an autoionizing level is usually determined by the time for a nonradiative transition of excited core electrons to less energetic orbitals of the ionized atom, with the simultaneous ejection of a free electron. This behavior is affected by the Coulomb interaction between electrons. For sufficiently slow autoionization times, even small oscillator-strength transitions can give rise to relatively large peak cross sections. Bekov et al. [Bekov, 1978b] have observed a long-lived (0.075 ns) autoionizing level in atomic gadolinium accessed in a three-step photoionization process. The measured peak cross section of 8×10^{-16} cm^2 corresponds to a transition oscillator strength of 1.5×10^{-5}. Results indicate that long-lived autoionizing states, with excitation cross sections comparable to those for excitation of bound high-lying states, exist in atoms with complex spectra. This is very important for the variety of applications discussed in the next section. The position and strength of the autoionizing levels accessed from excited atomic states are generally very difficult to calculate, and reliance is made on experimental data. Indeed, one of the applications of resonance photoionization is the identification and characterization of these levels.

3.2.9B. Ionization by Electric Fields

It is also worth noting that variations of the direct photoionization methods discussed above can have important advantages in the formation of ions. Field ionization can be an attractive technique [Littman, 1976; Letokhov, 1977b] in cases where the ionization cross sections are quite small but where excitations to states just below the ionization limit have large transition probabilities. Multistep excitation brings the atom or molecule to a highly excited state, which is then converted to an ion by the application of an externally applied field of the order of 1 kV/cm (see Fig. 3.9). If the high-lying state is at an energy where the excited electron is unbound in the combined field of the core and the externally applied dc Stark field, the electron is free to leave and an ion is produced. For a Rydberg state of effective principal quantum number n^*, the critical field is given by

$$\epsilon_c = \frac{1}{16n^{*4}} \text{ (in atomic units)} = 3.21 \times 10^8 \frac{1}{n^{*4}} \text{ V/cm} \qquad (3.28)$$

This formula is easily derived by considering a hydrogen-like atom in an electric field. The potential seen by the Rydberg electron, given in atomic units,

$$V = -\frac{e}{z} - ez \tag{3.29}$$

has a saddle point (maximum) at

$$V_s = -2\epsilon^{1/2} \tag{3.30}$$

Ionization can occur when the Rydberg-state energy,

$$E_n = -\frac{1}{2n^2} \tag{3.31}$$

exceeds the saddle-point energy. In practice, the experimentally determined threshold fields (i.e., to 50% ionization) can be somewhat larger (by ~10 to 20%) than that calculated using the classical expression given in Eq. (3.28). This small discrepancy may occur because of the experimental details of the pulsed electric field used in these experiments. Unlike published results for other elements, such as uranium and sodium [Paisner, 1976; Letokhov, 1977b], exact agreement with the classical expression has been observed in atomic rubidium, as shown in Fig. 3.9 [Jacquinot, 1977]. Field ionization has been used in such novel applications as (1) the detection of transitions among Rydberg states in sodium atoms induced by background 300-K blackbody radiation [Cooke, 1980]; (2) the observation, in strong magnetic fields, of Landau levels above the ionization limit [Castro, 1980]; and (3) has been suggested as a method of obtaining high-quantum-efficiency (0.1 to 1%), narrowband detection of microwaves using Rydberg states [Haroche, 1977]. For the blackbody experiments, transition rates out of level n^* are found to scale as ~6.79×10^4 T/n^{*4}, with T in K and n^* the effective principal quantum number. Accompanying ac Stark shifts for high-lying Rydberg states are independent of n^* but vary as T^2 (2.2 kHz at 300 K). For low-lying states, the ac Stark shifts vary as T^4 [Gallagher, 1981].

Pillet et al. [Pillet, 1984] have analyzed a set of microwave ionization experiments performed with sodium Rydberg atoms. Ionization occurs by the atom making a succession of Landau-Zener transitions to higher-lying levels. If the microwave frequency is properly chosen (15.4 GHz in their experiment), the critical microwave field amplitude for ionization is given by

$$
\epsilon_c \ (15.4 \ \text{GHz}) = \begin{cases} \dfrac{1.41 \times 10^9}{n^5} \ \text{V/cm} & ns \ (|m| = 0) \\[2em] \dfrac{1.53 \times 10^9}{n^5} \ \text{V/cm} & np \ (|m| = 0, \ 1) \\[2em] \dfrac{1.51 \times 10^9}{n^5} \ \text{V/cm} & nd \ (|m| = 0, \ 1) \\[2em] \dfrac{5.47 \times 10^8}{n^5} \ \text{V/cm} & nd \ (|m| = 2) \end{cases} \tag{3.32}
$$

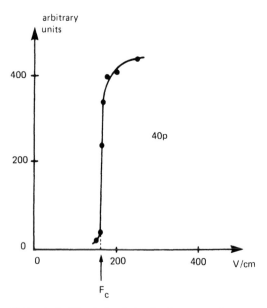

FIG. 3.9 Electric-field ionization of Rydberg states in rubidium:
(a) variation of ion current as a function of electric field for the laser-
excited 40p level; (b) variation of the critical field as a function of the
effective principal quantum number for s (▲), p (●), and d (■) states
[Jacquinot, 1977].

This is to be compared with classically derived expressions given by Pillet et al.:

$$\frac{1.71 \times 10^{9}}{n^{5}} \text{ V/cm} \tag{3.33}$$

$[1/(3n^{5})$ in atomic units] for $|m| = 0,1$ and

$$\frac{5.71 \times 10^{8}}{n^{4}} \text{ V/cm} \tag{3.34}$$

$[1/(9n^{4})$ in atomic units] for $|m| = 2$.

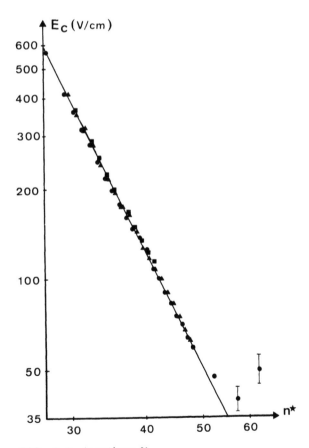

FIG. 3.9 (continued)

The origin of this behavior arises because the $|m| = 0, 1$ states of different principal quantum number n can mix in the electric field due to the same core effects that give rise to a nonzero quantum defect. Consequently, the "blue" Stark state of the n manifold at energy

$$E_n(\text{blue}) = -\frac{1}{2n^2} + \frac{3}{2n^2}\,\epsilon \qquad (3.35)$$

can make a transition to the "red" Stark state of the n + 1 manifold at energy

$$E_{n+1}(\text{red}) = -\frac{1}{2(n+1)^2} - \frac{3}{2(n+1)^2}\,\epsilon \qquad (3.36)$$

when

$$E_n(\text{blue}) = E_{n+1}(\text{red})$$

This occurs at field

$$\epsilon_c = \frac{1}{3n^5} \qquad (3.37)$$

Note the linear Stark effect that arises because of ℓ mixing in the electric field. Once the microwave field amplitude reverses its sign, the n + 1 red Stark state becomes the n + 1 blue Stark state. This state needs a smaller field than ϵ_c to make the transition to the n + 2 red Stark state. In this way the process continues until the atom ionizes. Since the $|m| = 2$ states have zero quantum defects, they ionize classically at

$$\epsilon_c = \frac{1}{9n^4} \qquad (3.38)$$

that is, when the red Stark state energy,

$$E_n(\text{red}) = -\frac{1}{2n^2} - \frac{3}{2n^2} \qquad (3.39)$$

exceeds the saddle-point energy,

$$V_s = -2\epsilon^{1/2} \qquad (3.40)$$

It is important to recognize that the $|m| = 0$, 1 states ionize at very low microwave fields (note the $1/n^5$ dependence), whereas the $|m| = 2$ states ionize at microwave fields somewhat higher than given by the dc field analog equation (3.28). A fuller description of the underlying physics is given in the article by van Linden van den Heuvell [van Linden van den Heuvell, 1984].

3.2.9C. Ionization via Collisions

Laser excitation followed by collisional ionization is another method exploited in many of the applications discussed in the next section. Ionization can occur via collisions with electrons, with buffer-gas atoms, as well as with ground or excited atoms of the same species. These areas of atomic and molecular collision physics are very diverse and are actively being investigated in many laboratories. Here we give a brief summary of the magnitudes of collisional mechanisms often encountered in resonance photoionization experiments.

3.2.9D. Excitation and Ionization by Electrons

Percival et al. [Percival, 1975; Lodge, 1976] have obtained an analytic semiempirical expression for the excitation of hydrogen-atom Rydberg states by monoenergetic electrons. These cross sections represent averages over all magnetic sublevels and the accuracy is claimed to be better than 15%, provided that the principal quantum numbers satisfy

$$n, n' > 5$$

and

$$E > \frac{4}{n^2}$$

where E is the energy of the incident electron in Rydbergs (13.6 eV). The cross sections are given by

$$\frac{E\sigma(n \rightarrow n')}{n^4 \pi a_0^2} = ADL + FGH \tag{3.41}$$

where

$$A = \frac{8}{3s}\left(\frac{n'}{ns}\right)^3\left(0.184 - \frac{0.04}{s^{2/3}}\right)\left(1.0 - \frac{0.2s}{nn'}\right)^{1+2s}$$

$$s = n' - n$$

$$D = \exp\left(\frac{-1}{nn'E^2}\right)$$

$$L = \ln\left[\frac{1 + 0.53E^2 n(n' - 2/n')}{1 + 0.4E}\right]$$

$$F = \left(1 - 0.3s\frac{D}{nn'}\right)^{1+2s}$$

$$G = 0.5\left(\frac{En^2}{n'}\right)^3$$

$$H = C_2(x_-, y) - C_2(x_+, y)$$

$$C_2(x,y) = \frac{x^2 \ln(1 + 2x/3)}{2y + 3x/2}$$

$$x_\pm = \frac{2}{\{En^2[((2 - n^2)/n'^2)^{1/2} \pm 1]\}}$$

$$y = \frac{1}{1 - D \ln(18s)/4s}$$

The cross section for deexcitation (n' → n) may be obtained from the detailed balance relation

$$E'n'^2 \sigma(n' \to n) = En^2 \sigma(n \to n') \tag{3.42}$$

The essential point to note is that these cross sections scale as n^4 and fall off rapidly as $|n - n'|$ increases.

Experiments in helium [Schiavone, 1977] yield excitation cross sections at an incident electron energy of 100 eV of

$$\sigma(1 \to n) = 9 \pm 5 \times 10^{-17} n^{-3} \text{ cm}^2 \tag{3.43}$$

In this experiment, angular momenta ($\Delta \ell$) changing collisions were shown to be very large ($\sim 10^6$ Å^2) and to scale geometrically (i.e., as n^4) for large n:

$$\sigma^e_{\Delta \ell}(E) \cong 5 \times 10^{-(15\pm0.3)} \text{ cm}^2 \ln[100\, E(eV)n^2]n^4[E(eV)]^{-1} \tag{3.44}$$

$\Delta \ell$ changing collisions of low ℓ-Rydberg excited atoms are important in the subsequent ionization process because the resultant high ℓ states do not suffer appreciable radiative decay.

Electron-impact-ionization cross sections for Rydberg states can be taken from a semiempirical formula derived by Percival [Percival, 1966]. These are total cross sections, integrated over all possible energies of the ejected electron, consistent with energy conservation. The formula is

$$\frac{\sigma_i^e(n,\varepsilon)}{n^4\pi a_0^2} = \frac{[(1.28/n)\ln(\varepsilon/n^2) + 6.67](\varepsilon - 1)}{\varepsilon^2 + 1.67\varepsilon + 3.57} \tag{3.45}$$

where $\varepsilon = E/U$ and U is the ionization energy of state n. These cross sections also scale as n^4. They become more and more sharply peaked about twice the threshold energy as n increases. Mansbach and Keck [Mansbach, 1969] treat the detailed collisional excitation and ionization cascade induced by thermal electrons. Their work is complementary to that of Percival and agrees in regions of overlap.

It is useful to make a comparison between electron-impact ionization and photoionization of ground-state atoms and molecules by conventional light sources. An important advantage of electron beams is their ability to be generated at very high energies. Thus they are beneficial when high-ionization-limit species are involved. The energy resolution of electron beams is quite poor, at best several milli-electron volts (several tens of wave numbers). This is done at a considerable sacrifice in current. Usually, spectroscopic electron-beam sources, such as the filaments on mass spectrometers, deliver only milliamperes of current at best. The number of electrons generated by a filament then is about the same as the photon flux from a conventional light source filtered through a monochromator. Using Eq. (3.45), which is only approximate for low-lying Rydberg states, the peak ionization cross section for ground-state hydrogen is 4.5×10^{-17} cm^2 and occurs at 1.6 times the threshold energy. Peak cross-section values of 10^{-17} to 10^{-16} cm^2 at several times the threshold energy are characteristic of most elements. Consequently, the electron-impact-ionization rates of ground-state atoms can be comparable to, or even larger than, the photoionization rates attainable using powerful incoherent light sources. The advantages of tunable coherent laser sources are evident.

3.2.9E. Ionizing Collisions with Neutrals

We briefly review several ionization processes of laser-populated high-lying Rydberg states that are induced by collisions with neutrals:

$$
\left.\begin{array}{l}
X^{**}(n\ell) + X \rightarrow X_2^+ + e^- \\[2em]
X^{**} + X^{*\prime} \rightarrow X_2^+ + e^-
\end{array}\right\} \quad \text{Associative ionization (AI)} \quad (3.46)
$$

$$
X^* + X^{*\prime} \rightarrow X^+ + X + e^- \quad \text{Penning ionization (PI)} \quad (3.47)
$$

$$
\left.\begin{array}{l}
X^{**}(n\ell) + A \rightarrow X^+ + A + e^- \\[2em]
X^{**}(n\ell) + A \rightarrow X^+ + A^-
\end{array}\right\} \quad \text{Collisional ionization (CI)} \quad (3.48)
$$

These ionization mechanisms are used in many of the applications discussed in the next section. For example, chemi- or associative ionization (AI) has been shown by Liu and Olson [Liu, 1978] to explain the observation of ion signals in the experimental series of Esherick et al. [Esherick, 1977]. Using a simple analysis, Liu and Olson show that the cross section for associative ionization depends linearly on the effective principal quantum number n^*:

$$
\sigma_{AI} = \pi P n^* \alpha_d^{1/2} a_0^{1/2} \tag{3.49}
$$

where the static polarizability α_d is in cm^3 and a_0 is the Bohr radius (5.29177×10^{-9} cm). The analysis assumes that the long-range portion of the ionic potential is determined by the point-charge-induced dipole interaction given in terms of the dipole polarizability of the neutral partner. P is the probability that the particles follow the diabatic potential into the continuum and may itself have an n and ℓ dependence. The static dipole polarizabilities have been tabulated by Teachout and Pack [Teachout, 1971]. Liu and Olson estimate an associative ionization cross section of 1×10^{-14} cm^2 for Ca** (4s20s) and 2×10^{-15} cm^2 for Ca** (4s20d) using an $\alpha_d = 170.4a_0^3$. More recently, Weiner [Weiner, 1985] has measured associative ionization cross sections and rate constants $k_{AI}(k_{AI} = \langle \sigma_{AI} v \rangle)$ for atomic sodium s, p, and d Rydberg states having $n^* = 5$ to 28. These rate constants increase rapidly with principal quantum numbers to a maximum (2×10^{-9} cm^3/s) at $n = 9$ to 11, after which they gradually decline. The corresponding associative ionization cross sections ($\sim 10^{-14}$ cm^2) are of the same order of magnitude as that deduced using Eq. (3.49). Weiner et al. have compared their experimental findings with more recent theories [Weiner, 1983]. In addition, an excited-state Na(3p)-excited-state Na(3p) associative ionization cross section of $1.5 \pm 0.6 \times 10^{-16}$ cm^2 has been measured by Bonanno et al. [Bonanno, 1983].

The cross section for Penning ionization for colliding Na(5s) and Na(3p) atoms has been measured to be 1.1×10^{-12} cm^2 [Carre, 1984]. Collisional ionization of a Rydberg atom by a ground-state atom of the same element should have a cross section similar to that given in Eq. (3.49). Associative ionization into the continuum is possible whenever the excited state is within $kT/2$ of the ionization limit, where k is Boltzmann's constant. The energy defect is compensated by the thermal energy of the colliding atoms. Cross sections as large as those observed for Penning ionization are certainly possible. Collisions of Rydberg atoms with molecules having large electron affinities have been studied by Hildebrandt et al. [Hildebrandt, 1978]. Ionization rate constants of $\sim 10^{-7}$ cm^3/s, corresponding to cross sections of $\sim 10^{-12}$ cm^2, are observed for SF_6 and some halogenated hydrocarbons. The ionization process can be understood in terms of a model in which the Rydberg electron behaves essentially as a free particle, the ionic core playing the role of a spectator. The variation of ionization rate with principal quantum number derives from the dependence of the electron attachment rate on the kinetic energy of the Rydberg electron.

Angular momentum ($\Delta \ell$) mixing collisions within high Rydberg states also play a role in collisions with buffer-gas atoms (He, Ne, and Ar) as first observed in atomic sodium by Gallagher et al. [Gallagher, 1975]. These cross sections rise sharply as n^4 at n = 5 to 10, reflecting the geometric size of the electron orbit, and then decline at higher n at decreasing density of the sodium atom wavefunction. This behavior is consistent with the short-range interaction between the Rydberg-atom electron and the colliding rare-gas atom. The peak $\Delta \ell$ cross sections range between 10^{-13} and 10^{-12} cm^2. In the experiments of Gallagher et al., the observed decay times τ_n of the Rydberg excited atoms exhibited the hydrogenic $n^{4.5}$ dependence. In fact, the results were in exact agreement with the predictions of the hydrogenic theory after averaging over the manifold of ℓ states for a given n Rydberg level:

$$\frac{1}{\tau_n} = 1.61 \times 10^{10} n^{-4.5} \text{ s}^{-1} \tag{3.50}$$

3.2.9F. Laser-Driven Ionization

Perhaps the most general mechanism discovered during the last 15 years is laser-driven ionization first reported in 1977 [McIlrath, 1977]. It was experimentally observed that a dense (> 10^{14} at./cm^3) atomic vapor, irradiated by a saturating pulse of resonance radiation, will ionize almost completely (>95%) within 1 µs. Laser-driven ionization has been observed in lithium, sodium, calcium, strontium, and barium. Essentially, the ionization process involves superelastic scattering and heating of seed electrons. These seed electrons are generated by a host of mechanisms: dimer photoionization, energy-pooling collisions, associative ionization, laser-induced inelastic collisions, and stimulated Raman scattering. The magnitude and interplay of these mechanisms are discussed in a review paper by Lucatorto and McIlrath [Lucatorto, 1980].

3.3 APPLICATIONS

3.3.1 Introduction

The previous sections introduced the fundamental concepts and techniques commonly used in resonant laser photoionization experiments
and applications found in the literature. In Secs. 3.3.2 to 3.3.7, examples are given of the major applications of resonant photoionization
in (1) atomic spectroscopy, (2) molecular spectroscopy, (3) mass spectrometry, (4) impurity and selective trace-element detection, (5) studies in chemical physics, and (6) isotope separation. Potential future
applications are described in Sec. 3.3.8.

3.3.2 Atomic Spectroscopy

An important application of resonance ionization spectroscopy is in the
identification and assignment of high-lying atomic states in heavy elements such as the actinides and lanthanides. These complex atomic
species have traditionally been studied using emission spectra photo-
graphed at high resolution. The emission spectra are generated using
a variety of sources. Electrodeless discharge lamps are commonly
used in the study of the neutral and singly ionized atomic species.
These lamps consist of some moderate-vapor-pressure compound (a
halide of some metal, for example) of the element to be studied, contained in a quartz tube with a few torr of noble gas. The encapsulat-
ed material is usually excited by a microwave source. The resulting
emission lines are dispersed by a large spectrometer and are recorded
photographically. After the resulting emission lines are cataloged, it
is usually possible to assign the lowest-lying states in the spectrum.
This is accomplished using coincidence methods supplemented by Zeeman spectra [Fred, 1967]. This standard method of assigning complex
atomic spectra does not work well for the higher-lying states of the
atom, however. The upper reaches of the spectra are dense (in some
cases over one state per wave number). Furthermore, only a very
minute fraction of the atoms in such discharges is in these highly excited states. These states decay to a very large number of lower-lying atomic levels, so that very few photons per allowed transition are
ever emitted. In the most complex atoms, the highest-lying states can
only be assigned by resonance excitation methods.

The early work of Carlson et al. [Carlson, 1976] used three lasers
to stepwise excite and ionize atomic uranium. The resonant multistep
laser-photoionization spectrometer used in this experimental series is
shown schematically in Fig. 3.10. The first laser was tuned to a
known transition near 6000 Å, exciting the atom from the ground state.
The second laser was tuned in wavelength so that it would further excite atoms, already excited by the first laser, to higher-lying levels,
from which they could subsequently be ionized by the third laser.

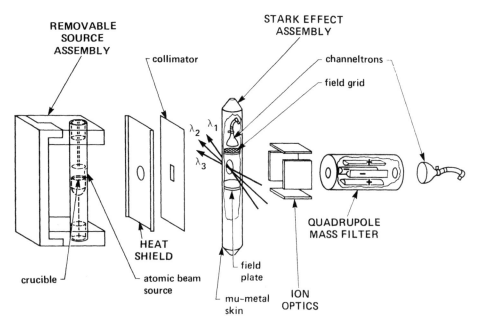

FIG. 3.10 Schematic of typical laboratory apparatus used in resonance photoionization experiments. The atomic vapor of interest is generated using a resistively heated tungsten crucible.

The third laser beam, which arrived after the first and second lasers had excited the vapor, was used to ionize atoms. Ions were detected as a function of wavelength of the second laser. In this way, levels of uranium near 4 eV of excitation were identified. These states were assigned a J value by sequentially accessing the level from a number of low-lying 2-eV states of different J and using $\Delta J = 0, \pm 1$ selection rules. The assignments were verified by examining the hyperfine spectra under high-resolution methods [Hackel, 1979] and by using polarization methods. Any number of lasers of any wavelength, including very long wavelength 10.6-μm CO_2 light, can be combined to map out any region of the spectrum. In this way the Rydberg states of many heavy atoms were observed for the first time [Solarz, 1976; Worden, 1978a].

An example of an autoionizing Rydberg spectrum of atomic iron taken with this apparatus is shown in Fig. 3.11. The observation of Rydberg progressions has led to the accurate determination of ionization potentials for a host of atomic systems [Worden, 1978a, 1979, 1984], which in turn has led to the discovery of systematic variation in the limit values [Rajnak, 1978]. Two- and three-step photoionization pathways have been used to study channel interactions between

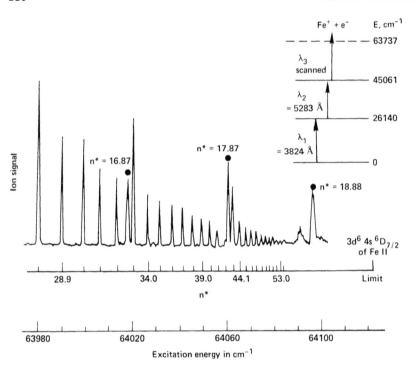

FIG. 3.11 Autoionizing Rydberg series of iron from the 45,061.327-cm^{-1} $3d^6 4s5s \, ^5D_3$ level that converges to the 384.77-cm^{-1} $3d^6 4sD_{7/2}$ level of the ion. The excitation sequence is shown in the figure. The dots and effective quantum numbers 16.87, 17.87, and 18.88 identify members of a series converging to the 667.64-cm^{-1} level of the ion [Worden, 1984].

Rydberg levels above and below the first ionization limit in the alkaline earth atoms [Esherick, 1977]. The elegant multichannel quantum defect theory (MQDT) and its associated graphical analysis [Lu, 1970] have been very successful in predicting and calculating the energies and wavefunctions of bound and autoionizing states [Armstrong, 1979]. MQDT has also been used to study the effects of increasing nuclear charge on autoionizing Rydberg resonances along the isoelectronic series Xe, Cs^+, and Ba^{2+} [Hill, 1982]. The Ba^{2+} spectra were obtained by employing two-stage laser-driven ionization of Ba to create a homogeneous column density of Ba^{2+} ions suitable for photoabsorption measurements. The analysis of the spectra illustrates the relationship between electron-electron correlations, term dependences, and autoionization widths in complex spectra.

Resonant multistep photoionization has allowed identification of narrow autoionizing levels accessible only from excited states in the noble gases [Rundel, 1975], uranium [Solarz, 1975; Coste, 1982], the lanthanides [Worden, 1978a, 1980], gadolinium [Bekov, 1978b], the alkaline earths [Safinya, 1979; Kachru, 1984], ytterbium [Bekov, 1980], neptunium [Worden, 1979], lutetium [Miller, 1982], and iron [Worden, 1984]. An autoionizing transition of width 0.07 cm^{-1} and 8×10^{-16} cm^2 cross section has been observed in gadolinium using a three-step photoionization process [Bekov, 1978b]. This contrasts with the very broad (hundreds of cm^{-1}), very weak (typically, 10^{-18} cm^2) autoionizing transitions accessible from atomic ground states [Garton, 1968, 1974]. Strong and narrow autoionizing transitions from excited states appear to be characteristic of the heavy atoms with several optically active electrons [Bekov, 1980]. The position and strength of autoionizing resonances are important for resonance ionization mass spectrometry, laser isotope separation, and efficient production of photoion beams, as discussed later in this chapter. Stark-field broadening of autoionizing levels, which is important for these applications, has been reported in only a few of these elements, such as gadolinium [Mishin, 1985] and barium [Safinya, 1980].

Autoionizing levels are also important in two-photon resonantly enhanced four-wave mixing (see three-photon sum generation in Chapter 2) [Hanna, 1979]. The presence of the autoionizing level further enhances the conversion efficiency of the up-converted light [Sorokin, 1975]. Four-wave mixing has allowed generation of continuously tunable narrowband radiation in ranges to wavelengths as short as 1174 Å [Tomkins, 1982]. The phase-matching requirements in four-wave mixing have been suggested and demonstrated [Wynne, 1981] as a method of accurately determining transition oscillator strengths. Before the advent of laser techniques, tabulated values for transition oscillator strengths (or cross sections) were 10% at best and usually significantly worse [Corliss, 1962]. Huber and Sandeman have recently reviewed transition probabilities and their accuracy [Huber, 1980].

Techniques for measuring cross sections both for bound-bound transitions and bound-free transitions have been developed using resonance photoionization spectroscopy. A very simple method in which time-resolved photoionization is used to obtain absolute values of oscillator strengths accurate to 5% has been reported by Carlson et al. [Carlson, 1977]. The principle of the method is that determination of the lifetime of a state and its fractional probability of decay back to the relevant lower-lying state (i.e., the emission branching ratio) is sufficient to calculate the transition probability or oscillator strength for the resonance of interest. A beam of atoms is stepwise excited using the general-purpose apparatus of Fig. 3.10. The first laser excites the atoms to a state near 2 to 3 eV of excitation, from which it can radiatively decay. The lifetime can be measured by delaying the second and/or third laser in time from the first laser and recording the ion signal as a function of delay time. This method was first described by Janes et al. [Janes, 1975]. The branching ratio can be ob-

tained either from conventional emission spectra or from a time-resolved photoionization scheme. The latter method works as follows.
First, a set of lasers resonantly excites the lower level of interest.
This is called the population sequence. Then a set of probe lasers
resonantly photoionizes the populated level via a strong autoionizing
transition. A saturating storage laser pulse is used to take atoms from
the lower excited level of interest to the higher-lying state of interest.
This state is not ionized by the probe-laser sequence since its cross
section for ionization at these wavelengths is small by design. We now
delay the ionizing probe lasers with respect to the storage laser and
plot the ion signal as a function of this delay time. The behavior
shown in Fig. 3.12a is observed. Initially, the signal height is reduced to about half of its maximum value since the storage pulse places
approximately half of the atoms in the more highly excited state. The
ion signal begins to recover as the delay time is increased since a certain fraction of the atoms in these highly excited states decays back to
the initial level. If all the atoms decayed directly to the initial state
(branching ratio of unity), we would recover the initial ion signal entirely. Clearly, the ratio of $K[S(t) - S/(0)]/[S(-) - S(0)]$ in Fig.
3.12a in the limit of many lifetimes $(t > 5\tau)$ of the decaying atoms is
simply the branching ratio. Here K accounts for level degeneracy and
light polarizations. It can be calculated using angular-momentum algebra and is ~1 for transitions involving high J states. The Einstein
transition probability for decay is then simply

$$A_{\ell u} = \frac{\beta_{\ell u}}{\tau_u} \tag{3.51}$$

where $\beta_{\ell u}$ is the branching ratio and τ_u is the excited-state lifetime.
Some data are shown in Fig. 3.12b. The analysis relies on spontane-
ous radiative emission being the only relevant decay process and is
not sensitive to the details of the excitation itself. The method has
the virtue of being applicable to Doppler-broadened transitions because each atom gives rise to the same relative signal independent of
its velocity.

The difficult problem of bound-free cross-section measurements is
also easily handled using resonance photoionization spectroscopy.
Due to the very high intensity available from the light source, saturation methods can be used. These methods remove the difficult task of
measuring experimental path lengths and atomic number densities.
When the ionizing laser pulse length is shorter than the radiative decay time of the lower level of the transition and the line width of the
laser is narrow compared to the homogeneous line width of the autoionizing feature, the ionization cross section is given by $h\omega$/saturation
fluence, as discussed earlier. The experiment consists of generating
a laser beam of smooth and constant fluence over some well-defined

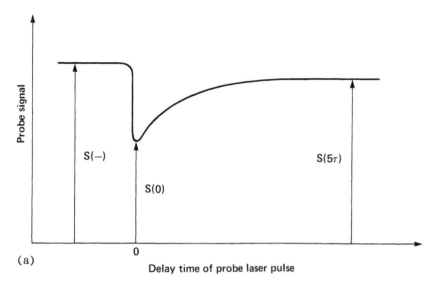

(a)

Probe signal

S(−)

S(0)

S(5τ)

0

Delay time of probe laser pulse

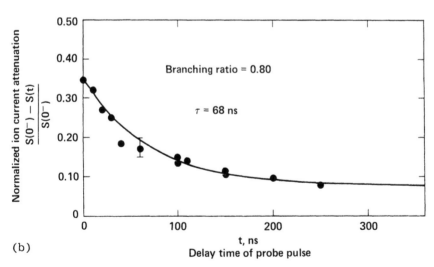

(b)

Normalized ion current attenuation $\frac{S(0^-) - S(t)}{S(0^-)}$

Branching ratio = 0.80

τ = 68 ns

t, ns
Delay time of probe pulse

FIG. 3.12 Cross-section measurement using lifetime-branching ratio technique: (a) theoretical probe signal versus time delay between resonant pulse and probe pulse; (b) lifetime-branching ratio data for the 0- to 22,919-cm^{-1} (4362.1 Å) transition in atomic uranium [Carlson, 1977].

area, generally by clipping the initial beam with an aperture and measuring relative photoion yield versus photon fluence. In Fig. 3.13 we
show data from the saturation measurement of the large autoionization
cross section in gadolinium referred to previously [Bekov, 1978b].
Cross-section measurements accurate to perhaps 25% can be obtained
in this manner. Excited atomic-state photoionization cross sections
have been reported in rubidium [Ambartzumian, 1976], in magnesium
[Bradley, 1976], and in gadolinium [Bekov, 1978b] using this technique.

Finally, owing to the sensitivity of the technique, multistep resonance photoionization has been used to measure atomic isotope shifts
of short-lived (<10 s) radioactive europium isotopes ($^{141-144}$Eu)
[Fedoseyev, 1984]. These on-line experiments performed at Gatchina
in the USSR, combined with off-line measurements of longer-lived
europium isotopes [Alkhazov, 1983], allowed investigation of nuclear
shell deformation effects about the magic neutron number of 82.

3.3.3 Molecular Spectroscopy

Resonant photoionization spectroscopy is useful in atomic systems for
several reasons. It is a sensitive tool that can be used to study highly
excited states that are not accessible using many standard methods.
It can be augmented with time-resolution and polarization methods that
can be used to greatly simplify complex atomic spectra. The same advantages are of great value in molecular systems. Resonant photoionization has recently become widely used. Two recent reviews of the
field are those of Johnson and Otis [Johnson, 1981] and that of Parker
[Parker, 1983].

The major advantage of resonant photoionization for molecular systems is that it permits the study of states not normally observable using absorption or fluorescence spectroscopy. This unique capability
derives from two effects. First, resonance photoionization spectroscopy of molecular systems generally involves the excitation of some excited molecular state by the absorption of two or more photons from
the ground state. Once the excited state is reached, additional photons from the same laser or from yet another laser system can be used
to ionize the molecule. Clearly, different selection rules are operative
than in the case of single-photon studies. Even in molecules that lack
a center of symmetry, states not observed in single-photon experiments can be studied. The resonance ionization spectroscopy of ammonia is a good example [Glownia, 1980].

The second effect of importance derives from the time evolution of
the laser-excited molecule. Once the excited state is populated, the
ionization step from this state must compete with other depopulating
mechanisms, such as radiative and nonradiative decay. Most valence
states undergo some form of rapid energy transformation either by the

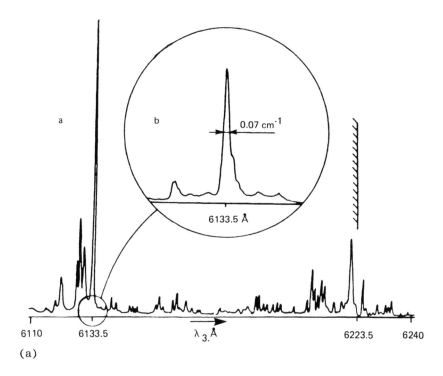

(a)

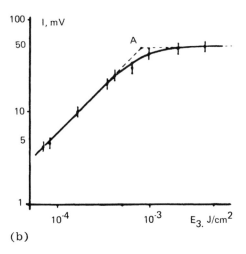

(b)

FIG. 3.13 Autoionization spectra from an excited state in atomic gadolinium: (a) ion signal versus third-step wavelength; (b) dependence of ion signal on energy fluence at maximum of autoionization resonance, 6133.5 Å [Bekov, 1978b].

release of radiation or by other processes, such as electronic-to-vibrational energy conversion. Rydberg states of molecules and atoms are less susceptible to these energy-conversion channels since their high-principal-quantum-number electrons are largely removed from strong interactions with the nucleus and remaining electrons. The persistence of energy in such states explains the ease with which Rydberg states are subsequently ionized and observed in photoionization spectra. By appropriate experimental design exploiting new selection rules and intramolecular kinetics, one can assign new molecular states or verify previous assignments from conventional studies.

Earlier in this chapter it was seen that a focused laser beam can produce a photon flux density of about 10^{29} photons/cm^2-s and that for typical molecular nonresonant processes, three-photon nonresonant processes were nearly as likely as one- and two-photon processes. As a result, tunable radiation between 3 and 4 eV permits access of states in a nonresonant mode up to about 9- to 12-eV excitation. The ionization process usually must be accomplished via a single-photon mechanism in order to compete with other energy-loss modes. Nonetheless, molecules with ionization potentials up to 12 eV constitute a very substantial fraction of those of spectroscopic interest [see Table 3.2(b)].

A recent study in the field of molecular photoionization spectroscopy is the work on ammonia by Glownia et al. [Glownia, 1980]. In this study, extensive analysis resulted in the reassignment of three states and the identification of seven new states. Only five band systems had been assigned previously. Figure 3.14 is reproduced from this work. It compares a portion of the multiphoton ionization spectrum obtained with a molecular beam with that observed using conventional vacuum ultraviolet absorption techniques. This work provides an excellent pedagogical example of the assignment of states from multiphoton ionization spectra.

Extensive and successful application has also been made of polarization effects in multiphoton ionization spectra. It is useful to compare the intensities of multiphoton ionization spectra in which the excitation laser is first circularly and then linearly polarized. The ratio of the spectral intensities can be an important clue to the assignment of the rovibronic structure [McLain, 1971]. Consider the case of n-photon excitation to a bound molecular Rydberg state in which the nonradiative relaxation from the state is slower than the ionization step. It is obvious that the ratio of the intensities is simply the ratio of the bound-state two-photon excitation probabilities for the two polarizations, if the ionization step is saturated. This last step has no effect since all molecules that reach the bound state are ionized. If the ionization step is not saturated, the ratio above must also be multiplied by the ratio for the cross sections for ionization for the two different polarizations. Examples of this type of analysis used in the labeling of spectra are given in the ammonia studies [Glownia, 1980] and by Parker [Parker, 1983].

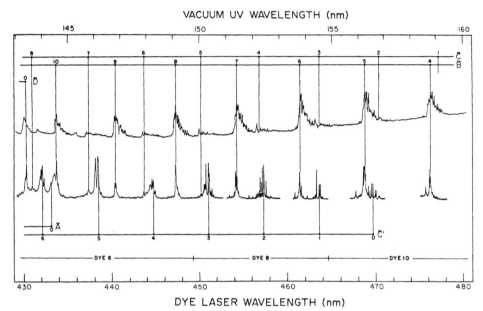

FIG. 3.14 Comparison of the single-photon vacuum ultraviolet (upper trace) and multiphoton ionization (lower trace) of ammonia. States not seen in the single-photon absorption spectrum can be identified using multiphoton ionization. The appearance of new states is due to selection rules that govern the multiphoton excitation and ionization of the molecular species [Glownia, 1980].

Evidence of the growing importance of this field of spectroscopy is given in the recent analysis of carbon monoxide by Loge et al. [Loge, 1983] and in the analysis of the methyl radical spectrum by Hudgens et al. [Hudgens, 1983]. Resonantly enhanced multiphoton ionization has been used to measure accurately and sensitively nitric oxide densities with good spatial and temporal resolution [Cool, 1984]. A few studies of multiphoton ionization in the liquid phase have been reported [Vaida, 1978; Braun, 1981]. Unfortunately, the technique has been, and is expected to be, less useful as a spectroscopic tool because of the absence of Rydberg spectra in the condensed phase. Multiphoton ionization is more useful as an impurity detector in such environments.

3.3.4 Multiphoton Ionization Mass Spectrometry

A very appealing concept is the use of laser photoionization in tandem with mass spectroscopic analysis of fragments as a universal two-dimensional tool for detecting trace amounts of complex molecules. By combining the state and isotopic specificity of the highly sensitive resonance photoionization technique with mass spectrometry, industrially important molecules could be fingerprinted. A number of fine reviews of this concept have appeared. We will draw upon a subset of these works [Antonov, 1981; Robin, 1980; Lubman, 1982; Bernstein, 1982] in this section.

It may be expected that the fragmentation patterns resulting from multiphoton-laser-induced dissociation and ionization of large molecules would vary with the power and possibly the wavelength of the exciting laser. In particular, Bernstein has sought to classify types of multiphoton-induced fragmentation behavior and to systematize the resulting observations by considering the energetics of the sequential physical steps that make up the process. These steps can be broken down as follows:

1. Excitation $\qquad$ $M + n - \text{photons} = M^*$

2. Ionization $\qquad$ $M^* + q - \text{photons} = M^+ + e^-$ $\qquad$ (3.52)

3. Photodissociation $\quad M^+ + r - \text{photons} = F^+ + R$

Here we recognize that the excited molecule can absorb photons to ionize. The ion itself or its fragments can further absorb photons, resulting in additional fragmentation. Indeed, it is the potential for further absorption of photons by the initial fragments that complicates the information that can be obtained in these experiments. This behavior is often observed. Figure 3.15, which is taken from Bernstein, illustrates diagrammatically the scheme used to systematize these experiments. Case (c) is an interesting example. Here n is 2 and the excited state can now absorb q + r of 2 photons. The last photon brings the molecule to an energy state that is sufficient not only to form the ion M^+ but also to generate the F^+ fragment. The branching ratio for the formation of M^+/F^+ is expected to depend on the frequency of the dissociating laser beam. The shorter the wavelength, the greater the excess energy of M^{**} and the greater is its tendency to form F^+ instead of M^+.

The fragmentation or dissociation steps themselves can also be examined. Case (e) diagrams a situation in which the parent ion M^+ can either absorb one photon further to form the fragment ion F^+ or internally relax quickly so that two more photons from the laser field are required to form F^+. Here the daughter-fragment-to-parent-ion ratio (d/p) can be strongly dependent on laser power. The opposite behavior is obtained in case (f). In this case a single photon in either the presence or absence of fast radiationless decay gives rise to the fragment ion. At the high laser powers that are often used in reso-

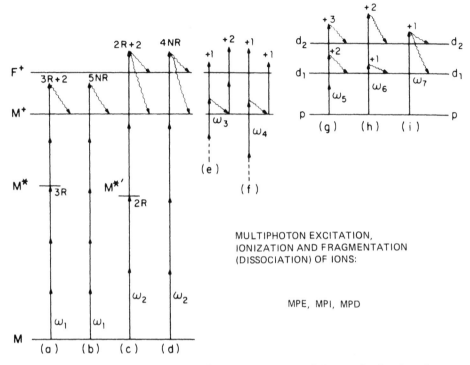

FIG. 3.15 Scheme to systematize observed multiphoton ionization-fragmentation behavior [Bernstein, 1982].

nantly enhanced multiphoton ionization (REMPI) mass spectroscopy, case (e) has a somewhat power-independent value for d/p in the limit that the two-photon ionization rate of the parent ion is saturated. Under those conditions there would be very little parent ion around to observe and d/p would be large.

Depending on the wavelength selected and the photodissociation dynamics of the parent and fragment ions, significantly different kinds of behavior can be achieved even within the same molecule. This is particularly important since, in general, multiphoton as opposed to single-photon fragmentation patterns are obtained at the laser powers generally employed in REMPI mass spectrometry.

Attention must also be paid to the polarization of the laser used [Robin, 1980], since as we saw in the preceding section, this can have a significant effect on the multiphoton excitation pathway. Nevertheless, given sufficient attention to laser intensity, wavelength, and polarization, useful and reproducible results can be obtained. Relevant examples are the fragmentation patterns obtained for the isomers

n-propyl and isopropyl iodide [Parker, 1982]. The results are reproduced in Fig. 3.16. It is clear that the two isomers are easy to distinguish. The different fragmentation patterns indicate that the parent n-propyl ions do not isomerize within the 5-ns laser pulse used in
the experiment. Instead, they absorb additional 368-nm radiation,
while the isopropyl cations are transparent at this wavelength and
survive further destruction. Although the process can be quite complicated, there does appear to be a place for REMPI mass spectrometry
as a supplement to fragmentation patterns obtained with 70-eV electrons (the current mass spectrometer standard).

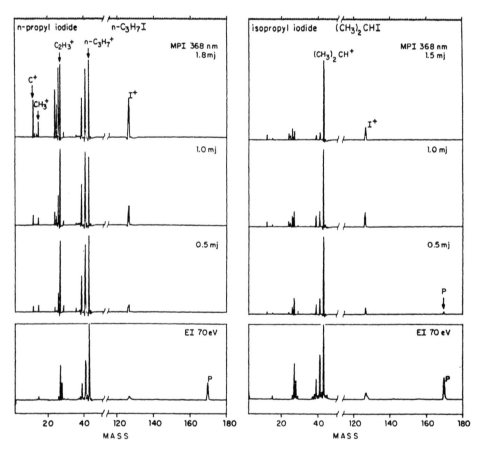

FIG. 3.16 Comparison of multiphoton ionization mass spectroscopy
fragmentation patterns for the isomers *n*-propyl iodide and isopropyl
iodide. The bottom panels show electron impact mass spectra [Parker,
1982].

Recent work using photoionization mass spectrometry by Rohlfing et al. [Rohlfing, 1983, 1984a, 1984b] on nickel and iron clusters produced by laser vaporization has allowed the measurement of the ionization potentials of Ni_x and Fe_x (x = 2 to 25) clusters. Tunable lasers were used to measure the appearance potential of each parent ion. Unexpected oscillations in the curve of ionization potential versus cluster size are observed. Some of these results are reproduced in Fig. 3.17. The authors conclude that this highlights the unique electronic and chemical, or catalytic, nature of metal clusters. Additional insight into the spectral dynamics of species important to a host of areas in physics and chemistry can be expected to come from experiments in this field.

3.3.5 Highly Selective Detection of Trace Elements

The clear advantage of highly selective photoionization of atoms leads naturally to the concept of the selective detection of trace elements and even single atoms [Hurst, 1977, 1979; Bekov, 1978a]. The ability to detect ultralow fractions of long-lived radioisotopes, such as ^{10}Be, ^{26}Al, and ^{36}Cl, with natural abundances of 1 part in 10^{10}, 10^{14}, and 10^{17}, respectively, may have applications in nuclear physics and geophysics [Kudriavtsev, 1982]. The long lifetimes of these isotopes, of order 10^6 years, makes them difficult to detect by usual nuclear radiation counting methods. Resonance ionization spectroscopy could greatly aid in their detection.

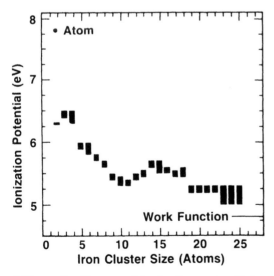

FIG. 3.17 Measured ionization potentials of iron clusters versus cluster size [Rohlfing, 1984b].

Testing of the fundamental models of solar fusion processes by the detection of solar neutrinos using resonance photoionization of the neutrino-formed atoms or their decay products has been proposed by Bahcall [Bahcall, 1978]. Further improvement using resonance ionization spectroscopy (RIS) has been suggested by Hurst et al. [Hurst, 1979]. The experiment proposed by Bahcall [Bahcall, 1969] and improved by Hurst relies on the reaction $\nu + {}^7Li \rightarrow e^- + {}^7Be$. The ${}^7Be$ has a 53-day half-life and decays by electron capture to produce a 53-eV ${}^7Li$ recoil nucleus and an Auger electron. These nuclei create a small amount of ionization as they are brought to rest in a position-sensitive proportional counter. The ionization can be used to trigger a set of lasers to photoionize the single ${}^7Li$ atom, which is coincidence and position detected along with the Auger electron. Only about 100 ${}^7Be$ atoms are produced, so that efficient single-atom detection methods are required. The 1978 proposal of Bahcall exploits the reaction $\nu + {}^{71}Ga \rightarrow e^- + {}^{71}Ge$. Kudriavtsev and Letokhov [Kudriavtsev, 1982] have proposed a similar experiment for solar neutrino detection that relies on the detection of ${}^{41}Ca$ to 1 part in 10^{21}. Proposals have also been made to use resonance ionization spectroscopy for detecting exotic atoms such as fission isomers and quark atoms [Hurst, 1979].

Resonance photoionization of atoms has also been proposed as a method of detecting Al impurities in Ge (with a detection limit of $10^{-9}\%$ demonstrated) and as a method for the detection of trace elements such as Al in water or in human blood [Bekov, 1983]. Resonantly enhanced three-photon ionization has been employed to detect ground-state atomic hydrogen and deuterium. Concentrations as low as 4×10^9 at./cm^3 in the presence of 10^{17} at./cm^3 buffer-gas atoms have been measured with a dynamic range of 10^4 and a time resolution of 10 ns [Bjorklund, 1978].

Laser photoionization has also been suggested as a method for the monitoring of atmospheric pollutants [Brophy, 1979]. An experimental detection of 1 part in 10^5 of aniline (C_6H_7N) in air was demonstrated and an ultimate detection limit of 1 part in 10^{11} was projected. Besides its high sensitivity, the method is subject to very little background absorption. The presence of background absorption limits the sensitivity of detection of atmospheric pollutants using standard techniques.

The ability of high-repetition-rate intense laser beams to produce intense ion beams of pure substances has also been demonstrated [Zherikhin, 1983]. A 2.5-μA beam of gallium ions was produced and indirectly demonstrated to be pure to about 1 part in 10^{10}. The method has the limitation, however, that the photoions will quickly charge-exchange with other species present in the beam having similar or smaller ionization potentials. Nevertheless, this method eliminates interaction with vessel walls to produce beam contaminants, which is a major drawback of discharge-produced beams. Caution must still be applied in interpreting experimental results before ultrahigh-purity beams can be unambiguously claimed. The work of Zherikhin et al., in fact, demonstrated that the ion content of the beam was photoion related but could not establish that it was solely gallium ions. Intense high-purity photoion beams have been suggested as useful for a range of materials technologies [Letokhov, 1977b], particularly for the formation of thin films of complicated stoichiometry and for ion implantation in semiconductors of elements such as phosphorus, arsenic, antimony, and boron. Letokhov et al. [Letokhov, 1978] have also devised a simple method for obtaining a high-intensity beam of highly polarized protons via multistep resonant photoionization of metastable atomic hydrogen.

Resonance ionization (RI) of trace or low-abundance isotopes has also been combined with mass spectroscopic (MS) techniques by several groups. The resulting RIMS method has demonstrated that fewer than 1000 atoms of ^{81}Kr can be counted directly without waiting for radioactive decay (2×10^5-year half-life) [Chen, 1984]. This is important for groundwater polar-ice-cap dating as well as for neutrino research. The RIMS technique has been applied to the precision measurement of isotopic ratios, particularly in the presence of isobaric interferences [Nogar, 1984]. Using continuous-wave dye-laser excitation to improve the instrumental duty cycle and signal-to-noise ratio, typical uncertainties of 10% at a ^{174}Lu/^{175}Lu ratio of 3.6×10^{-5} in the presence of Yb isobaric interferences have been demonstrated [Miller, 1983].

RIMS of atomic technetium has also been demonstrated by the same group [Downey, 1984b]. Technetium is a rare element with no stable isotopes. It is useful as a geologic tracer. The isotope ratios of $^{97\text{-}99}$Tc in deep underground molybdenite ore have been suggested as a measure of the integrated effects on the earth of solar neutrinos. Isobaric interferences from $^{94\text{-}100}$Mo and $^{96\text{-}102}$Ru are expected to mask the Tc isotopic ratios in the absence of the RIMS technique.

RIMS has also been actively studied by a group at the National Bureau of Standards in Gaithersburg, Maryland. They have demonstrated RIMS on carbon with detection limits of 10^7 at./cm^3 [Moore, 1985; Clark, 1983] and beryllium [Clark, 1985] and offer a novel opposed-beam two-photon approach to attain high selectivity (1 part in 10^{14}) for determining the isotope ratio ^{88}Sr/^{90}Sr [Lucatorto, 1984]. Of course, carbon is useful in radioactive dating and strontium isotopic abundances are relevant to environmental monitoring of nuclear-era contamination. RIMS has recently been demonstrated on atomic uranium using intracavity laser ionization [Downey, 1984a].

Two-step resonant photoionization has been suggested as a method to separate isomeric nuclei [Letokhov, 1973]. This has been realized in a three-step photoionization experiment on Eu [Fedoseyev, 1984]. Separation of radioactive waste material based on isomeric shifts in their atomic spectra has been proposed [Letokhov, 1977b].

Finally, we have already mentioned several methods for improving
the selectivity of the resonance photoionization process in the case of
small isotope shifts: by using a two-photon opposed-beam technique,
by polarization selection rules, and by nozzled expansion. It has also
been proposed that the atoms be collinearly excited in an atomic beam
that has accelerated the nonselectively formed ions in a potential field
[Kudriavtsev, 1982]. These ions will have differing velocities; they
can then be neutralized to atoms and selectively rephotoionized. We
note that infrared excitation can be used in one of the resonant steps
in the excitation chain. The reduced Doppler width of the infrared
transition will greatly enhance the selectivity of that step of the photoionization ladder.

3.3.6 Chemical Physics

The excellent sensitivity of photoionization spectroscopy has been
proposed in a variety of chemical physics studies. Laser photoionization has been used to improve the efficiency of detection of atoms generated by sputtering from ion-bombarded surfaces [Winograd, 1982].
An improvement over other postionization methods of several orders of
magnitude was easily established. Photoionization detection of stimulated Raman scattering in a sodium atomic beam has also been demonstrated [Beterov, 1983]. Using 10^5 to 10^6 atoms of sodium in the
laser interaction region, the authors were able to detect the presence
of stimulated Raman scattering as a dip in the photoionization signal.

Resonant photoionization has been used to investigate the chemistry of highly excited atoms [Solarz, 1980], such as association ionization of high Rydberg states in the alkaline earths and the binding energy of their dimer ions [Worden, 1978b; Hermann, 1980]. Data from
the former work are reproduced in Fig. 3.18 and illustrate the importance of collisional processes. Optically pumped amplified spontaneous
emission and stimulated Raman scattering were observed in Ca^+ and
Ba^+ in the latter work, which was performed using a high-density
space-charge-limited thermionic diode detector.

In addition, resonant photoionization has opened up to experimental study a whole new class of chemical reactions that can be induced
by the presence of intense tunable-wavelength exciting lasers. Laser-
induced ionization and associative ionization have been studied in the
alkali and mixed alkali-alkaline earth systems [Keller, 1984a, 1984b].
The cross sections for the laser-induced step in these reactions can
be reasonably strong. For example, in the Na-Ba system, the peak
cross section is calculated as 6.51×10^{-20} cm^2 I(W/cm^2). Weiner has
suggested laser-induced harpoon reactions [Weiner, 1980]. For the
reaction $Hg + Cl_2 + \hbar\omega \rightarrow HgCl + Cl$, he has estimated a cross section
of 5.03×10^{-26} cm^2 I(W/cm^2) at 193 nm. A laser intensity of 10^9 W/
cm^2 at 193 nm is necessary to obtain a gas-kinetic cross section. An
intensity of 1 GW/cm^2 at 193 nm is certainly attainable using available
ArF laser systems.

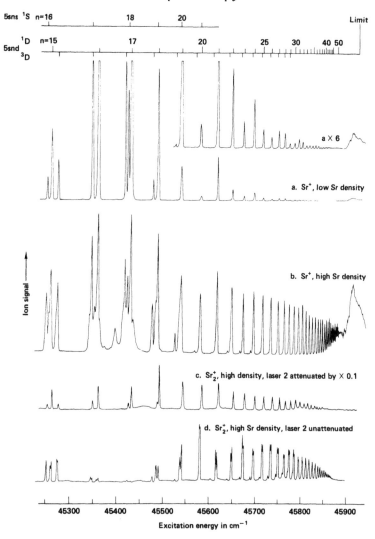

FIG. 3.18 Ion signals at mass 88 [Sr^+, scans (a), (b)] and at mass 176 [Sr_2^+, scans (c), (d)] versus excitation energy at various Sr densities. All scans are for two-step excitation with laser 1 at 4607 Å and laser 2 scanned from 4244 to 4125 Å. Sr ion signals in scan (a) are produced by photoionization of the laser-populated levels by laser 2 radiation, while those in scan (b) are produced mainly by collisional (atom and electron) ionization of the Rydberg levels. The Sr_2^+ signals in (c) and (d) result from associative ionization reactions. The reversal effect (dip at line center) in (d) from n = 15 to 35 results from electron ionization of these Rydberg levels before associative ionization collisions occur [Worden, 1978b].

An entirely unrelated process, which illustrates the wide-ranging
importance of photoionization phenomena, relates to its potential use in
the development of optical storage of information [Pokrowsky, 1983;
McFarlane, 1979]. Color centers in $LiF:Mg^{2+}$ can be photoionized by
GaAlAs diode lasers at liquid-helium temperatures. Only a small fraction of the inhomogeneously broadened zero-phonon feature is excited
at any time by the narrow-line-width diode laser. Typical ratios of
inhomogeneous to homogeneous line widths are of the order of 10^3.
Thus up to 10^3 resolvable data points can be encoded within each spatial location of the sample. Spatial storage densities of 10^{11} bits/cm^2
are estimated to be feasible.

3.3.7 Laser Isotope Separation

Isotope separation through n-step selective photoionization of atoms
has been actively and extensively pursued. Early recognition of the
potential of atomic-vapor laser isotope separation is generally attributed to Levy and Janes [Levy, 1973] and Ambartzumian and Letokhov
[Ambartzumian, 1972] following their experimental work on the nonselective photoionization of rubidium [Ambartzumian, 1971]. Laser
isotope separation of calcium [Brinkmann, 1974] and several lanthanides (Nd, Sm, Eu, Gd, Dy, and Er) have been demonstrated [Karlov,
1976, 1977, 1978]. Numerous papers have been published on the enrichment of ^{235}U by multistep laser ionization of the atomic vapor
[Snavely, 1975; Janes, 1976; Davis, 1977; Bohm, 1978].

For the very light atoms, the resonance-line isotope shifts are
generally governed by mass shift effects, while for the heavy elements the resonance-line isotope shifts are governed by nuclear volume effects [Kopfermann, 1958; Golovin, 1968]. The isotope shifts on
each successive excitation step can be much greater than the resonance-line absorption width due to the combined inhomogeneous effects of Doppler broadening and hyperfine structure. In atomic
uranium, for example, isotope shifts exceeding several gigahertz are
observed between transitions in ^{235}U and ^{238}U. This is typical of the
heavy elements. A possible situation for the heavy elements is for
the bound-bound peak Doppler cross sections to be $\sim 10^{-14}$ cm^2 and for
the final-step bound-free cross section to be $\sim 10^{-15}$ cm^2. The laser
fluence required to saturate each transition will then be between 0.1
and 1 mJ/cm^2 as long as the excitation time is short compared with the
radiative lifetime of the states. The radiative lifetimes of the bound
states are typically several hundred nanoseconds. A modest laser
fluence can then efficiently ionize the isotope to which it is resonantly
tuned.

Power broadening, or alternately the Rabi frequency, of the
transition can be significantly less than the isotope shift under these
conditions. In general, the selectivity for each step of the excitation

sequence has maximum selectivity (i.e., resonance transitions in desired isotopes unsaturated) of

$$S = \left(\frac{2 \Delta \nu_{IS}}{\Delta \nu_L} \right)^2 \tag{3.53}$$

where $\Delta \nu_L$ is the homogeneous natural line width (full width at half
maximum) and $\Delta \nu_{IS}$ is the isotope shift [see Eq. (3.15) and Table
3.3]. Selectivity can exceed 10^4 for each bound-bound transition.

Detailed calculation of a three-step scheme for selective photoionization of atomic ytterbium has recently been published [Karlov,
1983]. In essence, the resonance multistep photoionization pathway is
an optical distillation column where work is done only on the isotope
of interest. After one isotope in a mixture is selectively excited and
ionized, the ions are extracted onto a collector by an externally applied electric field. The applied field is required to accelerate the
ions before they kinetically charge-exchange with the unwanted isotopes. Once this requirement is met, effective separation of the two
isotopes is accomplished. In general, the method is not as useful for
the isotopically selective separation of elements from a molecular feed
material, owing to the combined effects of high molecular ionization potentials (to 16 eV), the generally small molecular photoionization cross
sections ($< 10^{-17}$ cm^2), and the reduction in selectivity due to the
added spectral complexity and confusion introduced by the molecular
fine structure.

Research by the U.S. Department of Energy over the last decade
at the Lawrence Livermore National Laboratory in Livermore, California, and Martin Marietta Energy Systems in Oak Ridge, Tennessee,
has demonstrated the feasibility of this approach for the enrichment
of uranium isotopes (0.7% ^{235}U to 3.2% ^{235}U) for fuel used in light-
water nuclear power reactors. A schematic of the uranium system is
illustrated in Fig. 3.19. The process is a three-step photoionization
scheme. Each step wavelength is generated by copper-vapor-laser-
pumped dye lasers configured in master oscillator/power amplifier
configurations (see Chapter 2). The laser system operates at a pulse-
repetition frequency ($\sim$10 kHz) necessary to irradiate the full atomic
uranium flow generated by an electron-beam vapor source. Table 3.4
summarizes the laser parameter requirements for U-AVLIS (Uranium
Atomic Vapor Laser Isotope Separation). The required product rate
necessary to displace the existing gaseous diffusion plants is $\sim 5 \times 10^6$
kg of 3.2% enriched uranium per year. Assuming 10% light utilization,
the tunable power necessary to process this amount of material is
$\sim$125 kW (or $\sim$0.75 MJ/kg).

Although quantitative details of the process physics remain classified (AVLIS for defense programs is also under development), pro-

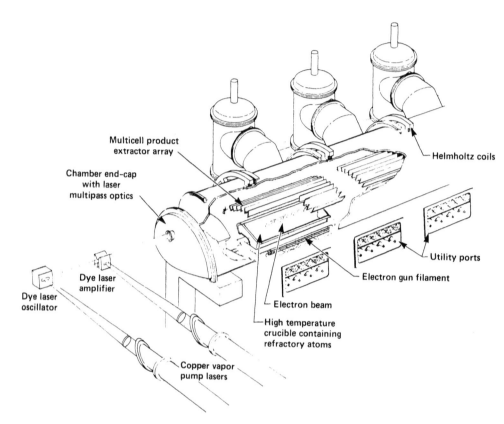

FIG. 3.19 Schematic illustration of Atomic Vapor Laser Isotope Separation (AVLIS) system being built at the Lawrence Livermore National Laboratory. Metallic uranium is melted and vaporized by an electron beam. The vapor is illuminated by laser light that photoionizes the isotope of interest. The ion is then electromagnetically extracted.

ponents of AVLIS have stated that the laser photoionization process is economically far superior (from 2 to 10 times) to both gaseous-diffusion and advanced gas-centrifuge methods [Davis, 1982a, 1982b, 1982c]. The present enrichment cost of a kilogram of 3.2% enriched material is ~ \$600. The AVLIS program is presently constructing a full-scale development module at Livermore to demonstrate technical and economic projections that an AVLIS plant can be deployed in the late 1980s or early 1990s and that the technology will be widely applicable to AVLIS of other elements and to industrial-scale photochemistry.

TABLE 3.4 Nominal Laser Parameters for a Practical Atomic-Vapor Laser Isotope Separation Process

Laser characteristic	AVLIS requirement	Reason for requirement
Pulse duration	20-200 ns	Laser pulse must be short in comparison with atomic-state lifetime
Pulse-repetition frequency	$\sim$10 kHz	Repetition rate must be adequate to illuminate all the flowing vapor
Pulse energy	0.1-1 J	Intensity must be sufficient to saturate atomic transitions
Frequency	Tunable	Frequency must be tunable to allow overlap with atomic transitions
Bandwidth	$\sim$1 to 3 GHz	Bandwidth should match inhomogeneously broadened atomic line
Frequency stability	$\sim\pm$30 MHz	Good spectral overlap of the laser and atomic line should be maintained
Phase front	$\sim 3\lambda/D$	Beams must propagate through long columns of vapor for efficient photon ultilization
Uniformity	$\sim$30%	Vapor must be fairly uniformly illuminated

Source: Davis et al. [Davis, 1982b].

3.3.8 Future Applications

The applications of resonant laser photoionizations are diverse. In the future, some of these applications will mature into major areas of scientific and industrial activity. Applications of lasers in chemistry are listed in Table 3.5 [Davis, 1982a] in order of economic potential. The first three entries have been demonstrated using atomic-vapor laser isotope separation. AVLIS is generic and is not restricted to the nuclear fuel cycle. As seen in Fig. 3.1, most of the elements can be photoionized by the first, second, and third harmonic outputs of the copper-vapor-laser-pumped dye lasers being developed for uranium

TABLE 3.5 Potential Applications of Lasers in Chemistry

Application	Economic potential
Isotope separation	Demands today a high-value product $\sim\$10/kg$
Cleanup of radioactive waste	Extremely high payoff if, and only if, a fully integrated reprocessing cycle is realized
Trace impurity removal	Cannot alone support technology development but will be major spin-off
Selective chemical reactions	Demands net gain in reaction (number particles/number photons) $>>1$
Photochemical activation or dissociation of gases	Potential high leverage in many applications (e.g., reactive etching, coatings)
Control of combustion particulates	Will probably happen slowly as a spin-off
Crystal and powder chemistry	Unique and interesting
Laser-induced biochemistry	Inevitable but a decade at least before some clarification

Source: Adapted from Davis [Davis, 1982a].

enrichment. AVLIS can be used, for example, to separate medical isotopes for skeletal (^{177}Lu, ^{153}Sm, ^{171}Er, ^{157}Dy) or myocardial (^{129}Cs) imaging. Perhaps the most remarkable potential application is enriching mercury for fluorescent lamps. Maya et al. [Maya, 1984] at General Telephone and Electronics (GTE) Lighting Products have demonstrated a 5% increase in lamp efficiency by simply increasing the ^{196}Hg content from 0.146% (natural) to 3%. The improved isotopic mix leads to a reduction in radiation trapping of the 2537-Å resonance emission and the resultant improvement in overall efficiency. Maya estimates that the annual consumption of electrical energy for fluorescent lighting in the United States is about 1.5×10^{11} kWh. At $0.05 per kWh, the cost savings in national consumption at 50% deployment could approach $200 million annually. Isotopically selective photoionization of mercury atoms has been reported in the literature [Dyer, 1983].

Finally, laser photoionization may find wide application in generating new laser sources. An autoionization-pumped laser was first observed in a resonant two-step photoionization scheme [Bokor, 1982]. The laser-pumped doubly excited autoionizing states of configuration

$6p_{3/2}$ np decay preferentially to the $6p_{1/2}$ state of the ion, resulting in a population inversion with respect to the ground $6s_{1/2}$ state and metastable $5d_{3/2}$ state. Amplified spontaneous emission at 493 and 650 nm, corresponding to transitions to these states, was observed. The process of selective autoionization and stimulated emission may possibly be extended to the vacuum ultraviolet using the appropriate atomic systems.

An exciting prospect for photoionization lasers derives from the recent work of Luk et al. [Luk, 1983]. By focusing the 193-nm output of a 10-ps 4-GW ArF laser to achieve intensities to 10^{14} W/cm^2, they observe collisionless multiphoton absorption in atoms spanning the region Z = 2 (He) to Z = 92 (U). High efficiency multiple atomic ionization is observed as shown in Fig. 3.20. The highest ion state observed was U^{10+}, corresponding to the absorption of 99 quanta (~633 eV)! This remarkable behavior, in this high-laser-intensity regime, is not predicted by the accepted nonresonant multiphoton ionization theory discussed in Sec. 3.2.7. The authors have correlated the maximum observed stage of ionization with the atomic shell structure of the heavier elements. This leads them to suggest strong coupling of the light field to a collective electron excitation (atomic plasmon) of the d or f shells. As in the autoionization-pumped laser system, this technique has the potential of creating population inversions in ionic states. However, in this process, large amounts of energy can be deposited into the atom, so that very-short-wavelength lasing is possible from various stages of ionization. Indeed, lasing in Kr at 93 nm has been observed in Rhodes' laboratory [Boyer, 1984]. The lasing transition is believed to occur via an autoionizing transition in Kr I: $4s4p^64d$ to $4s^24p^5n\ell$ ($n\ell$ = 4d, 4s). Lasing action occurs at a wavelength approximately equal to that corresponding to the 4p → 4s transition in the ion. The 4d or 4s electron probably only serves as a spectator in the emission process. These transitions are similar to those observed in earlier experiments on barium [Kachru, 1984]. This photoionization technique has opened up a new area of atomic physics to theoretical and experimental study.

Recently, Matthews et al. [Matthews, 1985] have reported the first observation of laboratory-produced soft-x-ray-amplified spontaneous emission using a high-power optical laser to form and heat a plasma. The heating occurs through the inverse Bremsstrahlung effect. This process may be loosely thought of as a form of nonresonant photoionization involving free-free transitions. The output from a frequency-doubled Nd-glass laser was used to form neon-like selenium and yttrium plasmas. X-ray emissions were observed at 206.3 and 209.6 Å in selenium and 154.9 and 157.1 Å in yttrium. The emissions are due to population inversions of the $2p^53p$-$2p^53s$ levels in the neon-like ions. These soft x rays are the shortest laser wavelengths ever produced in the laboratory.

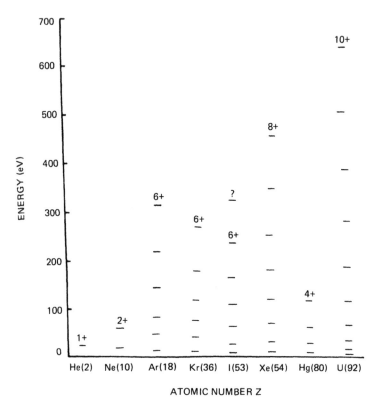

FIG. 3.20 High-intensity ($\sim 10^{14}$ to 10^{15} W/cm^2) ionization under collisionless conditions using the output of a 10-ps ArF laser at 194 nm:
(a) observed charge state for various atomic species; (b) relative
strength of ion signals for various atoms [Boyer, 1984].

3.4 SUMMARY

As discussed in this chapter, the applications of resonance photoionization spectroscopy include the identification of high-lying atomic
levels [Carlson, 1976], the measurement of transition cross sections
[Ambartzumian, 1976; Carlson, 1977], and the identification of atomic
species [Bekov, 1983]. Molecular spectroscopic applications also permit the assignment of high-lying states [Glownia, 1980; Loge, 1983],
using as in the atomic case, polarization studies [Parker, 1983], angular distributions of ejected electrons, or other labeling methods. The
extraordinary sensitivity and versatility of resonance photoionization
has already seen applications in the studies of chemical reactions
[Worden, 1978b], studies of electron binding energies in metal clus-

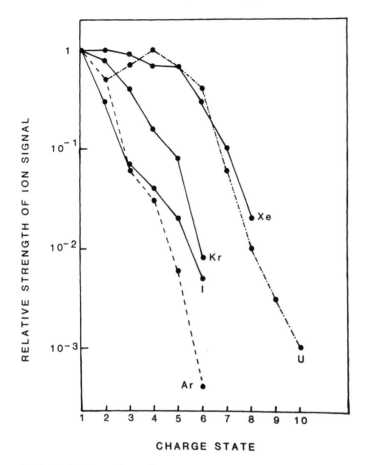

FIG. 3.20 (continued)

ters [Rohlfing, 1983], the precision measurement of ionization potentials [Janes, 1975; Solarz, 1976; Worden, 1978a, 1979, 1980, 1984], and in the new field of resonantly enhanced multiphoton ionization (REMPI) mass spectroscopy [Bernstein, 1982; Parker, 1983]. REMPI has yielded information on fragmentation systematics and mechanisms and may see increasing importance as an analytical tool in the chemical industry. Novel applications of resonance photoionization spectroscopy include its use as an impurity detector [Bekov, 1983], its use in the study of sputtering phenomena [Winograd, 1982], its proposed use in the detection of ultrarare isotopes [Kudriavtsev, 1982], the possibility of its use in the detection of solar neutrinos and exotic atoms [Hurst, 1979], and its emergence as a feasible method for the separation of uranium and other commercially valuable isotopes [Davis, 1982a, 1982b,

1982c; Emmett, 1984]. Given its simplicity and its applicability in a wide range of physical and chemical studies [Wynne, 1984], it is clear that resonance photoionization will remain as one of the most significant and widely used methods in spectroscopy for the foreseeable future.

ACKNOWLEDGMENT

We acknowledge with pleasure the members of the Process Development Group in the Laser Isotope Separation Program at the Lawrence Livermore National Laboratory who have kindly assisted us during the preparation of this manuscript.

REFERENCES

[Ackerhalt, 1977] J. R. Ackerhalt and B. W. Shore. Rate equations versus Bloch equations in multiphoton ionization, *Phys. Rev.*, *A 16*(1), 277 (1977).

[Alkhazov, 1983] G. D. Alkhazov, A. E. Barzakh, E. I. Berlovich, V. P. Denisov, A. G. Dernyatin, V. S. Ivanov, A. N. Zherikhin, O. N. Kompanets, V. S. Letokhov, V. I. Mishin, and V. N. Fedoseev. Measurement of isotopic variations in the charge radii of europium nuclei by the method of three-stepped laser photoionization of atoms, *JETP Lett.*, 37(5), 274 (1983).

[Allen, 1975] L. Allen and J. H. Eberly. *Optical Resonance and Two Level Atoms*, Wiley, New York, 1975.

[Ambartzumian, 1971] R. V. Ambartzumian, V. N. Kalinin, and V. S. Letokhov. Two-step selective photoionization of rubidium atoms by laser radiation, *JETP Lett.*, *13*(6), 217 (1971).

[Ambartzumian, 1972] R. V. Ambartzumian and V. S. Letokhov. Selective two step photoionization of atoms and photodissociation of molecules by laser radiation, *Appl. Opt.*, *11*(2), 354 (1972).

[Ambartzumian, 1976] R. V. Ambartzumian, N. P. Furzikov, V. S. Letokhov, and A. A. Puretsky. Measuring photoionization cross-sections of excited atomic states, *Appl. Phys.*, *9*, 335 (1976).

[Antonov, 1981] V. S. Antonov and V. S. Letokhov. Laser multiphoton and multistep photoionization of molecules and mass spectrometry, *Appl. Phys.*, *24*, 89 (1981).

[Armstrong, 1979] J. A. Armstrong, J. J. Wynne, and P. Esherick. Bound, odd-parity J = 1 spectra of the alkaline earths: Ca, Sr, and Ba, *J. Opt. Soc. Am.*, *69*(2), 211 (1979).

[Bahcall, 1969] J. N. Bahcall. What next with solar neutrinos? *Phys. Rev. Lett.*, *23*(5), 251 (1969).

[Bahcall, 1978] J. N. Bahcall. Solar neutrino experiments, *Rev. Mod. Phys.*, *50*(4), 881 (1978).

[Balling, 1976] L. C. Balling and J. J. Wright. Use of angular-momentum selection rules for laser isotope separation, *Appl. Phys. Lett.*, *29*(7), 411 (1976).

[Bayfield, 1974] J. E. Bayfield and P. M. Koch. Multiphoton ionization of highly excited hydrogen atoms, *Phys. Rev. Lett.*, *33*(5), 258 (1974).

[Bebb, 1966] H. B. Bebb and A. Gold. Multiphoton ionization of hydrogen and rare-gas atoms, *Phys. Rev.*, *143*(1), 1 (1966).

[Bekov, 1978a] G. I. Bekov, V. S. Letokhov, O. I. Matveev, and V. I. Mishin. Single atom detection of ytterbium by selective laser excitation and field ionization from Rydberg states, *Opt. Lett.*, *3*(5), 159 (1978).

[Bekov, 1978b] G. I. Bekov, V. S. Letokhov, O. I. Matveev, and V. I. Mishin. Observation of a long-lived autoionization state in the spectrum of gadolinium atom, *JETP Lett.*, *28*(5), 283 (1978).

[Bekov, 1979] G. I. Bekov, E. P. Vidolova-Angelova, V. S. Letokhov, and V. I. Mishin. Selective detection of single atoms by multistep excitation and ionization through Rydberg states, in *Laser Spectroscopy*, Vol. 4 (H. Walther and K. W. Rothe, eds.), Springer-Verlag, Berlin, 1979, p. 283.

[Bekov, 1980] G. I. Bekov, E. P. Vidolova-Angelova, L. N. Ivanov, V. S. Letokhov, and V. I. Mishin. Double-excited narrow autoionization states of ytterbium atom, *Opt. Commun.*, *35*(2), 194 (1980).

[Bekov, 1983] G. I. Bekov and V. S. Letokhov. Laser atomic photoionization spectral analysis of element traces, *Appl. Phys.*, *B30*, 161 (1983).

[Berkowitz, 1966] J. Berkowitz and W. A. Chupka. Photoionization of high-temperature vapors: I. The iodides of sodium, magnesium, and thallium, *J. Chem. Phys.*, *45*(4), 1287 (1966).

[Bernstein, 1982] R. B. Bernstein. Systematics of multiphoton ionization-fragmentation of polyatomic molecules, *J. Phys. Chem.*, *86*, 1178 (1982).

[Berry, 1966] R. S. Berry. Ionization of molecules at low energies, *J. Chem. Phys.*, *45*(4), 1228 (1966).

[Beterov, 1983] I. M. Beterov, V. P. Chebotayev, and N. V. Fatey-
ev. Photoionization detection of stimulated Raman scattering on
electron transitions of Na in an atomic beam, *Appl. Phys. B31*,
135 (1983).

[Bjorklund, 1978] G. C. Bjorklund, C. P. Ausschnitt, R. R. Freeman, and R. H. Storz. Detection of atomic hydrogen and deuterium by resonant three-photon ionization, *Appl. Phys. Lett.*,
33(1), 54 (1978).

[Bloembergen, 1976] N. Bloembergen and M. D. Levenson. Doppler
free two photon absorption spectroscopy, in *Topics in Applied
Physics, Vol. 13; High Resolution Laser Spectroscopy* (K. Shimo-
da, ed.), Springer-Verlag, Berlin, 1976, p. 55.

[Bohm, 1978] H.-D. V. Bohm, W. Michaelis, and C. Weitkamp. Hyperfine structure and isotope shift measurements on ^{235}U and
laser-separation of uranium isotopes by two-step photoionization,
Opt. Commun., *26*(2), 177 (1978).

[Bokor, 1982] J. Bokor, R. R. Freeman, and W. E. Cooke. Autoionization-pumped laser, *Phys. Rev. Lett.*, *48*(18), 1242 (1982).

[Bonanno, 1983] R. Bonanno, J. Boulmer, and J. Weiner. Determination of the absolute rate constant for associative ionization in
crossed-beam collisions between Na 3 $^{2}P_{3/2}$ atoms, *Phys. Rev.*,
A28(2), 604 (1983).

[Boyer, 1984] K. Boyer, H. Egger, T. S. Luk, H. Pummer, and
C. K. Rhodes. Interaction of atomic and molecular systems with
high intensity ultraviolet radiation, *J. Opt. Soc. Am.*, *B1*(1), 3
(1984).

[Bradley, 1976] D. J. Bradley, C. H. Dugan, P. Ewart, and A. F.
Purdie. Absolute photoionization cross-section measurement of
selectively excited magnesium, *Phys. Rev.*, *A13*(4), 1416 (1976).

[Braun, 1981] C. L. Braun, T. W. Scott, and A. G. Albrecht. Multiphoton ionization of N,N,N',N'-tetramethyl-p-phenylenediamine:
the condensed phase two-photon ionization spectrum, *Chem. Phys.
Lett.*, *84*(2), 248 (1981).

[Brinkmann, 1974] U. Brinkmann, W. Hartig, H. Telle, and H. Walther. Isotope selective photoionization of calcium using two-step
laser excitation, *Appl. Phys.*, *5*, 109 (1974).

[Brophy, 1979] J. H. Brophy and C. T. Rettner. Laser two-photon
ionization spectroscopy: a new method for real-time monitoring of
atmospheric pollutants, *Opt. Lett.*, *4*(10), 337 (1979).

[Camus, 1983] P. Camus, ed. Optogalvanic spectroscopy and its applications, in *Colloq. Int. CNRS*, Aussois, France, June 20-24 [*J.
Phys. (Paris)*, 44, Colloq. C7 Suppl. 11, Nov. 1983].

[Carlson, 1976] L. R. Carlson, J. A. Paisner, E. F. Worden, S. A. Johnson, C. A. May, and R. W. Solarz. Radiative lifetimes, absorption cross sections, and the observation of new high lying odd levels of ^{238}U using multistep laser photoionization, *J. Opt. Soc. Am.*, *66*(8), 846 (1976).

[Carlson, 1977] L. R. Carlson, S. A. Johnson, E. F. Worden, C. A. May, R. W. Solarz, and J. A. Paisner. Determination of absolute atomic transition probabilities using time resolved optical pumping, *Opt. Commun.*, *21*(1), 116 (1977).

[Carney, 1982] T. E. Carney and T. Baer. Translational energies of fragment ions in the multiphoton ionization of benzene, *J. Chem. Phys.*, *76*(12), 5968 (1982).

[Carre, 1984] B. Carre, G. Spiess, J. M. Bizau, P. Dhez, P. Gerard, F. Wuilleumier, J. C. Keller, J. L. LeGouet, J. L. Picque, D. L. Ederer, and P. M. Koch. Electron spectrometry study of associative ionization and Penning ionization in laser excited sodium vapor, *Opt. Commun.*, *52*(1), 29 (1984).

[Castro, 1980] J. C. Castro, M. L. Zimmerman, R. G. Hulet, D. Kleppner, and R. R. Freeman. Origin and structure of the quasi-Landau resonances, *Phys. Rev. Lett.*, *45*(22), 1780 (1980).

[Chen, 1981] H. L. Chen. Laser cleanup of Pt-group metals, in *1980 Laser Program Annual Report*, UCRL 50021-80, Vol. 3, Lawrence Livermore National Laboratory, Livermore, Calif., 1981, p. 10-52.

[Chen, 1984] C. H. Chen, S. D. Kramer, S. L. Allman, and G. S. Hurst. Selective counting of krypton atoms using resonance ionization spectroscopy, *Appl. Phys. Lett.*, *44*(6), 640 (1984).

[Chupka, 1968] W. A. Chupka and J. Berkowitz. Photoionization of the H_2 molecule near threshold, *J. Chem. Phys.*, *48*(12), 5726 (1968).

[Clark, 1983] C. W. Clark. Isotope shifts of C I spectral lines and their application to radioactive dating by laser-assisted mass spectrometry, *Opt. Lett.*, *8*(11), 572 (1983).

[Clark, 1985] C. W. Clark, J. D. Fassett, T. B. Lucatorto, L. J. Moore, and W. W. Smith. Observation of autoionizing states of beryllium by resonance ionization mass spectrometry, *J. Opt. Soc. Am. B.* (in press).

[Cooke, 1980] W. E. Cooke and T. F. Gallagher. Effects of blackbody radiation on highly excited atoms, *Phys. Rev.*, *A21*(2), 588 (1980).

[Cool, 1984] T. A. Cool. Quantitative measurement of NO density by resonance three-photon ionization, *Appl. Opt.*, *23*(10), 1559 (1984).

[Corliss, 1962] C. H. Corliss and W. R. Bozman. *Experimental Transition Probabilities for Spectral Lines of Seventy Elements*, NBS Monogr. 53, U.S. Government Printing Office, Washington, D.C., 1962.

[Coste, 1982] A. Coste, R. Avril, P. Blancard, J. Chatelet, D. Lambert, J. Legre, S. Liberman, and J. Pinard. New spectroscopic data on high-lying excited levels of atomic uranium, *J. Opt. Soc. Am.*, *72*(1), 103 (1982).

[Cowan, 1981] R. D. Cowan. *The Theory of Atomic Structure and Spectra*, University of California Press, Berkeley, Calif., 1981, p. 1.

[Davis, 1977] J. I. Davis and R. W. Davis. Some aspects of the laser isotope separation program at Lawrence Livermore Laboratory, in *Developments in Uranium Enrichment*, AIChE Symp. Ser., Vol. 73, No. 169 (M. Benedict, ed.), American Institute of Chemical Engineers, New York, 1977, p. 69.

[Davis, 1982a] J. I. Davis. *Lasers in Chemical Processing*, UCRL-53276, Lawrence Livermore National Laboratory, Livermore, Calif., 1982, p. 1.

[Davis, 1982b] J. I. Davis, J. Z. Holtz, and M. L. Spaeth. Status and prospects for lasers in isotope separation, *Laser Focus*, Sept. 1982, p. 49.

[Davis, 1982c] J. I. Davis and E. Rockower. Lasers in material processing, *IEEE J. Quantum Electron.*, *18*(2), 233 (1982).

[Delone, 1972] G. A. Delone, N. B. Delone, and G. K. Piskova. Multiphoton resonance ionization of atoms, *Sov. Phys. JETP*, *35*(4), 672 (1972).

[Dickie, 1973] L. O. Dickie, F. M. Kelly, T. K. Koh, M. S. Mathur, and F. C. Suk. Lifetime of the $5s5p\,^1P$ level of strontium, *Can. J. Phys.*, *51*, 1088 (1973).

[Downey, 1984a] S. W. Downey, N. S. Nogar, and C. M. Miller. Resonance ionization mass spectrometry of uranium with intracavity laser ionization, *Anal. Chem.*, *56*, 827 (1984).

[Downey, 1984b] S. W. Downey, N. S. Nogar, and C. M. Miller. Resonance ionization mass spectrometry of technetium, *Int. J. Mass Spectrom. Ion Phys.*, *61*(3), 337 (1984).

[Dunning, 1974] F. B. Dunning and R. F. Stebbings. Absolute cross sections for the photoionization of He $(n^{1,3}P)$ atoms, *Phys. Rev. Lett.*, *32*(23), 1286 (1974).

[Dyer, 1983] P. Dyer, G. C. Baldwin, G. Kittrel, D. G. Imre, and
E. Abramson. Isotopically selective photoionization of mercury
atoms, *Appl. Phys. Lett.*, *42*(4), 311 (1983).

[Eberly, 1977] J. H. Eberly and B. Karczewski. *Multiphoton Bibliography 1970-1976.*

[Eberly, 1979a] J. H. Eberly and S. V. O'Neil. Coherence versus
incoherence: time-independent rates for resonant two-photon
ionization, *Phys. Rev.*, *A19*(3), 1161 (1979).

[Eberly, 1979b] J. H. Eberly, J. W. Gallagher, and E. C. Beaty,
eds. *Multiphoton Bibliography 1977*, NBS LP-92 (1979).

[Eberly, 1979c] J. H. Eberly, J. W. Gallagher, and E. C. Beaty,
eds. *Multiphoton Bibliography 1978*, Suppl. 1, NBS LP-92
(1979).

[Eberly, 1980] J. H. Eberly, J. W. Gallagher, and E. C. Beaty,
eds. *Multiphoton Bibliography 1979*, Suppl. 2, NBS LP-92
(1980).

[Eberly, 1981] J. H. Eberly and J. W. Gallagher, eds. *Multiphoton
Bibliography 1980*, Suppl. 3, NBS LP-92 (1981).

[Eberly, 1984] J. H. Eberly, N. D. Piltch and J. W. Gallagher,
eds. *Multiphoton Bibliography 1981-1982*, Suppl. 4, NBS LP-92
(1984).

[Emmett, 1984] J. L. Emmett, W. F. Krupke, and J. I. Davis. Laser
R&D at the Lawrence Livermore National Laboratory for fusion
and isotope separation applications, *IEEE J. Quantum Electron.*,
QE-20(6), 591 (1984).

[Esherick, 1977] P. Esherick, J. J. Wynne, and J. A. Armstrong.
Multiphoton ionization spectroscopy of the alkaline earths, in
Laser Spectroscopy, Vol. 3 (J. L. Hall and J. L. Carlsten, eds.),
Springer-Verlag, Berlin, 1977, p. 170.

[Fano, 1968] U. Fano and J. W. Cooper. Spectral distribution of
atomic oscillator strengths, *Rev. Mod. Phys.*, *40*(3), 441 (1968).

[Fedoseyev, 1984] V. N. Fedoseyev, V. S. Letokhov, V. I. Mishin,
G. D. Alkhazov, A. E. Barzakh, V. P. Denisov, A. G. Dernyatin, and V. S. Ivanov. Atomic lines isotope shifts of short lived
radioactive Eu studied by high-sensitive laser resonance photoionization method in "on-line" experiments with proton beams,
Opt. Commun., *52*(1), 24 (1984).

[Feneuille, 1977] S. Feneuille and J. C. Lehmann, eds. Etats
atomique et moléculaires couples a un continuum atomes et molecules hautement excites, in *Proc. Aussois Conf.*, Aussois, France,
June 13-17, 1977.

[Foote, 1925] P. D. Foote and F. L. Mohler. Photo-electric ionization of caesium vapor, *Phys. Rev.*, *26*, 195 (1925).

[Fox, 1971] R. A. Fox, R. M. Kogan, and E. J. Robinson. Laser triple-quantum photoionization of cesium, *Phys. Rev. Lett.*, *26*(23), 1416 (1971).

[Fred, 1967] M. Fred. The significance of the Zeeman effect to the structural analysis of spectra, *Physica 33*, Zeeman Centennial Conf., Amsterdam, 1967, p. 47.

[Gallagher, 1975] T. F. Gallagher, S. A. Edelstein, and R. M. Hill. Collisional angular momentum mixing in Rydberg states of sodium, *Phys. Rev. Lett.*, *35*(10), 644 (1975).

[Gallagher, 1981] T. F. Gallagher, W. Sandner, K. A. Safinya, and W. E. Cooke. Blackbody-radiation-induced Stark shifts in perturbed Rydberg series, *Phys. Rev.*, *A23*(4), 2065 (1981).

[Garton, 1968] W. R. S. Garton, G. L. Grasdalen, W. H. Parkinson, and E. M. Reeves. Analysis of autoionization resonance structure in the $5s^2$ ($^1S_0$)-4dnp,nf spectrum of Sr I, *J. Phys.*, *B2*(1), 114 (1968).

[Garton, 1974] W. R. S. Garton and W. H. Parkinson. Series of autoionization resonances in Ba I converging on Ba II 6 ^{2}P, *Proc. R. Soc. London, A341*, 45 (1974).

[Geiger, 1928] H. Geiger and W. Muller. Electron counting tube for the measurement of weak activities, *Naturwissenschaften*, *16*, 617 (1928).

[Geltman, 1963] S. Geltman. Double-photon photo-detachment of negative ions, *Phys. Lett.*, *4*(3), 168 (1963).

[Glownia, 1980] J. H. Glownia, S. J. Riley, S. D. Colson, and G. C. Nieman. The MPI spectrum of expansion-cooled ammonia: photophysics and new assignments of electronic excited states, *J. Chem. Phys.*, *73*(9), 4296 (1980).

[Golovin, 1968] A. F. Golovin and A. R. Striganov. The isotope effect in the spectra of the heavy elements, *Sov. Phys. Usp.*, *10*(5), 658 (1968).

[Green, 1976] R. B. Green, R. A. Keller, G. G. Luther, P. K. Schenck, and J. C. Travis. Galvanic detection of optical absorptions in a gas discharge, *Appl. Phys. Lett.*, *29*(11), 727 (1976).

[Greinacher, 1926] H. Greinacher. Eine neue Methode zur Messung der Elementarstrahlen, *Z. Phys.*, *36*, 364 (1926).

[Hackel, 1979] L. A. Hackel, C. F. Bender, M. A. Johnson, and M. C. Rushford. Hyperfine structure measurement of high-lying levels of uranium, *J. Opt. Soc. Am.*, *69*(2), 230 (1979).

[Hanna, 1979] D. C. Hanna, M. A. Yuratich, and D. Cotter. *Nonlinear Optics of Free Atoms and Molecules*, Springer-Verlag, Berlin, 1978.

[Haroche, 1977] S. Haroche, C. Fabre, and P. Goy. Millimeter spectroscopy in Rydberg levels, in *Etats atomique et moléculaires couples a un continuum atomes et molecules hautement excites*, Proc. Aussois Conf., Aussois, France, June 13-17, 1977, p. 207.

[Held, 1973] B. Held, G. Mainfray, C. Manus, J. Morellec, and F. Sanchez. Resonant multiphoton ionization of a cesium atomic beam by a tunable-wavelength Q-switched neodymium-glass laser, *Phys. Rev. Lett.*, *30*(10), 423 (1973).

[Hermann, 1980] J. P. Hermann and J. J. Wynne. Ionization studies of laser-excited alkaline-earth vapors, *Opt. Lett.*, *5*(6), 236 (1980).

[Hertz, 1923] G. Hertz. Uber die Anregungs-und Ionisierungsspannungen von Neon and Argon und ihren Zusammenhang mit den Spektren dieser Gase, *Z. Phys.*, *18*, 307 (1923).

[Hildebrandt, 1978] G. F. Hildebrandt, F. G. Kellert, F. B. Dunning, K. A. Smith, and R. F. Stebbings. Ionization of xenon atoms in selected high Rydberg states by collision with CH_3I, C_7F_{14}, C_6F_6, and CH_3Br, *J. Chem. Phys.*, *68*(4), 1349 (1978).

[Hill, 1982] W. T. Hill, K. T. Cheng, W. R. Johnson, T. B. Lucatorto, T. J. McIlrath, and J. Sugar. Influence of increasing nuclear charge on the Rydberg spectra of Xe, Cs^+ and Ba^{++}: correlation, term dependence, and autoionization, *Phys. Rev. Lett.*, *49*(22), 1631 (1982).

[Hodgkinson, 1981] D. P. Hodgkinson and D. J. H. Wort. Doppler-free two-photon excitation of ^{238}U, *J. Phys.*, *B14*, 3891 (1981).

[Hofmann, 1971] W. Hofmann and G. L. Weissler. Measurement of the photoionization cross section in the resonance continuum of C I using a wall-stabilized arc, *J. Opt. Soc. Am.*, *61*(2), 223 (1971).

[Huber, 1980] M. C. Huber and R. J. Sandeman. Transition probabilities and their accuracy, *Phys. Scr.*, *22*, 373 (1980).

[Hudgens, 1983] J. W. Hudgens, T. G. DiGiuseppe, and M. C. Lin. Two photon resonance enhanced multiphoton ionization spectroscopy and state assignments of the methyl radical, *J. Chem. Phys.*, *79*(2), 571 (1983).

[Hurst, 1977] G. S. Hurst, M. H. Nayfeh, and J. P. Young. A demonstration of one atom detection, *Appl. Phys. Lett.*, *30*, 229 (1977).

[Hurst, 1979] G. S. Hurst, M. G. Payne, S. D. Kramer, and J. P. Young. Resonance ionization spectroscopy and one-atom detection, *Rev. Mod. Phys.*, *51*(4), 767 (1979).

[Jacquinot, 1977] P. Jacquinot, S. Liberman, and J. Pinard. High resolution spectroscopy of Rydberg states of rubidium, in *Etats atomique et moléculaires couples a un continuum atomes et molecules hautement excites*, Proc. Aussois Conf., Aussois, France, June 13-17, 1977, p. 215.

[Janes, 1975] G. S. Janes, I. Itzkan, C. T. Pike, R. H. Levy, and L. Levin. Post-deadline paper at CLEO, Washington, D.C., Two photon laser isotope separation of atomic uranium—spectroscopic studies, excited state lifetimes, and photoionization cross sections, *AVCO Everett Rep. 408*, 1975.

[Janes, 1976] G. S. Janes, I. Itzkan, C. T. Pike, R. H. Levy, and L. Levin. Two-photon laser isotope separation of atomic uranium: spectroscopic studies, excited-state lifetimes, and photoionization cross sections, *IEEE J. Quantum Electron.*, *QE-12*(2), 111 (1976).

[Johnson, 1976] P. M. Johnson. The multiphoton ionization spectrum of benzene, *J. Chem. Phys.*, *64*(10), 4143 (1976).

[Johnson, 1981] P. M. Johnson and C. E. Otis. Molecular multiphoton spectroscopy with ionization detection, *Annu. Rev. Phys. Chem.*, *32*, 139 (1981).

[Kachru, 1984] R. Kachru, N. H. Tran, H. B. van Linden van den Heuvell, and T. F. Gallagher. Enhancement of the autoionization rate of two-photon excited states of Ba $(6d_{5/2}\,nd_{5/2}(3/2))_{J=4}$ near the Ba^+ $6d_{3/2}$ limit, *Phys. Rev.*, *A 30*, 667 (1984).

[Karlov, 1976] N. V. Karlov, B. B. Krynetskii, V. A. Mishin, A. M. Prokhorov, A. D. Savelev, and V. V. Smirnov. Separation of samarium isotopes by two-step photoionization method, *Sov. J. Quantum Electron.*, *6*, 1363 (1976).

[Karlov, 1977] N. V. Karlov, B. B. Krynetskii, V. A. Mishin, A. M. Prokhorov, A. D. Savelev, and V. V. Smirnov. Isotope separation of some rare-earth elements by two-step photoionization, *Opt. Commun.*, *21*(3), 384 (1977).

[Karlov, 1978] N. V. Karlov, B. B. Krynetskii, V. A. Mishin, and A. M. Prokhorov. Laser isotope separation of rare earth elements, *Appl. Opt.*, *17*(6), 856 (1978).

[Karlov, 1983] N. V. Karlov, B. B. Krynetskii, V. A. Mishin, A. M. Prokhorov, and O. M. Stel'makh. Calculation of a three-stage scheme for selective photoionization of atoms for laser separation of isotopes, *Sov. Phys. Tech. Phys.*, *28*(8), 937 (1983).

[Keldysh, 1965] L. V. Keldysh. Ionization in the field of a strong electromagnetic wave, *Sov. Phys. JETP*, *20*(5), 1307 (1965).

[Keller, 1984a] J. Keller and J. Weiner. Multiphoton ionization spectroscopy of the sodium dimer, *Phys. Rev.*, *A 30*(1), 213 (1984).

[Keller, 1984b] J. Keller and J. Weiner. Production of Ba^+ in a two-photon radiative collision in crossed Na-Ba beams, *Phys. Rev.*, *A 29*(4), 2230 (1984).

[Kingdon, 1923] K. H. Kingdon. A method for the neutralization of electron space charge by positive ionization at very low gas pressures, *Phys. Rev.*, *21*, 408 (1923).

[Kopfermann, 1958] H. Kopfermann. *Nuclear Moments*, Academic Press, New York, 1958, p. 124.

[Korff, 1958] S. A. Korff. Geiger counters, in *Encyclopedia of Physics*, Vol. 45, *Nuclear Instrumentation: II* (S. Flügge, ed.), Springer-Verlag, Berlin, 1958, p. 52.

[Kudriavtsev, 1982] Y. A. Kuriavtsev and V. S. Letokhov. Laser method of highly selective detection of rare radioactive isotopes through multistep photoionization of accelerated atoms, *Appl. Phys.*, *B 29*, 219 (1982).

[Kurz, 1979] E. A. Kurz. Channel electron multipliers, *Am. Lab.* *11*(3), 67 (1979).

[Lambropoulos, 1974] P. Lambropoulos. Theory of multiphoton ionization: near-resonance effects in two-photon ionization, *Phys. Rev.*, *A 9*(5), 1992 (1974); errata: *Phys. Rev.*, *A 10*(6), 2516 (1974).

[Lambropoulos, 1980] P. Lambropoulos. Reaching VUV transitions with multiphoton processes, *Appl. Opt.*, *19*(23), 3926 (1980).

[Lecompte, 1974] C. Lecompte, G. Mainfray, C. Manus, and F. Sanchez. Experimental demonstration of laser temporal coherence effects on multiphoton ionization processes, *Phys. Rev. Lett.*, *32*(6), 265 (1974).

[Letokhov, 1973] V. S. Letokhov. Possibility of the optical separation of the isomeric nuclei by laser radiation, *Opt. Commun.*, *7*(1), 59 (1973).

[Letokhov, 1977a] V. S. Letokhov and V. P. Chebotayev. *Nonlinear Laser Spectroscopy*, Springer-Verlag, Berlin, 1977.

[Letokhov, 1977b]   V. S. Letokhov, V. I. Mishin, and A. A. Puretzky. Selective photoionization of atoms by laser radiation and its applications, *Prog. Quantum Electron.*, 5, 139 (1977).

[Letokhov, 1978]   V. S. Letokhov, V. M. Lobashev, V. G. Minogin, and V. I. Mishin. Method of obtaining polarized protons by laser radiation, *JETP Lett.*, 27(5), 284 (1978).

[Levy, 1973]   R. H. Levy and G. S. Janes. Method and apparatus for the separation of isotopes, U.S. patent 3,772,519 (1973).

[Littman, 1976]   M. G. Littman, M. L. Zimmerman, T. W. Ducas, R. R. Freeman, and D. Kleppner. Structure of sodium Rydberg states in weak to strong electric fields, *Phys. Rev. Lett.*, 36(14), 788 (1976).

[Liu, 1978]   B. Liu and R. E. Olson. Potential energies for Ca_2^+: cross sections for collisions of Ca^+ and Rydberg Ca^{**} with Ca, *Phys. Rev.*, A18(6), 2498 (1978).

[Lodge, 1976]   J. C. Lodge, I. C. Percival, and D. Richards. Semi-empirical cross sections for excitation of hydrogen atoms by protons, *J. Phys.*, B9(2), 239 (1976).

[Loge, 1983]   G. W. Loge, J. J. Tiee, and F. B. Wampler. Multiphoton induced fluorescence and ionization of carbon monoxide (B $^1\Sigma^+$), *J. Chem. Phys.*, 79(1), 196 (1983).

[Lu, 1970]   K. T. Lu and U. Fano. Graphic analysis of perturbed Rydberg series, *Phys. Rev.*, A2(1), 81 (1970).

[Lubman, 1982]   D. M. Lubman and M. N. Kronick. Mass spectrometry of aromatic molecules with resonance-enhanced multiphoton ionization, *Anal. Chem.*, 54, 660 (1982).

[Lucatorto, 1980]   T. B. Lucatorto and T. J. McIlrath. Laser excitation and ionization of dense atomic vapors, *Appl. Opt.*, 19(23), 3948 (1980).

[Lucatorto, 1984]   T. B. Lucatorto, C. W. Clark, and L. J. Moore. Possibilities for ultrasensitive mass spectrometry based on two photon, sub-Doppler resonance ionization, *Opt. Commun.*, 48(6), 406 (1984).

[Luk, 1983]   T. S. Luk, H. Pummer, K. Boyer, M. Shahidi, H. Egger, and C. K. Rhodes. Anamolous collision-free multiple ionization of atoms with intense picosecond ultraviolet radiation, *Phys. Rev. Lett.*, 51(2), 110 (1983).

[Mainfray, 1980]   G. Mainfray and C. Manus. Resonance effects in multiphoton ionization of atoms, *Appl. Opt.*, 19(23), 3934 (1980).

[Mansbach, 1969] P. Mansbach and J. Keck. Monte Carlo trajectory calculations of atomic excitation and ionization by thermal electrons, *Phys. Rev.*, *181*(1), 275 (1969).

[Matthews, 1985] D. L. Matthews, P. L. Hagelstein, M. D. Rosen, M. J. Eckart, N. M. Ceglio, A. V. Hazi, H. Medecki, B. MacGowan, J. E. Trebes, B. L. Witten, E. M. Campbell, C. W. Hatcher, A. M. Hawryluk, R. L. Kauffman, L. D. Pleasance, G. Rambach, J. H. Scofield, G. Stone, and T. A. Weaver. Demonstration of a soft x-ray amplifier, *Phys. Rev. Lett.*, *54(2)*, 110 (1985).

[Maya, 1984] J. Maya, M. W. Grossman, R. Lagushenko, and J. F. Waymouth. Energy conservation through more efficient lighting, *Science*, *226*, 435 (1984).

[McFarlane, 1979] R. M. McFarlane and R. M. Shelby. Photochemical and population hole burning in the zero-phonon line of a color center -F_3^+ in NaF, *Phys. Rev. Lett.*, *42*(12), 788 (1979).

[McIlrath, 1977] T. J. McIlrath and T. B. Lucatorto. Laser excitation and ionization in a dense Li vapor: observation of the even parity, core excited autoionizing states, *Phys. Rev. Lett.*, *38*(24), 1390 (1977).

[McLain, 1971] W. M. McLain. Excited state symmetry assignment through polarized two-photon absorption studies in fluids, *J. Chem. Phys.*, *55*(6), 2789 (1971).

[Meyerand, 1963] R. G. Meyerand and A. F. Haught. Gas breakdown at optical frequencies, *Phys. Rev. Lett.*, *11*(9), 401 (1963).

[Miller, 1982] C. M. Miller and N. S. Nogar. Autoionizing and high-lying Rydberg states of lutetium atoms, in *Proc. Topical Meet. Laser Tech. Extreme Ultraviolet Spectrosc.*, Boulder, Colo., March 3-10, 1982.

[Miller, 1983] C. M. Miller, N. S. Nogar, and S. W. Downey. Selective laser ionization for mass spectral isotopic analysis, *SPIE J.*, *426*, 8 (1983).

[Mishin, 1985] V. I. Mishin, G. Lombardi, D. E. Kelleher, and J. W. Cooper. Effects of very low electric fields on autoionizing states in gadolinium, *J. Opt. Soc. Am.* (in press).

[Mizushima, 1970] M. Mizushima. *Quantum Mechanics of Atomic Spectra and Atomic Structure*, Benjamin-Cummings, Menlo Park, Calif., 1970, p. 237 and references therein.

[Moore, 1985] L. J. Moore, J. D. Fassett, J. C. Travis, T. B. Lucatorto, and C. W. Clark. Resonance ionization mass spectrometry of carbon, *Int. J. Mass Spectrom.* (in press).

[Morellec, 1982] J. Morellec, D. Normand, and G. Petite. Nonreso-
nant multiphoton ionization of atoms, in *Advances in Atomic and
Molecular Physics*, Vol. 18 (D. Bates and B. Bederson, eds.),
Academic Press, New York, 1982, p. 97.

[Murad, 1964] E. Murad and M. G. Inghram. Photoionization of
aliphatic ketones, *J. Chem. Phys.*, *40*(11), 3263 (1964).

[Nogar, 1984] N. S. Nogar, S. W. Downey, R. A. Keller, and C. M.
Miller. Resonance ionization mass spectrometry at Los Alamos Na-
tional Laboratory, in *Analytical Spectroscopy* (W. S. Lyon, ed.),
Elsevier, Amsterdam, 1984, p. 155.

[Paisner, 1976] J. A. Paisner, L. R. Carlson, E. F. Worden, S. A.
Johnson, C. A. May, and R. W. Solarz. *D.C. Electric Field Be-
havior of High Lying States in Atomic Uranium*, UCRL-78034,
Lawrence Livermore National Laboratory, Livermore, Calif., 1976.

[Panock, 1977] R. L. Panock and R. J. Temkin. Interaction of two
laser fields with a three-level molecular system, *IEEE J. Quantum
Electron.*, *QE-13*(6), 425 (1977).

[Parker, 1982] D. H. Parker and R. B. Bernstein. Multiphoton ionization—fragmentation patterns of alkyl iodides, *J. Phys. Chem.*,
86, 60 (1982).

[Parker, 1983] D. H. Parker. Laser ionization spectroscopy and
mass spectrometry, in *Ultrasensitive Laser Spectroscopy* (D. S.
Kliger, ed.), Academic Press, New York, 1983, p. 233.

[Parr, 1968] A. C. Parr and F. A. Elder. Photoionization of ytterbium: 1350-2000 Å, *J. Chem. Phys.*, *49*(6), 2665 (1968).

[Penning, 1928] F. M. Penning. Demonstrate van een nieuw photo-
electrisch effect, *Physica*, *8* (old series), 137 (1928).

[Percival, 1966] I. C. Percival. Cross section for collisions of electrons with hydrogen atoms and hydrogen like ions, *Nucl. Fusion*,
6(3), 182 (1966).

[Percival, 1975] I. C. Percival and D. Richards. The theory of
collisions between charged particles and highly excited atoms, in
Atomic and Molecular Physics, Vol. 11 (D. Bates and B. Bederson, eds.), Academic Press, New York, 1975, p. 1.

[Pillet, 1984] P. Pillett, H. B. van Linden van den Heuvell, W. W.
Smith, R. Kachru, N. H. Tran, and T. F. Gallagher. Microwave
ionization of Na Rydberg atoms, *Phys. Rev.*, *A30*, 280 (1984).

[Pokrowsky, 1983] P. Pokrowsky, W. E. Moerner, F. Chu, and G.
C. Bjorkland. Reading and writing of photochemical holes using
GaAlAs-diode lasers, *Opt. Lett.*, *8*(5), 280 (1983).

[Popescu, 1974] D. Popescu, C. B. Collins, B. W. Johnson, and I. Popescu. Multiphoton excitation and ionization of atomic cesium with a tunable dye laser, *Phys. Rev.*, *A 9*(3), 1182 (1974).

[Rajnak, 1978] K. Rajnak and B. W. Shore. Regularities in the s-electron binding energies in $\ell^n s^m$ configurations, *J. Opt. Soc. Am.*, *68*(3), 360 (1978).

[Robin, 1980] M. B. Robin. Multiphoton fragmentation and ionization, *Appl. Opt.*, *19*(23), 3941 (1980).

[Rohlfing, 1983] E. A. Rohlfing, D. M. Cox, and A. Kaldor. Photoionization measurements on isolated iron atom clusters, *Chem. Phys. Lett.*, *99*(2), 161 (1983).

[Rohlfing, 1984a] E. A. Rohlfing, D. M. Cox, and A. Kaldor. Photoionization of isolated nickel atom clusters (Ni_x), *J. Phys. Chem.*, *88*, 4497 (1984).

[Rohlfing, 1984b] E. A. Rohlfing, D. M. Cox, A. Kaldor, and K. H. Johnson. Photoionization spectra and electronic structure of small iron clusters, *J. Chem. Phys.*, *81*(9), 3846 (1984).

[Rothberg, 1981] L. J. Rothberg, D. P. Gerrity, and V. Vaida. Effects of non-resonant ionization on multiphoton ionization lineshapes, *J. Chem. Phys.*, *75*(9), 4403 (1981).

[Rundel, 1975] R. D. Rundel, F. B. Dunning, H. C. Goldwire, Jr., and R. F. Stebbings. Near-threshold photoionization of xenon metastable atoms, *J. Opt. Soc. Am.*, *65*(6), 628 (1975).

[Rutherford, 1908] E. Rutherford and H. Geiger. An electrical method of counting the number of α-particles from radioactive sources, *Proc. R. Soc. London*, *A 81*, 141 (1908).

[Safinya, 1979] K. A. Safinya and T. F. Gallagher. Observation of interferences between discrete autoionizing states in the photoexcitation spectrum of barium, *Phys. Rev. Lett.*, *43*(17), 1239 (1979).

[Safinya, 1980] K. A. Safinya, J. F. Delpech, and T. F. Gallagher. Effects of electric fields on doubly excited autoionizing Rydberg states of barium, *Phys. Rev.*, *A 22*(3), 1062 (1980).

[Schiavone, 1977] J. A. Schiavone, D. E. Donohue, D. R. Herrick, and R. S. Freund. Electron-impact excitation of helium: cross sections, n and ℓ distributions of high Rydberg states, *Phys. Rev.*, *A 16*(1), 48 (1977).

[Shenstone, 1931] A. G. Shenstone. Ultra-ionization potentials in mercury vapor, *Phys. Rev.*, *38*, 873 (1931).

[Shore, 1978] B. W. Shore and M. A. Johnson. Coherence vs. incoherence in stepwise excitation, *J. Chem. Phys.*, *68*(12), 5631
(1978).

[Shore, 1980] B. W. Shore. Coherence in laser excitation, in *Proc.
II Int. Conf. Multiphoton Process.*, Budapest, Hungary, April
1980; UCRL-84260, Lawrence Livermore National Laboratory,
Livermore, Calif., 1980.

[Snavely, 1975] B. B. Snavely, R. W. Solarz, and S. A. Tuccio.
Separation of uranium isotopes by selective photoionization, in
Laser Spectrosc., *43*, Proc. Second Int. Conf., Megeve, France,
June 23-27, 1975 (S. Haroche, J. C. Pebay-Peyroula, T. W.
Hänsch, and S. E. Harris, eds.), Springer-Verlag, Berlin, 1975,
p. 267.

[Solarz, 1975] R. W. Solarz, J. A. Paisner, L. R. Carlson, C. A.
May, and S. A. Johnson. *Multi-Step Laser Spectroscopy in
Atomic Uranium*, UCRL-77590, Rev. 1, Lawrence Livermore Nation-
al Laboratory, Livermore, Calif., 1975.

[Solarz, 1976] R. W. Solarz, C. A. May, L. R. Carlson, E. F. Worden, S. A. Johnson, J. A. Paisner, and L. J. Radziemski. Detection of Rydberg states in uranium using time resolved stepwise
laser photoionization, *Phys. Rev.*, *A14*(3), 1129 (1976).

[Solarz, 1980] R. W. Solarz, E. F. Worden, and J. A. Paisner. Reactive collisions of laser excited atoms, *Opt. Eng.*, *19*(1), 85
(1980).

[Sorokin, 1975] P. P. Sorokin, J. A. Armstrong, R. W. Dreyfus,
R. T. Hodgson, J. R. Lankard, L. H. Manganaro, and J. J.
Wynne. Generation of vacuum ultraviolet radiation by nonlinear
mixing in atomic and ionic vapors, in *Laser Spectrosc.*, *43*, Proc.
Second Int. Conf., Megeve, France, June 23-27, 1975 (S. Haroche, J. C. Pebay-Peyroula, T. W. Hänsch, and S. E. Harris,
eds.), Springer-Verlag, Berlin, 1975, p. 46.

[Teachout, 1971] R. R. Teachout and R. T. Pack. The static dipole
polarizabilities of all the neutral atoms in their ground states,
At. Data, *3*, 195 (1971).

[Tomkins, 1982] F. S. Tomkins and R. Mahon. Generation of continuously tunable narrowband radiation from 1220 to 1174 Å in
Hg vapor, *Opt. Lett.*, *7*(7), 304 (1982).

[Vaida, 1978] V. Vaida, M. B. Robin, and N. A. Kuebler. A note on
the elusive $^1E_{2g}$ state in the two-photon spectrum of benzene,
Chem. Phys. Lett., *58*(4), 557 (1978).

[van Linden van den Heuvell, 1984] H. B. van Linden van den Heuvell, R. Kachru, N. H. Tran, and T. F. Gallagher. Excitation spectrum of Na Rydberg states in a strong microwave-field: a connnection between two points of view, *Phys. Rev. Lett.*, 53(20), 1901 (1984).

[Vasilenko, 1970] L. S. Vasilenko, V. P. Chebotayev, and A. V. Shishaev. Line-shape of two-photon absorption in a standing wave field in a gas, *JETP Lett.*, 12, 113 (1970).

[Voronov, 1965] G. S. Voronov and N. B. Delone. Ionization of the xenon atom by the electric field of ruby laser emission, *Sov. Phys. JETP*, 1(2), 66 (1965).

[Wainfan, 1955] N. Wainfan, W. C. Walker, and G. L. Weissler. Photoionization efficiencies and cross sections in O_2, N_2, CO_2, A, H_2O, H_2, and CH_4, *Phys. Rev.*, 99(2), 542 (1955).

[Weast, 1982] R. X. Weast, ed. *CRC Handbook of Chemistry and Physics*, CRC Press, Boca Raton, Fla., 1982, p. E71.

[Weiner, 1980] J. Weiner. On laser-induced harpooning reactions, *J. Chem. Phys.*, 72(10), 5731 (1980).

[Weiner, 1983] J. Weiner, R. Bonanno, and J. Boulmer. Absolute cross section measurements of associative ionization in Na(3s) + Na*(np) collisions, *J. Phys.*, B16, 3015 (1983).

[Weiner, 1985] J. Weiner. Associative ionization cross section vs. principal quantum for Na* (ns,p,d) + Na (3s) collisions with 16 < n < 28 (submitted for publication).

[Weissler, 1962] G. L. Weissler. Cross section measurements for some photon-gas interaction processes, *J. Quant. Spectrosc. Radia. Transfer*, 2, 383 (1962).

[Weissler, 1972] G. L. Weissler. Photoionization, a survey of methods of measurements, including plasma arc spectroscopy, *Sci. Light*, 21(1), 89 (1972).

[Wentzel, 1927] G. Wentzel. Uber strahlungslose Quantensprunge, *Z. Phys.*, 43, 524 (1927).

[Winograd, 1982] N. Winograd, J. P. Baxter, and F. M. Kimock. Multiphoton resonance ionization of sputtered neutrals: a novel approach to materials characterization, *Chem. Phys. Lett.*, 88(6), 581 (1982).

[Wiza, 1979] J. L. Wiza. Microchannel plate detectors, *Nucl. Instrum. Methods*, 162, 587 (1979).

[Worden, 1978a] E. F. Worden, R. W. Solarz, J. A. Paisner, and
J. G. Conway. First ionization potentials of lanthanides by laser
spectroscopy, *J. Opt. Soc. Am.*, *68*(1), 52 (1978).

[Worden, 1978b] E. F. Worden, J. A. Paisner, and J. G. Conway.
Associative ionization of laser-excited Rydberg states in strontium
vapor, *Opt. Lett.*, *3*(4), 156 (1978).

[Worden, 1979] E. F. Worden and J. G. Conway. Laser spectroscopy
of neptunium; first ionization potential, lifetimes and new high-
lying energy levels of Np I, *J. Opt. Soc. Am.*, *69*(5), 733 (1979).

[Worden, 1980] E. F. Worden and J. G. Conway. Multistep laser photoionization of the lanthanides and actinides, in ACS Symp. Ser.,
131, *Lanthanide and Actinide Chemistry and Spectroscopy* (N. M.
Edelstein, ed.), 1980, p. 381.

[Worden, 1984] E. F. Worden, B. Comaskey, J. A. Densberger, J. J.
ịChristensen, J. M. McAfee, J. A. Paisner, and J. G. Conway.
The ionization potential of neutral iron, Fe I, by multistep laser
spectroscopy, *J. Opt. Soc. Am.*, *B*1, 314 (1984).

[Wynne, 1981] J. J. Wynne and R. Beigang. Accurate measurement
of relative oscillator strengths by phase matched nonlinear optics,
Phys. Rev., *A*23(5), 2736 (1981).

[Wynne, 1984] J. Wynne, ed. *Current Trends in Atomic Spectroscopy*, report on a workshop of the National Research Council held
in Tucson, Ariz., Oct. 24-26, 1982, National Academy Press,
Washington, D.C., 1984.

[Zakheim, 1980] D. S. Zakheim and P. M. Johnson. Rate equation
modelling of molecular multiphoton ionization dynamics, *Chem.
Phys.*, *46*, 263 (1980).

[Zherikhin, 1983] A. N. Zherikhin, V. S. Letokhov, V. I. Mishin,
M. E. Muchnik, and V. N. Fedoseyev. Production of photoionic
gallium beams through stepwise ionization of atoms by laser radiation, *Appl. Phys. B*30, 47 (1983).

4

Applications of Laser Absorption Spectroscopy

Hao-Lin Chen   *Lawrence Livermore National Laboratory, Livermore, California*

4.1 INTRODUCTION

The advent of lasers, in particular tunable lasers, has greatly expanded the field of spectroscopy in many areas. As discussed in the previous chapters, tunable laser oscillations have now been obtained at wavelengths covering the range from the far-infrared to the ultraviolet. The special characteristics of laser light—its high intensity, directionality, coherence, wavelength tunability, monochromaticity, and temporal flexibility—have made lasers a particularly useful tool for spectroscopy. The recent development of lasers and associated technologies such as optical detectors and signal-averaging electronics has enabled many new kinds of spectral analyses to be made of chemical media and chemical-reaction dynamics studies. These analyses and studies could not be accomplished with conventional spectrometric techniques using spectral lamps or blackbody light sources. In this chapter we explore a number of promising applications of laser-absorption spectroscopy.

Absorption spectroscopy is probably one of the most widely used analytical tools in physics, chemistry, and industry. Infrared and far-infrared spectroscopy are commonly used for gas analysis and identification of chemical structures, while visible and ultraviolet spectroscopy are extensively used for quantitative analysis of atoms, ions, and chemical species in solution.

In addition to the characteristic frequencies absorbed by a given chemical species, another important characteristic is the absorption strength or coefficient. In conventional absorption spectroscopy, a cell containing the material under investigation is placed between a source of continuous radiation and the spectrometer. Radiation from the source coincident with frequencies at which the substance absorbs is removed from the continuum, producing the substance's absorption spectrum. As radiation of a given frequency from a source traverses a homogeneous absorbing medium, under certain conditions it is reduced in intensity by the same fractional amount through each succeeding unit length of path, given by the exponential relationship commonly known as Lambert's law:

$$I = I_0 \exp(-\alpha \ell) \tag{4.1}$$

where I_0 is the incident light intensity, I the intensity after traversing a thickness ℓ of the medium, and α is a constant known as the absorption coefficient.

It is well known that the low brightness of blackbody light sources and the attainable dispersion of monochromators impose limits on the resolution and sensitivity that can be achieved by conventional absorption spectroscopy. Conventional spectrometers are based on attaining wavelength selection by the mechanical motion of dispersive elements such as prisms or gratings, or, in the case of the Fourier-transform spectrometer, by the mechanical displacement of mirrors. Over the years, such devices have been developed into reliable and accurate systems with reasonable transmission and wavelength accuracy. Unfortunately, the resultant spectral resolution and sensitivity remain relatively poor compared to laser methods. The limitations are largely due to the trade-off between resolution and sensitivity. That is, the intensity of light from a blackbody source is always decreased as the bandpass of the monochromator is decreased, since the spectral brightness must be constant during measurement.

The introduction of lasers to absorption spectroscopy has significantly improved both the sensitivity and the spectral resolution of absorption techniques. This improvement is principally due to the attainable spectral purity and brightness of lasers, which are many orders of magnitude higher than those of conventional incoherent light sources coupled with dispersion elements. By using lasers it is now possible to achieve spectroscopic resolution that is about 1×10^4 to 1×10^5 × higher than that of conventional spectrometers. Today, the limit of resolution is no longer the instrumental line width, but more likely the natural width of the transition under observation or, in the case of long-lived excited states, the line width determined by the transit time of the absorber across the radiation field. The intense photon fluxes provided by the laser sources have also improved the signal-to-noise ratio of absorption spectra. The coherence, directionality, and temporal characteristics of lasers permit us to make a variety of different absorption measurements with high resolution in frequency, space, and time. Examples are saturation and polarization spectroscopy and excited-state absorption spectroscopy using optical double-resonance techniques.

There are many laser techniques that can be used to measure the densities of ground and excited states of molecules, atoms, and ions. We will discuss only the absorption technique here. Other techniques, such as laser-induced fluorescence and laser-photoionization techniques, will be discussed in other chapters. Because of the vastly increasing number of publications in the field of laser spectroscopy, we will not give a complete review of laser-absorption spectroscopy in this book. However, we will try to give an informative survey of some of the most popular and promising techniques and their developments.

Section 4.2 explains the fundamentals of absorption spectroscopy, comparing laser-absorption spectroscopy to conventional spectrometric techniques. Section 4.3 review the development of laser-absorption spectroscopy and, in particular, the techniques employing tunable diode lasers and tunable dye lasers. Section 4.4 reviews the past and current applications of laser-absorption techniques in physics, chemistry, and industry.

4.2 FUNDAMENTALS OF ABSORPTION SPECTROSCOPY

4.2.1 General Considerations

Absorption spectroscopy is the study of the attenuation of electromagnetic radiation by matter. In absorption, a photon of electromagnetic radiation is absorbed by a particle that is in a particular quantum state characterized by a certain energy level. After absorption, the particle makes a transition to a higher energy level. Atomic (or atomic ion) spectra involve only transitions of electrons from one electronic energy level to another. On the other hand, molecular (or molecular-ion) spectra may involve transitions between rotational and vibriational energy levels in addition to electronic transitions. Thus, if an isolated molecule is originally in the ground electronic energy level (the zero-point vibrational energy level and a particular rotational energy level), absorption of a photon may excite the molecule to a higher electronic vibrational and/or rotational energy level. There are many transitions that might give rise to absorption, but only those that satisfy certain selection rules are allowed. The origin of the specific selection rules was discussed in Chapter 1. Electronic, vibrational, and rotational state changes may all occur upon the absorption of a single photon. From the law of energy conservation [Herzberg, 1950],

$$h\nu = (E^h_{ele} - E^\ell_{ele}) + (E^h_{vib} - E^\ell_{vib}) + (E^h_{rot} - E^\ell_{rot}) \qquad (4.2)$$

where E^h and E^ℓ are the energy of the higher and lower energy states and ν is the frequency of radiation. Since only small quanta of energy are required to change from one rotational level to another, these changes may be isolated in the far-infrared and microwave regions of the spectrum. Vibrational energy-level changes result from absorption in the infrared region. However, for each vibrational level there are many rotational levels, so bands of closely spaced absorption lines are obtained. Electronic energy-level changes produce absorption in the visible and ultraviolet regions. Again because of the vibrational and rotational levels associated with each electronic level, bands of very closely spaced absorption lines may be obtained. As described in Chapter 1, detailed analysis of the electronic vibrational and rotational absorption spectra not only provides valuable information about the spacings and location of energy levels, but also allows us to determine the configuration and structure of the atomic and molecular species. A brief discussion of molecular energy levels pertaining to absorption spectroscopy is given in Sec. 4.2.3.

Figure 4.1 shows a conventional absorption spectrometer. The radiation from a source (glow bar, high-pressure arc, flashlamp, etc.) emitting a spectral continuum traverses the absorption cell. The transmitted intensity is recorded as a function of wavelength after passing through a spectrograph or an interferometer. The spectral resolution is limited by the resolving power of these instruments (resolving power is a measure of the ability of the spectrometer to distinguish closely spaced spectral lines). For example, if the minimum observable separation is $\delta\lambda$ at wavelength λ, the resolving power is given the dimensionless quantity $\lambda/\delta\lambda$. This can range up to about 6×10^5 in large grating spectrographs. Over the years the resolution of conventional absorption spectrometers has been improved using larger and larger grating spectrographs or high-dispersion interferometers such as the Fourier spectrometers, which may approach the Doppler-limited resolution of molecular absorption lines.

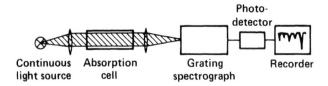

Continuous   Absorption   Grating   Recorder
light source   cell   spectrograph

Fig. 4.1 Conventional absorption spectroscopy. The spectral resolution is limited by the resolving power of the grating spectrograph.

The sensitivity of the conventional absorption spectrometer is determined by the lowest measurable light attenuation in the absorption cell, which is usually set by intensity fluctuations of the light source and by detector noise. Because of the divergence of the spectral lamp, light collection and propagation along a long-path absorption cell are often difficult. To measure weak absorption lines or low-concentration species, one therefore usually has to increase the signal-to-noise level by increasing the gas pressure rather than the cell length. This results in pressure broadening of the lines and therefore decreases the spectral resolution.

The most widely used conventional technique for measuring atomic densities in vapor is atomic-absorption spectrometry using a spectral line source [Robinson, 1966]. Figure 4.2 is a schematic diagram of the atomic-absorption spectrometer using hollow-cathode lamps and flame atomizers. The source radiation, which consists of the line spectrum of the desired element, is passed through the flame atomizer and the amount of radiation absorbed by the ground-state atoms in the atomizer is measured. The monochromator is used only to isolate the desired source radiation. Three types of spectral line sources are commonly used in atomic-absorption spectrometry: hollow-cathode discharge tubes, metal-vapor arc lamps, and electrodeless discharge tubes. Since dispersive elements are not required, the sensitivity of

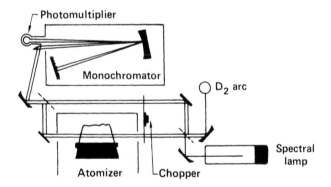

Fig. 4.2 Atomic absorption spectroscopy using spectral lamps. Energy from the spectral lamps is divided into two beams, the sample beam and the reference beam. The sample beam travels through the sample compartment while the reference beam travels around it. The beams recombined before entering the monochromator and the ratio of the two intensities is measured. The double-beam arrangement compensates for any drift in lamp intensity, detector sensitivity and electronic gain, and provides ultimate results in detectivity and precesion of the measurement.

this technique is significantly higher than that of infrared IR absorption spectrometry using gratings. However, it can only be applied to measure the concentration of atomic species for which a spectral line source of the same element is available. Also, to ensure the accuracy of this analytical technique, calibration runs using samples of similar concentration must be made before and after each measurement. Although numerous other disadvantages (such as radiation interference, etc.) have also been noted, the simplicity, convenience, and relatively low cost of this atomic-absorption technique make it very attractive for routine chemical analysis.

Molecular absorption spectra are commonly obtained with a grating or prism spectrometer and more recently with computer-operated Fourier-transform spectrometers. For chemical identification, a low-resolution grating or prism spectrometer as shown in Fig. 4.1 is adequate. However, for resolving complex interfering rotational lines, one normally uses a high-resolution Fourier-transform infrared spectrophotometer. Figure 4.3 is a simplified schematic of the Fourier-transform spectrophotometer using a Michelson interferometer. During the absorption measurement, the infrared radiation emitted from a thermal source is monitored by a mercury-cadmium-telluride detector or a deuterated triglycine sulfate detector. A He-Ne laser is used to provide the exact position and direction of the moving mirror, establishing the exact frequency of the infrared radiation. Data pro-

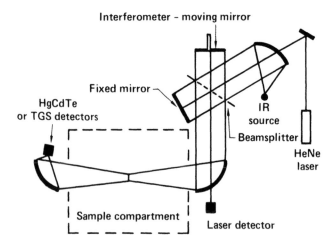

Fig. 4.3 Fourier transform infrared spectrometer—optical arrangement. The interferometer moving mirror is electronically controlled. The position of the moving mirror is monitored at all times.

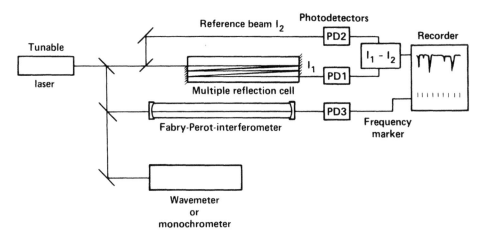

Fig. 4.4 Absorption spectroscopy using lasers. The spectral resolution is limited by the line width of the absorbing species. The Fabry-Perot interferometer is used as a frequency marker which enable one to determine the spectral separation of absorption peaks and line widths of absorption lines. The wavemeter is used to determine the exact wavelength where the laser absorption takes place.

cessing is achieved with a microprocessor coupled with a specially designed hardware multiplier, providing accurate instrumental control and fast Fourier transformation. The Fourier-transform infrared spectrometer can cover a spectral range of 10 to 15,000 cm^{-1} with a maximum resolution of 0.08 cm^{-1}, which is 10× better than the grating spectrometer. The frequency precision is better than 0.01%, which is many orders of magnitude better than grating spectrographs.

Figure 4.4 shows a typical setup for absorption spectroscopy using lasers. Because of the monochromaticity of laser light, the laser itself also acts as the monochromator and hence the spectral resolution is limited only by the line width of the absorbing species. The directionality and attainable spatial collimation of laser beams also allow us to propagate the beam along a long-path absorption cell, which increases the signal-to-noise ratio of the measurement and consequently increases the sensitivity. More important, this also allows us to perform absorption measurements at low pressure, avoiding pressure broadening of absorption lines. Finally, because of the high intensity of lasers, detector noise plays a minimal role in the absorption measurement. The signal-to-noise level and hence the sensitivity of the laser technique is often many orders of magnitude higher than that for conventional absorption spectroscopy. In Secs. 4.2.2 and 4.2.3 we

briefly review the physics of atomic- and molecular-absorption spectroscopy.

4.2.2 Atomic-Absorption Spectroscopy

When spectrally continuous light is passed through an atomic vapor,
the emergent light is no longer continuous because of the interaction
between the incident electromagnetic radiation and the atomic vapor.
Certain details of the atomic vapor, such as its density, thermal velocity, and internal state population, can be obtained by spectrally
analyzing the emergent light and comparing the resultant spectrum
with that of the incident light source.

In a typical absorption experiment the absorption spectrum of the
atomic vapor is obtained by interposing the vapor of interest between
the source and the entrance to the spectrograph. The result is a
series of characteristic absorption lines. The energy and sequence of
the lines from one type of atom are typical of that atom and differ
from the spectrum of any other atom. Hence we can use the footprint
of the atomic-absorption spectra to distinguish one atom from another
[Herzberg, 1944; Kuhn, 1962].

To correlate the experimentally measurable absorption intensity of
the atoms with the theoretical oscillator strength and transition moment, it is necessary to derive equations of transition probabilities of
absorption and spontaneous emission and to determine selection rules.
Details of these derivations and selection rules are given in Chapter 1.
We will not derive the Einstein transition probability here. However,
we will briefly discuss the correlation between the Einstein transition
probability and the integrated absorption of atoms and emphasize some
of the parameters pertaining to the physics of absorption spectroscopy.

Consider a parallel beam of light distributed in energy from $\bar{\nu}$ to
$\bar{\nu} + d\bar{\nu}$ and of intensity $I_{\bar{\nu}}$ passing through a layer of atoms between ℓ
and ℓ and $\ell + d\ell$. The decrease in intensity of the beam due to the
absorption of light between $\bar{\nu}$ and $\bar{\nu} + d\bar{\nu}$ is given as [Barrow, 1962;
Okabe, 1978]

$$-dI_{\bar{\nu}} \, d\bar{\nu} = d\ell \, ch\bar{\nu}\rho_{\bar{\nu}}(N_1 B_{12} - N_2 B_{21}) \qquad (4.3)$$

where B_{12} is the Einstein transition probability of absorption from
state 1 to state 2 and B_{21} is the transition probability of spontaneous
emission from state 2 to state 1. The nomenclature used in this section is described in Chapter 1. The intensity $I_{\bar{\nu}}$ is given in units of
erg cm^{-2} s^{-1} per wave number and N_1, N_2 are the number of atoms
per cubic centimeter at states 1 and 2 that are capable of absorbing
or emitting light between $\bar{\nu}$ and $\bar{\nu} + d\bar{\nu}$.

The intensity $I_{\bar{\nu}}$ is related to the energy density $\rho_{\bar{\nu}}$ by

$$I_{\bar{\nu}} = c\rho_{\bar{\nu}} \tag{4.4}$$

Thus we have

$$-dI_{\bar{\nu}} \, d\bar{\nu} = d\ell \, h\bar{\nu}I_{\bar{\nu}}(N_1 B_{12} - N_2 B_{21}) \tag{4.5}$$

The absorption coefficient, $\alpha_{\bar{\nu}}$ of N atoms per cubic centimeter, is defined by

$$I_{\bar{\nu}} = I_0 e^{-\alpha_{\bar{\nu}}\ell} \tag{4.6}$$

where I_0 and $I_{\bar{\nu}}$ are incident and transmitted intensities at frequency $\bar{\nu}$ and ℓ is the absorption path length in centimeters.

Differentiating, we find for the intensity decrease over the range from $\bar{\nu}$ to $\bar{\nu} + d\bar{\nu}$ and the path length between ℓ and $\ell + d\ell$:

$$-dI_{\bar{\nu}} \, d\bar{\nu} = d\ell \, I_{\bar{\nu}}\alpha_{\bar{\nu}} \, d\bar{\nu} \tag{4.7}$$

From the above we obtain

$$\int_0^\alpha \alpha_{\bar{\nu}} \, d\bar{\nu} = h\bar{\nu}(N_1 B_{12} - N_2 B_{21}) \tag{4.8}$$

Integration over the whole absorption line leads to

$$\int_0^\alpha \alpha_{\bar{\nu}} \, d\bar{\nu} = h\bar{\nu}(N_1 B_{12} - N_2 B_{21})$$

$$= h\bar{\nu}N_1 B_{12} \quad \text{(when } N_2 = 0\text{)} \tag{4.9}$$

Using the relationship

$$B_{21} = \frac{A_{21}}{8\pi hc\bar{\nu}^3} \qquad A_{21} = \frac{1}{\tau} \qquad g_1 B_{12} = g_2 B_{21} \tag{4.10}$$

we find that

$$\int_0^\alpha \alpha_{\bar{\nu}} \, d\bar{\nu} = \frac{g_2}{g_1} \frac{A_{21}N_1}{8\pi c\bar{\nu}^2} = \frac{g_2}{g_1} \frac{N_1}{8\pi c\bar{\nu}^2\tau} \tag{4.11}$$

where A_{21} is the Einstein transition probability of spontaneous emission from state 2 to state 1, g_1 and g_2 are the degeneracies, c is the velocity of light, $\bar{\nu}$ is the wavenumber of the absorption line, and τ is the radiative lifetime of the upper state. The intensity of the absorption line is sometimes expressed as a dimensionless quantity called oscillator strength f, which is the ratio of the observed absorption coefficient $\int \alpha_{\bar{\nu}} \, d\bar{\nu}$ to that of N electrons bound to the atom. Treating N bound electrons as N harmonic oscillators, we obtain as the integrated absorption coefficient of N electrons the quantity

$$N \frac{\pi e^2}{mc^2} \tag{4.12}$$

Thus the oscillator strength [Barrow, 1962] is

$$f = \frac{\int \alpha_{\bar{\nu}} \, d\bar{\nu}}{N\pi e^2/mc^2} = \frac{mc^2}{N\pi e^2} \int \alpha_{\bar{\nu}} \, d\bar{\nu} \tag{4.13}$$

where m and e are the mass and charge of the electron, respectively, and $\int \alpha_{\nu} \, d\bar{\nu}$ is in units of cm^{-2}.

In a typical absorption measurement there are several line-broadening mechanisms one must consider [Mitchell, 1961; Okabe, 1978]. A brief description of line shapes was given in Chapter 1. In this section we discuss the mechanisms of line broadening observed in absorption spectroscopy. If the pressure of absorbing atoms inside the sample cell is kept low (below 10^{-2} torr) and the foreign gas pressure is also kept below a few torr, the line width of the atomic absorption will be governed mainly by the lifetime of the excited states and by Doppler broadening. Doppler broadening is caused by motion of the atom. If an atom absorbs light while it is moving with a velocity v whose component along the line of sight is v_x, the absorbed wavelength would appear to be shifted by $\lambda_0 v_x/c$, where λ_0 is the absorbed wavelength from the atom at rest. This is called the Doppler shift. Because of the Doppler effect, spectral lines absorbed by atoms are broadened according to the Maxwell-Boltzmann velocity distribution of atoms. As described in Chapter 1, the Doppler line shape is best described by a Gaussian function with a Doppler width (FWHM) $\Delta \bar{\nu}_D$, given by

$$\Delta \bar{\nu}_D = \frac{2\sqrt{2k \ln 2}}{c} \bar{\nu}_0 \sqrt{\frac{T}{M}}$$

$$= 7.16 \times 10^{-7} \bar{\nu}_0 \sqrt{\frac{T}{M}} \quad cm^{-1}$$

$$= \frac{214}{\lambda} \sqrt{\frac{T}{M}} \quad MHz \tag{4.14}$$

where k is the Boltzmann constant, $\bar{\nu}_0$ the wavenumber at the center of absorption line, T the gas temperature in K, and M the atomic weight. In conventional spectroscopy, Doppler broadening is of comparative magnitude with instrumental broadening at very high resolution.

If the absorption-line profile is determined by Doppler broadening alone [Mitchell, 1961], the integrated absorption coefficient can be expressed in terms of α_0, the absorption coefficient at the peak, and $\Delta\bar{\nu}_D$, the Doppler width at the half-maximum points. For a Doppler-broadened absorption line,

$$\alpha_{\bar{\nu}} = \alpha_0 \, \exp\left(-\left\{\left[\frac{2(\bar{\nu} - \bar{\nu}_0)}{\Delta\bar{\nu}_D}\right]\sqrt{\ln 2}\right\}^2\right) \qquad (4.15)$$

we obtain

$$\int \alpha_{\bar{\nu}} \, d\bar{\nu} = \frac{1}{2}\sqrt{\frac{\pi}{\ln 2}}\,\alpha_0 \, \Delta\bar{\nu}_D$$

$$\alpha_0 = \frac{2}{\Delta\bar{\nu}_D}\sqrt{\frac{\ln 2}{\pi}}\,\frac{\pi e^2}{Mc^2}Nf$$

$$= \frac{0.832 \times 10^{-12}}{\Delta\bar{\nu}_D}Nf \qquad (4.16)$$

where $\Delta\bar{\nu}_D$ is the Doppler width in cm^{-1}. Therefore, by measuring α_0 and the Doppler width $\Delta\bar{\nu}_D$, one can determine the density of atoms in the sample cell.

In the low-pressure region there is another mechanism that could also broaden the absorption line—natural line broadening, resulting from the finite lifetime of excited atoms. This natural line broadening is usually very small, of the order of 10^{-3} cm^{-1}, which is 100 times narrower than the instrumental resolution of conventional spectroscopy. The relationship between the mean lifetime τ of the excited state and the line broadening $\Delta\bar{\nu}_0$ is

$$\Delta\bar{\nu}_0 = \frac{1}{2\pi c\tau} = \frac{5.30 \times 10^{-12}}{\tau} \text{ cm}^{-1} \qquad (4.17)$$

where c is the velocity of light in cm s^{-1} and τ is in seconds. Thus, when the pressure of a gas in the absorption cell is low, $<10^{-2}$ torr, the absorption line width is mainly controlled by natural and Doppler-broadening mechanisms.

When the gas pressure is higher than 10^{-2} torr, the absorption
line will be further broadened due to collisions with foreign gas atoms.
This line-broadening mechanism is commonly called Lorentzian broadening [Mitchell, 1961]. The broadening due to collisions with electrons or ions is called Stark broadening. The Lorentzian line-shape
function $f_L(\bar{\nu})$ and the Lorentzian line width (FWHM) $\Delta \bar{\nu}_L$ have been
described earlier. The collisional-broadening line width $\Delta \bar{\nu}_L$ in a
multicomponent gas can be written as

$$\Delta \bar{\nu}_L = \sum P_i \, \Delta \bar{\nu}_{L,i} \tag{4.18}$$

where $\Delta \bar{\nu}_{L,i}$ is the collision or Loretnzian-broadened width per unit
pressure of the ith gas species and P_i is the partial pressure in the
sample being analyzed. $\Delta \bar{\nu}_L$ increases linearly with the increase of
the foreign gas pressure. Taking atomic absorption of Hg as an example, when the background pressure is low, the line width of the
Hg 2537-Å line at 300 K is controlled mainly by the Doppler-broadening mechanism. However, as the pressure of foreign gas such as He
or Ne is increased to about 100 torr, the amount of line broadening
resulting from Lorentzian collisional broadening also becomes equal to
that of the Doppler broadening. Lorentzian broadening is characterized by asymmetry and a shift of the maximum of the absorption line.
The first attempt to explain these changes in an absorption line caused
by collisions with foreign gases was made in 1931 [Mitchell, 1961]. It
was thought that as an atom and molecule collide, a quasi-molecule
would form during the moment of close impact. Therefore, the excitation transition of an atom that is undergoing both optical absorption
and molecular collision can also be interpreted as a transition between
the ground and excited states of the short-lived quasi-molecule. By
applying the statistical theory of density fluctuations, one can calculate the shift of an absorption line in terms of the pressure of the
foreign gas and of atomic constants.

Increasing the pressure of the absorbing atoms will also cause
line-broadening effects. The self-broadening due to collisions of the
absorbing atoms with atoms of the same kind is commonly called Holts-
mark broadening. The line width of self broadening $\Delta \bar{\nu}_{SB}$ is proportional to the density of the absorbing gas [Hutcherson, 1973]:

$$\Delta \bar{\nu}_{SB} = \kappa \frac{e^2 f N_1}{c^2 n \bar{\nu}_0} \tag{4.19}$$

where κ is a constant, e and m are the charge and mass of an electron,
respectively, c is the velocity of light, f is the oscillator strength,
and $\bar{\nu}_0$ is the wavenumber at the line center. Self-broadening be-

comes appreciable at a pressure greater than 1 torr and the line shape becomes asymmetric as a result of molecular dimer or trimer formation.

In a typical absorption measurement at moderate density, the line shape will be controlled by both Doppler and collisional broadening effects. The simultaneous broadening of a transition by the Doppler and collisional mechanisms (Lorentzian) is mathematically described by the Voigt function [Penner, 1959; Armstrong, 1967] formed by the convolution of Gaussian and Lorentzian line shapes. A good approximation of the Voigt absorption profile can be obtained by assuming that every velocity group of atoms in the Gaussian-shaped Doppler profile is broadened to the same extent by the Lorentzian profile. The Lorentzian line shape for a group of atoms having a specified velocity v can be written as

$$\alpha_{\bar{\nu}}(v) = \frac{\Delta \bar{\nu}_L}{2\pi\{[\bar{\nu} - \bar{\nu}_0 + \bar{\nu}_0(v/c)]^2 + (\Delta \bar{\nu}_L/2)^2\}} \qquad (4.20)$$

If we further assume that the velocity of atoms has a Maxwell Boltzmann distribution function, we can write the fraction of atoms with velocity between v and v + dv as

$$\frac{dN}{N} = \left(\frac{M}{2\pi kT}\right)^{1/2} \exp\left(-\frac{Mv^2}{2kT}\right) dv \qquad (4.21)$$

By multiplying the two equations above and integrating over all velocities, we obtain the Voigt profile function:

$$\alpha_{\bar{\nu}} = \left(\frac{M}{2\pi kT}\right)^{1/2} \int_{-\infty}^{\infty} \frac{\Delta \bar{\nu}_L}{2\pi\{[\bar{\nu} - \bar{\nu}_0 + \bar{\nu}_0(v/c)]^2 + (\Delta \bar{\nu}_L/2)^2\}}$$

$$\exp\left(\frac{-Mv^2}{2kT}\right) dv \qquad (4.22)$$

The value of $\bar{\nu}_0(v/c)$ is included to account for the shift of absorption peak due to the Doppler effect. The profile of the Voigt function can be rearranged and written as

$$\alpha_{\bar{\nu}} = \alpha_{\bar{\nu}_0} \frac{a}{\pi} \int_{-\infty}^{\infty} \frac{\exp(-y^2)\, dy}{a^2 + (\chi - y)^2} = \alpha_{\bar{\nu}_0} V(a, \chi) \qquad (4.23)$$

where $\alpha_{\bar{\nu}_0}$ is the maximum absorption coefficient when only Doppler broadening is important. The value of a and χ are defined as (Omenetto, 1979)

$$a = \frac{\Delta \bar{\nu}_L}{\Delta \bar{\nu}_D} \sqrt{\ln 2}$$

$$\chi = \frac{\bar{\nu} - \bar{\nu}_0}{\Delta \bar{\nu}_D} 2 \sqrt{\ln 2} \qquad\qquad (4.24)$$

The values of $V(a, \chi)$ have been tabulated by Young [Young, 1965] and Mitchell and Zemansky [Mitchell, 1961] and have been used directly to calculate $\alpha_{\bar{\nu}}$. The parameter χ is the spectral distance from the line center normalized by the Doppler width. The Voigt parameter a is a measure of the relative magnitude of Doppler and collisional broadening line widths.

There are many other mechanisms for the broadening of spectral lines. For instance, the absorption line is also broadened by the interaction of atoms with moving ions and electrons. Thus in a high-density plasma, Stark broadening may be important. In a solid the local crystalline field may also shift and split the energy levels due to the Stark effect.

Many experimental methods for the measurement of absorption profiles have been developed. Although many absorption profiles have been measured directly with an ordinary light source using the Zeeman scanning technique [Kopfermann, 1958] or by using a continuum source with a synchronized monochromator-interferometer scan [Wagenaar, 1974], these techniques are experimentally complicated. They also suffer from limited wavelength resolution that requires deconvoluting the experimental results with an instrument function to obtain the true absorption-line profile.

Importantly, one of the advantages of a narrowband tunable laser is its ability to scan across an absorption line to observe directly the entire profile of atomic absorption in a sample cell. Advantages of observing the entire profile during a chemical-analysis measurement using laser-absorption spectroscopy include the resolution of overlapping atomic absorption lines, absolute density determination, and thermal analysis of atoms. Specific examples of these applications are discussed in Secs. 4.4.1 to 4.4.5.

4.2.3 Molecular-Absorption Spectroscopy

As discussed earlier, the absorption of light in molecular systems occurs between two quantum states, just like in atomic systems. However, these states consist not only of electronic but also vibration and rotation states. The molecular-absorption spectrum is much more complicated than that observed in the atomic system. Because the energy separation between successive rotation levels is much smaller than those between vibration levels, we can observe the gross vibration structure in the electronic band using a low-resolution grating or prism spectrograph. At room temperature, most molecules are in the lowest vibration level of the ground electronic state. The absorption bands would consist of transitions from the ground vibration state ($v'' = 0$) to various vibration levels of the upper electronic state ($V' = 0, 1, 2, \ldots$), which is commonly called a progression. The intensity distribution of the progression $v'' = 0$ of the lower electronic state to $v' = 0, 1, 2, 3$ is not uniform and usually has a maximum, following the Franck-Condon principle [Herzberg, 1950; King, 1964]. This principle assumes that the electronic transition occurs so fast compared with the nuclear motion that the relative position and velocity of the nuclei remain approximately the same during the electronic transition. Hence the maximum intensity should appear for the v'' to v' transitions while the internuclear distance remains unchanged.

For pure vibrational-rotational transitions where the molecule remains in the same electronic state, the spectrum is much simpler. With a conventional spectrograph of high resolution, we can observe the fine structure in vibrational-rotational bands [Barrow, 1962; King, 1964]. For a given vibration transition observed in the IR absorption spectra, we obtain transitions between various rotation levels of the upper and lower vibration states with energies in wavenumbers:

$$\bar{\nu} = \bar{\nu}_{v'-v''} + B_{v'}J'(J' + 1) - B_{v''}J''(J'' - 1) \tag{4.25}$$

where $\bar{\nu}_{v'-v''}$ is the wavenumber for the transition between the two vibration levels v' and v'' in the absence of rotation. B' and B'' are the rotation constants in the upper and lower vibration states, respectively. By measuring the precise wavenumber of each vibrational-rotational transition, we determine the rotation constants and hence the internuclear distance of each vibration and rotation level. In general, the average internuclear distance increases slightly with v for an anharmonic oscillator. Therefore, the B value usually decreases with increasing v.

Vibrational-rotational transitions, taking place in an electronic state with no electronic angular momentum, are possible only for $\Delta J = \pm 1$. This is true for the ground electronic states of the majority of stable diatomic molecules. For a molecule in an electronic state with electronic angular momentum greater than zero (such as NO), the vibrational-rotational transition with $\Delta J = 0$ is also possible [Herzber, 1971].

In addition to the electronic and vibration transitions described above, molecules with a permanent dipole moment may undergo purely rotation transitions. The relatively large moments of inertia of poly-atomic molecules result in a close spacing of their rotation levels [Hertzberg, 1966a]. The transitions between these rotation levels normally fall in the microwave region. Usually, the rotation absorption spectrum is measured for a molecule remaining in the ground electronic and vibration state, which in most cases has a $^1\Sigma$ symmetry. The rotation selection rules for transitions between the rotation levels in such states are $\Delta J = \pm 1$. The transition between the level J and J + 1 gives a line at frequency

$$\bar{\nu} = 2B_0(J + 1) - 4D_0(J + 1)^3 \tag{4.26}$$

where B_0 and D_0 are rotational constants for the ground electronic and vibrational states. Again, by measuring the exact wavenumber of each rotational transition and spectral separation, we can deduce the rotational constants and hence detailed information regarding molecular structures.

Quantitative determination of molecular density by absorption spectroscopy is generally more difficult than the measurement of atomic density using atomic-absorption techniques described in the preceding section. This is mainly due to the complexity of the absorption spectra observed in molecular systems. For polyatomic molecules, many low-lying vibrational and rotational energy levels are populated. Transitions originating from these low-lying levels (called hot bands) introduce additional complexity to the resulting spectra.

The absorption-line profiles of molecular transitions are also more complex than those of the atomic systems. In addition to the line-broadening mechanisms described earlier for the atomic transitions, the line widths of molecular transitions are strong functions of rotational energy levels for a variety of reasons. Collisions of molecules with electric-dipole or higher-multipole moments, such as HF, have enhanced resonance broadening. Collisional line broadening of HF, DF, HI, and so on, has been studied thoroughly [Meredith, 1974]. It was found that the variation in line widths with J is often considerable and complex. For example, the line widths of the transitions in the 1 → 0 band of HCl vary by a factor of 6 when broadened by Ar, a factor of 2 when broadened by He, and almost no variation is observed when broadened by He [Tipping, 1970]. Figure 4.5 shows measurements of the v = 2 → 0 band of HF broadened by N_2 [Smith, 1974]. The interaction here is dipole-quadrupole and there is no falloff at low J. For collisions between molecules with large dipole moments such as HF by HF or DF by DF, the half-widths decrease at

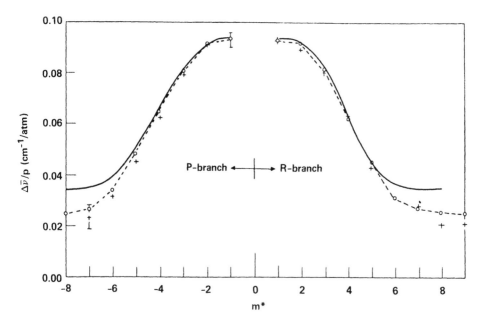

Fig. 4.5 Measured line widths of the first overtone band of HF broadened by N_2 at 373 K. (After Smith and Meredith [Smith, 1974].)

small J and consequently the collision-broadened half-widths show a falloff at low J. Thus, to determine the density of molecular species by absorption spectrometry, one must measure the entire line profile.

Most of our present knowledge about molecular structure has been provided by molecular spectroscopy using conventional spectrophotometric techniques. The accuracy and sensitivity of conventional spectroscopic techniques are commonly limited by the equipment used, in particular by the intensity and spectral purity of the light source. This is especially true for absorption measurements using grating or prism spectrometers. The introduction of lasers to molecular-absorption spectroscopy has remarkably improved both the sensitivity and spectral resolution of the measuring techniques, enabling us to make unambiguous identification of the vibrational, rotational, and electronic transitions of chemical species. Lasers have also permitted sensitive detection of spurious chemical species in the presence of large amounts of other chemical components. Several reliable tunable laser devices are now commercially available for spectroscopic applications. The spectral range, tuning methods, and applications are discussed below.

4.3 ABSORPTION SPECTROSCOPY WITH LASERS

4.3.1 Linear Laser Absorption Spectroscopy—Overview

There are two commonly used methods of laser-absorption spectroscopy: direct transmission measurements and opto-acoustic measurements [Letokhov, 1977]. Both methods are shown in Fig. 4.6. In a
typical laser transmission measurement, the wavelength dependence
of laser energy transmitted by the absorber is measured. This method is especially useful for measuring intense absorption lines when
the absorbed energy is an appreciable portion ($>10^{-3}$) of the incident
beam. However, if the absorbed energy is only a very small portion
of the incident beam, it would be more accurate to directly measure
the amount of laser energy absorbed by the sample. In an opto-
acoustic measurement, the wavelength dependence of the magnitude
of pressure changes resulting from sample heating after laser absorption is recorded. In some cases, atomic or molecular absorption of
laser light is also followed by subsequent reemission of light, which
is called laser-induced fluorescence or phosphorescence. The measurement of fluorescence intensity as a function of laser wavelength
is discussed in more detail in Chapter 9. In performing these absorption measurements, one has to keep the intensity of the laser at an
appropriate level so that the level population of the absorber will not
be significantly altered by the laser and multiquantum transition will
not occur. Under these conditions, the amount of laser absorption
and the density of the absorber will have a linear relationship. Sections 4.3.2 to 4.3.4 describe the linear laser-absorption techniques
developed at various regions of wavelength. Absorption measurements
based on nonlinear properties of the absorber at higher laser intensity
are discussed in Sec. 4.3.5.

During laser-absorption measurements, it is sometimes difficult to
detect and measure weak absorption lines. To overcome this difficulty,
one often monitors the first- or higher-order derivatives of the absorption spectrum. The laser wavelength is periodically scanned and
the corresponding intensity modulations of the transmitted laser beam
are detected. Modulation of the laser wavelength can be achieved by
dithering either the laser dc current (which will be discussed in the
section below, as in the case of semiconductor diode lasers) or magnetic field (as in the case of spin-flip Raman lasers) and the laser cavity length and etalons (as in the case of dye lasers and frequency-
mixing sources). The derivative method allows us to separate weak and
narrow absorption lines from a background of more intense wide-band
absorption. The sensitivity of the derivative method is about 100×
higher than that of the direct-transmission measurement. Figure 4.7
is a sample of the first derivative absorption spectrum of NO near
the 3724 cm^{-1} region obtained by the derivative method [Eng, 1982].

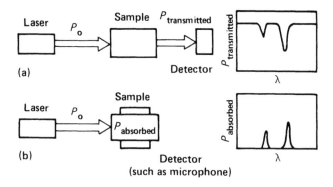

Fig. 4.6 Principal methods of linear laser-absorption spectroscopy using a tunable laser: (a) direct absorption; (b) opto-acoustic method using a microphone. (After Letokhov, [Letokhov, 1977].)

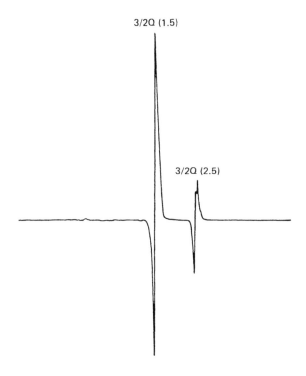

Fig. 4.7 Sample of the first derivative absorption spectrum of NO near the 3724-cm^{-1} region obtained by the derivative method, showing that the Λ-doubling splitting of about $0.012\ \text{cm}^{-1}$ can easily be resolved. (After Eng and Ku [Eng, 1982].)

The derivative method is particularly useful in line-position measurements and in line-width measurements. In locating the center of absorption lines, the derivative signal is significantly sharper than that obtained from the direct-absorption measurement. In line-width measurements, the separation between the positive and negative peaks in the first and second derivative signals can be related to the absorption line width, provided that its line shape is determined.

Another laser-absorption technique that can be used to measure the concentration of absorbers is based on the effect of Faraday rotation, in which the polarization plane of laser radiation is found to rotate as the light is propagating through the absorbing medium placed in a longitudinal magnetic field. During the measurement, the wavelength of the propagating laser light is tuned to the center of the absorption line. By determining the degree of angular rotation of the polarization planes, one deduces the concentration of absorbers in the absorption path. The effect of polarization rotation described here is similar to the Faraday rotation observed in ferromagnetic materials [Yariv, 1967]. Although this method has not been widely applied in absorption spectroscopy, it could be used to measure the absorption coefficients of atoms and molecules in a broad range of pressure and laser intensities [Gus'kov, 1977].

One of the major difficulties in tunable laser-absorption spectroscopy is the determination of the absolute wavelengths and the exact separation of absorption lines. As shown in the figures above, the most commonly used technique to obtain wavelength separation is to send part of the laser output through a precalibrated Fabry-Perot interferometer, which gives frequency marks when the laser frequency is tuned across the transmission maximum of the interferometer.

To get the absolute wavelengths of absorption lines, one could use a calibration spectrum from molecules with known absorption lines and absorption frequencies simultaneously measured with the sample of interest. However, one of the most accurate methods used for frequency calibration is the heterodyne technique [Hillman, 1979; Frerking, 1977]. The basic principle of the heterodyne technique is to mix the laser output in a nonlinear crystal (detector) with a reference laser of fixed frequency and wavelength. The difference frequency generated by these lasers is directly counted to yield the exact frequency of the laser beam of interest. In the following sections we shall separately review these absorption spectroscopic techniques using lasers.

4.3.2 Infrared and Microwave Regions

As mentioned earlier, all known molecules or molecular ions have their vibrational-rotational spectra in the range between 2 and 30 μm. Up, to now, the most widely used tunable lasers for infrared spectroscopy have been the semiconductor diode laser, spin-flip Raman laser, op-

tical parametric oscillators, difference frequency IR and submillime-
ter-wavelength lasers, and high-pressure gas lasers.

4.3.2A. Absorption Spectroscopy Using Semiconductor Diode Lasers

Tunable diode lasers have many attractive features for applications in absorption spectroscopy [Butler, 1976; Hinkley, 1976;
McDowell, 1981]. These unique features include high spectral purity,
line widths of 10^{-4} cm^{-1}, high spectral brightness, broad and continuous tunability, ease of rapid frequency and amplitude modulation,
and wide spectral coverage (3 to 30 µm, a region of great interest
to the molecular spectroscopist).

Laser action in the Pb salt family of semiconductors was first demonstrated by Butler et al. [Butler, 1964] at the MIT Lincoln Laboratory. Radiation at 6.6 µm was observed from a PbTe diode laser. Diode laser action was also achieved in other binary and ternary Pb
salts [Butler, 1965]. By 1966 it was demonstrated that the entire
spectral range of 2.7 to 33 µm could be covered by lasers using various Pb salt compounds [Melngailis, 1968]. The feasibility of using
diode lasers for high-resolution infrared spectroscopy was first demonstrated on SF$_6$ [Hinkley, 1970; Ralston, 1974]. Subsequently,
Hinkley et al. [Hinkley, 1976] perfected the measurement techniques
for general applications in IR molecular spectroscopy. As we discussed earlier in Chapter 2, laser action in semiconductors [Eng, 1980]
results from optical transitions across the energy gap between the
nearly empty conduction band and nearly full valence band. Figure
4.8 shows the dependence of laser-emission wavelength on composition for commonly used Pb salt semiconductors, which provide a class
of laser spanning the IR region from 2.7 to over 30 µm.

The spectral region of a typical diode laser is governed by the
band gap and the precise spectral-mode positions are determined by
the refractive index of the semiconductor and the beam cavity length.
In general, the lasers will emit in several spectral modes simultaneously
when the current is significantly above threshold. The spectral width
of a single mode depends on factors such as the stability of the op-
erating temperature and bias current. Line widths of less then 3×10^{-6} cm^{-1} have been measured under controlled conditions. The tuning rate of a spectral mode is determined primarily by the variation of
refractive index with temperature.

The attainable continuous-wave (CW) power output from a Pb salt
diode laser over all modes is typically on the order of 3 mW for wavelengths less than 8.5 µm, 0.5 mW for 8.5 µm $< \lambda <$ 20 µm, and 0.2
mW for $\lambda >$ 20 µm [Butler, 1976]. Figure 4.9 is a schematic diagram
of an absorption spectrometer using a tunable diode laser. Figure 4.10
shows the infrared spectra of the 000 → 010 band of N$_2$O taken by a
tunable diode laser and by a conventional Fourier tranform spectro-

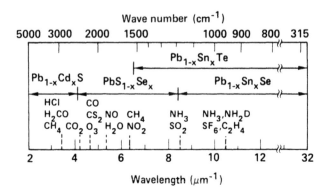

Fig. 4.8 Dependence of laser emission wavelength on composition of
Pb-salt semiconductors. The tunable diode laser emission occurs in
several spectral modes separated by about 1 cm^{-1}. The emission is
generally tuned by varying the temperature. Control of the heat-sink
temperature provides a course-tuning capability, while the Joule heating associated with the bias current is used as a fine-tuning control.
The tuning range for current tuning alone is typically 15 cm^{-1}. Tuning ranges greater than 50 cm^{-1} are achievable using both temperature and current tuning. Also shown are some of the gases in which
diode laser absorption spectroscopic studies have been performed.
(After Eng et al. [Eng, 1980].)

meter grating [Reisfeld, 1979]. The spectral resolution of 3×10^{-4}
cm^{-1} achieved by the tunable diode laser is 1000× better than that of
conventional spectroscopic instruments. Because of the widespread
availability of diode lasers in the last few years, a large number of
molecules has been studied using high-resolution absorption spectroscopy. Table 4.1 is a brief list of the molecules studied.

With a semiconductor diode laser of a given composition, wavelength tuning can be accomplished by any method that alters the energy gap, such as varying the diode temperature [Harman, 1969] or
applying a magnetic field [Nill, 1972a] or external pressure [Corcoran,
1970] to the semiconductor. However, the laser frequency generally
does not follow continuously the tuning of the spectral gain profile
over the entire range, but jumps after a few cm^{-1} of continuous tuning by some cm^{-1}. The mode-hopping problem has been overcome by
using an external cavity with a length that can be separately controlled. Recently, Forrest and Hall [Forest, 1983] have developed a fully automated tunable diode laser system for long-term precesion spectroscopic measurements of gases, liquids, and solids. The newly developed mesa-stripe geometry lasers [Butler, 1983] of $Pb_{1-x}Sn_xSe$

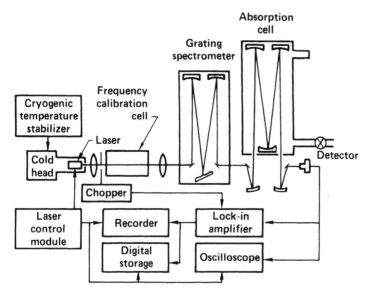

Fig. 4.9 Schematic diagram of absorption spectrometer using a tunable diode laser. The absolute emission frequency can be determined to about 0.5 cm^{-1} with a grating spectrometer. A higher precesion determination can be achieved by comparing the laser frequency to a frequency standard using calibrated molecular absorption lines. (After Eng and Mantz [Eng, 1979a].)

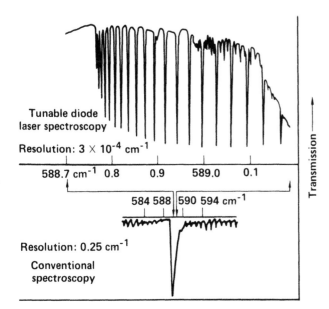

Fig. 4.10 Infrared spectra of the Q branch by the $00^{\circ}0 \to 01'0$ transition of N_2O taken by a tunable diode laser and by a conventional Fourier transform spectrometer (After Reisfeld and Flicker [Reisfeld, 1979].)

TABLE 4.1 Molecules Studied Using Semiconductor Diode Lasers

Molecule	Spectral region	Measurement result	Reference
$^{14}N^{16}O$	1876-1880 cm^{-1} 1903-1904 cm^{-1}	Λ-doubling, nuclear hyperfine	[Nill, 1972b; Blum, 1973]
$^{14}N^{16}O$	1876 cm^{-1}	Nonlinear Zeeman spectrum	[Zeiger, 1973]
$^{14}N^{16}O$	1902-1915 cm^{-1}	Regular and satellite bands sub-Doppler spectroscopy with molecular-beam collimator	[Pine, 1979a]
$^{14}N^{16}O$	1876-1881 cm^{-1}	Sub-Doppler spectroscopy with molecular beam produced by supersonic nozzle	[Gough, 1978]
$^{14}N^{16}O^{+}$	2280-2300 cm^{-1}	P and R branch line position of several vibrational bands	[Bien, 1978]
CO	2190 cm^{-1}	Doppler width measurement	[Nill, 1971]
CO	1880-1900 cm^{-1}	Gain profile in a CO laser discharge	[Blum, 1972b]
CO	2075-2105 cm^{-1}	Combustion and shock-tube temperature and species concentration determination	[Hanson, 1978]

CO	$2100-2150$ cm^{-1}	Air-broadened width and line strengths	[Ku, 1975]
HBr	$2660-2680$ cm^{-1}	Line intensity and Ar and N$_2$ broadened widths	[Mantz, 1978]
HI	$2400-2500$ cm^{-1}	Nuclear quadrupole and line profile	[Strow, 1980]
CS	$1180-1266$ cm^{-1}	Line positions of several isotopic species, band analysis	[Todd, 1979]
CS	$1250-1304$ cm^{-1}	Line positions and band analysis	[Yamada, 1979]
C1O	Band centered at 850 cm^{-1}	Line positions and band analysis	[Menzies, 1977]
H$_2$	1034.67 cm^{-1}	Collisional narrowing for the S(3) quadrupole line	[Rogowski, 1978 Reid, 1978c]
H$_2$O	$1879-1880$ cm^{-1}	High-J absorption-line-width atmospheric pressure-line intensity	[Blum, 1972a]
H$_2$O	1879 cm^{-1}	Collision narrowing of a high-J H$_2$O vapor line	[Eng, 1972]
H$_2$O	$1800-1960$ cm^{-1}	Self-broadening, N$_2$ broadening, air-broadened width, line intensity, pressure shift	[Eng, 1973]

TABLE 4.1 (continued)

Molecule	Spectral region	Measurement result	Reference
H_2O	1800-1960 cm^{-1}	General absorption-line-parameter determination; heterodyne experiment	[Eng, 1974]
H_2O	10-15 μm	Absorption-line-parameter determination for H_2O lines	[Eng, 1979a]
D_2O	9.4 μm	Line-position measurement using heterodyne method	[Worchesky, 1978]
OCS	859-cm^{-1} and 1711-cm^{-1} bands	Line-position measurement using heterodyne technique	[Maki, 1980; Wells, 1979]
SO_2	1115-1158 cm^{-1}	Line-position, intensity, and air-broadened widths	[Hinkley, 1972]
SO_2	1176-1266 cm^{-1}	Line-position determination	[Allario, 1975]
CO_2	618 cm^{-1}	Line position of several lines near the Q-branch head	[Aldridge, 1975a]
CO_2	10-15 μm	Line intensity, width and temperature-dependent width parameter; line shape near 618 Q branch	[Eng, 1979a]

Species	Frequency/Range	Measurement	Reference
CO_2	$667\ cm^{-1}$	Q-branch line-position measurement	[Planet, 1975]
CO_2	ν_2 at 15 μm	Line-position, intensity, and width measurement	[Plante, 1979]
$^{16}O\,^{12}C\,^{18}O$ $^{16}O\,^{12}C\,^{17}O$ $^{17}O\,^{12}C\,^{17}O$	$662{-}$ $666\ cm^{-1}$	Q-branch line-position determination	[Flicker, 1979]
$^{13}CO_2$	$648\ cm^{-1}$	Q-branch line position	[Reisfeld, 1978]
$^{14}CO_2$	ν_3	Line position and strength	[Eng, 1977]
CO_2	9.4 μm	Line position	[Reid, 1978b]
CO_2	9.4 μm	Weak-band line position	[Nereson, 1977]
$H^{12}CN$	$712{-}737\ cm^{-1}$	Line position	[Devi, 1979a]
$H^{13}CN$		Line position	
O_3	$1034{-}1068\ cm^{-1}$	Line position	[El-sherbing, 1979]
O_3	$683{-}717\ cm^{-1}$	Line position	[Devi, 1979b]
N_2O	$550{-}619\ cm^{-1}$	Line position and molecular constants	[Reisfeld, 1979]

TABLE 4.1 (continued)

Molecule	Spectral region	Measurement result	Reference
NH_3	$1545-1595\ cm^{-1}$	Line position and Stark spectroscopy	[Weber, 1974]
NH_3	$9.5\ \mu m$	Relative positions for multiplet components	[Sattler, 1978b]
NH_3	$942-956\ cm^{-1}$	Line positions using CO_2 laser lines as reference	[Nereson, 1978]
NH_3	$800-1200\ cm^{-1}$	Line positions and relative intensity	[Cappelani, 1979]
NH_3	$10\ \mu m$	General measurements	[Urban, 1980]
C_2H_2 $^{12}C^{13}CH_2$	$13.7\ \mu m$	Line positions of Q branch	[Reddy, 1979; Hawakami, 1978]
$^{12}C_2D_2$	$9.5\ \mu m$	Line positions and self- and helium-broadened line width	[Rutt, 1978]
CH_4	$1243-1337\ cm^{-1}$	Several line positions	[Fox, 1979]
CH_4	$1292-1307\ cm^{-1}$	Line positions determined	[Restelli, 1979a]

CH_4	Q, R branches of ν_4 band	Intensity of seven transitions	[Restelli, 1979b]
CH_4	ν_4	Line-intensity measurement	[Jennings, 1980]
OsO_4	961 cm^{-1}	Line-position measurement	[McDowell, 1978a]
CF_4	ν_3 band near 1282	Line-position measurement	[McDowell, 1980]
CH_3F	ν_3 band Q branch	Line positions displayed on	[Sattler, 1977]
CH_3F	1183 cm^{-1}	Line positions of Q branches	[Hirota, 1979]
CH_3I	824-862 cm^{-1} ν_6 band	Line positions of Q branches	[Das, 1980]
CCl_2F_2	10.8 μm	High-resolution scan	[Jennings, 1978a]
$C^{35}Cl_2F_2$	10.8 μm	High-resolution scan and pressure-broadened absorption profile	[Nordstrom, 1979]
CCl_2F_2	10.8 μm	Line strength, line position, and broadening coefficients	[Restelli, 1978]
HNO_3	11.3 μm 5.9 μm	Line positions and broadening coefficients	[Brockman, 1978]

TABLE 4.1 (continued)

Molecule	Spectral region	Measurement result	Reference
C_3O_2	1565-1600 cm^{-1}	Line-measurement positions	[Weber, 1976]
C_3O_2	819-866 cm^{-1}	Line-position measurements	[Jennings, 1978b]
HCCCN	2066-2095 cm^{-1}	Line position and B' value	[Yamada, 1979]
C_2H_{44}	947.1-951.9 cm^{-1} 1350-1400 cm^{-1}	Line position	[Montgomery, 1975]
CH_3OH	1027-1068 cm^{-1}	Stark spectroscopy and heterodyne position measurement	[Sattler, 1978a]
$F_2C = CH_2$	10-μm region	Heterodyne technique for line-position measurement	[Sattler, 1980]
H_2SO_4	880, 1225 cm^{-1}	Absorption coefficient and dissociation constant at elevated temperatures	[Eng, 1978]
H_2SO_4	1225 cm^{-1}	Absorption coefficient	[Montgomery, 1978]

Species	Band	Description	Reference
SF$_6$	ν_3 band near 948 cm^{-1}	Line position using heterodyne technique	[Aldridge, 1975b]
SF$_6$	ν_4 band 615 cm^{-1}	Line-position measurement and line assignment	[Aldridge, 1975b]
SF$_6$	ν_3 band 948 cm^{-1}	Line position and band analysis	[McDowell, 1978b]
SF$_6$	ν_4 band 581-641 cm^{-1}	Line position using a long-path cell and band analysis, also other isotopic species	[Person, 1978; Kim, 1979]
MoF$_6$	ν_3 band 713 cm^{-1}	Molecular-beam method for low-temperature data	[Cummings, 1980]
MF$_6$	ν_3 band 625-630 cm^{-1}	Molecular-beam method for low-temperature data	[Travis, 1977]
UF$_6$	ν_3 band	Molecular-beam method for low-temperature data	[Travis, 1977]

Source: After Eng et al. [Eng, 1980].

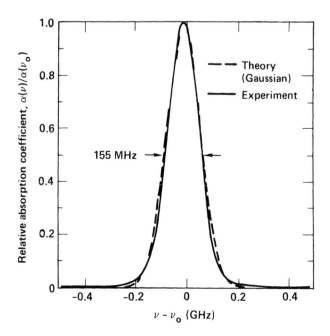

Fig. 4.11 Measured and calculated Doppler-broadened absorption line
shapes of CO at 2115.63 cm^{-1}, v = 0 → 1, P(7). The Doppler width
for this line is found to be 155 MHz. A theoretical Gaussian line shape
having this width is indicated by the dashed trace. (After Melngailis
[Melngailis, 1972] and Hinkley et al. [Hinkley, 1976].)

and PbS$_{1-x}$Se$_x$ have very impressive single-mode tuning ranges with
high output power.

   As discussed earlier, three of the most fundamental properties of
an isolated absorption line are its frequency, intensity, and shape.
Tunable diode lasers have contributed to our understanding of each of
these properties. The intensity and shape of a spectral line can be
measured directly with a tunable laser. The key to such measurements is the laser's narrow line width and intensity. Examples of the
measured Doppler-broadened and collisional-broadened line shapes are
shown in Fig. 4.11. The Doppler-broadened P(7) line of CO at
2115.63 cm^{-1} was taken with a Pb salt tunable diode laser [Melngailis,
1972]. The CO gas pressure was 0.1 torr and the optical absorption
length was 10 cm. The Doppler width for this line was found to be
155 MHz. The theoretical Gaussian line shape having this FWHM is
also plotted. Figure 4.12 shows an example of a collision-broadened
absorption line in the v$_2$ band of water vapor present in the atmo-

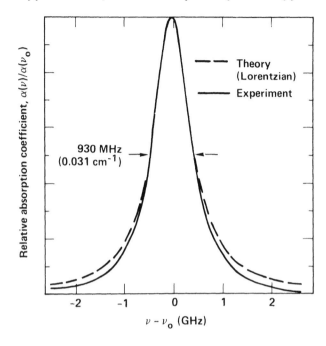

Fig. 4.12 Measured and calculated Doppler-broadened line shape of H_2O at 1879.6 cm^{-1}, atmospheric pressure, $v_2 = 0 \rightarrow 1$. The line has a measured width of 870 MHz. A theoretical Lorentzian line shape having a width of 930 MHa is indicated by the dashed trace. (After Blum et al. [Blum, 1972a] and Hinkley et al. [Hinkley, 1976].)

sphere observed at 1879.6 cm^{-1}, also taken with a tunable diode laser [Blum, 1972a]. The absorption measurment was made over an optical path 7.4 m long. The absorption line has a measured with of 870 ± 90 mHz. A calculated Lorentzian line shape using a width of 930 MHz and normalized absorption peak is also displayed in this figure. The agreement between measured and calculated line width as observed in both cases was excellent.

Figure 4.13 shows the experimentally measured line width (half width at half maximum) for NH_3 at 939.21 cm^{-1} as a function of NH_3 and air pressure [Hinkley, 1976]. At low pressure the line width is dominated by the Doppler effect with a width of 84 MHz. However, as the pressure of NH_3 increases beyond 4 torr, the line width increases linearly with the pressure of NH_3 at a rate of 25 MHz/torr due to self-broadening. As the pressure of air increases beyond 10 torr, the line width increases at a rate of 6.3 MHz/torr of air, which is four times slower than that of self-broadening.

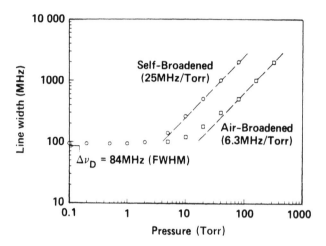

Fig. 4.13 Measured absorption line widths for the Q(9,3) transition of NH₃ at 939.21 cm⁻¹ region as a function of NH3 and air pressure. At low pressures, the line width asymptotically approaches the Doppler width of 84 MHz. The line width starts to increase with increasing pressure at approximately 4 torr at a rate of 25 MHz/torr for self-broadening and at a rate of 6.3 MHz/torr for air broadening. (After Hinkley et al. [Hinkley, 1976].)

Although semiconductor lasers have not been applied to absorption spectroscopy as widely as conventional grating spectrometers, they have much higher resolution and spectral brightness. Regarding their total cost, a complete semiconductor laser system is still much less expensive than any other infrared instrument of comparable resolution.

4.3.2B. Absorption Spectroscopy Using Spin-Flip Raman Lasers

The spin-flip Raman lasers can deliver considerably more power than semiconductor lasers. However, the output is not as widely tunable as that of the semiconductor lasers described above. It is only tunable around the wavelength of pump sources such as pulsed or CW CO_2 lasers at 10.6 μm, CO lasers at 5.1 μm, or HF chemical lasers in the 2.7-μm region.

The first spin-flip scattering experiment was performed by Slusher, et al. [Slusher, 1967] on n-type InSb. The excitation source was a CO_2 laser operating at 10.6 μm. To generate light from a spin-flip Raman laser, one places a cryogenically cooled semiconductor in a magnetic field and irradiates it with a focused laser beam. Just like conventional Raman scattering, when the incident laser radiation is inelastically scattered by electrons in the semiconductor material, it can excite the electron from a lower energy level onto the higher Landé level. The resulting scattered radiation has an energy equal to the pump energy minus the excitation energy, which is commonly called Stokes radiation. Scattering by electrons in the upper levels would produce anti-Stokes radiation with a corresponding energy of pump energy plus excitation energy. Thus the output frequency of the spin-flip Raman laser can be tuned by the magnetic field B according to the relation [Hinkley, 1976]

$$\bar{\nu} = \bar{\nu}_p \pm g\mu_B \frac{B}{H} \qquad (4.27)$$

where $\bar{\nu}_p$ is the wave number of the pump laser, g the Landé g factor, and μ_B the Bohr magneton. The + and − signs correspond to the interactions of the magnetic field with the magnetic moment of the electron spin, which is parallel or antiparallel (with respect to the external B field). The tuning range depends on the Landé g factor and is about 2.3 cm^{-1} per kilogauss for InSb [Hinkley, 1976]. To obtain wider tuning ranges, high magnetic fields are necessary and superconducting magnets are often needed to obtain fields as high as 200 kG. The tuning range also depends on semiconductor materials [Patel, 1976] such as InSb, CdS, InAs, and so on. The line width is limited mainly by frequency jitter due to instabilities of the pump laser intensity, which causes thermal drifts, and by acoustic vibrations of the resonator system. In order to achieve continuous tuning without mode hopping, one has to use external cavity mirrors instead of the end faces of the semiconductor crystal [McKenzis, 1975].

Figure 4.14 shows the absorption spectrum of a portion of the H_2O/NO mixture around 1886 cm^{-1}, obtained with a spin-flip Raman laser [Hinkley, 1976]. A sample of NO (20 ppm) was contained in the cell at 76 torr total pressure. The NO lines are indicated by the arrows. The pump laser was at 1892.25 cm^{-1}. Figure 4.15 is a schematic diagram of the experimental arrangement for absorption spectroscopy using a CW spin-flip Raman laser [Mozolowski, 1979]. Although the spin-flip Raman laser was invented at about the same time as the tunable diode laser, its use in absorption spectroscopy has been comparatively limited. A number of spectroscopic measurements performed using CW spin-flip Raman lasers are listed in Table 4.2. In comparison with conventional spectrometers, the spin-flip lasers are more costly to operate and cumbersome to use because of the need for a highly stable pump laser and a cryogenic cooling system for the superconducting magnet. However, its potential of wide-range tunability, high output power, and good mode quality makes the spin-flip Raman laser attractive for infrared spectroscopy.

TABLE 4.2 Molecules Studied with Spin-Flip Raman Lasers

Molecule	Vibration band	Spectral coverage	Results	References
^{14}NO	1-0	$1887.5\ cm^{-1}$	Stratospheric NO concentration	[Patel, 1974]
^{14}NO	1-0	Q branch	Hyperfine structure	[Butcher, 1975]
^{14}NO	1-0	Q branch	High sensitivity (intercavity)	[Dutta, 1977]
^{14}NO	1-0	Q branch	Highest resolution SFRL data	[Mozolowski, 1979]
^{14}NO	1-0	Q branch	Zeeman modulation technique demonstrated	[Urban, 1978]
^{14}NO	1-0	P and R branch lines	NO, N_2, Ar^+, and He broadening coefficients	[Rohrbeck, 1980]
^{15}NO	1-0	R branches	Λ-doubling	[Patel, 1978a]
^{14}NO	2-1	P and R branches	CO-pumped double-resonance experiment	[Patel, 1977]

^{15}NO	2-1	P and R branches	Vibrational energy transfer between ^{14}NO and ^{15}NO	[Patel, 1978b]
H_2O	ν_2	1884.57 cm^{-1} and 1889.58 cm^{-1}	H_2O, N_2, O_2, Ar, and Xe broadening coefficients	[Guerra, 1975]
H_2O	ν_2	1885.2 cm^{-1}	Air broadened width	[Patel, 1972]
OCS	Combination bands	1882.5-1886.5 cm^{-1}	Line position	[Butcher, 1975]
OCS	Combination bands	1875-1882 cm^{-1}	Line position (intracavity)	[Dutta, 1977]
OCS	Combination bands	1865-1905 cm^{-1}	Line position, 300-MHz resolution	[Fayt, 1977]
C_2H_4	$\nu_7 + \nu_8$	1877-1884	Molecular constant	[Hager, 1976]
SbH_3	Sb-H stretching bands	Q branch 1880 cm^{-1}	Hyperfine structure	[Butcher, 1975]

[a]*Source:*   After Eng et al. [Eng, 1980]

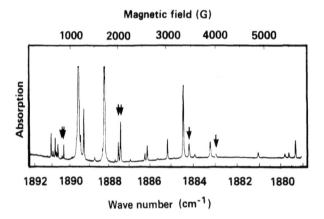

Fig. 4.14 Absorption spectrum of a portion of the H_2O/NO bands around 1886 cm^{-1}, obtained with a spin-flip Raman laser—20 ppm of NO were contained in the cell with 76 torr total pressure. The NO lines are indicated by the arrows. (After Hinkley et al. [Hinkley, 1976].)

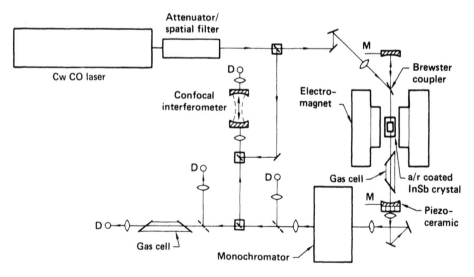

Fig. 4.15 Schematic diagram of the absorption experiment using CW spin-flip Raman laser. (After Mozolowski et al. [Mozolowski, 1979].)

4.3.2C. Absorption Spectroscopy Using Difference-Frequency
Laser Sources

One can also generate tunable IR and millimeter-wavelength laser
light by frequency-mixing techniques. By mixing outputs from two
single-mode lasers with frequencies ν_1 and ν_2 in a nonlinear optical
crystal, one can generate light at a different frequency $\nu_3 = \nu_1 - \nu_2$
with high efficiency if the phase-matching condition is fulfilled. If
one of the lasers is chosen to be tunable, the frequency of ν_3 would
also be tunable as long as the phase-matching condition is fulfilled
continuously.

Pine [Pine, 1974] constructed a CW tunable difference-frequency
infrared laser source operated in the region 2.2 to 4.2 μm by mixing
the radiation from an argon ion laser with a tunable CW dye laser.
With single-mode operation of both lasers and frequency mixing in
$LiNbO_3$, Pine [Pine, 1975, 1976] achieved a tunable infrared output
with microwatt power and a spectral resolution of 0.001 cm^{-1}.

It was later reported by Byer [Byer, 1974] that frequency mixing
in nonlinear crystals of AgGaSe should allow continuous tuning of ν_3
from 3 to 12 μm, in CdSe from 10 to 25 μm, in GaP from 20 to 200 μm,
and in $LiNbO_3$ from 170 μm to 1 cm. Thus with nonlinear crystals the
whole IR and millimeter range from 1 μm to 1 cm can be covered by the
mixing technique for applications in absorption spectroscopy. Figure
4.16 is a schematic diagram of the optical layout of an absorption measurement using a frequency-mixing technique. Table 4.3 lists molecules studied by absorption techniques using frequency mixing.

One disadvantage of the frequency-mixing technique using nonlinear crystals is the severe restriction of tuning ranges imposed by
the phase-matching condition. This difficulty has recently been overcome by four-wave parametric mixing in alkali metal vapors [Wynne,
1974]. Pulsed dye lasers independently tuned to ν_1 and ν_p regions
are combined in a collinear beam and focused into a heat-pipe oven
containing alkali metal atoms. One of the lasers (ν_1) is tuned to the
vicinity of an alkali resonance line and serves as a pump for stimulated electronic Raman scattering at ν_s. The frequency ν_p of the second laser mixes with ν_1 and ν_s in the alkali vapor to generate the difference frequency $\nu_{IR} = \nu_1 - \nu_s - \nu_p$. This technique also allows
the generation of IR pulses tunable in the range 2 to 25 μm.

4.3.2D. Absorption Spectroscopy Using Optical Parametric
Oscillators

Tunable IR laser pulses can also be generated by optical parametric oscillations in nonlinear-birefringent crystals [Giordmaine, 1965;
Harris, 1969]. If laser light of sufficient intensity is focused onto a
nonlinear birefringent crystal, a parametric interaction between the
pump light and the molecules in the crystal may take place that splits

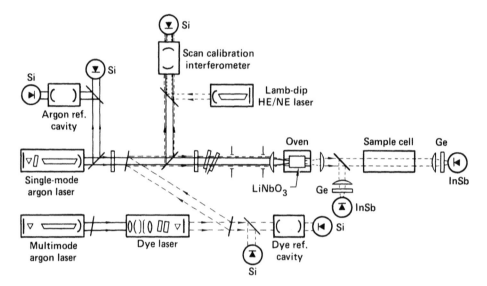

Fig. 4.16 Schematic diagram of the absorption experiment using frequency-mixing technique. (After Eng et al. [Eng, 1980] and Pine
et al. [Pine, 1977a, 1977b].)

the photon of energy ν_p into two photons of energy ν_i and ν_s with
$\nu_p = \nu_i + \nu_s$, where ν_i and ν_s are the frequencies of the idler wave
and signal wave, respectively. By rotating the nonlinear crystal or
by varying its refractive index through temperature changes, a tunable output at ν_s and ν_i can be achieved. For example, pumping a
$LiNbO_3$ crystal with a frequency-doubled YAG laser at $\nu_p = 0.56$ μm
generates tunable laser light at 0.6 to 0.85 μm for the signal wave and
at 1.6 to 3.6 μm for the idler wave by varying the temperature of the
crystal from 200 to 430°C. Commercial optical parametric oscillator devices are now available with an output peak power of 300W near 2.5 μm.
Byer [Byer, 1974] has recently extended the tunable range from 1.4
to 4.4 μm with an energy conversion efficiency of 40%. The line width
of the optical parametric oscillator of 0.001 cm^{-1} was achieved using a
tilted etalon within the resonator. Figure 4.17 shows an absorption
spectrum of the first overtone (v = 0 → v = 2) of CO taken with a
$LiNbO_3$ parametric oscillator. The pressure of CO was 600 torr [Hink-
ley, 1976].

4.3.2E. Absorption Spectroscopy Using High-Pressure Gas Lasers

Tunable high-intensity IR pulses from 9 to 11 μm can also be generated by operating high-power CO_2 and N_2O lasers to oscillate on

Molecule	Vibration band	Spectral region (cm^{-1})	Results	Reference
$^{12}CH_4$, $^{13}CH_4$	ν_3 ν_3	2841–3166	Position, intensity, and line width	[Pine, 1975, 1976; Dang-Nhu, 1979; Bray, 1979]
SO_2	$\nu_1 + \nu_3$	2463–2524	Line intensity, width, and position	[Pine, 1977a,b]
H_2CO	$\nu_3 + \nu_4$, $\nu_3 + \nu_6$, ν_1, ν_5, $\nu_2 + \nu_4$, $2\nu_3$, $\nu_2 + \nu_6$	2700–3000	Position, intensity, and width	[Pine, 1978; Brown, 1979]
CO_2	ν_3	R-branch head	Line position and intensity	[Pine, 1980a]
HF	1-0	P-branch lines	Line width and shape	[Pine, 1980b]
NO	2-0	3850–3950	Λ-doubling splitting	[Pine, 1979b]
D_2	1-0 band	2915–3492	Line position, collisional narrowing, and pressure shift	[McKeller, 1978]
PH_3	$3\nu_2 \leftarrow 0$	2930–3900	Line position and centrifugal splitting	[Bernard, 1979]
H_3^+	$\nu_2 \leftarrow 0$	2450–2950	Line position	[Oka, 1980]

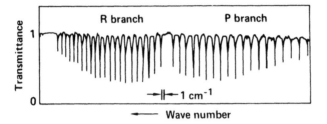

Fig. 4.17 Absorption spectrum of the first overtone ($v = 0$ to $v = 2$) of CO taken with a $LiNbO_3$ parametric oscillator. The pressure of CO was 600 torr. (After Hinkley et al. [Hinkley, 1976].)

many closely spaced vibrational-rotational lines [Basov, 1975]. One could pressure broaden these vibrational-rotational lines to such an extent that the width of each individual vibrational-rotational line becomes larger than the line spacing and thus the laser becomes continuously tunable within the vibrational-rotational energy band.

For example, Harris et al. [Harris, 1976] achieved tunable laser operations in electron-beam pumped CO_2 laser systems at 15 atm with a tunable range of 70 cm^{-1}, with 100 mJ per pulse and a pulse duration of 100 ns. A line width of 0.03 cm^{-1} was achieved by line narrowing with two intracavity etalons. The first spectroscopic experiments using high-pressure gas lasers were performed in SF_6 [Beterov, 1973].

As described above, the entire IR and microwave regions between 2 and 30 μm can be covered by tunable laser sources. Currently, the most widely used laser sources in these regions are semiconductor diode lasers. Fully automated tunable-diode laser systems for long-term high-resolution spectroscopic measurement of gas, liquid, and solid samples have recently been developed. Although the other laser systems are not as convenient and easy to use and as tunable as the semiconductor diode laser, they can deliver considerably more power than semiconductor diode lasers.

4.3.3 Visible and Ultraviolet Regions

As described earlier, electronic transitions of atoms, molecules, and ions take place mainly in the visible and ultraviolet regions. In this spectral region the most widely used tunable laser systems are CW dye lasers, pulsed dye lasers, and Raman-shifted pulsed dye lasers. The entire ultraviolet to the visible region (from 0.267 to 1 μm) can be completely covered by commercially available single-frequency dye lasers using different dyes or dye mixtures and pump lasers.

4.3.3A. Absorption Spectroscopy Using CW Dye Lasers

Traveling-wave ring-resonator CW dye lasers (pumped by argon or krypton lasers) that efficiently provide tunable single-frequency radiation over the entire visible spectrum from 0.4 to 1 μm are now commercially available [Divens, 1982; Jarrett, 1979] for absorption spectroscopy. The output power of ring dye lasers is typically from a few hundred milliwatts to a few watts, depending on pump laser power and dye efficiency. The root-mean-square (RMS) line widths are 150 kHz and frequency drifts are typically less than 50 MHz/h, which is ideal for high-resolution absorption spectroscopy.

Frequency doubling of the ring-dye-laser output has also been achieved using nonlinear optical crystals such as KDP, ADP, LiO_3, or urea crystals inside the laser cavity [Gabel, 1972; Wagstaff, 1979; Mariella, 1979; Clough, 1980]. Second-harmonic generation in these nonlinear crystals enables us to expand the tuning range of the single-frequency dye lasers from 0.26 to 0.412 μm, with a RMS line width of 1 MHz and output power of a few milliwatts to 20 mW. Figure 4.18 shows the wavelength tuning curves of the single-frequency ring dye lasers. Pump laser power and dye used are also indicated.

Figure 4.19 shows a typical absorption spectrum taken by a tunable CW dye laser pumped by an Ar ion laser. During a typical measurement using a tunable CW dye laser, three signals were monitored.

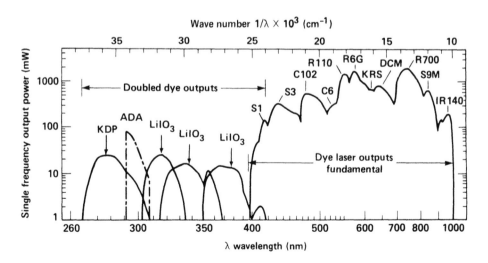

Fig. 4.18 Wavelength tuning range obtained with a single-frequency CW ring dye laser. (After Johnston et al. [Johnston, 1982].)

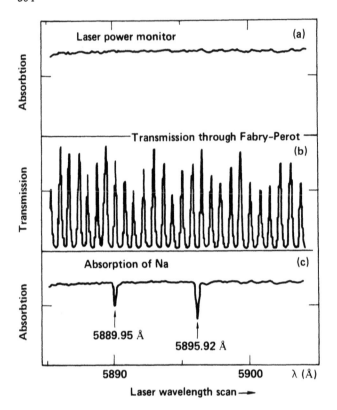

Fig. 4.19 Typical absorption spectrum taken with a tunable CW dye laser.

Laser power (I_0) before entering the absorption cell is displayed in Fig. 4.19a. Laser power transmitted through the Fabry-Perot interferometer is shown in Fig. 4.19b, and the laser power (I) after absorption by atomic vapor (which is Na in this example) is displayed in Fig. 4.19c. During the signal processing the transmitted signal (I) is normalized to the laser power (I_0) and its logarithm is calculated. A Fabry-Perot interferometer is used as a wavelength standard for determining the separation and width of the absorption lines.

4.3.3B Absorption Spectroscopy Using Pulsed Dye Lasers

The four most commonly used pulsed dye lasers are: (1) flash-lamp-pumped dye lasers [Snavely, 1969; Marowsky, 1973] with pulse durations of 0.1 to 100 μs, peak power in the kilowatt range, and repetition rates up to 100 pps; (2) N_2 laser-pumped dye lasers [Basting, 1974; Dunning, 1974] with pulse lengths of 20 ns, peak power up to MW, and repetition rates of a few hundred pps; (3) excimer-laser-pumped dye lasers [Sutton, 1976; Uchino, 1979] with pulse durations of less than 100 ns, peak power of many megawatts, and repetition rates of a few hundred pps; and (4) frequency-doubled YAG-laser-pumped dye lasers [Lavi, 1979], with pulse length of 5 ns, peak power of 10 MW, and repetition rate of up to 30 pps. Like the CW dye lasers, the line width of pulsed dye lasers can be narrowed by inserting etalons inside the laser cavity. Coarse wavelength tuning is achieved by tuning the gratings or prisms. Fine tunings are made by tilting etalons and by changing refractive indexes through the changing of gas pressure inside the optical cavity. In the UV region, wavelength tuning can be obtained by frequency doubling of visible laser pulses using nonlinear optical crystals. Since the efficiency of second-harmonic generation increases with the intensity of the fundamental wave, the conversion efficiency of nonlinear crystals such as KDP, ADP, or RDP is typically 10 to 50% [Kuhl, 1975], which is many orders of magnitude higher than that of CW dye lasers. Narrow-bandwidth high-energy tunable pulsed dye lasers are now commercially available for wavelengths from 0.19 to 4.5 μm. Line widths of the pulsed dye lasers are about 200 MHz, limited mainly by the duration of laser pulses.

During a typical absorption measurement using pulsed dye lasers, the fluence of each laser pulse before and after propagating through the sample cell is continuously monitored by a pair of photodiodes or photomultipliers. The output of the detectors is sent to boxcar integrators for signal averaging. The transmitted signal is normalized to the initial reference signal and its logarithm is calculated.

Since most nonlinear optical crystals absorb strongly at wavelengths below 0.2 μm, harmonic generation below this wavelength cannot be achieved by the simple doubling techniques described above. Recently, Sorokin et al. [Sorokin, 1975] demonstrated that tunable vacuum-ultraviolet light could be generated by frequency-summing techniques, where one mixes two visible dye-laser pulses in atomic and ionic vapors. At present, the wavelength coverage in the vacuum-ultraviolet region is still very limited.

4.3.3C. Absorption Spectroscopy Using Raman-Shifted Pulsed Dye Lasers

As described above, continuously tunable UV light can be generated by frequency doubling the output of a continuously tunable dye laser operating in the visible region. This technique is generally effective only at wavelengths longer than 0.22 μm. In 1979 a simple versatile method based on stimulated Raman scattering was developed [Paisner, 1979; Wilke, 1979]. This technique uses a tunable dye laser in the visible region and converts the laser light to both shorter

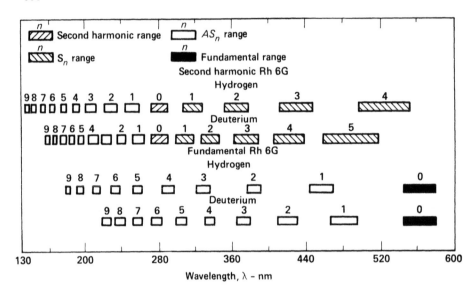

Fig. 4.20 Wavelength coverage between 0.130 and 0.60 μm using a Raman-shifted rhodamine 6G dye laser. We have observed AS_8 to S_2 using the fundamental and AS_5 to S_6 using the second harmonic of the rhodamine 6G dye laser. (After Paisner and Hargrove [Paisner, 1979].)

and longer wavelengths by stimulated Raman scattering in molecular hydrogen or deuterium gas. Tunable coherent beams covering the electromagnetic spectrum from 0.171 μm in the vacuum ultraviolet to above 1 μm in the IR were generated. Figure 4.20 shows the wavelength coverage possible between 0.13 and 0.6 μm using a Raman-shifted rhodamine 6G dye laser pumped by an Nd-YAG laser [Paisner, 1979]. Both anti-Stokes and Stokes radiation are continuously tunable by tuning the visible dye lasers. The line width of the anti-Stokes and Stokes emission follows closely the spectral line width of the pump laser. The optical resolution $\Delta\lambda/\lambda$ of the Raman-shifted light is about 10^{-5} to 10^{-6}. Power levels of Raman-shifted beams are between several μJ to 20 mJ, depending on the orders of the anti-Stokes or Stokes transitions.

Figure 4.21 is a nitric oxide absorption spectrum taken around 0.215 μm using a tunable stimulated Raman laser source and optoacoustic technique. The upper trace (a) shows the absorption spectrum taken by the conventional absorption spectrometer at low resolution. The three major bands (0,0), (1,0), and (2,0) correspond to transitions between the v = 0 vibrational level of the ground electronic state $X^2\pi_{1/2}$ and the v = 0, 1, 2 vibrational levels of the excited elec-

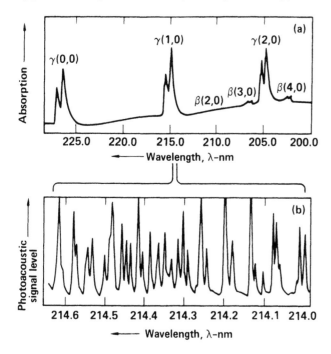

Fig. 4.21 Nitric oxide absorption spectrum taken around 215 nm using a tunable stimulated Raman laser source and opto-acoustic technique. (After Paisner and Hargrove [Paisner, 1979].)

tronic state $^2\Sigma^+$. The lower trace (b) is a portion of the resolved rotational bands of the $v = 0 \to v = 1$ transitions taken photoacoustically with the anti-Stokes stimulated Raman source described above.

4.3.4 Extreme and Vacuum Ultraviolet Regions

Traditionally, most of our spectroscopic knowledge in the XUV and VUV regions has been derived by using vacuum spectographs in combination with spectral lamps with a resolving power of less than 3×10^5 at 0.15 μm. Such resolution is sufficient for Doppler-limited spectroscopy of light molecules. However, for heavier molecules, higher resolution would be necessary. A common problem in conventional spectroscopy is the photon flux available when such instruments are used at their highest resolution. The flux is often very low and inadequate for high-resolution spectroscopy studies under these conditions. The most commonly used laser technique to generate tunable XUV and VUV radiation is by wavelength conversion via nonlinear

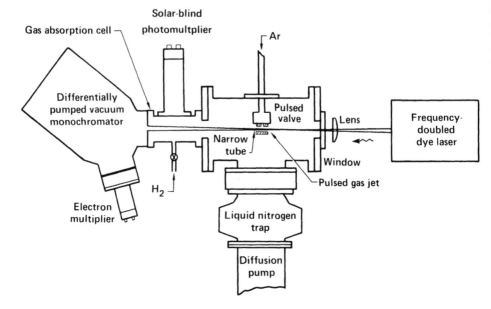

Fig. 4.22 Schematic diagram of the optical arrangement for XUV harmonic generation in a pulsed gas jet. (After Boker et al. [Boker, 1983].)

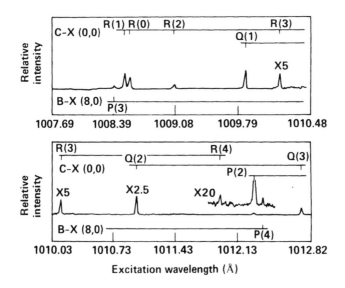

Fig. 4.23 Fluorescence spectrum of H_2 observed during irradiation of H_2 using the XUV laser described in Fig. 4.22. (After Rettner et al. [Rettner, 1983].)

processes with excimer lasers [Rettner, 1983; Kung, 1983]. As we have discussed earlier, the most important technical problems that must be addressed in XUV and VUV harmonic generation are the lack of a suitable transparent window material and the self-absorption of the generated harmonic by the nonlinear medium. Boker et al. [Boker, 1983] and Kung [Kung, 1983] developed a relatively simple technique for generating radiation at these regions using a pulsed rare-gas jet source as the nonlinear medium. Kung has frequency tripled the 3550-Å light in Xe, producing 4.7×10^{11} photons/pulse at 1182 Å, while Boker et al. generated the third, fifth, and seventh harmonics of a picosecond KrF excimer laser to achieve radiation at 828, 497, and 355 Å. Figure 4.22 is a schematic diagram of the optical arrangement for the observation of XUV harmonic generation in a pulsed gas jet. Recently, Rettner et al. [Rettner, 1983] extended the latter technique to the generation of tunable XUV radiation by frequency tripling the second harmonic of a tunable dye laser using argon as the nonlinear medium. They used the resulting radiation to excite the Lyman bands $(B'\Sigma_u^+ - X')_g^+\Sigma$ and Werner $(C'\pi_u - x'\Sigma_g)$ of H_2. Figure 4.23 shows a portion of the excitation spectrum obtained, covering the region from 1007.7 to 1012.8 Å. Such laser-based XUV and VUV sources should permit new absorption spectroscopy experiments to be performed in which the coherence and temporal and polarization characteristics in the XUV and VUV regions are exploited. The development of the tunable XUV and VUV lasers is still in its infancy. The spectral coverage of lasers at this range is currently very limited.

4.3.5 Nonlinear Laser-Absorption Spectroscopy

As described in the previous sections, lasers have enormously enhanced the sensitivity and application range of conventional absorption spectroscopy. At the same time, lasers have also made it possible to observe many nonlinear spectroscopic phenomena with unprecedented detail. For example, intense laser light can significantly change level populations and bleach the absorbing transitions. By using a second laser as a probe beam, it is now possible to see optical spectra without Doppler broadening. Based on this general concept, many Doppler-free techniques have been developed, such as saturation spectroscopy [Lamb, 1964; Szoke, 1963; Hall, 1976], polarization spectroscopy [Hänsch, 1975], optical double-resonance spectroscopy [Fields, 1975; Kaminsky, 1976], and two-photon absorption spectroscopy [Vasilenko, 1970]. In atomic spectroscopy and molecular spectroscopy in particular, the electronic transitions often exhibit a high density of overlapping lines where the spacing between neighboring lines is smaller than the width resulting from Doppler broadening. These laser-based Doppler-free nonlinear absorption methods permit us to eliminate Doppler broadening in gas samples without any need for cooling and to

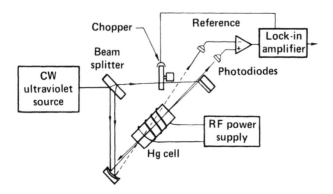

Fig. 4.24 Schematic of the saturation absorption spectrometer using CW UV laser as the saturation and probe source. The metastable level (6P $^3P_0$) of Hg was populated by running RF discharge in Hg vapor. (After Bloomfield et al. [Bloomfield, 1983].)

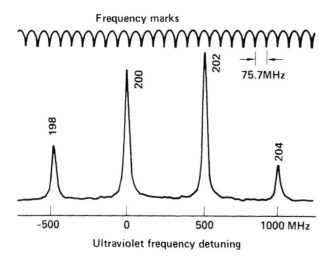

Fig. 4.25 Ultraviolet Hg absorption line 6P $^3P_0 \to$ 9s $^3S_1$ at 2465-Å region recorded using a CW saturation absorption spectrometer. The top trace shows the frequency calibration made by a reference interferometer with a free spectral range of 75.7 MHz. (After Bloomfield et al. [Bloomfield, 1983].)

simplify and unravel complex absorption spectra for detailed studies of atomic and molecular structures. We briefly discuss below a few useful nonlinear absorption techniques. These methods have been used mainly for high-resolution spectroscopy of atoms and molecules and are now generating new applications.

4.3.5A Saturation and Polarization Spectroscopy

Saturation spectroscopy was first realized by Lamb [Lamb, 1964]. This Doppler-free spectroscopic technique uses two laser beams, a strong saturation beam and a weak probe beam that are sent by mirrors in nearly opposite directions through the absorber. When the saturation beam is on, it bleaches a path through the cell and the probe signal is received more strongly at the detector. As the saturation beam is alternately stopped and transmitted by a chopper, the probe signal is also modulated. However, this will occur only when the pump and probe lasers are tuned near the center of the Doppler-broadened absorption line and both beams are interacting with the same velocity group of atoms or molecules.

Figure 4.24 shows an experimental arrangement for taking the saturation spectra of Hg at 2465 Å [Bloomfield, 1983]. It resembles the standard setup except that a pair of probe beams is used to increase the measurement sensitivity. The metastable $6P\,^3P_0$ level of Hg was populated by running a radio-frequency (RF) discharge in a cylindrical quartz cell filled with Hg vapor. Continuous-wave tunable UV radiation near 2465 Å was generated by mixing the output of a single-mode Ar^+ ion laser at 3638 Å with the output of a ring dye laser at 7644 Å in a crystal of ammonium dihydrogen phosphate (ADP). During the measurement a portion of the UV output is used as the saturation beam while the two weaker fractions, which reflect off both surfaces of an uncoated quartz plate, serve as the counter-propagating probe beams. The two probe beams are detected by photodiodes and the signals are sent to a lock-in amplifier. The saturating beam is focused and modulated by a chopper. A 2-GHz portion of the $6P\,^3P_0 \rightarrow 9s\,^3S_1$ spectrum is shown in Fig. 4.25. Absorption signals corresponding to transitions from the four stable even isotopes of Hg are displayed. Note that the peak heights correspond closely with the natural abundances of Hg. The line width of each absorption line is only 34 MHz (FWHM), in good agreement with the natural line width as determined by the lifetime of the $9s\,^3S_1$ state of Hg.

Polarization spectroscopy is a more sensitive Doppler-free absorption method. It is very similar to the saturation absorption method described above except that the nonlinear interaction of the two counterpropagating laser beams is monitored via changes in light polarization instead of intensity. Figure 4.26 shows the apparatus for Doppler-free polarization spectroscopy. A circularly polarized beam from

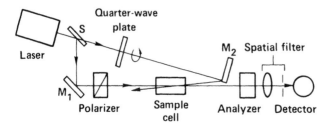

FIG. 4.26 Apparatus for Doppler-free polarization spectroscopy. A circularly polarized beam from the pump laser orients the rotation axes of atoms in a narrow range of velocities. These atoms can be detected with high sensitivity by a counter-propagating probe beam because they change its polarization, causing light to pass through the analyzer into the detector. (After Weiman and Hänsch [Weiman, 1976].)

the pump laser labels atoms or molecules of selected axial velocity by orienting them through optical pumping. Normally, atoms or molecules in a given gas have their rotation axes distributed in random directions. Because the absorption cross section for circularly or linearly polarized light depends on the atomic or molecular orientation, the saturating beam will therefore preferentially deplete atoms or molecules with particular orientations, leaving the remaining ones polarized. Those atoms or molecules can then be detected with high sensitivity because they can change the polarization of a counterpropagating probe beam, causing light to pass through the crossed polarizer into the detector. Polarization spectroscopy has an important advantage in its signal-to-noise ratio. There is almost no transmission of the probe beam until its polarization is changed by the absorber pumped by the saturating beam. With essentially no background noise, the signal is not easily obscured by noise or intensity fluctuations of the laser. The high sensitivity of polarization spectroscopy has made it possible to work with a highly monochromatic CW dye laser of low power.

4.3.5B. Excited-State Absorption Spectroscopy

When a large fraction of a population is excited into a selected upper energy level by the high irradiance of a laser beam, it becomes feasible to observe absorption lines that originate from the selected excited state. A completely different series of energy levels become accessible for sensitive spectral measurements. This two-step absorption technique is useful in analytical work when the first absorption

step happens to be in a wavelength region where spectral interferences are a serious problem, while the absorption from the laser-excited
upper states is relatively free from interference. Absorption spectra
of the laser-excited atoms, such as Ba, Ca, Rb, and Mg, and so on
have been thoroughly studied since 1973.

In a two-photon absorption process, each photon contributes one
half of the energy necessary to excite the upper level. If the excitation of an upper energy level is achieved by the simultaneous absorption of two photons traveling in opposite directions to each other, it
would also provide a way to minimize Doppler broadening. Here absorption occurs when the sum of the energies of the two photons
equals the energy difference between the energy levels. Because the
photons are traveling in opposite directions, any Doppler wavelength
shift of the absorber with respect to one photon is compensated by the
opposite Doppler shift of the absorber with respect to the other photon.
Consequently, all atoms in the total population are equally influenced
by the lasers, irrespective of any velocity component they have. This
is in contrast to saturation spectroscopy as described above when only
atoms with zero Doppler velocity component are observed by the probe
laser beam. The power required for the two-photon absorption process to pump a significant population to the upper energy level is much
higher than if a one-photon absorption process were used. The required power is reduced when the energy of one of the photons is in
near resonance with an intermediate energy level. Two-photon absorption spectroscopy and excited-state absorption spectroscopy have
already become valuable complements to single-photon absorption
spectroscopy because of their different selection rules and because
they permit one to reach high-lying excited states and to obtain high-
resolution spectroscopic information about the excited states of atoms
and molecules that could not be measured by conventional techniques.

4.3.6 Intracavity Absorption

Another particularly promising technique to improve the sensitivity
of atomic and molecular absorption is to place the absorption cell in
the cavity of a laser [see Fig. 4.27 for a comparison of external (Fig.
4.27a) and intracavity (Fig. 4.27b) absorption measurements]. One
finds that the output power of the laser can be changed dramatically
even for weak absorption. An enhancement in absorption signal of 10
to 100 is obtained by using the intracavity method. A number of physical phenomena contribute to the enhancement of intracavity absorption. The most obvious contribution is that because the absorber is
placed inside a multipass optical resonator, the extent of absorption
increases with the number of passes inside the optical cavity. The
enhancement also depends on how close the laser is operated to thres-

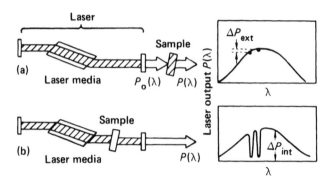

FIG. 4.27 Comparison of the methods of detecting weak absorption lines when the absorption cell is inside or outside the optical cavity. (After Letokhov [Letokhov, 1977].)

hold. If the laser operates near threshold where the gain is only slightly larger than the losses due to absorption, the output power becomes a very sensitive function of the absorption losses. Although the enhancement factor becomes very large when one is operating near threshold, the stability of the laser becomes rather poor. Many workers have studied the enhancement of the detectability of trace absorption using the intracavity technique. It was found that the enhancement is equal to the gain in optical path length caused by the multiple passes of the laser beam and becomes larger as the laser operates close to threshold. Datta et al. [Datta, 1975] used the intracavity absorption technique to enhance the chemical reactions between Cl_2 and $CH_2 = CH_2$ for chlorine isotope separation. Although this technique improves the absorption enhancement factor, because of its nonlinear nature it is not clear to what extent it improves the signal-to-noise ratio of the determined absorbance.

4.4 APPLICATIONS OF LASER-ABSORPTION SPECTROSCOPY

Laser spectroscopy has had a revolutionary impact in the areas of chemical-reaction diagnostics, chemical analysis, and chemical-process control. We will briefly review the achievements of laser-absorption spectroscopy in these areas where truly remarkable improvements in sensitivity, detectivity, and temporal, spatial, and energy resolution are obtained over conventional techniques. In a laser laboratory it is now possible to disect completely a complex chemical reaction, to follow the rate of appearance or disappearance of starting materials or products, and to carry out such measurments in real time with resolution down to less than 10^{-12} s. These measurements can also be performed with a high degree of spatial resolution. We can identify and monitor the appearance of trace reactive components in a complicated mixture and selectively interact with individual components with tunable narrow-line-width lasers. Lasers have also been used to study some state-selected chemical reactions, new reaction pathways, and transition-state spectroscopy. We are now on the threshold of the last frontier of reaction dynamics, where we move from the spectroscopy of reactants and products to the spectroscopy of reaction intermediates.

It may be possible to further develop laser spectroscopy from its current role as a research or measurment tool in the analytic laboratory to become an integral part of process-control technology. Much progress has been made along these lines in the past few years. Success in this area allowed us to understand many chemical processes in intimate detail and to optimize process-control technology via process parameters that could not have been measured previously. In this chapter we briefly review applications of laser-absorption spectroscopy in chemical-reaction diagnostics, chemical-concentration monitoring, chemical-process control, vapor-density monitoring, and isotope separation.

4.4.1 Chemical-Concentration Monitoring

The earliest experiments with lasers in absorption spectroscopy were performed by Geritsen [Geritsen, 1966] with a HeNe laser as a source at 3.39 µm to measure vapor concentrations of hydrocarbons. Hydrocarbons such as CH_4, CH_3F, CH_3Cl, C_2H_4, and C_3H_8 all have strong absorption at 3.39 µm, which corresponds to the excitation of C-H stretching vibration. Alobaidi et al. [Alobaidi, 1975] further improved this laser-absorption technique to measure alcohol-vapor concentrations. By using a double-beam arrangement and phase-sensitive detection as is commonly used in conventional absorption spectrometers, they were able to detect alcohol concentration to ppm levels—a technique particularly useful for rapid analysis of alcohol concentration in the breath.

Another good example of concentration monitoring using coincidental resonance absorption is the monitoring of NH_3 in flue gases using two $^{13}CO_2$ laser lines. This technique was suggested [Kaldor, 1982] for in situ absorption measurements of NH_3 concentration at the ppm level. Figure 4.28 is a proposed optical layout for such a measurement. The beam from a $^{13}CO_2$ laser is directed through the flue gas via optical ports in the duct wall and then reflected back into a detector. Figure 4.29 illustrates the complexity of a portion of the IR spectrum and frequencies of the proposed $^{13}CO_2$ laser transitions, R(18) and

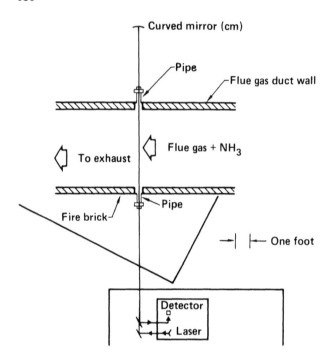

FIG. 4.28 Optical layout for NH3 monitoring in flue gases. The beam from a diode laser is directed through the flue gas via optical ports in the duct wall and then reflected back again and into a detector. (After Kaldor and Woodin [Kaldor, 1982].)

R(14), at which the measurements are made. It is obvious that one could use a tunable diode laser to perform similar measurements with ease.

Another important application of laser-absorption spectroscopy is in trace-gas monitoring. The first successful industrial pollution monitoring instrument using tunable diode lasers was developed by Ku et al. [Ku, 1975]. The method is based on the resonance absorption by pollutant molecules in the atmosphere. By tuning the diode laser in resonance with the P(4) line of the $(v = 1 \rightarrow 0)$ band of CO and using the first-derivative technique with a 10-kHz modulation laser signal, Ku et al. were able to detect the amount of CO in air to about the 5-ppb level. The total integration time for such measurements was about 1 s. This technique was later adapted by the U. S. Environmental Protection Agency (EPA) for a Regional Air Pollution Study. The sensitivity of such a technique used in the industrial environment is limited primarily by the turbulence of the gas sample and by the me-

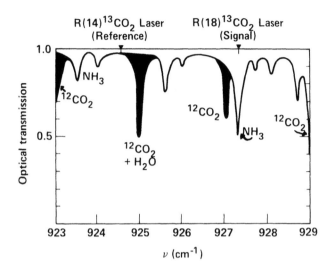

FIG. 4.29 High-resolution spectrum of flue gas in which NH3 mea-surements are made. The blackened peaks are interferences due to CO_2 and H_2O. Indicated at the top are two $^{13}CO_2$ lines, which may be used for reference (no absorption) and signal (NH3 absorption) mea-surements. The reference beam is necessary to take into account op-tical distortions caused by turbulence in the gas. (After Kaldor and Woodin 1982 [Kaldor, 1982].)

chanical stability of the optical components. One could easily improve the sensitivity of the laser-absorption technique by using a long-path white cell. Figure 4.30 shows a schematic of a long-path low-pressure laser-absorption spectrometer developed by Reid et al. [Reid, 1978a]. A detection sensitivity of 0.003 ppb for molecular CO using a tunable diode laser near 2120 cm^{-1} was achieved. The overall optical path length was 200 m and the integration time used to achieve such sensi-tivity was 100 s. Similar laser-absorption techniques were used to monitor SO_2 concentrations to gas. By tuning the diode laser to a strong absorption line of SO_2, Reid et al. were able to detect SO_2 down to the 0.1-ppb level. They measured an absorption coefficient of 3×10^{-8} cm^{-1} using a 200-m optical path and a 100-s integration time constant. It is obvious that the laser-absorption technique above can be used for monitoring many other molecular species provided that they have strong absorption lines. For molecules having narrow and resolvable absorption features, one could use the derivative technique for improving the sensitivity of the measurement. For molecules that have broadband absorptions, the derivative technique is not applica-

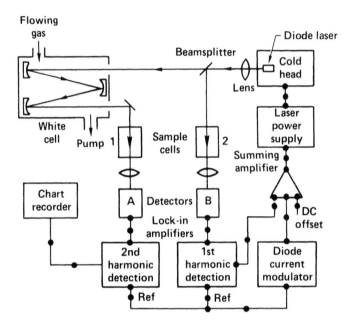

FIG. 4.30 Optical layout of a long-path low-pressure laser absorption
spectrometer for monitoring SO_2 in gas. A sample of gas of known
SO_2 concentration is put in sample cell 1 for absolute calibration. A
pure sample of SO_2 at low pressure is place at sample cell 2. By tuning the laser to an absorption line of SO_2 and frequency modulating
the laser, one uses the first derivative signal from detector B to sta-
bilize the laser to the SO_2 line center. The SO_2 concentration in the
white cell is determined from the second derivative signal monitored
by detector A. (After Reid et al. [Reid, 1978a].)

ble. One such molecule is H_2SO_4, an important smokestack-emission
air pollutant. In 1979 Eng et al. [Eng, 1979b] developed a dual-laser
differential absorption technique to measure the H_2SO_4 concentration
in smokestacks. Figure 4.31 is a schematic diagram of such a measurement system. Lasers A and B are frequency locked at 880 and 962
cm^{-1} using OCS and NH_3 gases in the reference cell. The absorption
of H_2SO_4 at 880 cm^{-1} is about nine times stronger than at 962 cm^{-1}.
(The exact choice of frequency was dictated by the strongly interfering species, such as H_2O and CO_2, in the smoke.) The two laser
beams are combined and sent across the stack. The intensity of each
before and after passing through the stack gas was monitored and ra-
tioed for cancellation of any laser-beam amplitude fluctuations. This
monitoring technique is capable of detecting 1 ppm of H_2SO_4 and has

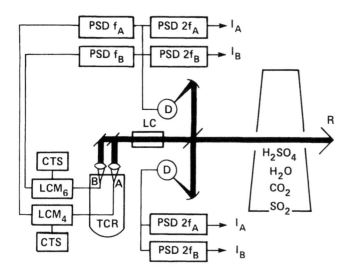

FIG. 4.31 Schematic diagram of a dual-laser differential absorption monitor of gaseous sulfuric acid in a smokestack. PSD, phase-sensitive detector; D, detector; LC, lock cell; R, retros; CTS, cryogenic temperature stabilizer; LCM, laser-control module; TCR, temperature-controlled refrigerator. (After Eng et al. [Eng, 1979b].)

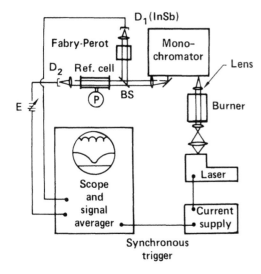

FIG. 4.32 Schematic diagram of a laser absorption system for flame-temperature measurements. The temperature can be accurately determined from the line profile information. (After Hanson [Hanson, 1977b].)

successfully been used as a real-time H_2SO_4 concentration monitor in power-plant smoke stacks. We anticipate a large growth in field and industrial applications of such devices in the future.

A tunable-diode laser as a source of radiation also offers vastly differnt concentration measurement techniques never before possible. Recently tunable-diode laser-absorption spectroscopy was applied to the post-flame region of a flat flame burner [Hanson, 1977b] and in a shock-tube experiment [Hanson, 1977a] for temperature and gaseous species measurements. The accurate determination of combustion temperature and concentration of gaseous species in flames is quite important for the fundamental understanding of combustion chemistry and for improving combustion efficiency. Figure 4.32 is a schematic diagram of an in situ measurement of flame combustion [Hanson, 1977b] By analyzing the line profile of absorption lines and relative intensities, one obtains accurate information regarding flame temperatures. The output beam of a diode laser is collimated to about 3 mm in diameter as it passes through the combustion flame above the burner surface. A grating spectrometer, a Ge etalon, and a 60-cm cell are used for mode isolation, providing a laser frequency calibration. The absorption profile can be displayed or processed with a signal averager to yield spectroscopic data such as line width and intensity of absorption at line center, which enables the analyst to determine the combustion temperature and concentration of CO and NO, etc. Figure 4.33 shows a comparison of the calculated and measured absorption profiles for the experiment shown in Fig. 4.32. The absorption line width (full width at half maximum) of CO measured at 2077.0 cm^{-1} $v = 1 \rightarrow 2$, P(10) in an atmospheric pressure propane-air burner was found to be 0.042 cm^{-1}. The theoretical curve was calculated based on Voigt profile using a Voigt parameter a of 2.47 and gas temperature of 2066 K.

In short, the ability to tune a diode laser or dye laser to a specific infrared or visible frequency and its high spectral brightness, combined with the improved optical detection techniques, have made lasers a powerful tool for examining trace impurities in many industrial environments.

4.4.2 Chemical-Reaction Diagnostics

There are vast amounts of literature on laser diagnostics of chemical-reaction kinetics. In this section we selectively review a few outstanding examples where laser-absorption spectroscopy has been used to selectively monitor the concentration of reactants and products, their internal state energy distribution, and kinetic energy, providing the analyst with a powerful means for determining the dynamics of chemical reactions. One such example is the measurement of the chemical reaction rates of HNO_3 with NO and HNO_2 with O_3 using a tunable

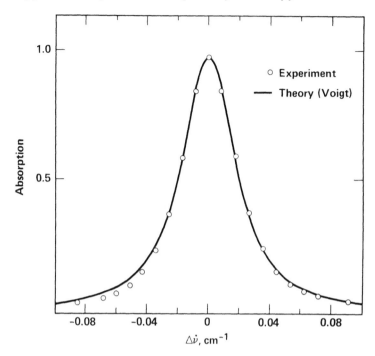

FIG. 4.33 Comparison of calculated and measured absorption profiles of CO at 2077.0 cm^{-1}, $v = 1 \rightarrow 2$, P(10) for the experiment shown in Fig. 4.32. The absorption line width (full width at half maximum) is 0.0042 cm^{-1}. (After Hanson [Hanson, 1977b].)

diode laser [Streit, 1979]. These chemical reactions are of great importance to the understanding of atmospheric chemistry. Figure 4.34 is a block diagram of the experimental arrangement used by Streit et al. The reaction chamber is a 12-liter Pyrex sphere that has a relatively low surface-to-volume ratio. Two gas-phase reactions were monitored:

$$HNO_3 + NO \rightarrow HNO_2 + NO_2 \text{ and } HNO_2 + O_3 \rightarrow HNO_3 + O_2 \quad (4.28)$$

During the experiment, low-pressure gases of HNO_3 and NO_2 (or HNO_2 and O_3) were independently fed into the sphere as gas A and gas B. The concentration of HNO_3 in the first reaction is monitored by a tunable-diode laser operated at 6 μm. In a similar experimental arrangement, the concentration of HNO_2 in the second reaction is monitored. Both HNO_3 and HNO_2 have strong absorptions in the 6-μm region,

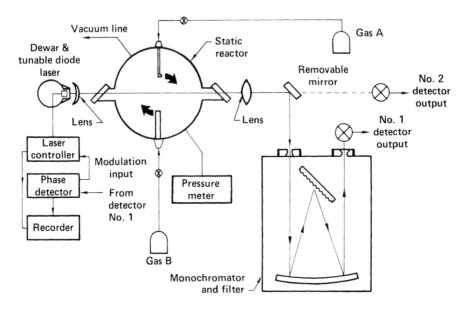

FIG. 4.34 Experimental arrangement for measuring the chemical reaction rate using a tunable diode laser. (After Streit et al. [Streit, 1979].)

corresponding to excitation of the ν_2 band of the HNO_2 and HNO_3 molecules [Bair, 1979].

The high brightness and temporal properties of the laser have also enabled us to make many new kinetics measurements that are not possible with conventional techniques. Laguna [Laguna, 1982] has successfully demonstrated a technique based on tunable-diode laser absorption for a direct time-resolved detection of methyl radicals in the gas phase. Methyl radicals play an important role in most combustion systems and are particularly important in the oxidation of methane. A diagram of the kinetic apparatus is shown in Fig. 4.35. Methyl iodide (CH_3I) is photolyzed at 248 nm using a 30-ns pulsed KrF excimer laser. The density of the CH_3 radical is monitored as a function of time after the photolysis laser pulse by measuring the transient absorption of the diode laser light tuned to the ν_2 mode vibrational-rotational transition of the radical at 570 to 630 cm^{-1}. The absorption signal is strong enough to do real-time kinetics, to measure the recombination rate of CH_3 radicals, and to measure the three-body reaction rate constant of CH_3 with O_2 at room temperature. The sensitivity and selectivity of the tunable laser absorption make this a particularly powerful technique for detailed studies of gas-phase re-

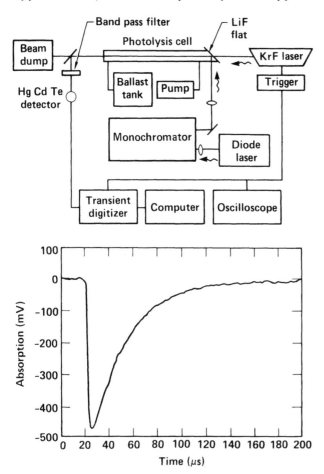

FIG. 4.35 Experimental arrangement for the detection of CH_3 radicals in the gas phase by diode laser absorption at 608 cm^{-1}. The lower trace shows the absorption signal of CH_3 radical as a function of time after photolysis. (After Laguna and Baughcum [Laguna, 1982].)

action kinetics and energy-transfer processes of polyatomic free radicals or molecules.

Based on the same measurement concept (infrared flash kinetics spectroscopy), Petek et al. [Petek, 1983] measured the transient concentration of CH_2 radicals in gas. The transient infrared vibrational-rotational spectrum of singlet methylane ($^1CH_2$) produced by pulsed XeCl laser photolysis of CH_2CO was taken by time-resolved absorption spectroscopy using a difference frequency infrared laser source,

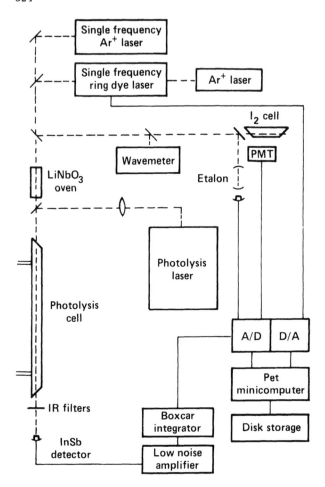

FIG. 4.36 Schematic of the high-resolution flash kinetic IR spectrometer used to obtain the spectra of $^1CH_2$ symmetric (ν_1) and asymmetric (ν_3) stretch. (After Petek et al. [Petek, 1983].)

tuned through the range 2740 to 2940 cm^{-1} with 0.0013-cm^{-1} instrumental resolution. A schematic of the high-resolution flash kinetic infrared spectrometer used to obtain the spectra of the $^1CH_2$ symmetric
(ν_1) and asymmetric (ν_3) stretches is shown in Fig. 4.36. Figure
4.37 is a sample spectral scan of $^1CH_2$, demonstrating a typical signal-to-noise ratio obtained in these experiments. The observed IR
line (0.01 cm^{-1}) is Doppler limited, which is much broader than the
intrinsic resolution of the tunable difference frequency laser (0.0007

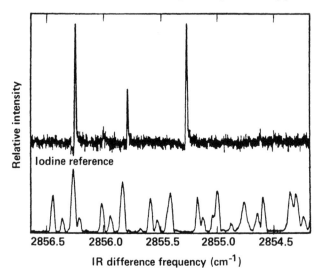

FIG. 4.37 Sample spectral scan of $^1CH_2$ demonstrating typical S/N ratio obtained in the flash kinetic spectroscopic experiment. The observed IR line widths (0.01 cm^{-1}) are Doppler limited and much broader than the intrinsic resolution of the tunable difference frequency laser (0.0007 cm^{-1}). (After Petek et al. [Petek, 1983].)

cm^{-1}). This work demonstrated that infrared flash kinetic spectroscopy using either a tunable diode laser or a difference frequency laser as a probe is a feasible method of obtaining high-resolution and high-sensitivity spectra of reactive chemical transients.

Langford et al. [Langford, 1983] have further extended flash kinetic spectroscopy from the infrared to the visible region using similar measurement concepts. Using a tunable dye laser, the technique of resonance absorption was used to monitor the time evolution of individual CH_2 ($^1A_1$) rotational levels following the excimer laser photolysis of CH_2CO. The experimental arrangement is very similar to Fig. 4.36 except that the tunable IR laser is replaced with a tunable dye laser. Figure 4.38 is a typical absorption spectrum of several rotational lines near $16,929$ cm^{-1}. The assignments are from Herzberg et al. [Herzberg, 1966a], [Herzberg, 1966b]. The temporal evolution of the 4_{14} rotational level of CH_2 ($^1A_1$) can be closely followed. One could deduce the production and removal rates of the specific level by careful analysis of the history of absorbance. Valuable information regarding the production, relaxation, and chemical reaction rates was obtained by this method.

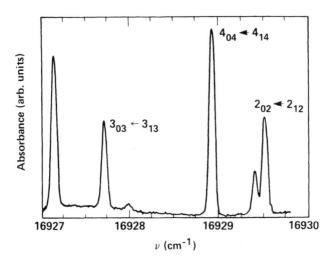

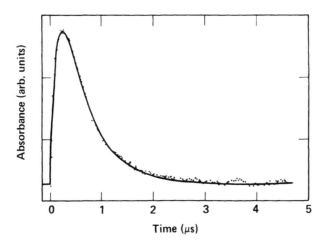

FIG 4.38 Absorption spectrum of several rotational lines near 16,929 cm^{-1} region obtained from the photolysis of 0.5 torr CH_2CO in 4.0 torr of He. The lower trace shows the temporal evolution of one of the rotational levels of CH_2 ($^1A_1$) in 0.2 torr of CH_2CO with 3.0 torr of He. (After Langford et al. [Langford, 1983].)

4.4.3 Chemical-Processing Control

A logical extension of laser-absorption spectroscopy from the laboratory to industry is to use it for chemical-processing and quality control. A few practical chemical-processing and monitoring techniques
using tunable lasers recently have been developed. One such application was developed by Schweid et al. [Schweid, 1983] for the manufacture of room-temperature vulcanized (RTV) silicone products. By
analyzing the content of methyltriacetoxysilicone (MTAS) in RTV using a tunable diode laser, the researchers were able to improve both
the process efficiency and product quality, resulting in demonstrable
cost savings. One-component RTV sealants are made by compounding
the silicone polymers with other components. The chemical and physical properties of RTV sealant such as curing rate and product
strength can be fine-tuned over a wide range by varying the formula-
tions. The manufacture of RTV one-component sealant adhesives is
a simple compounding operation in which the essential ingredients such
as dimethyl siloxane polymers, fillers, and methyltriacetoxysilane
cross-linker are continuously and thoroughly mixed in the reactor. In
order to perform the analysis and actively control the amount of methyltriacetoxysilane in RTV, the product is continuously flowed
through the laser absorption cell. Figure 4.39 shows the absorption
spectra of RTV in the region of 1600 to 1900 cm^{-1}, which shows a
strong peak at 1750 cm^{-1}, a characteristic absorption of methyltriacetoxysilane. In this application the absorbance of RTV in the sample
cell is measured and the concentration of cross-linked material is calculated by an on-line computer and fed back to the process-control
loop.

There is always the question of whether the same measurement or
control can be done with a more conventional commercially available
and cheaper nonlaser process-control instrument. In this case the
high intensity, narrow line width, and tunability of the laser emission enabled a more timely and representative sampling of the product.

Another potential application using the high brightness of tunable-diode lasers was to accurately determine the distribution of SiO
in silicon crystal wafers—an important material widely used in the
semiconductor industry. Silicon oxide is strongly absorbing near
9.04 μm, so that one could map the SiO concentration by focusing and
collimating a narrow IR beam through the silicon wafer and measuring
its transmittence. This measurement technique is very important in
further understanding the annealing process in silicon crystal growth
[Ohsawa, 1980] and could be applied to in situ silicon-wafer process
control.

Another interesting application of laser absorption is to monitor
the quality of plastics during the plastic-extruding process [Forrest,

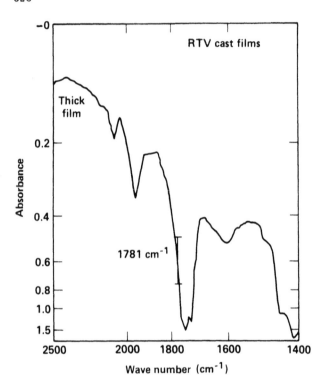

FIG. 4.39 Low-resolution IR spectrum of room-temperature-vulcanized (RTV) silicone rubber. (After Schweid et al. [Schweid, 1983].)

1983]. Since most plastics and additives have strong and character-istic absorptions in the IR and far-IR regions, one could monitor the absorptions in those regions simultaneously and determine the product composition with high precision. The on-line nature of the measure-ment means more uniform quality material can be manufactured and lower-quality product can be effectively segregated from higher-qual-ity product.

Multiwatt single-frequency CW ring dye lasers have been avail-able commercially for years. However, applications using these sourc-es had been limited to research environments until recently because the laser systems were relatively difficult to handle and were not re-liable under industrial conditions. Nevertheless, dye-laser absorption spectroscopy can be very useful in chemical processing. One example reported by Stuke et al. [Stuke, 1978] involves the isotopically se-lective photoaddition of iodine chloride (ICl) to acetylene. This la-ser-driven reaction is a potential technique for chlorine or iodine iso-

Absorption

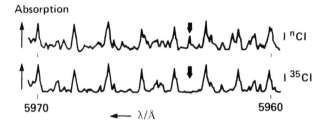

5970   ⟵ λ/Å   5960

FIG. 4.40 Absorption spectra of natural I^nCl and pure $I^{35}Cl$ over 10 Å around 5965 Å, recorded with a 1.5-m grating spectrometer having a resolution of 2×10^{-2} Å . The arrows mark the transmission window for $I^{35}Cl$. (After Stuke and Marinero [Stuke, 1978].)

tope enrichment. A computer-controlled CW dye laser was used to excite selectively the molecules of interest and enhance the photoaddition reaction between the laser-excited ICl molecule and acetylene forming C_2H_2ICl. Figure 4.40 shows the absorption spectra of natural I^nCl and pure $I^{35}Cl$ over a 10-Å range around 5965 Å recorded with a 1.5-m grating spectrometer having a resolution of 20×10^{-3} Å (1.7 GHz). The arrow marks the transmission window for $I^{35}Cl$, showing absorption lines for $I^{37}Cl$. This wavelength region is ideal for selective excitation of the $I^{37}Cl$ molecule using narrow-bandwidth tunable dye lasers. During the photoaddition chemical process, the tunable dye laser is tuned to a region that has a maximum absorption for one isotopic molecule and a minimum absorption for the other. The laser is locked to this specific region by the computer-controlled electronics. Figure 4.41 shows a simple schematic of this device. This laser system was applied to the selective photoaddition of iodine chloride to acetylene:

$$I^iCl^* + HC \equiv CH \rightarrow \underset{I}{\overset{H}{\diagdown}}C = C\underset{{}^iCl}{\overset{H}{\diagup}} \qquad (4.29)$$

giving *cis*-1,2-iodochloroethylene enriched in either ^{37}Cl or ^{35}Cl with an isotopic purity greater than 97%. The described on-line computer-controlled CW dye-laser system is a useful tool for industrial applications because it can be tuned to the wavelength of maximum excitation with reasonable stability.

4.4.4 Vapor-Properties Monitoring

The emergence of the atomic vapor laser isotope separation and vapor deposition technologies has generated considerable interest in the di-

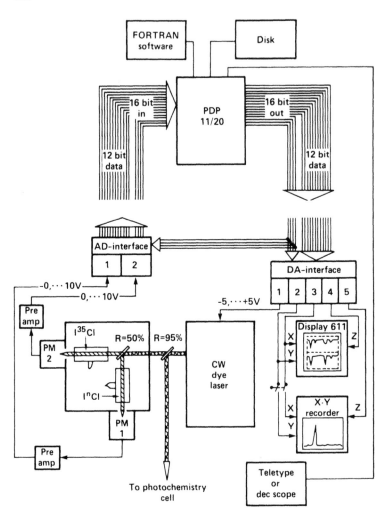

FIG. 4.41 On-line computer-controlled CW-dye-laser spectrometer for photochemical separation of chlorine isotopes. The laser is automatically tuned to a region with maximum excitation selectivity for isotope separation. (After Stuke and Marinero [Stuke, 1978].)

agnosis of vapor concentrations and flow properties during the deposition process. Spatially and temporally resolved species concentration, flow velocity, and temperature are important process parameters that must be probed remotely. Vapor-flow information can be obtained by absorption spectroscopy using tunable lasers. Chen et al.

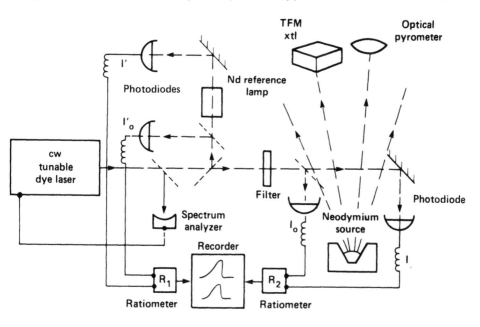

FIG. 4.42 Apparatus for atom-absorption measurement using CW dye laser. The atomic vapor is produced by surface evaporation of metallic neodymium by impact with high-energy electrons. Transient evaporation rates were monitored using a pair of quartz-crystal deposition-rate thin-film monitors. (After Chen et al. [Chen, 1978].)

[Chen, 1978] used a laser resonance absorption technique to determine the density of neodymium vapor during an evaporation process. Figure 4.42 shows the setup for an atomic absorption experiment using CW tunable dye lasers. During the evaporation process, a CW dye laser was tuned to sweep across the 5620-Å, 5676-Å, 5785-Å, and 5827-Å region to monitor the ground, 1128-cm^{-1}, 2367-cm^{-1}, and 3682-cm^{-1} states of Nd atoms. Line widths and percentage absorptions for each transition were measured at various distances from the evaporation source. The absolute density of neodymium at each state was then calculated using published gA values, where g is the degeneracy factory and A is the Einstein transition of probability as defined earlier. Figure 4.43 shows a typical absorption trace observed during the evaporation experiment. The CW dye laser was tuned to sweep across the 5729-Å region for monitoring excited-state atom density. The absorption profile of all the even isotopes of Nd can be clearly identified. Similar resonance absorption techniques have been applied to determine densities of different elements in the vapor for which the gA values are known.

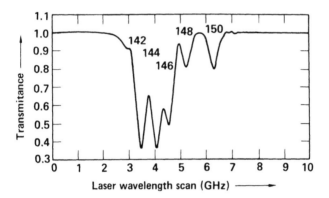

FIG. 4.43 Typical atomic-absorption trace observed during the Nd evaporation process. The CW dye laser was tuned to sweep across the 5729-Å region to monitor the density of the second metastable state at 2367 cm^{-1}. (After Chen et al. [Chen, 1978].)

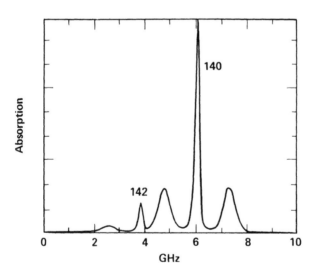

FIG. 4.44 Atomic-absorption trace observed during the vapor flow velocity measurement. The absorption peaks observed at the two sides of the 140 peak result from the Doppler shifts from which vapor flow velocity is calculated. (After Chen and Abdallah [Chen, 1985].)

Vapor flow velocity can also be measured by Doppler shifting the resonance absorption spectra. By sending the laser probe in at 45° with respect to the vapor flow direction and reflecting it back on itself, Chen et al. [Chen, 1985] observed a symmetric split of the absorption spectra about v_0 (corresponding to zero velocity). The frequency of the split yields the vertical velocity component $V = \lambda \, \Delta v / 2 \cos \theta$, where θ is the angle between the velocity component and the laser probe (45°), and the factor of 2 accounts for the Doppler shift in both directions about zero velocity. Figure 4.44 shows the laser absorption trace. The laser scan width was calibrated with a 300-MHz interferometer. The multiple peaks on each of the two Doppler-shifted manifolds are due to the two isotopes in Ce. Using the values of wavelength and measured Doppler shift, we obtained a vapor flow velocity of 9×10^4 cm/s.

4.4.5 Other Applications

There are many other applications of laser-absorption spectroscopy that warrant discussion. Tunable diode and tunable dye lasers have been used to probe the dynamics and small-signal gain coefficients of many laser systems. Reid [Reid, 1983] used a tunable lead salt diode laser operating in the region 2140 to 2310 cm^{-1} to probe the vibrational populations in a CW CO_2 laser discharge. By operating the tunable diode laser at 12 μm, Reid was able to measure the gain coefficient at 12 μm in an optically pumped CW NH_3 Raman laser. Pulsed and CW tunable dye lasers are commonly used for measuring the small-signal gain of electronic transition lasers. In the HF chemical laser system, the tunable IR lasers or pulsed HF lasers operating at 2.7 μm have been used to detect the amount of ground-state HF molecules generated prior to the initiation of the $H_2 + F_2$ chemical-reaction chains [Wilson, 1973]. The tunability, high resolution, and temporal properties of IR, visible, and UV lasers enable experimental measurements in regimes unattainable using conventional techniques. Significant progress has also been made in the diagnostics of molecular and atomic ions. Haese et al. [Haese, 1983] have recently developed an IR laser-absorption spectroscopic method for in situ ArH^+ ion drift velocity and mobility measurements in a dc glow discharge. IR laser detection of ions in plasmas is nonintrusive and can be carried out with relatively high spatial resolution. Similar resonance-absorption techniques will be used to monitor the drift velocity of atomic ions in plasma using tunable dye lasers. Traditionally, ion-transport properties are measured in a time-of-flight drift-tube experiment using mass spectroscopy detection. The laser technique provides detailed information on the motions of ions in plasmas. In addition, it can also measure the internal energy distribution of ions.

Laser-absorption spectroscopy has also been used extensively in isotope analysis. Isotope shifts in spectral lines can be as large as several hundredths of a nanometer in atomic systems and several tenths of a nanometer in molecular systems. For elements having an isotope splitting greater than the Doppler width, isotope-ratio analysis can be made easily with a single-frequency tunable laser. For atoms or molecules having an isotope shift smaller than the Doppler width, the Doppler-free nonlinear absorption techniques can be used. In the latter case, line-narrowed methods such as saturation spectroscopy and two-photon spectroscopy permit isotope-ratio analysis for many elements.

4.5 SUMMARY

As described in this chapter, tunable lasers with narrow spectral bandwidths and high brightness have made possible many new types of analytical absorption measurements. Atomic, molecular, and ionic species in the ground state as well as in excited states can be monitored under conditions that are difficult or impossible to measure using a conventional spectrometer and blackbody or spectral lamps. In analytical chemistry it is now possible to obtain spatial and temporal maps of density, temperature, and state distributions for species of interest. High laser intensity enables atoms, molecules, and ions to be pumped into excited energy levels in a step wise fashion, opening up new atomic spectral regimes to study. Two-photon absorption techniques offer the possibility of minimizing Doppler line broadening. The method offers great promise for isotope analysis where spectral lines are very close together. Laser-absorption spectroscopy has set new standards in detection. In recent years the analytical techniques using laser-absorption spectroscopy have been sufficiently developed to be of immediate and widespread use in industry. Real-time monitoring of industrial-scale chemical processes has provided prompt and accurate measurements of reactant and product species, including the direct determination of temperature, pressure, and flow velocity. Using the laser-absorption technique, it is now possible to examine the course of many chemical reactions in detail, allowing for active chemical process control. It is clear that laser-absorption spectroscopy is well suited for certain specialized industrial chemistry applications. However, there is some disagreement on projections as to how widespread such applications can become in large-scale bulk chemical processing. Nevertheless, innovative concepts for producing laser light at lower cost to optimize laser performance, reliability, and system lifetimes will be the future direction for laser-absorption spectroscopy.

REFERENCES

[Aldridge, 1975a] J. P. Aldridge, R. F. Holland, H. Flicker, K. W. Nill, and T. C. Harmon, High resolution Q-branch spectrum of CO_2 at 618 cm^{-1}, *J. Mol. Spectrosc*, 54, 328 (1975).

[Aldridge, 1975b] J. P. Aldridge, H. Filip, H. Flicker, R. F. Holland, R. C. McDowell, N. G. Nereson, and K. Fox, Octahedral fine-structure splittings in ν_3 of SF_6, *J. Mol. Spectrosc.*, 58, 165 (1975).

[Allario, 1975] F. Allario, C. H. Bair, and J. F. Butler, High-resolution spectral measurements of SO_2 from 1176.0 to 1265.8 cm^{-1} using a single PbSe laser with magnetic and current tuning, *IEEE J. Quantum Electron.*, QE-11, 205 (1975).

[Alobaidi, 1975] T. A. Alobaidi and D. W. Hills, A helium-neon laser infrared analysis for alcohol vapour in the breath, *J. Phys.*, E8, 30 (1975).

[Armstrong, 1967] B. H. Armstrong, Spectrum line profiles, the Voigt function, *J. Quant. Spectrosc. Radiat. Transfer*, 7, 61 (1967).

[Bair, 1979] C. H. Bair and P. Brockman, High-resolution spectral measurement of HNO_3 5.9-μm band using a tunable diode laser, *Appl. Opt.*, 18, 4152 (1979).

[Barrow, 1962] G. M. Barrow, *Introduction to Molecular Spectroscopy*, McGraw-Hill, New York, 1962.

[Basov, 1975] N. G. Basov, E. M. Blenov, V. A. Danilychev, and A. F. Suckhov, High pressure carbon diode electrically excited preionization lasers, *Sov. Phys.*, 17, 705 (1975).

[Basting, 1974] D. Basting, F. P. Schafer, and B. Steyer, New laser dyes, *Appl. Phys.*, 3, 81 (1974).

[Bernard, 1979] P. Bernard and T. Oka, A study of the K = 3n A_1-A_2 centrifugal splitting in the $3\nu_2$ and $4\nu_2$ states of PH_3 using a difference-frequency laser system, *J. Mol. Spectrosc.*, 75, 181 (1979).

[Beterov, 1973] I. M. Beterov, V. P. Chebotayer, and A. S. Provorov, High precision spectroscopy of SF_6 with CW high pressure tunable CO_2 lasers, *Opt. Commun.* 7, 410 (1973).

[Bien, 1978] F. Bien, Measurements of the nitric oxide ion vibrational absorption coefficient and vibrational transfer to N_2, *J. Chem. Phys.*, 69, 2631 (1978).

[Bloomfield, 1983] L. A. Bloomfield, B. Couvilland, E. A. Hildum, and T. W. Hänsch, CW ultraviolet saturation spectroscopy of the 6P $^3P_0$ − 9S $^3S_1$ transition in mercury at 246.5 nm, *Opt. Commun.*, 45, 87 (1983).

[Blum, 1972a] F. A. Blum, K. W. Nill, P. L. Kelley, A. R. Calawa, and T. C. Harmon, Tunable infrared laser spectroscopy of Atmospheric water vapor, *Science, 177*, 694 (1972).

[Blum, 1972b] F. A. Blum, K. W. Nill, A. W. Calawa, and T. C. Harmon, Measurement of the gain line shape of a gas laser using a tunable semiconductor laser, *Appl. Phys. Lett., 20*, 377 (1972).

[Blum, 1973] F. A. Blum, K. W. Nill, and A. J. Strauss, Line shape of the Doppler-limited infrared magnetic rotation spectrum of nitric oxide, *J. Chem. Phys., 58*, 4968 (1973).

[Boker, 1983] J. Boker, L. Eishner, R. Story, P. Bucksbaum, and R. P. Freeman, Wavelength conversion with excimer lasers, *AIP Conf. Proc. Excimer Lasers, 1983*, American Institute of Physics, New York, 1983.

[Brockman, 1978] P. Brockman, C. H. Bair, and F. Allario, High resolution spectral measurement of HNO_3 11.3-μm band using tunable diode lasers, *Appl. Oct., 17*, 91 (1978).

[Brown, 1979] L. R. Brown, R. H. Hunt, and A. S. Pine, Wave numbers, line strengths and assignments in the Doppler-limited spectrum of formaldehyde from 2700 to 3000 cm^{-1}, *J. Mol. Spectrosc., 75*, 406 (1979).

[Butcher, 1975] R. J. Butcher, R. B. Dennis, and S. D. Smith, The tunable spin-flip Raman laser: II, Continuous wave molecular spectroscopy, *Proc. R. Soc. London, A344*, 541 (1975).

[Butler, 1964] J. F. Bulter, A. R. Calana, R. J. Phelan, Jr., T. C. Harmon, A. J. Strauss, and R. H. Rediker, PbTe diode laser, *Appl. Phys. Lett., 5*, 75 (1964).

[Butler, 1965] J. F. Butler and A. R. Calawa, PbS diode laser, *J. Electrochem. Soc., 112*, (1965).

[Butler, 1976] J. F. Butler, Tunable diode laser instruments for advanced measurment and analysis, *Proc. Soc. Photo-Opt. Instrum. Eng., 82*, 33 (1976).

[Butler, 1983] J. F. Butler, R. E. Reeder, and K. J. Linden, Mesastripe $Pb_{1-x}Sn_xSe$ tunable diode lasers, *IEEE J. Quantum Electron., 10*, 1520 (1983).

[Byer, 1974] R. L. Byer, *Laser Spectroscopy* (R. G. Brewer and A. Mooradian, eds.), Plenum Press, New York, 1974.

[Cappellani, 1979] F. Cappellani and G. Restelli, Diode laser measurements of NH_3 ν_2-band spectral lines, *J. Mol. Spectrosc.*, 77, 36 (1979).

[Chen, 1978] H. L. Chen, R. Bedford, C. Borzileri, W. Brunner, and M. Hayes, Relaxation of neodymium in a weakly ionized expanding plasma, *J. Appl. Phys.*, 49, 6136 (1978).

[Chen, 1985] H. L. Chen and S. Abdallah. Private communication, 1985.

[Clough, 1980] P. H. Clough and J. Johnston, Laser-induced fluorescence measurement of the vibrational distribution of CS formed in the reaction $O + CS_2 \rightarrow CS + SO$, *Chem. Phys. Lett.*, 71, 253 (1980).

[Corcoran, 1970] V. J. Corcoran, R. E. Cupp, J. J. Gallagher, and W. T. Smith, Nonlinear optical effects using a CO_2 laser and klystron, *Appl. Phys. Lett.*, 16, 316 (1970).

[Cummings, 1980] J. C. Cummings, Diode-laser spectroscopy of the ν_3 α branch of MoF_6, *J. Mol. Spectrosc.*, 83, 417 (1980).

[Dang-Nhu, 1979] M. Dang-Nhu, A. S. Pine, and A. G. Robiette, Spectral intensities in the ν_3 bands of $^{12}CH_4$ and $^{13}CH_4$, *J. Mol. Spectrosc.*, 77, 57 (1979).

[Das, 1980] P. P. Das, V. M. Devi, and K. N. Rao, Diode laser spectroscopy of the ν_6 band of $^{12}CH_3I$, *J. Mol. Spectrosc.*, 84, 305 (1980).

[Datta, 1975] S. Datta, R. W. Anderson, and R. N. Zare, Laser isotope separation using an intracavity absorption technique, *J. Chem. Phys.*, 63, 5503 (1975).

[Devi, 1979a] V. M. Devi, P. P. Das, and K. N. Rao, Intercomparison of CO_2 and HCN wave numbers at 800-600 cm^{-1} with a diode laser spectrometer, *Appl. Opt.*, 18, 2918 (1979).

[Devi, 1979b] V. M. Devi, S. P. Reddy, K. N. Rao, J. M. Flaud, and C. Camy-Peyret, Interpretation of diode laser spectra of ozone at 14 μm: (010) and (020) states of ozone, *J. Mol. Spectrosc.*, 77, 156 (1979).

[Divens, 1982] W. G. Divens and S. M. Jarrett, Design and performance of a frequency-stabilized ring dye laser, *Rev. Sci. Instrum.*, 53, 1363 (1982).

[Dunning, 1974] F. B. Dunning and R. F. Stebbings, The efficient generation of tunable near UV radiation using a N_2 pumped dye laser, *Opt. Commun.*, 11, 112 (1974).

[Dutta, 1977] N. Dutta, R. T. Warner, and G. J. Wolga, Sensitivity enhancement of a spin-flip Raman laser absorption spectrometer through use of an intracavity absorption cell, *Opt. Lett.*, *1*, 155 (1977).

[El-sherbing, 1979] M. El-sherbing, E. A. Ballik, J. Shewchun, R. K. Garside, and J. Reid, High-sensitivity point monitoring of ozone and high resolution spectroscopy of the ν_3 band of ozone using a tunable semiconductor diode laser, *Appl. Opt.*, *18*, 1198 (1979).

[Eng, 1972] R. S. Eng, A. R. Calawa, T. C. Harmon, P. L. Kelley, and A. Javan, Collisional narrowing of infrared water-vapor transitions, *Appl. Phys. Lett.*, *21*, 303 (1972).

[Eng, 1973] R. S. Eng, P. L. Kelley, A. Mooradian, A. R. Calawa, and T. C. Harmon, Tunable laser measurements of water-vapor transitions in the vicinity of 5 μm, *Chem. Phys. Lett.*, *19*, 524 (1973).

[Eng, 1974] R. S. Eng, P. L. Kelley, A. R. Calawa, T. C. Harmon, and K. W. Nill, Tunable diode laser measurements of water-vapor absorption line parameters, *Mol. Phys.*, *28*, 653 (1974).

[Eng, 1977] R. S. Eng, K. W. Nill, and M. Wahlen, Tunable diode laser spectroscopy determination of ν_3 band center of $^{14}C^{16}O_2$ at 4.5 μm, *Appl. Opt.*, *16*, 3072 (1977).

[Eng, 1978] R. S. Eng, G. Petagna, and K. W. Nill, Ultra-high (10^{-4} cm^{-1}) resolution study of the 8.2 μm and 11.3 μm bands of H_2SO_4: accurate determination of absorbance and dissociation constants, *Appl. Opt.*, *17*, 1723 (1978).

[Eng, 1979a] R. S. Eng and A. W. Mantz, Tunable diode laser spectroscopy of CO_2 in the 10- to 15-μm spectral region—lineshape and Q branch head absorption profile, *J. Mol. Spectrosc.*, *74*, 331 (1979).

[Eng, 1979b] R. S. Eng, A. W. Mantz, and T. R. Todd, Improved sensitivity of tunable-diode-laser open-path trace gas monitoring systems, *Appl. Opt.*, *18*, 3438 (1979).

[Eng, 1980] R. S. Eng, J. F. Butler, and K. H. Linden, Tunable diode laser spectroscopy: an invited review, *Opt. Eng.*, *19*, 945 (1980).

[Eng, 1982] R. S. Eng and R. T. Ku, High-resolution linear laser absorption Spectroscopy—Review, *Spectrosc. Lett.*, *15*, 803 (1982).

[Fayt, 1977] A. Fayt, D. VanLerberghe, J. P. Kupfer, H. Pascher, and H. G. Hafele, Analysis of vibration-rotation bands of OCS in the 5.3 μm region using a tunable Q-switched spin-flip Raman laser, *Mol. Phys.*, *33*, 603 (1977).

[Fields, 1975] R. W. Fields, G. H. Capella, and M. A. Revelli, Optical-optical double resonance laser spectroscopy of BaO, *J. Chem. Phys.*, *63*, 3228 (1975).

[Flicker, 1979] H. Flicker, N. G. Hereson, and W. S. Benedict, Diode laser spectroscopy of Q branches at 662-666 cm^{-1} of CO_2 enriched in ^{18}O and ^{17}O, *Proc. Soc. Photo-Opt. Instrum. Eng.*, *190*, 275 (1979).

[Forrest, 1983] G. T. Forrest and D. L. Hall, Automated tunable diode laser monitoring systems, in *Tunable Diode Laser Development and Spectroscopy Applications*, Proc., Soc. Photo-Opt. Instrum. Eng., 438 (1983).

[Fox, 1979] F. Fox, M. J. Reisfeld, and R. S. McDowell, Tunable diode laser measurements of line strength in the ν_4 band of $^{12}CH_4$, *J. Chem. Phys.*, *71*, 1058 (1979).

[Frerking, 1977] M. Frerking and D. J. Muehlner, Infrared heterodyne spectroscopy of atmospheric ozone, *Appl. Opt.*, *16*, 526 (1977).

[Gabel, 1972] C. Gable and M. Hercher, A continuous tunable source of coherence UV radiation, *IEEE J. Quantum Electron.*, *QE-8*, 850 (1972).

[Geritsen, 1966] H. J. Geritsen, *Physics of Quantum Electronics*, McGraw-Hill, New York, 1966.

[Giordmaine, 1965] J. A. Giordmaine and R. C. Miller, Tunable coherent parametric oscillation in $LiNbO_3$ at optical frequencies, *Phys. Rev. Lett.*, *14*, 1973 (1965).

[Gough, 1978] T. E. Gough, R. E. Miller, and G. Scoles, Sub-Doppler resolution infrared spectroscopy of supersonic molecular beams of nitric oxide, *J. Mol. Spectrosc.*, *72*, 124 (1978).

[Gray, 1979] D. L. Gray, A. G. Robiette, and A. S. Pine, Extended measurement and analysis of the ν_3 infrared band of methane, *J. Mol. Spectrosc.*, *77*, 440 (1979).

[Guerra, 1975] M. A. Guerra, M. Ketahi, A. Sanches, M. S. Feld, and A. Javan, Water vapor spectroscopy at 5 µm using a tunable SFR laser, *J. Chem. Phys.*, *63*, 1317 (1975).

[Gus'kov, 1977] L. N. Gus'kov, B. I. Troshin, V. P. Chebotaev, and A. V. Sishaev, A laser absorption polarization spectrometer and measurement of small absorption coefficients, *Zh. Prikl. Spektrosk. (J. Appl. Spectrosc. USSR)*, *27*, 993 (1977).

[Haese, 1983] N. N. Haese and T. Oka, Doppler shift ion mobility measurements of ArH^+ in a He dc glow discharge by diode laser spectroscopy, in *Tunable Diode Laser Development and Spectro-*

scopy Applications, Proc. Soc. Photo-Opt. Instrum. Eng., 438,
165a (1983).

[Hager, 1976] J. Hager, W. Hinz, H. Walther, and G. Strey, High-
resolution spectroscopy of ethylene by means of a spin-flip-raman
laser, Appl. Phys., 9, 35 (1976).

[Hall, 1976] J. L. Hall, C. J. Borde, and K. Jehara, direct optical
resolution of the recoil effect using saturated absorption spectroscopy, Phys. Rev. Lett., 37, 1339 (1976).

[Hänsch, 1975] T. W. Hänsch, S. A. Lee, R. Wallenstein, and C. Weiman, Doppler-free two-photon spectroscopy of 1S-2S hydrogen,
Phys. Rev. Lett., 34, 307 (1975).

[Hanson, 1977a] R. K. Hanson, Shock tube spectroscopy: advanced
instrumentation with a tunable diode laser, Appl. Opt., 16,
1479 (1977).

[Hanson, 1977b] R. K. Hanson, High-resolution spectroscopy of combustion gases using a tunable IR diode laser, Appl. Opt., 16,
2045 (1977).

[Hanson, 1978] R. K. Hanson and P. K. Falcone, Temperature measurement technique for high-temperature gases using a tunable
diode laser, Appl. Opt., 17, 2477 (1978).

[Harman, 1969] T. C. Harman, Temperature and compositional dependence of laser emission in $Pb_{1-x}Sn_xSe$, Appl. Phys. Lett., 14,
333 (1969).

[Harris, 1969] S. E. Harris, Tunable optical parametric oscillators,
Proc. IEEE, 57, 2096 (1969).

[Harris, 1976] N. W. Harris, F. O. Neill, and W. T. Whitney, Wide-
band interferometer tuning of a multiatmosphere CO_2 laser, Opt.
Commun., 16, 57 (1976).

[Hawakami, 1978] H. Hawakami, Y. Izawa, and C. Yamanaka, A high
resolution infrared spectrum of C_2H_2 using a $Pb_{0.93}Sn_{0.07}$ diode
laser, Jpn. J. Appl. Phys., 17, 461 (1978).

[Hertzberg, 1944] G. Hertzberg, Atomic Spectra and Atomic Structure,
Dover, New York, 1944.

[Hertzberg, 1950] G. Hertzberg, Molecular Spectra and Molecular
Structure: I. Spectra of Diatomic Molecules, II. Infrared and
Raman Spectra of Polyatomic Molecules, Van Nostrand Reinhold,
New York, 1950.

[Hertzberg, 1966a] G. Hertzberg, Molecular Spectra and Molecular
Structure: III. Electronic Spectra and Electronic Structure of
Polyatomic Molecules, D. Van Nostrand, Princeton, N.J., 1966.

[Hertzberg, 1966b] G. Herzberg and J. W. C. Johns, The spectrum and structure of singlet CH_2, *Proc. R. Soc. London, A295*, 107 (1966).

[Hertzberg, 1971] G. Hertzberg, *The Spectra and Structures of Simple Free Radicals*, Cornell University Press, Ithaca, N.Y., 1971.

[Hillman, 1979] J. J. Hillman, D. E. Jennings, and J. L. Faris, Diode laser-CO_2 laser heterodyne spectrometer: measurement of 2sQ (1,1) in $2\nu_2 - \nu_2$ of NH_3, *Appl. Opt., 18*, 1808 (1979).

[Hinkley, 1970] E. D. Hinkley, High resolution infrared spectroscopy with a tunable diode laser, *Appl. Phys. Lett., 16*, 351 (1970).

[Hinkley, 1972] E. D. Hinkley, A. R. Calawa, P. L. Kelley, and S. A. Clough, Tunable-laser spectroscopy for the ν_1 band of SO_2, *J. Appl. Phys., 43*, 3222 (1972).

[Hinkley, 1976] E. D. Hinkley, K. W. Nill, and F. A. Blum, Infrared spectroscopy with tunable lasers, in *Laser Spectroscopy* (H. Walther, ed.), Springer-Verlag, Berlin, 1976.

[Hirota, 1979] E. Hirota, Diode laser spectroscopy of the ν_6 band of methyl fluoride, *J. Mol. Spectrosc., 74*, 209 (1979).

[Hutcherson, 1973] J. W. Hutcherson and P. M. Griffin, Self-broadened absorption linewidths for the krypton resonance transitions, *J. Opt. Soc. Am., 63*, 338 (1973).

[Jarrett, 1979] S. M. Jarrett and J. F. Young, High-efficiency single-frequency CW ring dye laser, *Opt. Lett., 4*, 176 (1979).

[Jennings, 1978a] D. E. Jennings, Diode laser spectra of Cl_2F_2 near 10.8 μm: air-broadening effects, *Geophys. Res. Lett., 5*, 241 (1978).

[Jennings, 1978b] D. E. Jennings, J. J. Hillman, and W. H. Weber, Diode laser analysis of the $3\nu_7^{1f} + \nu_2 \leftarrow 2\nu_7^0$ and $4\nu_7^0 + \nu_2 \leftarrow 3\nu_7^{1f}Q$ branches of C_3O_2, *Opt. Lett., 2*, 157 (1978).

[Jennings, 1980] D. E. Jennings, Absolute line strengths in ν_4, $^{12}CH_4$: a dual-beam diode laser spectrometer with sweep integration, *Appl. Opt., 19*, 2695 (1980).

[Johnston, 1982] T. F. Johnston, R. H. Brady, and W. Proffitt, Powerful single-frequency ring dye laser spanning the visible spectrum, *Appl. Opt., 21*, 2307 (1982).

[Kaldor, 1982] A. Kaldor and R. L. Woodin, Applications of lasers to chemical processing, *Proc. IEEE, 70*, 565 (1982).

[Kaminsky, 1976] M. E. Kaminsky, R. T. Hawkins, F. V. Kowalski, and A. L. Schawlow, Identification of absorption lines by modulated lower-level population spectrum of Na_2, *Phys. Rev. Lett.*, *36*, 671 (1976).

[Kim, 1979] K. C. Kim, W. B. Person, D. Seitz, and B. J. Krohn, Analysis of the v_4 (615 cm^{-1}) region of the Fourier transform and diode laser spectra of SF_6, *J. Mol. Spectrosc.*, *76*, 322 (1979).

[King, 1964] G. W. King, *Spectroscopy and Molecular Structure*, Holt, Rinehart and Winston, New York, 1964.

[Kopfermann, 1958] H. Kopfermann, *Nuclear Moments*, Academic, Press, New York, 1958.

[Ku, 1975] R. T. Ku, E. D. Hinkley, and J. O. Sample, Long-path monitoring of atmospheric carbon monoxide with a tunable diode laser system, *Appl. Opt.*, *14*, 854 (1975).

[Kuhl, 1975] J. Kuhl and H. Spitschan, A frequency doubled dye laser with a servo-tuned crystal, *Opt. Commun.*, *13*, 6 (1975).

[Kuhn, 1962] H. G. Kuhn, *Atomic Spectra*, Academic Press, New York, 1962.

[Kung, 1983] A. H. Kung, Third-harmonic generation in a pulsed supersonic jet of xenon, *Opt. Lett.*, *8*, 24 (1983).

[Laguna, 1982] G. A. Laguna and S. L. Baughcum, Real-time detection of methyl radicals by diode laser absorption at 608 cm^{-1}, *Chem. Phys. Lett.*, *88*, 568 (1982).

[Lamb, 1964] W. E. Lamb, Jr. Theory of an optical maser, *Phys. Rev.*, *A134*, 1429 (1964).

[Langford, 1983] A. Langford, H. Petek, and C. B. Moore, Collisional removal of CH_2 ($^1A_1$): absolute rate constants for atomic and molecular collisional partners at 295°K, *J. Chem. Phys.*, *78*, 6650 (1983).

[Levi, 1979] S. Lavi, L. A. Levir, and E. Miron, Efficient oscillator-amplifier dye laser pumped by a frequency-doubled Nd-YAG laser, *Appl. Opt.*, *18*, 525 (1979).

[Letokhov, 1977] V. S. Letokhov, Laser Spectroscopy: 3. Linear laser spectroscopy, *Opt. Laser Tech.*, *9*, 263 (1977).

[Maki, 1980] A. G. Maki, W. B. Olson, and R. L. Sams. High-resolution infrared spectrum of the 859- and 1711-cm^{-1} bonds of carbonyl sulfide (OSC), *J. Mol. Spectrosc.*, *81*, 122 (1980).

[Mantz, 1978] A. W. Mantz and R. S. Eng. Tunable laser measurements of line intensities and pressure-broadened linewidths of HBr:Ar on the HBr infrared fundamental region, *Appl. Spectrosc.*, *32*, 239 (1978).

[Mariella, 1979] R. P. Mariella. A study of laser-assisted surface ionization of lithium, *J. Chem. Phys.*, *71*, 94 (1979).

[Marowsky, 1973] G. Marowsky. Reliable single-mode operation of a flashlamp pumped dye laser, *Rev. Sci. Instrum.*, *44*, 890 (1973).

[McDowell, 1978a] R. S. McDowell, L. Radziemski, H. Flicker, H. W. Galbraith, R. C. Kennedy, N. G. Nereson, B. J. Krohn, J. P. Aldridge, and J. P. King. High-resolution spectroscopy of OsO_4 stretching fundamental, *J. Chem. Phys.*, *69*, 1513 (1978).

[McDowell, 1978b] R. S. McDowell, H. W. Galbraith, B. J. Krohn, N. G. Nereson, P. F. Moulton, and E. D. Hinkley. High-J assignments in the 10.5 μm SF_6 spectrum: identification of the levels pumped by CO_2 P(12) and P(22), *Opt. Lett.*, *2*, 97 (1978).

[McDowell, 1980] R. S. McDowell, M. J. Reisfeld, H. W. Galbraith, B. J. Krohn, A. Flicker, R. C. Kennedy, J. P. Aldridge, and N. G. Nereson. High-resolution spectroscopy of the 16-μm bending fundamental of CF_4, *J. Mol. Spectrosc.*, *83*, 440 (1980).

[McDowell, 1981] R. S. McDowell. Vibrational Spectroscopy using tunable lasers, in *Vibrational Spectra and Structure*, Elsevier, New York, 1981.

[McKellar, 1978] A. R. W. McKeller and T. Oka. A study of the electric quadruple fundamental band of D_2 using an infrared difference frequency laser systems, *Can. J. Phys.*, *56*, 1315 (1978).

[McKenzis, 1975] H. A. McKenzis, S. D. Smith, and R. B. Dennis. Tuning and mode characteristics of the CW InSb spin-flip Raman laser determined by use of a Fabry-Perot interferometer, *Opt. Commun.*, *15*, 151 (1975).

[Melngailis, 1968] I. Melngailis. Measurement of the fundamental vibration-rotation spectrum of IO, *J. Phys. Paris Colloq. C-4 (Suppl)*. *11*, 11 (1968).

[Melngailis, 1972] I. Melngailis. The use of lasers in pollution monitoring, *IEEE Trans. Geosci. Electron.*, *Ge-10*, 7 (1972).

[Menzies, 1977] R. T. Menzies, J. S. Margolis, E. D. Hinkley, and R. A. Toth. The use of lasers in pollution monitoring, *Appl. Opt.*, *16*, 523 (1977).

[Meredith, 1974] R. E. Meredith and F. G. Smith. Broadenings of hydrogen fluoride lines by H_2, D_2, and N_2, *J. Chem. Phys.*, *80*, 3388 (1974).

[Mitchell, 1961] A. C. G. Mitchell and M. W. Zemansky. *Resonance Radiation and Excited Atoms*, Cambridge University Press, Cambridge, 1961.

[Montgomery, 1975] G. P. Montgomery and J. C. Hill. High-resolution diode-laser spectroscopy of the 949.2 cm^{-1} band of ethylene, J. Opt. Soc. Am., 65, 579 (1975).

[Montgomery, 1978] G. P. Montgomery and R. F. Majkowski. Diode laser spectroscopy of foreign gas broadened sulfuric acid (H_2SO_4) vapor, Appl. Opt., 17, 173 (1978).

[Mozolowski, 1979] M. H. Mozolowski, R. B. Dennis, and H. A. Mackenzie. The external-resonator CW spin-flip Raman laser: output characteristics and spectroscopic applications, Appl. Phys., 19, 205 (1979).

[Nereson, 1977] N. G. Nereson and H. Flicker. Wavenumber measurement of weak CO_2 laser line around 10.6 µm, Opt. Commun., 23, 171 (1977).

[Nereson, 1978] N. G. Nereson. Diode laser measurements of NH_3 absorption lines around 10.6 µm, J. Mol. Spectrosc., 69, 489 (1978).

[Nill, 1971] K. W. Nill, F. A. Blum, A. R. Calawa, and T. C. Harmon. Infrared spectroscopy of CO using a tunable PbSSe diode laser, Appl. Phys. Lett., 19, 79 (1971).

[Nill, 1972a] K. W. Nill. High-resolution spectroscopy using magnetic-field-tuned semiconductor lasers, Appl. Phys. Lett., 21, 132 (1972).

[Nill, 1972b] K. W. Nill, F. A. Blum, A. R. Calawa, and T. C. Herman. Observation of Λ-doubling and Zeeman splitting in the fundamental infrared absorption band of nitric oxide, Chem. Phys. Lett., 14, 234 (1972).

[Nordstrom, 1979] R. J. Nordstrom, M. Morillon-Chapey, J. C. Deroche, and D. E. Jennings. A first study of the ν_6 fundamental of CF_2I_2, J. Phys. Lett., 40, L37 (1979).

[Ohsawa, 1980] A. Ohsawa, K. Honda, S. Ohkawa, and R. Ueda. Determination of oxygen concentration profiles in silicon crystals observed by scanning IR absorption using semiconductor lasers, Appl. Phys. Lett., 36, 147 (1980).

[Oka, 1980] T. Oka. Observation fo the infrared spectrum of H_3^+, Phys. Rev. Lett., 45, 531 (1980).

[Okabe, 1978] H. Okabe. Photochemistry of Small Molecules, Wiley, New York, 1978.

[Omenetto, 1979] N. Omenetto, Analytical Laser Spectroscopy, Wiley, New York, 1979.

[Paisner, 1979] J. Paisner and S. Hargrove. A tunable laser system for UV, visible, and IR regions, in Energy and Technology Re-

view, Lawrence Livermore National Laboratory, Livermore, Calif. 1979.

[Patel, 1972] C. K. N. Patel. Linewidth of tunable stimulated Raman scattering, *Phys. Rev. Lett.*, *28*, 649 (1972).

[Patel, 1974] C. K. N. Patel. Spectroscopic measurements of stratospheric nitric oxide and water vapor, *Science*, *184*, 1173 (1974).

[Patel, 1976] C. K. N. Patel, T. Y. Chang, and V. T. Nguyen. Spinflip Raman laser at wavelengths up to 16.8 μm, *Appl. Phys. Lett.*, *28*, 603 (1976).

[Patel, 1977] C. K. N. Patel, R. J. Kerl, and E. G. Burkhardt. Excited-state spectroscopy of molecules using opto-acoustic detection, *Phys. Rev. Lett.*, *38*, 1204 (1977).

[Patel, 1978a] C. K. N. Patel and R. J. Kerl. High-resolution opto-acoustic spectroscopy of ^{15}NO, Λ-doubling measurements, *Opt. Commun.*, *24*, 294 (1978).

[Patel, 1978b] C. K. N. Patel. Use of vibrational energy transfer for excited-state opto-acoustic spectroscopy of molecules, *Phys. Rev. Lett.*, *40*, 535 (1978).

[Penner, 1959] S. S. Penner. *Quantitative Molecular Spectroscopy and Gas Emissivities*, Addison-Wesley, Reading, Mass., 1959.

[Person, 1978] W. B. Person and K. C. Kim. Vibrational anharmonicity constants for SF_6: I. A 16 μm diode laser study of transitions in the ν_4 region, *J. Chem. Phys.*, *69*, 2117 (1978).

[Petek, 1983] H. Petek, D. J. Nesbitt, P. R. Ogllby, and C. B. Moore. IR flash kinetic spectroscopy: the ν_1 and ν_3 spectra of singlet methylene, *J. Phys. Chem.*, *87*, 5367 (1983).

[Pine, 1974] A. S. Pine. Doppler-limited molecular spectroscopy by difference-frequency mining, *J. Opt. Soc. Am.*, *64*, 1683 (1974).

[Pine, 1975] A. S. Pine. Doppler-limited spectra of the ν_3 vibration of $^{12}CH_4$ and $^{13}CH_4$, *J. Mol. Spectrosc.*, *54*, 132 (1975).

[Pine, 1976] A. S. Pine. High-resolution methane ν_3-band spectra using a stabilized tunable difference-frequency laser system, *J. Opt. Soc. Am.*, *66*, 97 (1976).

[Pine, 1977a] A. S. Pine and P. F. Moulton. Doppler-limited and atmospheric spectra of the 4 μm $\nu_1 + \nu_3$ combination band of SO_2, *J. Mol. Spectrosc.*, *64*, 15 (1977).

[Pine, 1977b] A. S. Pine, G. Dresselhaus, B. Palm, R. W. Davies, and S. A. Clough. Analysis of the 4 μm $\nu_1 + \nu_3$ combination band of SO_2, *J. Mol. Spectrosc.*, *67*, 386 (1977).

[Pine, 1978] A. S. Pine. Doppler-limited spectra of the C-H stretching fundamentals of formaldehyde, *J. Mol. Spectrosc.*, *70*, 167 (1978).

[Pine, 1979a] A. S. Pine and K. W. Nill. Molecular-beam tunable-diode-laser sub-Doppler spectroscopy of Λ-doubling in nitric oxide, *J. Mol. Spectrosc.*, *74*, 43 (1979).

[Pine, 1979b] A. S. Pine, J. W. C. Johns, and A. G. Robiette. λ-doubling in the v = 2 ← 0 overtone band in the IR spectrum of NO, *J. Mol. Spectrosc.*, *74*, 52 (1979).

[Pine, 1980a] A. S. Pine and G. Guelachvilli. R-branch head of the ν_3 band of CO_2 at elevated temperatures, *J. Mol. Spectrosc.*, *79*, 84 (1980).

[Pine, 1980b] A. S. Pine. Collisional narrowing of HF fundamental band spectral lines by neon and argon, *J. Mol. Spectrosc.*, *82*, 435 (1980).

[Planet, 1975] W. G. Planet, J. R. Aronson, and J. F. Butler. Measurements of the widths and strengths of low-J lines of the ν_2 branch of CO_2, *J. Mol. Spectrosc.*, *54*, 331 (1975).

[Planet, 1979] W. G. Planet and G. L. Tettmer. Temperature dependent intensities and widths of N_2 broadened CO_2 lines at 15 μm from tunable laser measurements, *J. Quant. Spectrosc. Radiat. Transfer*, *22*, 345 (1979).

[Ralston, 1974] R. W. RAlston, J. N. Walpole, A. R. Calawa, T. C. Harman, and J. P. McVitte. High single-ended CW output power in stripe-geometry PbS diode lasers, *J. Appl. Phys.*, *45*, 1323 (1974).

[Reddy, 1979] S. P. Reddy, V. M. Devi, A. Baldacci, W. Ivanic, and K. N. Rao. Acetylene spectra with a tunable diode laser: ($\nu_4 + \nu_5$)$^{0+} - \nu_4$^{1f} Q branches of $^{12}C_2H_2$ and $^{12}C^{13}CH_2$, *J. Mol. Spectrosc.*, *74*, 217 (1979).

[Reid, 1978a] J. Reid, J. Shewchun, B. K. Garside, and E. A. Ballike. Point monitoring of ambient concentrations of atmospheric gases using tunable lasers, *Opt. Eng.*, *17*, 56 (1978).

[Reid, 1978b] J. Reid, J. Shewchen, and B. K. Garside. Measurement of the transition strength of the 00°2 9.4 μm sequence band in CO_2 using a tunable diode laser, *Appl. Phys.*, *17*, 349 (1978).

[Reid, 1978c] J. Reid and A. R. W. McKeller. Observation of the $S_0(3)$ pure rotation quadruple transition of H_2 with a tunable diode laser, *Phys. Rev.*, *A18*, 224 (1978).

[Reid, 1983] J. Reid. Investigation of the dynamics of IR gas laser using tunable diode lasers, in *Tunable Diode Laser Devlopment and Spectroscopy Applications*, Proc. Soc. Photo-Opt. Instrum. Eng., *438*, 170 (1983).

[Reisfeld, 1978] M. J. Reisfeld and H. Flicker. The Q-branch spectrum of the (01'0-00°0) transition of $^{13}C^{16}O_2$, *J. Mol. Spectrosc.*, *69*, 330 (1978).

[Reisfeld, 1979] M. J. Reisfeld, and H. Flicker. The ν_2 band of N_2O as a frequency standard in the 17 μm region of the infrared, *Appl. Opt.*, *18*, 1136 (1979).

[Restelli, 1978] G. Restelli, F. Cappellani, and G. Meladrone. Evaluation of CF_2I_2 spectral parameters for atmospheric sensing, *Pure Appl. Geophys.*, *117*, 531 (1978).

[Restelli, 1979a] G. Restelli and F. Cappellani. High resolution spectroscopy of the ν_4 band of methane, *J. Mol. Spectrosc.*, *78*, 161 (1979).

[Restelli, 1979b] G. Restelli, F. Cappellani, and G. Melandrone. High-resolution measurements of absolute line intensities in the Q and R branches of the ν_4 fundamental of $^{12}CH_4$, *Chem. Phys. Lett.*, *66*, 454 (1979).

[Rettner, 1983] C. T. Rettner, E. E. Marinero, R. N. Zare, and A. H. Kung. XUV excitation of H_2 using the third harmonic of a frequency-doubled dye laser, in *AIP Conf. Proc. Excimer Lasers, 1983* (Rhodes, ed.), American Institute of Physics, New York, 1983.

[Robinson, 1966] J. W. Robinson. *Atomic Absorption Spectroscopy*, Marcel Dekker, New York, 1966.

[Rogowski, 1978] R. S. Rogowski, C. H. Bair, W. R. Wade, J. M. Hoell, and G. E. Copeland. Infrared vibration-rotation spectra of the ClO radical using tunable diode laser spectroscopy, *Appl. Opt.*, *17*, 1301 (1978).

[Rohrbeck, 1980] W. Rohrbeck, R. Winter, W. Herrmann, J. Wild, and W. Urban. Pressure broadening coefficients for nitric spectrometer, *Mol. Phys.*, *39*, 673 (1980).

[Rutt, 1978] H. N. Rutt and D. N. Travis. Diode laser measurements on dideuteroacetylene, *J. Phys.*, *B11*, L447 (1978).

[Sattler, 1977] J. P. Sattler and G. J. Simonis. Tunable diode laser spectroscopy of methyl fluoride, *IEEE J. Quantum Electron.*, *QE-13*, 461 (1977).

[Sattler, 1978a] J. P. Sattler, T. L. Worchesky, and W. D. Riessler. Diode laser spectra of gaseous methyl alcohol, *Infrared Phys.*, *18*, 521 (1978).

[Sattler, 1978b] J. P. Sattler and K. J. Ritter. Diode laser spectra of NH_3 in the 9.5 μm region, *J. Mol. Spectrosc.*, *69*, 486 (1978).

[Sattler, 1980] J. P. Sattler, T. L. Worchesky, K. J. Ritter, and W. J. Lafferty. Technique for wideband, rapid and accurate diode-laser heterodyne spectroscopy: measurements on 1.1 di-fluo-ethylene, *Opt. Lett.*, *5*, 21 (1980).

[Schweid, 1983] A. N. Schweid and B. B. Hardman. An industrial application of an IR tunable diode laser for the continuous and automatic analysis of a process stream, *Proc. Soc. Photo-Opt. Instrum. Eng.*, *438*, 61 (1983).

[Slusher, 1967] E. Slusher, C. K. N. Patel, and P. A. Fluery. Inelastic light scattering from Landau-level electrons in semiconductors, *Phys. Rev. Lett.*, *18*, 77 (1967).

[Smith, 1974] F. A. Smith and R. E. MEredith. Anomalous resonance effect in collision broadened spectral line widths, *J. Quant. Spectrosc. Radiat. Transfer*, *14*, 385 (1974).

[Snavely, 1969] B. B. Snavely. Flashlamp-excited organic dye lasers, *Proc. IEEE*, *57*, 1374 (1969).

[Sorokin, 1975] P. P. Sorokin. Generation of vacuum ultraviolet radiation by nonlinear mixing in atomic and ionic vapors, *Laser Spectroscopy* (S. Haroche, ed.), Springer-Verlag, Berlin, 1975.

[Streit, 1979] G. E. Streit, J. S. Wells, F. C. Fehsenfeld, and C. J. Howard. A tunable diode laser study of the reactions of nitric and nitrous acids: HNO_3 + NO and HNO_2 + O_3, *J. Chem. Phys.*, *70*, 3439 (1979).

[Strow, 1980] L. L. Strow. Observation of nuclear quadrupole hyperfine structure in the infrared spectrum of hydrogen iodide using a tunable-diode laser, *Opt. Lett.*, *5*, 166 (1980).

[Stuke, 1978] M. Stuke and E. E. Marinero. On-line computer controlled CW dye laser spectrometer for laser isotope separation, *Appl. Phys.*, *16*, 303 (1978).

[Sutton, 1976] D. G. Sutton and G. A. Capelle. KrF-laser-pumped tunable dye laser in the ultraviolet, *Appl. Phys. Lett.*, *29*, 563 (1976).

[Szoke, 1963] A. Szoke and A. Javan. Isotope shift and saturation behavior of the 1.15 μm transition of Ne, *Phys. Rev. Lett.*, *10*, 521 (1963).

[Tipping, 1970] R. H. Tipping and R. M. Herman. Impact theory for the noble gas pressure induced HCl vibration-rotation and pure rotation line width: I, *J. Quant. Spectrosc. Radiat. Transfer,* 10, 881 (1970).

[Todd, 1979] T. R. Todd and W. B. Olson. The infrared spectra of $^{12}C^{32}S$, $^{12}C^{34}S$, $^{13}C^{32}S$, $^{12}C^{33}S$, *J. Mol. Spectrosc.,* 74, 190 (1979).

[Travis, 1977] D. N. Travis, J. C. McGurk, D. McKeown, and R. G. Denning. Infrared spectroscopy of super-cooled gases, *Chem. Phys. Lett.,* 45, 287 (1977).

[Uchino, 1979] O. Uchino, T. Mizunami, M. Maeda, and Y. Miyazue. Efficient dye lasers pumped by a XeCl excimer laser, *Appl. Phys.,* 19, 35 (1979).

[Urban, 1978] W. Urban and N. Herrmann. Zeeman modulation spectroscopy with spin-flip Raman laser, *Appl. Phys.,* 17, 325 (1978).

[Urban, 1980] S. Urban, V. Spirko, P. Papousek, R. S. McDowell, N. G. Nereson, S. P. Belov, L. I. Gershstein, A. V. Maslovskij, A. F. Krupnov, J. Curtis, and K. N. Rao. Coriolis and 1-type interactions in the ν_2, $2\nu_2$ and ν_4 states of $^{14}NH_3$, *J. Mol. Spectrosc.,* 79, (1980).

[Vasilenko, 1970] L. S. Vasilenko, V. P. Chebotaev, and A. V. Shishaev. Line shape of a two-photon absorption in a standing-wave field in a gas, *JETP Lett.,* 12, 113 (1970).

[Wagenaar, 1974] H. C. Wagenaar, C. J. Pickford, and L. de Galan. The interferometric measurement of atomic absorption line profiles in flames, *Spectrochim. Acta,* 29B, 211 (1974).

[Wagstaff, 1979] C. E. Wagstaff and M. H. Dunn. A second-harmonic, ring dye laser for the generation of continuous-wave single-frequency UV radiation, *J. Phys.,* D12, 355 (1979).

[Weber, 1974] W. H. Weber, P. D. Maker, K. F. Yeung, and C. W. Peters. High-resolution Stark spectroscopy in the ν_4 band of NH_3 using a thin-film diode laser, *Appl. Opt.,* 13, 1431 (1974).

[Weber, 1976] W. H. Weber, P. D. Maker, a nd C. W. Peters. Analysis of tunable diode laser spectra of the ν_4 bands of $^{12}C_3$ $^{16}O_2$, *J. Chem. Phys.,* 64, 2149 (1976).

[Weiman, 1976] C. Weiman and T. W. Hänsch. Doppler-free laser polarization spectroscopy, *Phys. Rev. Lett.,* 36, 1170 (1976).

[Wells, 1979] J. S. Wells, F. R. Peterson, and A. G. Maki. Spectroscopic reference frequencies in the 9.5 μm band of carbonyl sulfide, *Appl. Opt.,* 18, 3567 (1979).

[Wilke, 1979] V. Wilke and W. Schmidt. Tunable coherent radiation source covering a spectral range from 185 to 880 nm, *Appl. Phys.*, *17*, 477 (1979).

[Wilson, 1973] J. Wilson, H. L. Chen, W. Frye, R. L. Taylor, R. Little, and R. Lowell. Electron beam dissociation of fluorine, *J. Appl. Phys.*, *44*, 5447 (1973).

[Worchesky, 1978] T. L. Worchesky, K. J. Ritter, J. P. Sattler, and W. A. Riessler. Heterodyne measurements of infrared absorption frequencies of D_2O, *Opt. Lett.*, *2*, 70 (1978).

[Wynne, 1974] J. J. Wynne, P. Sorokin, and J. R. Lankard, A tunable infrared coherent source for the 2 to 25 micron region and beyond, *Laser Spectroscopy* (R. G. Brewer and A. Moradian, eds.), Plenum Press, New York, 1974.

[Yamada, 1979] C. Yamada and E. Hirota. Infrared diode laser spectroscopy of carbon monosulfide, *J. Mol. Spectrosc.*, *74*, 203 (1979).

[Yariv, 1967] A. Yariv. *Quantum Electronics*, Wiley, New York, 1967.

[Young, 1965] C. Young. *Tables for Calculating the Voigt Profile*, Univ. Michigan Rep. 058637-T, Ann Arbor, Mich., 1965.

[Zeiger, 1973] H. J. Zeiger, F. A. Blum, and K. W. Nill. Observation of strong nonlinearities of the high field Zeeman spectrum of NO at 1876 cm^{-1}, *J. Chem. Phys.*, *59*, 39688 (1973).

5

Laser Plasmas for Chemical Analysis

DAVID A. CREMERS *Los Alamos National Laboratory, Los Alamos, New Mexico*

LEON J. RADZIEMSKI *New Mexico State University, Las Cruces, New Mexico*

5.1 INTRODUCTION

Laser plasma techniques of chemical analysis have some unique properties not found in standard laboratory analysis methods. These are (1) that bulk solid samples can be prepared rapidly in either the vapor or particulate state for analysis via an auxiliary technique, and (2) that species for analysis can be both prepared and excited in one step via atomic emission spectroscopy. The purpose of this chapter is to describe some recent uses of laser plasmas for chemical analysis and to review established applications. The discussion of pulsed laser plasmas is divided into two main sections according to whether the spark plasma is used only to ablate material from a sample for analysis by an auxiliary technique or whether the spark plasma is analyzed directly (direct spark analysis). Particular emphasis is placed on the analytical results and figures of merit for each method. Because of the extensive literature on laser plasmas, it is not possible to cover each topic in detail here, so the reader is referred to the general references listed at the end of the chapter and the references cited therein for additional information. The intention here is to give the reader a general idea of the wide range of possible applications of laser plasmas. It is emphasized that due to the high temperatures of laser plasmas all techniques described here provide an elemental analysis of materials. Although every attempt was made to have this presentation complete, the topics selected reflect somewhat the authors' own interests and experiences.

FIG. 5.1 Laser spark in air generated by focusing a Nd:YAG laser pulse with a 5-cm-focal-length lens. A transmission diffraction grating was placed in front of the camera lens to produce the spectra which radiate from the spark. The emission lines, mainly from once-ionized nitrogen atoms, are superimposed on the spectrally dispersed intense background radiation from the spark.

5.1.1 Physics of Laser Plasmas in Gases

When a sufficiently energetic laser pulse is focused in a gas, a spark plasma is formed at the focus which appears as a bright flash of intense light accompanied by a loud sound. Maker et al. (1964) were the first to report observation of this phenomenon called optically induced gas breakdown. A photograph of a spark generated by focusing a Nd:YAG laser pulse of 15-ns duration in ambient air is shown in Fig. 5.1. The high temperature of the plasma can vaporize solid material and electronically excite the resulting atoms. For this reason, the laser spark is useful to remove small amounts of material from a surface for analysis via another technique (Sec. 5.2), or the spark light can be monitored directly to determine its elemental content (Sec. 5.3). For example, the spectrally dispersed light radiating from the spark in Fig. 5.1 exhibits atomic line emission superimposed on the spectrally broadband continuum radiation from the spark. The line spectrum is due primarily to emission from once-ionized nitrogen atoms formed from nitrogen molecules by the high plasma temperature. In addition to gases and surfaces, the laser spark can be generated by focusing intense laser radiation directly into bulk liquids. The discussion in this section is confined to laser plasmas formed in gases (DeMichelis, 1969; Hughes, 1975; Raizer, 1977). Some characteristics of plasma formation on solid surfaces are discussed in Sec. 5.2.2.

There are two main steps leading to breakdown of a gas by a laser pulse. The first step involves generation of a few free electrons in the focal volume of the focused laser pulse. These "priming-electrons" are produced by multiphoton ionization of atoms, molecules, or even dust particles in the focal region. The energy of a single photon from lasers typically used to generate the laser spark is usually much less than the energy needed to ionize an atom. For example, the energies of photons from ruby and Nd:YAG lasers are 1.79 and 1.17 eV, respectively, whereas the ionization potential of inert gases is 12 eV or greater. However, because of the high power density (MW/cm^2) and large photon flux (photons/cm^2) of the focused laser pulses, there is a high probability that ionization will occur by the absorption of many laser photons during the laser pulse. The second step leading to breakdown is avalanche ionization of matter in the focal region to form a plasma. In the classical picture, the free electrons are accelerated by the electrical fields of the optical pulse in the time period between collisions with neutral atoms. The collisions act to produce an isotropic electron energy distribution. Eventually, if the applied electric fields are intense enough and exist long enough, the electron energy becomes sufficiently high to collisionally ionize an atom. This produces other free electrons that gain energy from the electric fields and cause further ionization. This process of electron multiplication continues throughout the laser pulse and results in significant ionization of the gas and breakdown. Strictly speaking, the avalanche process at optical frequencies requires a quantum mechanical description because the photon energy is much greater than the gain in energy of an electron per collision due to the impressed electric fields. In the quantum description, electrons acquire energy from these fields initially via absorption of photons during collisions with neutral atoms (inverse Bremsstrahlung absorption). As the avalanche develops and the electron density increases, absorption via electron-ion inverse Bremsstrahlung becomes important. Collisions between electrons and atoms are necessary to the absorption process because, as in the classical description, a free electron cannot directly absorb a photon. Despite the differences between the classical and quantum descriptions of the avalanche process, the quantitative predictions of both theories are similar (Hughes, 1975).

The onset of electron multiplication depends not only on the rate of increase of free electron energy due to the impressed fields but also the magnitude of mechanisms that act to decrease the electron energy and the number of free electrons in the avalanche region. A small fraction of the electron energy is lost during each elastic collision with an atom because of momentum conservation. Electron energy is also lost via inelastic collisions that result in electronic excitation of atoms, vibrational excitation of molecules, and especially excitation of low-lying molecular energy levels. The number of free electrons may decrease because of diffusion out of the avalanche region, recombination with positive atomic ions, and attachment to electronegative molecules. Taking these gain and loss mechanisms into account, the rate of change of electron density (n_e) may be written as (Hughes, 1975)

$$\frac{dn_e}{dt} = (\nu_i - \nu_a - \nu_d)n_e + \nu_r n_e^2 \qquad (5.1)$$

where $\nu_i n_e$ is the total ionization rate, $\nu_a n_e$ the rate of attachment to molecules, $\nu_d n_e$ the net rate of electron diffusion out of the focal volume, and $\nu_r n_e^2$ the rate of electron-ion recombination, all rates per unit volume. For optical pulses of short duration (τ_ℓ), recombination is negligible, so the last term in the equation can be ignored. If n_{eb} is defined as the electron density corresponding to breakdown, then Eq. (5.1) can be integrated to give the breakdown condition (Hughes, 1975)

$$\ln \frac{n_{eb}}{n_{eo}} \leqslant (\nu_i - \nu_a - \nu_d)\tau_\ell \qquad (5.2)$$

here n_{eo} is the initial electron density produced by multiphoton ionization. At optical frequencies n_{eb} is usually defined to be the electron density above which electron-ion inverse Bremsstrahlung absorption becomes more effective in heating the plasma than electron-neutral atom inverse Bremsstrahlung. Experimental and theoretical data indicate that breakdown begins when the ratio of the electron density to the neutral atom density is approximately 0.001 (Young and Hercher, 1967). Once n_{eb} is exceeded, ionization will continue rapidly because the absorption coefficient for the electron-ion process is much greater than that for the electron-neutral atom interaction.

The breakdown threshold is usually specified to be the minimum laser pulse power density at the focal spot (I_{th}) necessary to generate a visible spark. Based on avalanche theory, the breakdown threshold for a short-duration laser pulse is given by (Hughes, 1975)

$$I_{th} = \frac{n_a}{c_i} \left(\frac{\omega}{\nu_{ea}} \right)^2 \frac{1}{\tau_\ell} \ln(n_e V_f) \qquad (5.3)$$

which assumes that diffusion and recombination processes do not significantly deplete the electron population in the focal region. Here V_f is the focal volume, n_a the neutral atom density, c_i a gas-dependent parameter, ω the laser frequency, and ν_{ea} the neutral atom collision frequency. Since ν_{ea} varies as n_a, according to Eq. (5.3) the breakdown threshold should vary as p^{-1}. Numerous experiments have been conducted to test the predictions of avalanche theory. Detailed compilations of some results are presented by Hughes (1975) and Raizer (1977). In general, these data tend to verify the pressure dependence predicted by Eq. (5.3) over the range 200 to 10^4 torr, but they indicate a different dependence of the breakdown threshold on laser frequency. Instead of I_{th} decreasing monotonically with decreasing frequency, the breakdown threshold reaches a maximum value at a frequency near or within the visible spectral region dependent on the gas, and then decreases at higher and lower frequencies. The low breakdown threshold at the higher frequencies may be due to multiphoton effects not considered in avalanche theory. Typical breakdown thresholds are 10 MW/cm^2 for gas pressures around 760 torr using Q-switched ruby or Nd:YAG laser pulses of 15 to 50 ns duration. This corresponds to electric field strengths on the order of 10^5 V/cm.

Following the breakdown of the gas, the luminous plasma expands outward in all directions from the focal volume. In most gases, the rate of expansion is greatest toward the focusing lens, because this is the direction from which optical energy enters the plasma: a significant portion of the incident energy is absorbed by the plasma front moving toward the lens. Laser plasmas in hydrogen and helium gases at moderate pressures are somewhat transparent, so that a large fraction of the incident laser energy is absorbed by the interior of the plasma instead of the portion facing the lens. The initial rate of plasma expansion is about 10^5 m/s. As the plasma expands, this rate steadily decreases. The nonisotropic expansion of the plasma accounts for its general pear-shaped or cone-shaped appearance, which sometimes includes a complex substructure (Young et al., 1966). The loud sound that accompanies breakdown is due to the shock wave emanating from the focal volume. The electron temperature, electron density, and spectroscopic temperature of the plasma, important characteristics of the laser spark used as an excitation source, are discussed in Sec. 5.3.2.

5.1.2 Spectrochemical Figures of Merit

The figures of merit commonly used to describe the quantitative analytical performance of a method are the (1) limits of detection, (2) precision, and (3) dynamic range. Because these criteria are referred to extensively throughout this chapter, a brief explanation of each one is presented here. More detailed discussions of these figures of merit can be found in the text by Torok, et al. (1978).

The dynamic range of a method refers to the concentration or absolute mass range over which the method is useful to measure how much of the species of interest (the analyte) is present. The dynamic range is determined by constructing a calibration curve of analyte signal versus concentration or mass. Concentration is usually specified as weight percent, parts per million (ppm), or parts per billion (ppb) of analyte mass to total mass of other constituents. Ideally, the resulting curve will be a straight line that passes through the coordinate origin, indicating that the analyte signal is directly proportional to the amount of analyte present. Frequently, the curves will be linear over only a portion of the dynamic range, exhibiting nonlinear behavior at low or high concentrations or both. The slope of the curve at any point is called the sensitivity of the technique at that concentration. The useful dynamic range of a method is determined by how the sensitivity varies with concentration or mass and the measurement precision.

The limit of detection of a method is the minimum analyte mass or concentration that can be quantitatively measured with a specified reliability. One main factor determining the detection limit is the noise level of the analytical technique. Noise may arise from many sources, which include the method of sample introduction, variations in the excitation characteristics of an emission source, and fluctuations in the intensity of light sources used for absorption experiments. Typically, the limit of detection is specified to occur at the analyte concentration or mass that produces a signal that is three times larger than the relative standard deviation (σ_n) of the noise level. Using this definition, there is a 99.86% certainty that signals greater than $3\sigma_n$ are due to the analyte.

Precision refers to the degree of reproducibility of a measurement. Precision is usually defined to be the standard deviation (σ) of 16 replicate measurements of the analyte signal using identically prepared samples. The precision is sometimes specified in terms of the percent relative standard deviation (RSD), where RSD = ($\sigma/\bar{x}$) 100%, where $\bar{x}$ is the mean of the 16 measurements. An analytical method has good precision if the RSD is 1% or less for measurements near the detection limit.

5.2 LASER ABLATION AND ANALYSIS BY AN AUXILIARY TECHNIQUE

5.2.1 General Considerations

The use of the laser to ablate material from a surface dates back to the early 1960s and the development of the laser microprobe with cross excitation. In this method, high-power laser pulses (10 MW or greater) are focused on a surface and the resulting ejected material is subsequently excited by a conventional electrode spark. Recently, there has been renewed interest in laser ablation for use with other forms of analysis because it exhibits several attractive features.

1. In many cases, the ablated matter is sufficiently atomized to permit analysis by auxiliary methods requiring material in the atomic state, thus eliminating the time-consuming steps sometimes involved in sample preparation. This is important for situations requiring rapid analysis (e.g., real-time diagnostics) or analysis in the field, where sample preparation is not always practical. For other auxiliary methods in which the removed material is subsequently subjected to a hot vaporizing source (e.g., a plasma), complete atomization is not important and the material may be removed as particles.
2. The laser pulse can ablate material for analysis at a remote location. In this case, the ablated material would be transported, for example, in a flowing gas stream to the desired location. This would be important for applications in which the spectral analysis could not be carried out near the sampled medium.
3. All types of material can be sampled with the laser spark because ablation is accomplished by focused light energy and does not rely on the electrical properties of the material, as does ablation via conventional electrode sparks. Lasers can sample material that is conducting or nonconducting, transparent or opaque, or composed of organic or inorganic compounds.
4. The small spot size of a focused laser pulse (few tens of micrometers) can provide a spatially resolved microanalysis of a surface.

5. Laser ablation minimizes the amount of sample used in the analysis. Depending on the amount of sample required to achieve the desired precision or to obtain a representative sample from an inhomogeneous surface, typically only nanograms or less of material need to be consumed because of the microsampling capabilities of the laser spark.

6. Many of the auxiliary techniques discussed in this section have greater detection sensitivities, dynamic ranges, and freedom from matrix effects than direct spectroscopic analysis of the laser plasma. Combining these techniques with laser ablation permits use of these improved capacities together with the rapid sampling provided by the laser spark.

5.2.2 Characteristics of Laser Ablation

Basically, three types of laser pulses have been investigated in detail for laser ablation for analytical purposes. These are (1) the normal (free-running or non-Q-switched) laser pulse having an energy of up to several joules and a duration of about 1 ms, (2) the semi-Q-switched pulse containing up to 0.5 J with a duration of several tens of microseconds, and (3) the electro-optically Q-switched laser pulse with a duration of 10 to 50 ns and energies of up to 1 J/pulse. Pulses 1 and 2 actually consist of an envelope of many shorter duration pulses (due to relaxation oscillations in laser operation), whereas the Q-switched pulse is really a single short duration pulse of light energy. Almost exclusively, solid-state lasers, such as ruby (694.3 nm), Nd:glass (1060 nm), or Nd:YAG (1064 nm), have been used to ablate material for analytical purposes. The reasons for this are somewhat historical, because these lasers were among some of the earliest types commercially available, providing a convenient source of reliable high-power pulses. Over the past few years, small, rugged, and portable Nd:YAG lasers have been developed which are well suited for many analytical applications that require the unique sampling ability of the laser spark. The dependence of focal spot size and surface reflectivity on wavelength does not lead to significantly different ablation characteristics for the ruby and Nd lasers. Large, high-power, pulsed-CO_2 lasers (10.6 μm) can remove large amounts of material from surfaces, but these lasers are employed mainly in industrial applications and so are not considered here (Ready, 1971).

A photograph of a Nd:YAG laser pulse focused on a metal surface is shown in Fig. 5.2. The interaction between the laser pulse and solid surface is a complicated process dependent on many parameters of the laser pulse and the surface material (Ready, 1971, 1978; Scott and Strasheim, 1978; Laqua, 1979). Effects produced by the laser pulse range from minor surface heating to the ejection

FIG. 5.2 Laser spark generated by focusing a Nd:YAG laser pulse on to a solid metal surface. The laser plume extends up from the surface toward the focusing lens.

of material from the surface and formation of a crater. Because the
energy of a high-power laser pulse is delivered in a short time period there is very rapid surface heating and melting of material in the
focal volume. In a time Δt, energy deposited at the surface of a
material with thermal diffusivity (κ) penetrates to a depth (d) given
by

$$d = (4\kappa \, \Delta t)^{1/2} \qquad\qquad\qquad\qquad (5.4)$$

A laser pulse of 15-ns duration incident on carbon steel ($\kappa = 0.119$
cm^2/s) will only heat the sample to a depth of 0.8 μm during the
pulse. Irradiation of a surface with focused power densities of
$\sim 10^5$ W/cm^2 or less and pulse widths of a millisecond or less results
only in surface heating. At higher power densities, the heated
surface begins to melt and ejection of material is observed. If the
laser power density is $\sim 10^7$ W/cm^2 and the pulse width less than
200 μs (normal laser pulse), some vaporization occurs but only about
0.1% of the ablated matter is in the vapor state (Petukh and Yankovskii, 1978). High-speed photographs show that many particles are
ejected from the crater formed by the normal laser pulse (Scott and
Strasheim, 1970), and this is attributed to a flushing mechanism in
which molten material is pushed out toward the crater walls and
then from the surface by the high-vapor pressure in the laser plume
(Ready, 1978). The time required for the surface in the focal volume
to reach the vaporization temperature (t_v) depends on the thermal
conductivity (K), density (ρ), specific heat (c), and boiling temperature (T_b) and initial ambient temperature (T_0) of the pure material
and the absorbed power density (P) according to the equation (Ready,
1978)

$$t_v = \frac{\pi}{4} \, \frac{K\rho c}{P^2} \, (T_b - T_0)^2 \qquad\qquad\qquad (5.5)$$

Values of t_v depend on the element. For example, if $P = 10^7$ W/cm^2,
lead and tungsten will reach their respective boiling temperatures of
1788 and 6173 K within 10 and 1336 ns after the start of surface heating. For pulse durations less than about 1 μs, the laser power density required to raise a material to the vaporization temperature is
independent of the focal spot size because thermal conduction processes are negligible at such short times. At high-power densities,
vaporization occurs rapidly and only a small fraction of material in the
focal volume is molten. At power densities of $\sim 10^8$ W/cm^2 and pulse
widths of about 25 μs (the semi-Q-switched laser pulse) more material is vaporized and less solid and molten material is ejected from the
surface. At energies above 10^9 W/cm^2 and pulse durations of less
than 100 ns (Q-switched laser pulse), vaporization is the predominant

process, with only minor expulsion of molten material from the surface.
No particles are observed from surfaces irradiated with these high-
power pulses (Scott and Strasheim, 1970). At these higher powers,
solid material is quickly vaporized, so only a small fraction of the
material in the focal volume is molten at one time. The vaporized
material forms a luminous plume that extends up from the surface
(Fig. 5.2) and absorbs a large fraction of the remaining incident
pulse energy via the inverse Bremsstrahlung processes discussed in
Sec. 5.1.1. In this way, the plume effectively shields the surface
from further interaction with the laser pulse. Emissions from neutral,
singly, doubly, and triply ionized atoms have been observed from the
plume (Archbold et al., 1964) which exhibits high temperatures and
high electron densities. The analytical utility of these emissions is
discussed in Sec. 5.3.4. The maximum mass of material (M) that can
be vaporized by a laser pulse of energy E is given by the equation

$$M = \frac{E}{c(T_b - T_0) + \ell} \tag{5.6}$$

where ℓ is the latent heat of vaporization of the material. A typical
laser pulse will vaporize only a small amount of material. For example,
a 1-J pulse can vaporize a maximum of 2.6 μg of pure Fe. Partial
reflection of the laser pulse by the surface, shielding of the surface
by the plume, and conduction of some of the deposited energy away
from the focal region will keep the vaporized mass below that predicted by Eq. (5.6). On the other hand, because of the flushing
mechanism discussed above that produces airborne particles, the
amount of material ablated by pulses of moderate power may actually
be greater than predicted by Eq. (5.6).

The diameters of craters produced by laser ablation typically
range from 10 to 300 μm, depending on experimental conditions,
and are generally much larger than the laser pulse focal spot on the
surface. Crater depths range from a few micrometers (for electro-
optically Q-switched lasers) to 1 mm for the normal laser pulse. The
greater penetration depth of the longer laser pulse is desirable to
avoid only a surface analysis that may not be representative of the
bulk material. Attempts to relate the crater size to properties of the
surface and focused beam have met with only partial success (Ready,
1965; Klocke, 1969; Bar-Isaac and Korn, 1974).

There is some evidence that the composition of ablated material
is not completely representative of the bulk material. In a pair of
studies (Baldwin, 1970, 1973), material ablated from Zn/Cu alloys was
collected on a thin film placed directly above the surface and then
analyzed using x-ray fluorescence techniques. The Zn/Cu ratio of
vapors generated by a Q-switched laser pulse ($\approx 10^9$ W/cm^2) was
significantly greater than the Zn/Cu ratio of the alloy. In addition,

the Zn/Cu ratio of the vapors generated by the lower power normal laser pulse ($\sim 10^6$ W/cm^2) was much greater than that produced by the Q-switched laser pulse, but as the power density of the normal pulse increased, the Zn/Cu ratio decreased and approached the value obtained with the Q-switched pulse. In contrast, analysis of all material (vapor, liquid, and solid phases) removed from the alloy showed a close correspondence between the compositions of the ablated and bulk materials. Bingham and Salter (1976) evaluated the composition of material vaporized by the CO_2, ruby, and Nd:YAG lasers using a mass spectrometer. The focused power densities of these lasers were, respectively, 10^9, 10^9 to 10^{11}, and 10^8 to 10^{11} W/cm^2. The ruby and Nd:YAG lasers were Q-switched and the CO_2 laser pulses had an intense peak about 100 ns wide followed by a low intensity tail of 1 to 4 μs duration. Particles generated with the CO_2 laser showed the greatest difference in composition compared to the bulk material. For example, the concentrations of the refractory elements W, Zr, and Ta in particles ablated with the CO_2 laser were depleted by a factor of about 11, 15.5, and 106, respectively, whereas these concentrations in particles ablated with the ruby and Nd:YAG lasers were within a factor of 4 of the bulk sample. Recently, one of the authors (DAC) has examined the composition of material ablated using an acousto-optically Q-switched Nd:YAG laser (150-ns pulse width, 5000-Hz repetition rate, and 10 mJ/pulse). The ablated particles were transported through a tube 1 m long and were collected on filters positioned at the end of the tube. The particles were dissolved in acid to produce a solution which was analyzed using a conventional method of atomic emission spectrometry. The refractory elements Mo and W were depleted by a factor of about 4 in the particles, whereas the concentrations of Mn and Cu were enhanced by a factor of between 1.5 and 2 compared to the bulk material. These effects were calibrated out of the analysis if the experimental conditions (i.e., lens-to-sample distance) were kept constant. Significant changes in these conditions degraded the accuracy of the analysis.

If maximum atomization of the material is required by the auxiliary analysis technique, the highest available laser pulse power densities should be utilized for several reasons: (1) the total mass of vaporized material increases with laser power density, although the total amount of ejected material may decrease; (2) the ablation process becomes less dependent on the composition of the material and all materials can be vaporized; (3) at the higher powers the problem of selective volatization is minimized because the ejected material does not reside in a molten state for a significant period of time; and (4) high-power pulses from Q-switched lasers are more reproducible in terms of power and duration than those from a normal or semi-Q-switched pulsed laser. On the other hand, if the ablated material is to be analyzed by an auxiliary atomizing source, like the inductively coupled plasma (Sec. 5.2.7), it is preferable to have the ejected

material in the form of small particles, because these are more likely
to have the composition of the original substance than would the
vaporized material (Baldwin, 1970; Panteleev and Yankovskii, 1965).
In addition, more material is removed from a surface if the energy
is delivered in a laser pulse of lower power. For example, Carr and
Horlick (1982) found that a Q-switched pulse of 1.5 J generated about
25 μg of ejected material, whereas a normal laser pulse of about the
same energy removed 500 μg from the same metal surface.

5.2.3 Laser Microprobe with Cross Excitation

First developed in 1962, this is the oldest form of laser microprobe
analysis (Brech and Cross, 1962) and perhaps the most thoroughly
investigated laser plasma analytical technique. The developments
in this field up to 1973 are covered in detail by Moenke and Moenke-
Blankenburg (1973) and in several reviews (Harding-Barlow et al.,
1973; Keil and Snetsinger, 1973; Harding-Barlow and Rosan, 1973;
Margoshes, 1973; Harding-Barlow, 1974). A typical experimental
arrangement is shown in Fig. 5.3. A laser pulse is focused on the
surface to be analyzed using a microscope objective lens. The distance between the sample and lens is usually only 18 to 25 mm, so a
transparent protective plate is placed between them to prevent deposition of ablated material on the lens. A visual imaging system is
made collinear with the laser pulse focusing optics to permit accurate
alignment of the sample. Directly above the surface is a pair of
conventional spark electrodes which excites material ejected by the
laser pulse. (On some commercial instruments, the operator has the
choice of monitoring the laser spark light directly for analysis or
using the electrical spark.) The electrode spark is triggered either
by ionized material arriving between the electrodes or by external
circuitry timed to the laser pulse, the latter method yielding better
precision. The light from the electrical discharge is imaged on the
entrance slit of a spectrograph for analysis and the dispersed light
is detected either photographically or photoelectrically. Measurements
are typically restricted to repetition rates ⩽ 10 Hz because of the low
duty cycle of the laser and the need to position a fresh surface in the
focal volume after each shot. Microprobe instruments have been produced in the United States, Japan, Germany, and the USSR.

The advantages of using a laser pulse for sampling have already
been discussed. There are three main spectrochemical advantages of
combining conventional electrode spark excitation with laser spark
ablation.

1. The electrical spark is much more intense than the laser spark
    because the deposited energy is 100 to 1000 times greater than
    the energy of the laser pulse. This increases the analyte line

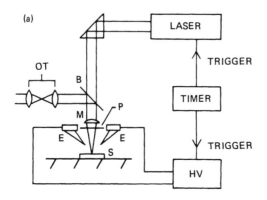

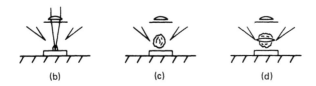

FIG. 5.3 (a) Schematic diagram of the major components of a laser microprobe system with cross excitation. (b) Material ablated by the laser pulse, (c) expands upward, and is (d) excited by an electrode spark. OT, observation telescope; B, beamsplitter; M, microscope objective (simplified); P, protective plate; E, electrodes; S, sample. The timer triggers the electrode spark a predetermined time after the laser pulse.

intensities (Rasberry et al. 1967a), and hence the sensitivity, of the combined technique by at least an order of magnitude over that obtained by monitoring the laser plume directly.

2. The emission lines excited by the electrical discharge are much narrower than the lines from the laser plume (due to the lower electron density of the electrical spark) and do not exhibit the marked self-reversal often observed from the laser plume. When electrical excitation is weak, the emission spectra appear as the superposition of the spectra produced by the laser spark and electrical discharge (Allemand, 1972).

3. The electrode spark does not contain the strong background continuum radiation characteristic of the laser spark (Moenke and Moenke-Blankenburg, 1973), due to ion-electron recombination and Bremsstrahlung processes, although in air the electrical

spark introduces cyanogen bands and some continuum (Harding-Barlow et al., 1973).

Laser microprobe analysis with cross excitation has been applied to a wider variety of materials than any other microprobe method. This includes minerals, meteorites, ceramics, glasses, biological samples, oils, crystals, welded joints, and powdered materials, as discussed in the references listed at the beginning of this section. Absolute elemental detection limits obtained with the microprobe depend on the source material. For example, the detection limits for Cu in steel, in pressed reference powders, and in alloys are, respectively, 50, 0.5, and 500 pg (Moenke and Moenke-Blankenburg, 1973). Detection limits for most elements lie in the range 0.1 to 500 pg, the lower limit being attained for powdered samples. Concentration detection limits are between 10 and 200 ppm for most elements in either the powdered or solid metal form. In principle, femtogram masses of some elements should be detectable, but incomplete vaporization and other plasma characteristics prevent analysis at these low levels (Klockenkamper and Laqua, 1977).

Despite high sensitivity, the laser microprobe with cross excitation has several disadvantages compared to other analysis methods: (1) because the technique uses a laser and electric spark it requires more complex instrumentation than analyses which monitor only the light from the laser plume; (2) the method has all the disadvantages of electrode sparks: possible spectral interference by electrode materials and wearing of the electrodes with use; and (3) in general, the precision of this method is poorer than other spectrochemical techniques. The precision of laser microprobe analyses with cross excitation ranges from 2.5% (Moenke and Moenke-Blankenburg, 1973) up to 39% (Rasberry, et al., 1967b) with the larger values apparently more common (Van Deijck et at., 1979). For comparison, more conventional methods of atomic emission spectrometry have precisions of 1% or less (Barnes, 1978). Attempts have been made to increase the precision of laser microprobe measurements by normalizing the analyte line intensities using an internal standard emission line and the depth and volume of the crater produced by ablation (Van Deijck et al., 1979). The precision obtained with these methods was no better than the 3.6 to 16.1% RSD obtained from the absolute analyte signal. The conclusion was that cross excitation of the laser plume results in a random source of error restricting the technique to mainly qualitative analysis. Comparisons of the laser microprobe with and without cross excitation may be found in Harding-Barlow et al. (1973) and Harding-Barlow and Rosan (1973). Routine applications of the laser microprobe with cross excitation have decreased in recent years along with the decreased availability of commercial instruments.

5.2.4 Laser Ablation for Atomic Absorption Spectrometry Analysis

At sufficiently high concentrations, the emission lines of some elements
in the plume of a laser pulse focused on a surface appear self-reversed,
indicating the presence of absorbing states of these species in the
cool outer sheath of the plume. Several studies have investigated the
analytical utility of these absorptions.

In the simplest experiments (Karyakin and Kaigorodov, 1968) the
sample surface was oriented at 45° to the axes of both the laser beam
and detection system, as shown in Fig. 5.4a. In this geometry, the
intense background light from the plasma served as the continuum
source for absorption experiments. A normal laser was used having
up to 20 J/pulse. The spectra were recorded photographically. A
linear calibration curve was obtained for Cu in a bricketted sample
over the range 0.3 to 10%. The reproducibility of the measurements
was 7% and a detection limit of 0.01% Cu was established. In a second
study using the same method (Krivchikova and Demin, 1971), materials examined included ruby and high-chrome steels. Linear calibration curves were made for Si, Mn, and Ni in steel covering about an
order of magnitude in concentration.

In other work, a primary light source was used to measure elemental absorptions (Fig. 5.4b). This has the advantage that species
can be observed at later times because the pulsed light from the
primary source lasts longer than the few microseconds of intense
background emissions and it can be triggered at a predetermined
time following plasma formation. Mossotti et al. (1967) obtained time-
resolved information about the evolution of analyte absorptions by
using a pulsed xenon lamp as the primary source. The lamp circuit
was designed so that the output intensity remained constant for a
350-μs period during which quantitative measurements of the absorption signal were made. An appreciable number of free atoms were
observed to outlast the 10 to 15-μs period of intense laser plume
emission following ablation. Detection limits of 0.0025, 0.0040, and
0.0035% were obtained for Ca, Ag, and Cu, respectively, contained in
a graphite matrix. Calibration curves for Ag and Ca were linear over
the ranges 0.0063 to 0.1% and 0.0024 to 0.01%, respectively, but the
curve for Cu showed a significant deviation from linearity at concentrations above 0.01%, which was attributed, in part, to insufficient
resolving power of the spectrometer combined with a spectrally broad
primary source. The detection sensitivity of the elements studied
was highly dependent on the matrix and could not be accounted for
by considering differences in the amount of material vaporized.

Pulsed hollow cathode lamps (PHCL) have been used to increase
the analytical performance of absorption measurements above that
obtained with continuum sources. The narrow atomic line widths
emitted by hollow cathode lamps permit the use of high-throughput
spectrometers since high resolution is provided by the lamps them-

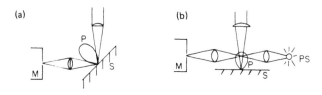

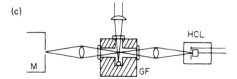

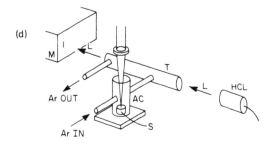

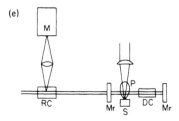

FIG. 5.4 Methods of performing atomic absorption spectrometry on laser-ablated material. (a) Observation of absorbing species against the background of the laser plume. (b) Observation of absorbing species using a primary light source (flash lamp or hollow-cathode lamp). (c) Laser ablation inside a graphite furnace to increase the lifetime of free atomic species. (d) Transport of laser-ablated material to an absorption cell for analysis. (e) Intracavity absorption of laser-ablated material to increase the sensitivity of atomic-absorption measurements. M, monochromator; P, laser plume; S, sample; PS, primary light source; GF, graphite furnace; HCL, hollow-cathode lamp; AC, ablation chamber; T, absorption chamber; L, light path; Mr, laser cavity mirror; DC, laser dye cell; RC, resonance fluorescence cell.

selves. Increased resolution leads to more nearly linear calibration
curves. Using single-shot absorbance measurements, calibration
curves were made for 0.002 to 0.15% Cu in an Al alloy and 0.013 to
0.124% Mn in a graphite matrix by Osten and Piepmeier (1973). The
precision was 20% for measurements of Cu absorption made at different areas on the bulk material. The lifetimes of free atoms increased
significantly as the pressure of the surrounding gas was reduced
below 400 torr. At 1 torr, free Cu atoms existed out to 9 ms after
the 60-ns laser pulse. The composition of some materials appeared
to change during repeated sampling by the laser spark. For example,
absorption signals from Cu in Al decreased with repetitive sampling
at the same site, indicating selective partitioning of the elements
between the vapor and molten states. On the other hand, no selective
vaporizaiton of Cu and other elements in pressed graphite powders
was observed, possibly because graphite sublimes rather than melts
when irradiated by intense laser radiation.

A normal pulsed laser (1 ms, 20 J/pulse) and a PHCL were used
to analyze powdered geochemical materials and laser glasses for Cu,
Mn, Zn, and Ag (Vul'fson et al., 1973; Karyakin et al., 1973).
Steel samples were analyzed by Manabe and Piepmeier (1979) using
a PHCL and a rhodamine 6B/ethanol flash-lamp-pumped dye laser
producing 1-μs pulses at about 590 nm. Precisions of 10 to 20% were
obtained for the Mn signal when fresh surfaces were used for each
measurement. However, 16 repetitive measurements at the same spot
gave a precision of 6% which is on the order of the expected shot
noise from the PHCL. The effects of surrounding gas pressure on
analyte signal and sampling were also examined in this study. In another study using a PHCL, Fe and Mn absorptions were monitored
from the laser plume produced on samples of algae, root crops, and
penicillin derivatives (Sukhov et al., 1976). Analysis of samples
prepared on glass plates gave linear calibration curves within the
limits 0.1 to 0.00001% and a measurement error of ±50% at 0.0001%.

Some studies have attempted to increase the analytical utility of
laser ablation-absorption measurements in different ways. Atomization produced by a CO_2 laser pulse focused on a sample positioned
in a graphite furnace (Fig. 5.4c) was investigated by Matousek and
Orr (1976). The sensitivity of the method increased as the furnace
temperature increased, due to more extensive atomization and longer
residence time of vaporized material in the atomic state. At the highest temperature (740°C) used, the absorption signal from vaporized
Ag was observed out to 10 ms after vaporization. The precision was
typically 5%, and more uniform analyte signals were obtained with
this method because absorption originated from samples contained in
a sealed chamber rather than from material in an inhomogeneous and
rapidly changing laser plume.

In another study, material was vaporized in one chamber and then transported in flowing Ar gas to an absorption cell using an apparatus similar to that shown in Fig. 5.4d (Ishizuka et al., 1977). Analytical curves were linear and the precision was 1 to 12%, and depended on the sample and data reduction method. Detection limits for elements in brass, in an Al alloy, and in steel were within the concentration range 2 to 330 ppm. Element masses of 1 to 10 pg were detected in these materials.

Some of the material vaporized in the foregoing scheme condensed on the walls of tubes and airborne particles, thereby reducing the concentration of material in the atomic state. At least two methods have been examined to revaporize the particles for atomic absorption measurements. In one method, based on flame atomic absorption, the analyte wavelength was monitored as ablated material was introduced into an acetylene-air flame (Ka'ntor et al., 1976, 1979). Detection limits for elements ablated from Ni alloys and thin metallic films were not as low as reported by Ishizuka et al. (1977), probably due to dilution of the analyte in the flame. In another method, detection limits comparable to those obtained with the method depicted in Fig. 5.4d were obtained by transporting the ablated material into a hot graphite tube for analysis via flameless atomic absorption spectometry (AAS) (Wennrich and Dittrich, 1982). Precision was 20 to 30%. Laser atomization and flameless AAS were also combined to analyze biological samples (Sumino et al., 1980) and geological specimens (Schron et al., 1983).

An acousto-optically Q-switched Nd:glass laser was used to vaporize material for atomic absorption analysis by Quentmeier et al. (1979, 1980). Each firing of the laser produced a series of 25 to 30 pulses with an interpulse spacing of about 8 μs. The intensity of the individual pulses varied by as much as 60%. However, because of the even spacing of the pulses over a period of $\sim$200 μs, material was vaporized more uniformly over a longer time period than with the single pulse from a Q-switched laser or the random spikes of a normal laser. Hollow-cathode lamps were used to measure absorptions of elements atomized from an Al alloy. Detection limits of 2.5, 3, 10, 1500, and 7000 ppm were obtained for Mg, Cu, Mn, Fe and Zn, respectively. The lowest measurable masses were 40 and 70 pg for Mg and Cu, respectively. Linear calibration curves (on a linear-log plot) were obtained over two or three orders of magnitude in concentration. Shot-to-shot variations in laser vaporization and in expansion of the plume limited precision.

In a study by Vul'fson et al. (1983), laser-intracavity absorption was used to increase the sensitivity of the laser ablation absorption method (Fig. 5.4e). This method relies on the strong effect that minute changes in absorption losses introduced into the laser cavity have on the output of a laser operating near threshold. The sample was positioned in the cavity of a pulsed dye laser and material was vaporized into the optical path of the dye laser by another pulsed laser. The dye-laser pulses were directed into a resonance cell outside the cavity containing the element to be detected. The dye-laser wavelength was tuned to excite a resonance transition in the element of interest and the resultant fluorescence from the cell was spectrally resolved and detected with a photomultiplier tube. The change in fluorescence intensity from the cell as the laser plume was generated was related to the concentration of the element in the plume. A detection limit of 0.000003% Mo in rock was established with this method, which is about an 80-fold increase in sensitivity compared with single-pass absorption measurements using a hollow cathode lamp. Projected gains in sensitivity for other elements range from 7 to 80. It may be possible to decrease these detection limits even more using a CW dye laser that is more sensitive to intracavity absorptions than a pulsed dye laser (Keller and Travis, 1979).

5.2.5 Laser Ablation for Laser-Induced Fluorescence Analysis

Laser-induced fluorescence (LIF) has been used to probe the atomic constituents of the laser plume. Compared to monitoring the laser plume light directly, some advantages of this method should be, (1) greater concentration and absolute mass detection sensitivity, (2) greater freedom from matrix effects, (3) wider dynamic range, and (4) isotope selectivity. In one study, Cr emission was monitored from steel, flour, and skim milk powder (Measures and Kwong, 1979a,b; Kwong and Measures, 1979). A Q-switched ruby laser ablated the sample and the resulting plume was interrogated by a synchronized dye-laser pulse using the geometry shown in Fig. 5.5. The sample was maintained in a vacuum to ensure rapid expansion of the ablated plume. The calibration curves were linear over the range investigated (10 to 1000 ppm), and no chemical matrix effects were observed due to Cr in different chemical states or due to additions of other compounds to the sample. The uncertainty in the measurements was 20 to 30% and was attributed mainly to correctable experimental difficulties instead of the method. The detection limit for Cr was 1 ppm, corresponding to 0.1 pg of atomized Cr. The main source of noise was Mie scattering of the dye-laser radiation into the detection system from particles ablated into the fluorescence volume. This was minimized by interrogating the atomized material with the dye-laser pulse before arrival of slower-moving particles. Mie scattering is only a problem for resonance fluorescence measurements in which the wavelengths of the fluorescence and exciting

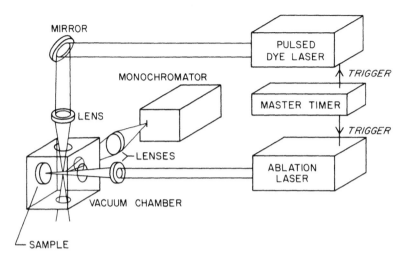

FIG. 5.5 Fluorescence detection of laser-ablated atomic species. The ablating laser pulse, dye-laser pulse, and observation axis of the fluorescence are mutually perpendicular. The dye-laser pulse is delayed with respect to the ablating laser pulse.

radiations are the same. In principle, the sensitivity of fluorescence measurements in the absence of background sources is limited by shot noise. However, with the sample prepared by laser ablation, variations in the amount and composition of material vaporized on consecutive laser shots would probably prevent the attainment of the lowest limit.

The temporal and spatial characteristics of the laser plume were studied in detail using LIF and an automated data collection system (Lewis et al., 1983; Lewis and Piepmeier, 1983). The effects of atmosphere, pressure, and time delay between atomizing and interrogating pulses were examined. Detection limits for Li (in a thin-film emulsion) and Cu (in an Al alloy) were 2 and 50 ppm, respectively.

Laser-induced fluorescence was used to study metal clusters formed in the material laser-vaporized from a solid sample. The vaporized species were entrained in a low-pressure and low-temperature (78 K) flowing He stream and subsequently interrogated by a tunable dye-laser pulse. Spectral information was obtained on diatomic metal molecules such as Pb_2 and Cu_2 (Bondybey and English, 1981, 1982) and the heteronuclear metal clusters CuGa and CuIn (Bondybey et al., 1983). The dynamics of metal cluster formation were examined by time-resolving the fluorescence intensity.

5.2.6 Laser Ablation into a Microwave-Induced Plasma

Using the laser microprobe with cross excitation, the ablated material
is sampled by the electrode spark immediately above the surface.
In some applications the sample surface may become contaminated by
the action of the spark. Also, the surface geometry may be irregular
so that repetitive sampling using identical laser plasma-electrode
spark geometry is not possible. In addition, the precision of the
analysis is often poor due to the interaction between the laser plume
and spark (Sec. 5.2.3). One method developed to overcome these
problems is to transport the ablated material to a continuous secondary plasma excitation source, such as the microwave-induced plasma,
for analysis.

Leis and Laqua (1978, 1979) examined microwave excitation of
material laser ablated from Al and Zn alloys in a chamber adjacent to
a microwave-sustained Ar plasma. The temperature of the plasma was
measured spectroscopically to be 4530 K at an ambient Ar pressure
of 800 torr. Detection limits for Al, Cd, Mg, Pb, and Cu from zinc
samples were 250, 20, 13, 150, and 150 ppm, respectively, using
photoelectric detection. Absolute detection limits were 0.5, 0.04, 0.03,
0.3, and 0.3 ng, respectively. Linear calibration curves were constructed covering about an order of magnitude in concentration.

In another study (Ishizuka and Uwamino, 1980), material was
ablated from brass, Al, and steel samples and carried into the microwave cavity for excitation using a flowing Ar stream at a pressure
of 0.5 to 7 torr. The transient signal measured as the material
passed through the cavity was integrated and the peak intensity was
recorded. Linear calibration curves were constructed for Ni and Fe
extending over one and two orders of magnitude, respectively. The
precision was element dependent and ranged from 1 to 14%. Detection limits for the 10 elements studied were in the range 1 to 25 pg
(absolute) and 1 to 25 ppm (concentration).

5.2.7 Laser Ablation into the Inductively Coupled Plasma

The inductively coupled plasma (ICP) is one state-of-the-art emission
spectrometric excitation source (Barnes, 1978). The high temperatures (5000 to 8000 K), high stability, and relative freedom from
matrix effects of this source yield low detection limits for most elements, calibration curves which are linear over four to five orders
of magnitude, and a precision of a few percent or less. Because
the ICP operates at atmospheric pressure, the introduction of ablated material is easier compared to low-pressure sources such as
some microwave plasmas. Combining the sampling capabilities of
laser ablation with the analytical performance of the ICP presents
some interesting possibilities. (Commercial units are available to

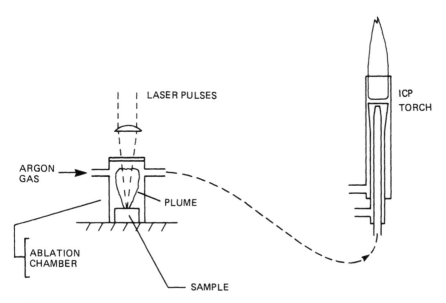

FIG. 5.6 Laser ablation of solids into the ICP.

ablate material into the ICP using an electrode spark, but these are usable only with conducting samples.) Most studies to date have concentrated on ablation of metals into the ICP, although some work has been done with airborne particles and powders (Abercrombie et al., 1978). A diagram of a typical setup is shown in Fig. 5.6. The sample to be vaporized is contained inside a small sealed ablation chamber and the laser pulses are focused on the sample through a window. The chamber is flushed with Ar at a flow rate of 0.5 to 1 liter/min. and the ablated particles are carried into the plasma torch through the central tube normally used for liquid nebulization. Argon is also used for the cooling and plasma gases supplied to the torch. For maximum transport efficiency of ablated material, the distance between the chamber and torch is kept short. However, with proper design, particles can be transported over distances of up to 30 m with only a minimal loss of material. The most reproducible results have been obtained with ablation chambers designed to minimize volume without causing losses to the chamber walls via spattering

There have been several detailed investigations of laser ablation combined with the ICP. Metal samples and silicate rock specimens were analyzed by Thompson et al. (1981). Reproducibility of the signals was element dependent. Relative standard deviations for Fe,

Cu, Ni, and Cr in steel were between 1.5 and 9%, whereas a value of
25% was obtained for Si. The large value for Si indicates the problems
associated with the microsampling characteristics of the focused laser
pulse since a different, freshly cleaned surface area was sampled
with each pulse (Si is known to segregate in steels). Normalizing
the element signals to Fe significantly increased the precision of the
measurements. Linear calibration curves were obtained for Cu, Cr,
Mn, Mo, V, P, and S in steel over concentrations ranging from less
than 0.01% up to 3.2%, depending on the element. The data for Si
were scattered, possibly due to vaporized silica adhering to the walls
of the chamber and transport tube. Concentration and absolute detection limits were established for several elements. The former were
found to be poorer, except for S and P, than those measured by
other methods, whereas the absolute detection limits were comparable to those obtained by pneumatic nebulization of solutions into
the ICP, except for S and P, which had significantly lower limits
using laser ablation. Results obtained with the geological samples
ablated directly into the ICP were not as good. Because of the
inhomogeneity of these materials, analyte signals varied widely with
the location sampled by the laser pulses. To minimize this effect,
the samples were reduced to powders and formed into pellets to increase their homogeneity. This resulted in more satisfactory calibration curves, but further work will be required to obtain results
equaling those obtained with the steel samples.

Metal samples were analyzed by Carr and Horlick (1982) using
an ablation chamber mounted directly under the torch to ensure
maximum material transport to the plasma. The laser was operated in
the Q-switched and normal pulse modes. About 25 and 500 µg of
material were removed, respectively, by each type of pulse. Analytical results were essentially the same with both types of pulses,
although the normal pulse was preferred because it removed more
material. Calibration curves, established for the normalized intensities of Mn/Si, Mn/Mg, and Si/Mg from high-Al-alloy samples, were
linear on a log-log plot, but the slopes were element dependent and
were much less than unity. This was attributed to some characteristic of laser ablation since the ICP is generally free of matrix effects.
The average precision was 5%. Calibration curves for low-Al alloys
were nonlinear. More satisfactory results were obtained with brass
samples; the curves were linear with a slope of 1 and the average
RSD was 3.3%.

In a third study, steel, brass, Al, and Ti samples were ablated
into the ICP (Ishizuka and Uwamino, 1983). The ablation characteristics of Q-switched and normal laser pulses were examined using
an apparatus similar to that shown in Fig. 5.6. The effects of laser
energy, laser pulse focusing, and length of tubing between the
ablation chamber and plasma were investigated and some optimum
operating conditions identified for each type of pulse. Measurement

precision was 3 to 11% and calibration curves were linear with unity
slope on a log-log plot over 0.007 to 1% Cr and 0.005 to 5% Cu using
either type of pulse. Analyte signals measured from samples ablated
with normal laser pulses were 10 to 30 times larger than those measured with the Q-switched laser, due to the greater mass ablated with
the former. Concentration detection limits were lowest using normal
pulses, whereas the absolute analyte masses detected were lowest
using Q-switched pulses.

In a study by Kawaguchi et al. (1982), ablation into the ICP
was accomplished with a normal laser (100 μs, 0.1 J/pulse) operating in the single-pulse or 10-Hz repetition rate mode. The mass of
ablated material was element dependent and ranged between 0.01 and
130 μg. The peak intensity and reproducibility of analyte signals
depended on the length and diameter of the tube transporting the
aerosol to the plasma. Optimum tube sizes were determined. Detection limits for the single-pulse operating mode were 0.017, 0.074, and
0.025%, respectively, for Cr, Mn, and Ni in steel. For the 10-Hz
mode the limits were 0.004, 0.047, and 0.004%, respectively, for
these same elements. The precision values ranged between 1.8 and
6.9% for both single-pulse and 10-Hz operation. Linear calibration
curves were obtained over the concentrations 0.036 to 1.16% Cr,
0.76 to 1.07% Mn, and 0.031 to 2.97% Ni.

The work done to date on combining the ICP with laser ablation
indicates the method shows promise as a convenient and rapid means
of sample analysis. The ICP offers significantly improved analytical
capabilities over the laser microprobe with cross excitation. However,
much work remains to be done. Studies are needed to determine the
size distribution of particles generated by laser ablation and the
maximum particle size that is vaporized by the ICP. Although the
ICP exhibits good precision, this may be degraded by the reproducibility of the laser ablation process. However, removal of sufficient
material using high-power or high-repetition-rate lasers or both may
improve the precision, sensitivity, and sampling reproducibility.

5.2.8 Laser Ablation for Mass Spectrometry Analysis

Interest is currently being directed to laser microprobe mass analysis
(LAMMA) because of the recent availability of a commercial instrument.
The technique uses a high-power laser pulse to vaporize a small amount
of solid sample. The resulting ions and ion clusters are then mass
analyzed with a time-of-flight or quadrupole mass spectrometer. The
advantages are extremely high sensitivity ($\approx 10^{-20}$ g), speed of analysis, applicability to both inorganic and organic (including biological)
samples, and microsampling capability (spatial resolution to 1 μm).
Recent reviews have discussed the basic principles, characteristics,
and limitations of the method (Denoyer et al., 1982), applications to
structural analysis (Hercules et al., 1982), and details of multiphoton
and plasma desorption techniques (Cotter, 1984).

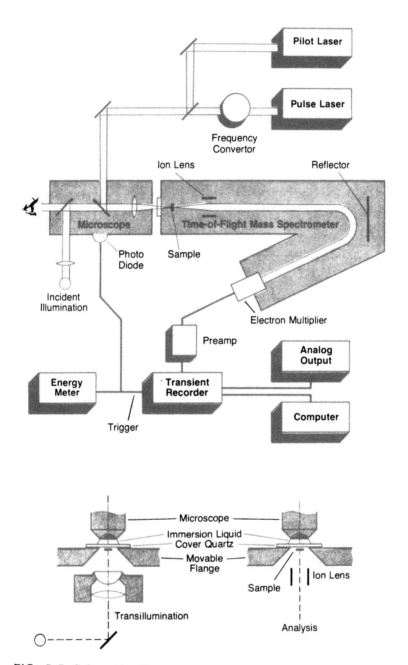

FIG. 5.7 Schematic diagram of LAMMA-500 laser microprobe mass spectrometer. (Reprinted with permission from E. Denoyer, R. Van Grieken, F. Adams, and D. F. S. Natusch, *Anal. Chem.*, *54*; 26A; copyright 1982, American Chemical Society.)

The first LAMMA studies were reported by Honig and Woolston (1963), but the broad kinetic energy spread of the resulting ions limited the utility of the method. The use of a Q-switched laser and a focusing system incorporating microscope optics led to extremely low detection limits, 10^{-19} g and 200 ppb of Li in epoxy resin films (Hillenkamp et al., 1975). Further improvements combined with Hillenkamp's technique led to a commercial instrument (Vogt et al., 1981), a schematic of which is shown in Fig. 5.7. The thin sample is mounted in vacuum with the laser beam incident from one side. The resultant ions emerge from the other side into a time-of-flight mass spectrometer. Complications arise from the requirement that the sample or the substrate on which it is mounted be extremely thin (less than a few micrometers), and the corresponding requirement on the accuracy of the focusing. Recent studies have looked at methods of simplifying the sample preparation and mounting problem. Holm et al. (1984) discuss mounting the sample on a thin polymer substrate in the ambient atmosphere. The substrate then provides the seal for the high vacuum required by the mass spectrometer. For the samples investigated, the mass spectrum was identical to that obtained using conventional LAMMA excitation. Grazing incidence irradiation has also been investigated, but leads to problems in reproducibility of focusing because of shot-to-shot sample erosion.

The LAMMA technique has been used to analyze metal atoms in organic (Jansen and Witmer, 1982) and inorganic samples (Nitsche et al., 1978). Michiels and Gijbels (1983) reported fingerprint spectra for TiO_2, Ti_2O_3, and TiO. The stoichiometry of these compounds was correlated with the relative intensity of the positive or negative ion clusters and atomic ions. Shankai et al. (1984) described the measurement of C, O, and N contained in metal specimens. Comparisons to standard reference materials indicated that ppb levels of these elements could be measured in a straightforward manner. Dutta et al. (1984) have applied LAMMA to mass analysis of refinery source emissions through the analysis of emitted particles. In a novel arrangement, Hardin et al. (1984), combined LAMMA with the effluent of a liquid chromatograph. Samples were sprayed on a moving stainless steel belt, which then went through a vacuum seal to be presented to the irradiating laser. Data were presented on the mass spectrum obtained from histoichine, erythromycin, and other organic samples.

The best analysis of the figures of merit for LAMMA is given by Jansen and Witmer (1982). They conclude that for organic samples the dynamic range extends from ppb levels to 100%. The wide dynamic range of mass spectrometric analysis is due to the absence of matrix effects at high sample concentrations, which usually show up in spectrometric-based techniques. Precision is on the average 20% and accuracy is good to 20% with standards and good to a factor of 2 without. In their opinion, LAMMA is a rapid semiquantitative analysis method with extremely low detection limits. The poor accuracy and precision of the technique reside primarily with sampling problems associated with the reproducibility of focusing and positioning optical elements.

5.2.9 Laser Ablation Combined with Other Techniques

5.2.9A. Laser Ablation onto a Substrate for Subsequent Analysis

The auxiliary techniques described above analyze the laser-ablated material immediately after removal. Several two-step methods have been developed to collect the ablated products on a substrate for subsequent analysis using a variety of techniques. Substrates include transparent films, carbon cylinders, and single and double electrodes used in the geometries shown in Fig. 5.8a-c (Petukh et al.,(1976b). The maximum ablated material is collected with the transparent film but collection directly on an electrode has the advantage of shortening the analysis time because the material can be excited directly on the electrodes. With optimum conditions as much as 30 to 40% of the ablated matter is directly transferred to the electrodes. Another advantage of a two-step analysis is that the ablating and excitation steps can be operated under optimum conditions, free of restrictions

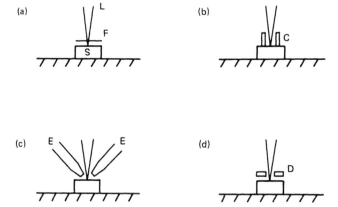

FIG. 5.8 Collection of laser-ablated material on a substrate for subsequent analysis. Collection devices include (a) a thin transparent film (F), (b) the interior of a graphite cylinder (C), (c) a pair of electrodes (E), and (d) a thin graphite disk with a hole in the center (D) for focusing the laser pulse. L, focused laser pulses: S, sample.

imposed by the other step, because they are done separately. For example, using the laser microprobe with cross excitation, the electrodes cannot be positioned too close to the sample because of arcing to the surface and contamination of the sample by electrode material. These problems are avoided by performing the ablating and excitation steps separately.

Using a two-step method, Boitsov and Zil'bershtein (1981) microanalyzed synthetic oxide monocrystals. Laser-vaporized material was condensed on a graphite rod positioned above the target surface area. The condensed material was then analyzed using atomic emission spectrometry by making the rod one electrode of either an ac arc or high-voltage impulse discharge. This method gave better analytical results than direct spectral analysis of the laser plume with or without cross excitation, which were also used to analyze the crystals. The composition of the collected material was found to be identical to that of the original sample. Using only ≈ 3 μg of sample, relative impurity concentrations of Mn, Al, Mg, and Ca down to 0.01 to 0.005% were determined in the crystalline samples. The precision for replicate analysis was 15 to 35%, depending on the element.

In another study, a thin graphite disk (Fig. 5.8d) with a hole in the center for focusing the laser beam onto the sample was investigated for collection of ablated samples (Rudnevsky et al., 1984). The disk collected about twice as much material as the graphite cylinder used in previous work. Collected material was analyzed by positioning the disk on a specially designed electrode of a dc arc discharge or by placing the disk in a demountable cathode of a hollow-cathode lamp. The lowest detection limits were obtained with a hollow-cathode lamp positioned in an external magnetic field: 0.05% (Ni), 0.03% (Mn), 0.01% (Ag), 0.01% (Al), and 0.05% (Cr).

5.2.9B. Laser Ablation and Resonance Ionization Spectrometry Analysis

Sensitive resonance ionization spectrometry (RIS) was used by Mayo et al. (1982) to monitor impurity Na atoms produced by laser ablation of material off a solid Si sample contained in a proportional counter. Previously, the RIS technique was used only with gases and vapors. Vaporized Na atoms were preferentially ionized using photons supplied by two counter-propagating dye laser pulses passing through the ablation region. The electrons generated by photoionization were detected as an indication of Na atoms because other atoms were not efficiently ionized by the narrow dye-laser wavelengths used. The density of Si atoms in the laser plume was estimated to be about 10^{14} cm^{-3}. The Na concentration measured from the purest Si samples was 5 x 10^{11} cm^{-3} for the bulk material. Because of the sensitivity of the method, calibration will be the main problem to its application.

5.3 DIRECT SPECTROSCOPIC ANALYSIS OF
LASER PLASMAS

5.3.1 General Considerations

The purpose of this section is to describe some of the recent developments in techniques and applications of the laser spark without
auxiliary analysis. Earlier work in this area is described in several
reviews and the reader is referred to them for additional information
(Moenke and Moenke-Blankenburg, 1973; Scott and Strasheim, 1978;
Petukh and Yankovskii, 1978; Laqua, 1979). The discussion here
is confined to laser pulses of 10 to 20 ns duration and powers of
0.1 to 100 MW produced by electro-optically Q-switched lasers, which
have been used in the majority of published experiments. The
characteristics of sparks produced by other types of laser pulses
(e.g., picosecond pulses) are discussed by Hughes (1975) and Raizer
(1977).

Direct spectroscopic analysis of the laser spark formed on a
surface or in a gas or liquid has several advantages over the methods
described in Sec. 5.2:

1. The method is simpler because the laser pulse vaporizes and
    excites the sample in one step; no auxiliary analysis equipment
    is needed.
2. The absence of auxiliary equipment makes this method more
    economical, especially in comparison to techniques requiring a
    microwave or inductively coupled plasma.
3. Being an emission technique, direct spark analysis provides
    simultaneous multielement analysis capabilities without increased
    instrumental complexity and cost, unlike LIF and absorption
    methods.
4. Because the spark can be generated in remote locations, it is
    useful in applications requiring noninvasive analysis. Only
    optical access to the medium being sampled is required at the
    laser frequency and the emission wavelengths of the elements
    being monitored.
5. For some materials, matrix effects are less pronounced using
    laser spark excitation compared to the electrical spark (Karya-
    kin and Kaigorodov, 1967).

5.3.2 Excitation Characteristics of the Laser Plasma

Because of the high temperature of the laser plasma, material in the
plasma volume is immediately reduced to elemental form and the
resulting atoms electronically excited. If local thermodynamic equilibrium (LTE) exists even approximately, the excitation characteristics
of the laser plasma are determined by the temperature and electron

density (Griem, 1964). If not, the relative populations of the levels determine the excitation properties. Because of the high electron density of the laser spark and because the plasma is in a state of decay following the laser pulse, the spark is probably in a state of LTE at later times when analytical measurements are usually performed. For a 20 ns laser pulse, this corresponds to times beyond 1 μs (Radziemski et al., 1983a). The medium interacting with the laser pulse exerts a strong influence on the plasma properties, and this is reflected in the temperature and electron density of the plasma. In liquids and on solid surfaces much of the laser energy goes into material vaporization, reducing the amount left for excitation, compared to the spark in gases. The laser spark is not homogeneous, being at most cylindrically symmetric, and is very small, with a volume of 0.03 to 0.1 cm^3 when generated by a Nd:YAG laser pulse of about 100 mJ. Hence, when the temperature and electron densities are quoted, these are often population-averaged quantities.

Repetitive laser-induced plasmas in gases have been studied extensively: H_2 (Litvak and Edwards, 1966), He (Braerman et al., 1969; George et al., 1971; Ahmad et al., 1969; Montgolfier et al., 1972), Ar (Stevenson, 1975), N_2 and O_2 (Stricker and Parker, 1982), and air (Mandel'shtam et al., 1964; Radziemski et al., 1983a) The temperature and electron density of the decaying plasma have been measured as a function of time and other parameters. It is difficult to compare the results of these studies because of the different experimental conditions, but the following general comments can be made. At early times ($\leqslant$100 ns) the electron density of plasmas produced in H_2, He, and O_2 near atmospheric pressure is on the order of 10^{18} cm^{-3} or higher, significantly greater than the electron density of electrode sparks. This corresponds to about 10% ionization of the gas. The high electron density broadens spectral lines, reducing the selectivity of direct laser spark analysis in some applications. Also at early times, the temperature of the plasma can be very high: at least 100,000 K for H_2 and He and a maximum of 30,000 K for O_2 at 15 to 20 ns. The temperature and electron density of the laser spark produced in air by a Q-switched Nd:YAG laser (Radziemski et al., 1983a) are shown in Figs. 5.9 and 5.10, respectively, as a function of time. Separate experiments are indicated by the different points, including measurements made using Saha and Boltzmann analyses which assume that the plasma is in LTE (Griem, 1964). Electron densities were obtained from the Stark widths of spectral lines. Recently, the temperature of the spark produced in air by CO_2 laser pulses of 500 mJ was measured two different ways (Radziemski et al., 1985): the electron temperature was determined using a double-floating-probe method and the spectroscopic temperature was measured from a Boltzmann analysis using oxygen lines. At early times the electronic and spectroscopic temperatures were widely different, indicating the

absence of equilibrium between the two species. As the plasma decayed, however, the two temperatures converged, and equilibrium was
attained via collisions in the cooling plasma at about 25 μs. Comparisons of the temperatures and electron densities of some different
plasmas are presented in Table 5.1.

A photograph of the laser spark produced in air is shown in Fig.
5.1. Superimposed on the spectrally dispersed intense background

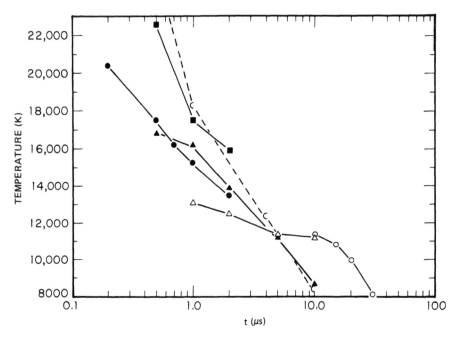

FIG. 5.9 Temporal variation of the temperature of a decaying laser
plasma produced in air at 580 torr. The temperature was determined using the Saha equation and the species (●) C II 251.2/C I
247.8 nm, (■) N II 399.5/N I 415.0 nm, (○) Be II 313.0/Be I 234.8
nm, and (△) Be II 313.0/Be I 234.8 nm spatially resolved by Abel
inversion. The temperature was also determined by a Boltzmann
plot using Be I lines (▲). The temperatures calculated using a computer hydrodynamic code are indicated by C. (Reprinted with permission from Radziemski et al. (1983a), *Anal. Chem.*, 55, 1246; copyright
1983, American Chemical Society.)

continuum of the spark are emissions from singly ionized nitrogen atoms. These emissions and those from other species can be separated in some cases by time-resolving the spark light. Because of the changing temperature and electron density of the cooling plasma, spectral emissions from neutral and ionized atoms and simple molecules (formed from recombining atoms) are observed to maximize during distinct periods after plasma formation. This is shown schematically

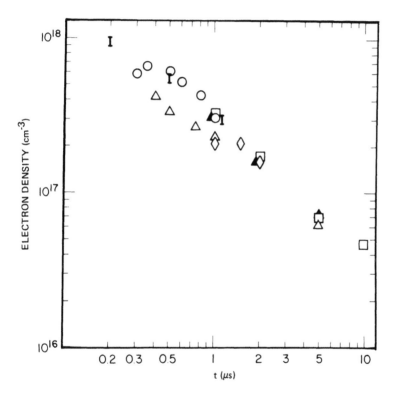

FIG. 5.10 Temporal variation of the electron density of a laser plasma produced in air at 580 torr. The electron density was determined from the Stark widths of the following lines: (o) F I 685.6 nm; (Δ) Ar I 706.7 nm; (□ and ▲) N I 415.0 nm; (I) N II 343.7, N II 395.5, and N II 399.5 nm; (◊) Cl I 837.5 nm. (Reprinted with permission from Radziemski et al. (1983a), *Anal. Chem.*, 55, 1246; copyright 1983, American Chemical Society.)

TABLE 5.1 Temperatures and Electron Densities of Some Plasmas

Plasma	Maximum temperature (K)	Electron density (cm^{-3})	Conditions
Dc arc	4000-8000	5×10^{14}-10^{16}	Air at 1 atm
Electrode spark	> 20,000	10^{16}	Air at 1 atm
Microwave-induced plasma	4000-6000	10^{14}	Ar at 1 atm
Inductively coupled plasma	5000-9000	10^{15}-10^{16}	Ar at 1 atm
Laser spark	> 20,000	10^{17}-10^{18}	Air at 1 atm
Continuous optical discharge	17,000	—	Air at 1 atm
	18,000	10^{17}-10^{18}	Ar at 2 atm

Source: Zil'bershtein (1980) and Raizer (1980).

in Fig. 5.11 for the spark in air. Immediately after plasma initiation,
the spark light is dominated by an intense background continuum due
to Bremsstrahlung processes in the plasma. Between 0.5 and 2 μs,
spectral features due to once-ionized species are observed. From
2 to 10 μs, emissions from neutral atoms occur as ions and free electrons combine. The transition between ionic and neutral nitrogen
atoms is shown in Fig. 5.12. For times beyond 10 μs, the intensities
of emissions from neutral atoms decrease steadily and emissions from
simple molecules are observed. Emissions from CN are readily observed from the spark in air. Some species exhibit behavior that
differs from this general scheme: emissions from Be II begin at early
times and continue out beyond 10 μs and the sodium D lines emit out
past 30 μs. Because of the temporal separation of emissions from
different species in the plasma, time-resolved detection of the plasma
light is useful to maximize the signal-to-background ratio and minimize
spectral interferences between atoms of different ionization stages
and between atoms and molecules.

Spectrochemical information has been obtained from laser plasmas generated directly in liquids, including water, methanol, ethanol, and acetone (Cremers et al., 1984). Electrical breakdown of a liquid produces a gaseous cavity at very high pressure (~1 Mbar) containing vapors characteristic of the bulk material. The temporal history of the spark is compressed to less than 2 μs, about a factor of 2 shorter compared to the spark in air and far less than would be expected given the large difference in density between air and liquids. Also, emissions from the background continuum, neutral and ionized atoms, and simple molecules appear simultaneously, minimizing the utility of time resolution to eliminate spectral interferences for liquid analysis. The temperature of the spark generated in water using pulses of 50 mJ and 15 ns duration was measured spectroscopically from the intensity ratio of Ca II (393.4 nm) to Ca I (422.7 nm), assuming LTE. Temperatures ranged from 11,600 K at 250 ns after plasma formation to 6,900 K at 1.5 μs. These are significantly less than the temperatures of the spark in air. The electron density, measured from the Stark broadening of Li I (670.8 nm) was 5 x 10^{18} cm^{-3} at 250 ns and 9 x 10^{17} cm^{-3} at 1 μs.

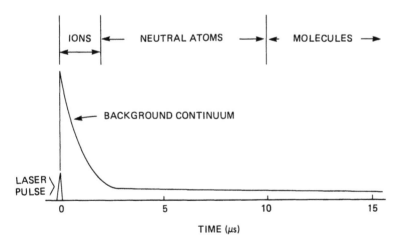

FIG. 5.11 Major spectrochemical periods during decay of the laser spark. The spark light is dominated by an intense spectrally broad background at early time.

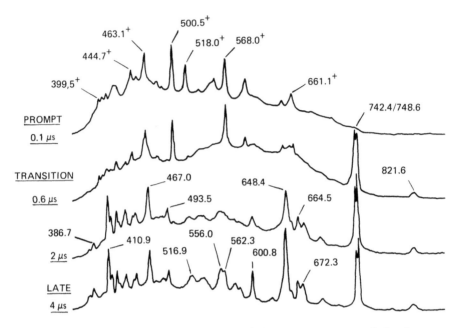

FIG. 5.12 Temporally resolved photodiode array spectra of the laser spark in air as a function of time. The spectra were recorded at the times listed at the left following plasma formation. Emissions from singly ionized nitrogen atoms appear at early times. As the plasma cools these lines decay and emissions from neutral atomic nitrogen become prominant.

5.3.3 Apparatus for Spectrochemical Analysis of Laser Plasmas

A typical apparatus used for the spectrochemical analysis of laser plasmas is shown in Fig. 5.13. The plasma is shown produced in ambient room air. The laser spark can also be generated in different gasses at various pressures, on solid surfaces in a vacuum or surrounded by specific atmospheres, and in bulk liquids using specially designed sample cells. Because the laser spark is generated by optical radiation, the sample cell can have a simple design with only one or two windows required to focus the laser pulse into the medium and to monitor the plasma light. According to Eq. (5.3), for a specific gas at a fixed pressure, the focal volume of the laser pulse and the radiation frequency determine the power density needed to induce gas breakdown. The power density decreases with the size of the focal volume. The focal volume in turn depends on the

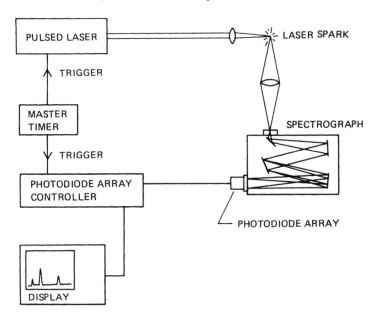

FIG. 5.13 Apparatus for spectral analysis of laser sparks. The laser spark can be generated in a gas, in a liquid, or on a surface. The plasma light is spectrally dispersed and detected in the same way for each type of sample. Detection using a photodiode array is shown here. The array is gated to record only the spark light during a certain period after the start of breakdown.

mode quality of the laser pulses and the focusing optics. In practice, air breakdown is obtained using a TEM_{00}-mode Q-switched Nd:YAG laser generating pulses of about 15 ns duration containing a few tens of millijoules energy and a lens of moderate quality with a 5-cm focal length. Using a multimode Nd:YAG laser, somewhat greater energies of about 100 mJ/pulse are required for breakdown.

Regardless of the medium being sampled, the method of analyzing the plasma light is the same. The spark light is spectrally resolved by imaging the plasma on the entrance slit of a scanning monochromator or a spectrograph. Spectrometers having higher throughput than those used to monitor conventional electrode sparks are required because of the lower integrated light levels from the laser spark. For example, a 1-m spectrograph with an f-number of 35, commonly used with the electrode spark, results in only very low light levels reaching the detectors when used with laser sparks generated by 250-mJ laser pulses at 10 Hz. Spectrometers with an f-number of about 8 are acceptable for laser spark analysis. The spectrally dispersed spark light can be recorded in many ways. Photographic plates provide wide spectral coverage, but exhibit low sensitivity and require time-consuming handling and analysis. Greater sensitivity and real-time analysis over a wide spectral range are possible using an optical multichannel analyzer which records the spectral intensity profile at the spectrograph focal plane using a vidicon camera or linear array of photodiodes. These devices are convenient for survey work, for optimization of operating parameters, and for nonrepetitive signals because the complete spectral signature can be obtained on a single shot. Photomultiplier tube (PMT) detection exhibits the ultimate sensitivity but requires that the spectral signal be scanned over to measure the emission line intensities or line widths, or that the on-line and off-line signal intensities be monitored simultaneously to determine the net line intensity. In the former case, the spectral signals must be repetitive so the spectrometer wavelength can be scanned over the line(s) of interest, whereas nonrepetitive signals can be analyzed with the latter method. Detection with a PMT would probably be employed in a practical device devoted to a specific set of elements because of its low cost. Each of these methods can be used to integrate or time-resolve the plasma light.

Time resolution of the laser spark is necessary to examine plasma evolution in detail, to eliminate the strong background emissions that occur at early times, and to prevent interferences between species with overlapping spectral signatures that emit within different periods after spark formation (Sec. 5.3.2). With film, time resolution is achieved using a streak camera to sweep the spectrally resolved spark light across the film plane to obtain a continuous record of events (Evtushenko et al., 1966). Also, rotating mirrors synchronized to the laser Q-switch can sweep the spark light across the entrance slit of the spectrograph at the desired moment to obtain information at a particular time period (Archbold et al., 1964; Piepmeier and Malmstadt, 1969). Time resolution is achieved electronically with image-converter-streak photography (Ahmad et al., 1969). With PMT detection, the simplest method is to display the tube signal on a simple storage oscilloscope, although this lacks signal-averaging capabilities. A better method is to electronically gate and integrate the PMT signals from many laser sparks (Schroeder et al., 1971; Treytl et al., 1971a; Allemand, 1972). Boxcar averagers provide a versatile method of time resolution with the PMT. However, with this method the PMT or the electronics of the boxcar or both can become saturated by the intense continuum radiation occurring at early times if the detection system gain is set high to maximize analyte signals occurring at late times or if the light intensity on the tube is high. Saturation can be avoided by gating the high voltage to the PMT so that the tube gain can be rapidly increased (rise time $\approx$20 ns) at a precise moment after the continuum intensity has decayed substantially. The boxcar is still necessary to provide a definite aperture period because the tube gain cannot be decreased readily: it decays slowly over several hundred microseconds. With diode array detection, time resolution is possible, as with film, by sweeping the image of the spark plasma across the spectrograph slit or by mechanically or electronically shuttering the light into the spectrometer. Perhaps the most convenient, but not most economical method, is to use a time-gated image-intensified diode array which consists of a gateable microchannel plate image intensifier situated in front of the diode array (Radziemski and Loree, 1981). The intensifier amplifies the incident light signal with a gain of about 25,000. By rapidly gating the intensifier on and off, the light reaching the diode array can be precisely controlled to monitor only specific time periods during plasma decay. Some rather complete direct spark analysis systems are described in the literature (Peppers et al., 1968; Marich et al., 1974).

5.3.4 Analysis of Surfaces

One of the first uses of the laser spark for spectrochemical analysis of solids was reported by Debras-Guédon and Liodec (1963). The pulses from a non-Q-switched ruby laser were used to excite solid material and the spark light was spectrally resolved and then recorded photographically. The feasibility of quantitative analysis using direct spark excitation of solid metals was demonstrated in 1964 with a Q-switched ruby laser (Runge et al., 1964). Linear calibration curves were obtained for the photographically measured integrated intensity ratios Ni/Fe and Cr/Fe. Precisions of 5.3% and 3.8% were obtained for Ni and Cr, respectively. In a subsequent study using similar equipment, molten metal was analyzed directly with the laser spark (Runge et al., 1966). Linear calibration curves were constructed using the intensity ratios of Ni/Fe and Cr/Fe, which closely approximated those obtained from solid metal. It was recognized that the intense background emission from the laser spark at early times interferes with detection of species at trace levels.

A detailed comparison was made between the normal, semi-Q-switched, and Q-switched laser pulses (Sec. 5.2.2) for spectral analysis by Scott and Strasheim (1970). The highest plasma temperatures were obtained with the Q-switched and semi-Q-switched pulses and the analyte emission lines in the plumes produced by these pulses were red-shifted in wavelength and exhibited significant self-reversal. By monitoring a selected region of the plume, the analytical capabilities of all three types of laser pulses were increased. However, due to differences in the temporal behavior of the background light from the plasmas, time resolution was useful only with the Q-switched pulse.

The interaction of Q-switched laser pulses with metals, quartz and Teflon was studied by Allemand (1972) and found to be strongly dependent on the physical properties of the materials. The value of spatially masking the plasma to mimimize background emission was demonstrated together with the use of time-resolved detection to separate emissions from neutral and ionized atoms. Subsequent excitation of material in the plasma plume with an electrode spark produced long-lived emitting species with narrower line widths than could be obtained with laser excitation alone. The absolute intensity of Fe emission from steel was reproducible to within 3.2 to 7.7%, depending on conditions. On the other hand, the Zn/Cu ratio from a brass sample was reproducible to within 0.3%. In the case of metals, the most reproducible results were obtained with polished surfaces.

Using a Q-switched ruby laser and photographic detection, Sc, Yb, Y, and Eu were detected in a NaCl matrix at concentrations down to $\sim$5 ppm using a single laser pulse (Ishizuka, 1973). Linear calibration curves were constructed for several elements over an order of magnitude in concentration.

An extensive investigation of steel analysis was carried out by Felske et al. (1972) using a Q-switched laser. Samples were positioned in a special holder that moved between laser shots so that fresh surfaces were analyzed by consecutive sparks. The measurement precision was 1%, which was three times smaller than the precision obtained using cross excitation of the laser plume. However, the detection sensitivity with laser spark excitation alone was an order of magnitude poorer.

In another study, the occurrence of self-reversed emission lines was minimized in laser spark analysis by using reduced pressures of the surrounding atmosphere and by viewing a certain portion of the laser plume (Petukh et al., 1976a). Also in this study, and in an earlier more detailed study (Piepmeier and Osten, 1971), the effects of atmosphere on the laser sampling process were described. Some characteristics of laser plume emissions from surfaces were discussed in a series of reports from one laboratory. Materials investigated included steels, Al, and metals in an organic matrix and biological tissues. Topics addressed included the influence of atmosphere (Treytl et al., 1971b), matrix effects (Marich et al., 1970), time-resolved detection (Treytl et al., 1971a, 1972), and the use of spatial differentiation to monitor plasma light (Treytl et al., 1975). The most significant gains in detection sensitivity were achieved using time resolution and by carefully choosing the region of the plasma being monitored.

The feasibility of solid steel analysis using a Q-switched ruby laser was reexamined by Ozaki et al. (1982a). The elements Mn, Si, and C were monitored from standard steel samples. The greatest absolute line intensities and signal-to-noise ratios were obtained using an atmosphere of Ar or Ar mixed with 3% H_2 compared to using either pure N_2 or air. Differences between analyses obtained on consecutive laser sparks were reduced by operating the laser above a certain threshold to maximize pulse reproducibility and by normalizing the analyte lines to the 271.4-nm Fe II line. Nonlinear calibration curves were obtained over the ranges 0.1 to 1.2% for C, 0.12 to 1.3% for Si, and 0.2 to 1.65% for Mn, using optimum experimental conditions. The reproducibility of five replicate measurements was 2.5 to 3.1% for the normalized line intensities. The reproducibility of conventional electrode spark data, also evaluated in this study, was better than that obtained with the laser spark using the same detection apparatus. The temporal decays of the analyte and background emissions indicated that the signal-to-noise ratio could be improved using time-resolved detection.

In a second study by Ozaki et al. (1982b) a Q-switched ruby laser was used to monitor C, Si, and Mn in liquid iron at about 1600°C. A 25-cm focal length lens and method of collinearly imaging the laser pulses and the spark light were used to minimize the effect of changes in lens-to-liquid surface distance. However, changes of a few millimeters in this distance affected the absolute analyte signals, but this was minimized by normalizing to the background light or the 271.4-nm Fe II line. Evolution of CO from the molten surface interfered with the determination of C, but this effect was minimized by using a sample chamber in which a rapid flow of Ar swept the CO out of the laser spark volume. Nonlinear calibration curves were prepared over the ranges 0.1 to 1.8% for Mn, 0.15 to 1.5% for Si, and 0.1 to 3.8% for C. However, at the higher Si and Mn concentrations the curves showed a significant decrease in sensitivity. Changes in the molten iron temperature between 1440 and 1600°C produced no change in spectral line intensities. The line intensities and precision measured by analyzing the liquid iron were lower than those obtained by monitoring solid samples.

Although solid-state lasers have been used most extensively for direct spark analysis, the UV pulses from a N_2 laser have been applied to steel analysis (Kagawa and Yokoi, 1982). The short (337.1 nm) wavelength of these pulses, which readily induces photoejection of electrons from the metal surface, permits the generation of surface plasmas at lower pulse energies than possible with the longer wavelengths. Also, the high repetition rates possible with N_2 lasers allow averaging of many laser shots. Reduction of the surrounding gas pressure to about 1 torr was found to increase the signal-to-background ratio and decrease the atomic line widths. The intensity of plasma emission was observed to decrease steadily if the same location on the sample was interrogated repeatedly by the laser pulses, complicating the analysis. Linear calibration curves were made over the range 0.1 to 3% Cr by measuring the ratio Cr/Fe from standard steel samples. The precision was a few percent. In addition, it was possible to monitor emissions from 10% B in glass with the laser spark Detection limits for Cu in Fe and Mg in Al were about 0.05%.

The laser spark has been applied to environmental monitoring. Inhalation of particles containing Be and its alloys represents a significant hazard to workers. In current monitoring practices, airborne Be is collected on air sampling filters and then analyzed by dissolving the filter to produce a solution compatible with standard atomic absorption or ICP analysis. This step is time consuming and so does not provide real-time monitoring of the workplace. The ability of the laser spark to vaporize and excite material in one step has been used to monitor directly the Be particles collected on filters (Cremers and Radziemski, 1985). A spark about 0.5 cm long was formed on the filter surface by focusing the repetitive pulses from a Nd:YAG laser with a cylindrical lens. The filter was rotated under the lens to interrogate a large area of the filter surface with the spark. The Be mass on the filter was determined semiquantitatively in only a few minutes because the filter was analyzed directly without time-consuming sample preparation. The limit of detection for Be particles of respirable size (0.5 to 5 μm diameter) was 0.45 ng/cm^2 of filter surface. Accordingly, if contaminated air containing Be at the maximum average 8-h exposure level of 2 μg/m^3 is passed through the filter at 40 liter/min, only a few seconds are required to collect a measurable sample, providing near real-time warning of hazardous Be concentrations. Calibration curves were constructed for Be masses of 0.001 to 10 μg on the filters, which covers the range anticipated for a monitoring instrument. Unfortunately, the laser spark did not completely vaporize Be particles larger than about 10 μm in diameter, so the method is restricted to particles below this size.

5.3.5 Analysis of Gases

Although the physics of laser plasmas produced in gases has been studied extensively, only recently has the plasma been used for quantitative spectrochemical analysis of materials contained in gases. Time-integrated photographic and diode array detection were used by Schmieder (1981) and Schmieder and Kerstein (1980) to demonstrate the laser spark as a diagnostic for monitoring the elemental constituents of combustion products. Using different mixtures of N_2 and O_2 the relative abundances of N and O atoms were accurately measured by monitoring the spark light. The method can be extended to include H and C for a complete combustion analysis. In another experiment, the fuel-to-air ratio in a methane flame was determined by measuring the C/N ratio. Spatial information was obtained by moving the spark within the flame to sample different regions. Also in this study, the relative abundances of H, C, N, O, and S from shale oil vapors were determined with the laser spark and the results were in agreement with the slower gaschromatographic method.

In a series of preliminary studies, time-integrated detection of the laser spark was used to measure the Na and K content of an experimental coal-gasification system (Loree and Radziemski, 1981a), the presence of beryllium particles in air, and the elements Cl, S, and P from organic vapors (Loree and Radziemski, 1981b). It was estimated that 510 ng of Na and 75 ng of K were detected with the laser spark formed directly in the hot gases of the coal-gasifier effluent. The acronym LIBS for "laser-induced breakdown spectroscopy" was coined to describe emission spectroscopy using the laser spark.

Time-resolved detection of the laser spark in gases was applied to monitor pure oxygen, Cl in air, and P and C from a phosphorus-bearing organic vapor in air and He (Radziemski and Loree, 1981). The temporal decays of neutral and ionized lines of C, P, and O were monitored using a time-gated optical multichannel analyzer or a PMT-boxcar detection system. Detection limits for P and Cl in air were projected to be 15 to 60 ppm, respectively.

In a subsequent study, the laser spark was investigated by Radziemski et al. (1983a) to detect aerosols in air. Solutions containing Na, P, As, and Hg were nebulized into the laser spark volume through a heat chamber to produce particles about 0.05 μm in diameter. Detection limits for these elements in air were 0.0006, 1.2, 0.5 and 0.5 μg/g, respectively, The ICP detection limits for these elements in Ar are about a factor of 100 lower. A calibration curve was established for Na/air extending over four orders of magnitude in concentration.

Because of the 20,000+ K initial temperatures of the laser spark, elements with high-lying excited states can be efficiently excited. Two examples are Cl and F, which exhibit strong emissions from energy levels lying 10.4 and 14.1 eV, respectively, above the ground state. Using the laser spark, detection limits for Cl and F in air were measured to be 8 and 38 ppm (w/w), respectively (Cremers and Radziemski, 1983). By injecting small amounts of Cl and F bearing gases into a small cell containing the laser spark in He, mass detection limits of 3 ng were established for both elements. The precision for replicate analysis was 8%. The feasibility of normalizing the Cl and F emissions to an internal standard (i.e., Ar in air) to increase precision was demonstrated. The slopes of calibration curves were proportional to the numbers of Cl and F atoms on the molecules (SF_6, CCl_4, $C_2Cl_3F_3$) used to construct the curves, indicating complete dissociation of these species by the high spark temperatures.

The laser spark has also been applied to monitor directly the concentration of airborne Be particles (Radziemski et al., 1983b). Beryllium particles in the spark volume are vaporized and the atoms are excited. Direct analysis of air provides a real-time warning of toxic levels in contrast to the filter method discussed in Sec. 5.3.4.

A detection limit of 0.6 ng/g of Be/air was established, which is one-third the permissible average concentration for an 8-h workday. A calibration curve was prepared extending over the range 0.5 to 20,000 ppb of Be/air. Because of insufficient sensitivity, this technique would probably not be usable as a routine air monitor, but it would provide immediate warning of dangerously high Be concentrations.

5.3.6 Analysis of Liquids

Although the physical properties of the laser spark in liquids and some of the accompanying phenomena have been examined extensively, only recently has consideration been given to its spectrochemical applications (Cremers et al., 1984). By focusing the 10-Hz repetitive pulses from a Nd:YAG laser into bulk liquids, detection limits were established for several elements in aqueous solution. The alkali elements were detectable at levels of 1 ppm or less, but the detection limits were somewhat higher for the other elements studied, indicating the method would be useful in general for monitoring minor and major species in liquids. The calibration curve prepared for Li/water is shown in Fig. 5.14. It is linear over about four orders of magnitude in concentration and exhibits a loss of sensitivity at

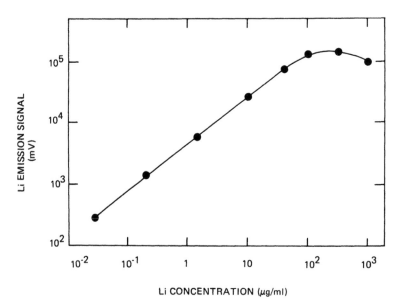

FIG. 5.14 Calibration curve for Li/water determined using the laser spark.

high concentrations due to self-absorption. Detection limits for some elements decreased using two sequential laser sparks to sample the same volume of liquid. The sparks were produced by identical pulses from separate Nd:YAG lasers operating at 10 Hz. The pulses were made collinear before entering the sample cell using a beam splitter, so the sparks were produced in the same volume of liquid to a high degree of accuracy. The emission intensities from B and Be were significantly increased using the sequential sparks. Maximum increase occurred for a separation of about 18 µs between consecutive sparks. The B detection limit decreased from 1200 ppm, obtained using the single repetitive spark, to 80 ppm using the sequential spark method. Much of the energy in the first spark goes into vaporizing the liquid, producing a gaseous cavity containing species characteristic of the liquid. A large portion of the energy of the second spark, formed inside the cavity, further excites and reexcites the vaporized material. A reduction in emission line width using the sequential spark method was also noted for the elements Be and B. Experiments with flowing samples showed that liquid velocities up to 90 cm/s through the spark volume had no effect on the analytical results.

5.4 THE CONTINUOUS OPTICAL DISCHARGE

To produce optical breakdown of gases, power densities of 10^7 W/cm^2 or greater are required. However, it has been shown (Generalov et al., 1970) that plasmas can be maintained continuously in high-pressure gases using power densities as low as 10^6 W/cm^2 from a focused continuous-wave (CW) CO_2 laser beam. The resulting continuous optical discharge (COD) resides near the focus of the laser beam as a very bright ball of intense light about 1 to 3 mm in diameter. A photograph of the COD is shown in Fig. 5.15. Since the CW powers are too low to induce breakdown, the COD is initiated by introducing an auxiliary spark from either a focused laser pulse or electrode spark into the focal volume of the CW CO_2 laser beam. A typical experimental setup is shown in Fig. 5.16. This plasma is unique because it is generated by focused high-frequency optical radiation (10^{12} Hz) independent of any physical support. Other plasmas used for chemical analysis are generated by fields at much lower frequencies supplied to the plasma via some physical device: arcs and sparks use electrodes, the ICP requires an induction coil, and microwave discharges use a waveguide or cavity. A detailed discussion of optical discharges, pulsed and continuous, is presented by Raizer (1980, 1981).

FIG. 5.15 Photograph of the COD produced in Xe gas at 2000 torr using 45 W of CO_2 laser power. The plasma is contained in a quartz tube 1.5 cm in diameter.

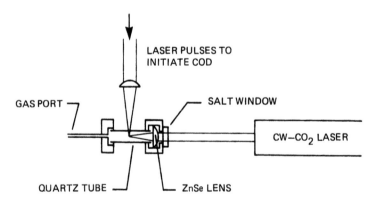

FIG. 5.16 Schematic diagram of the apparatus used to generate the COD.

The COD has been generated in a number of different gases, including Xe and Ar (Generalov et al., 1972), Xe, Ar, Kr (Franzen, 1973), air (Smith and Fowler, 1973), H_2, D_2, N_2, and air (Kozlov et al., 1979). The laser power needed to maintain the COD depends on many factors, including type of gas and pressure and quality of the laser beam. Only 20 W of power are required to produce the COD in Xe at 10 atm, whereas 2 kW are needed to generate the plasma in air. Typically, 0.5 to 1 kW of CW CO_2 power is needed to produce the plasma in molecular gases like N_2 and D_2 at pressures of 6 to 8 atm. For a given laser power and gas, the plasma can be sustained only within a certain pressure regime. As the gas pressure is increased above the lower limit, the maintenance threshold laser power decreases dramatically, but at sufficiently high pressures the COD is extinguished due to increased radiative losses from the plasma.

From the standpoint of chemical analysis, the most appealing aspect of the COD is its high temperature and electron density, which should provide good excitation for elements difficult to excite because of their high-lying energy levels. For example, the spectroscopic temperature of the COD in Ar, measured using relative line intensities and the Saha equation, was 13,000 K at 16 atm and increased to 18,000 K at 2 atm pressure. Also, the electron density remained high at 5×10^{17} cm^{-3} over the range 4 to 16 atm and decreased slightly below 4 atm (Generalov et al., 1972). The temperature of the plasma in air at 1 atm was about 17,000 K (Keefer et al., 1975). It is projected that in helium at 5 atm the temperature at the center of the COD will reach 30,000 K (Raizer, 1981). That such high temperatures can be attained in a continuous source is unusual. For example, the continuous ICP and microwave sources have maximum temperatures of about 8000 and 5000 K, respectively, whereas electrode and laser produced sparks exhibit temperatures in the range 20,000 K.

The COD has only recently been investigated as an analytical source (Cremers et al., 1985). Calibration curves were established for O_2 and Cl_2 injected into a small cell containing the COD. The curves were linear over most of the two orders of magnitude in concentration examined. Approximate detection limits were established for solid samples ablated by a pulsed laser into the COD volume. Limits of detection for Na and Cu were, respectively, 5.2 and 0.87 ng, assuming a measurement period of 1 s. Detectability could probably be increased significantly through development of an improved sample introduction system. Introduction of some gases into the COD perturbed the plasma because the gas absorbed the CO_2 laser wavelength or the decomposition products of the gases coated the optical elements or both. To avoid this, the COD should be operated as a plasmatron in which the carrier gas is flowed rapidly through the plasma volume to prevent contamination of the optics.

TABLE 5.2 Detection Limits of Some Laser Spark Methods

		Detection limits	
Technique	Sample [Reference]	Absolute (pg)	Concentration (ppm)
Laser microprobe	Al in steel [1]	2500	—
with cross excita-	V in steel [1]	750	—
tion (Sec. 5.2.3)	V in an alloy [1]	—	830
	Cu in steel [1]	50	—
	Cu in powder [1]	0.1	—
	Cu in zinc [1]	—	10
Laser ablation combined with:			
Atomic-absorption	V in steel [2]	20	22
spectrometry	Al in steel [2]	10	11
(Sec. 5.2.4)	Cu in aluminum [3]	70	3
	Mg in aluminum [3]	40	2.5
Laser-induced	Li on film [4]	0.01	2
fluorescence	Cr in flour [5]	0.1	1
(Sec. 5.2.5)			
Microwave-induced	Al in steel [6]	8.4	9.3
plasma (Sec.-	Zn in Al alloy [6]	0.7	0.9
5.2.6)			
Inductively	Elements ablated	—	—
coupled	from steel: [7]	—	—
(Sec. 5.2.7)	P	10	10
	S	15	15
	V	10	10
	Mn	80	80
	Elements ablated		
	from steel: [8]		
	V	20	20
	Mn	3	3
Mass spectrometry	Li in epoxy	10^{-7}	0.2
(Sec. 5.2.8)	resin [9]		
	Metals in organic	1	10^{-3}
	materials [10]		
	O, N, C in solids [11]	—	$\sim 10^{-3}$
Direct laser spark analysis of:			
Surfaces (Sec.-	Metals in an organic		
5.3.4)	matrix [12]		
	Cu	0.002	—
	Mg	0.002	—
	Hg	0.3	—

TABLE 5.2 (Continued)

Technique	Sample [Reference]	Detection limits Absolute (pg)	Detection limits Concentration (ppm)
Surfaces continued	Be on filter [13]	2	—
	Elements in steel [14]		
	Cr	—	200
	Ni	—	570
	Si	—	490
Gases (Sec.-5.3.5)	Cl in air [15]	3000	8
	F in air [15]	3000	38
	Aerosol in air containing: [16]		
	Be	—	0.0006
	Na	—	0.006
	P	—	1.2
	Hg	—	0.5
	As	—	0.5
	U [17]	—	10.0
Liquids (Sec.-5.3.6)	Aqueous solutions containing: [18]		
	Li	—	0.006
	Be	—	10.0
	Mg	—	100.0
	Ca	—	0.8
	Al	—	20.0

[1]   Cited in Moenke and Moenke-Blankenburg (1973).
[2]   Ishizuka et al (1977).
[3]   Quentmeier et al (1980).
[4]   Lewis et al. (1983).
[5]   Kwong and Measures (1979).
[6]   Ishizuka and Uwamino (1980).
[7]   Thompson et al. (1981).
[8]   Ishizuka and Uwamino (1983).
[9]   Hillenkamp et al. (1975).
[10] Jansen and Witmer (1982).
[11] Shankai et al. (1972).
[12] Treytl et al. (1972).
[13] Shankai et al. (1984).
[14] Felske et al. (1972).
[15] Cremers and Radziemski (1983).
[16] Radziemski et al. (1983a).
[17] Cremers, D. A., unpublished results.
[18] Cremers et al. (1984).

5.5 ANALYTICAL PERFORMANCE OF LASER PLASMA TECHNIQUES

5.5.1 Introduction

In each discussion of the laser plasma techniques, representative analytical figures of merit have been given that include detection limits, precision, and range over which the method has been usefully calibrated. This section discusses some of the factors influencing the analytical performance of the techniques.

5.5.2 Detection Limits

Representative detection limits for the laser spark methods are listed in Table 5.2. Both absolute mass and concentration detection limits are presented for comparison. Some of these limits were determined rigorously using calibration curves and precision measurements, whereas others are estimates. Except for mass spectrometry, all of the laser spark methods discussed above use optical spectrochemical detection, so the detection limits are a strong function of the element monitored. In general, the alkali metals (Li, Na, K, Rb, Cs) are detectable at low levels, whereas gases (e.g., O, C, F) and the semimetals (e.g., S, P) have somewhat higher detection limits. This variability is due to many factors, including (1) the intensity of the strong analytical line (related to the oscillator strength of the corresponding transition), (2) the upper energy of the emitting analytical line (the higher the level energy, the more difficult it is to populate with plasma sources in LTE), (3) the spectral region of the analytical line (the strong sulfur lines lie near the vacuum ultraviolet or beyond 900 nm, which are difficult spectral regions in terms of materials and detector sensitivity), and (4) the high reactivity of some free atoms which may readily combine with other elements, reducing the lifetime of the free emitting species. In general, concentration limits are in the range 1 to 80 ppm for most methods listed in Table 5.2, which are respectable values considering the minimal sample preparation required to achieve them. Conventional methods of atomic absorption spectrometry and atomic emission spectrometry have much lower limits for routine analysis but often require time-consuming sample preparation. Absolute masses measurable with the laser spark fall in the range 0.1 to 80 pg for most elements, which is comparable to conventional methods. Because of the small size of the laser spark, low absolute masses are detectable even though the concentration limits are at moderate levels. In mass spectrometry analysis, detectability depends weakly on the element being monitored because ionization efficiencies vary only within an order of magnitude, so the main factor determining detection limits is the sample introduction method. For this reason, detection limits obtained with laser microprobe mass analysis are uniformly low.

In addition to the purely element-based factors influencing detection limits, the matrix containing the analyte has a large influence
because of the different amounts of material removed and excited
by the laser spark from different samples. Excluding chemical matrix
effects and their effect on emission-line intensities, lower concentration detection limits will be obtained with samples from which more
material is removed by the laser spark. In fact, this effect must be
taken into account when using calibration standards that are sufficiently different from the samples to be analyzed in terms of hardness
and reflectively at the laser wavelength. This can be done by
changing the laser power between samples so that equal masses of
material are removed from each standard and unknown or by normalizing the analyte line to some reference element present at a known
concentration in each sample (Moenke and Moenke-Blankenburg, 1973).

5.5.3 Precision, Dynamic Range, and Accuracy

In the laser ablation methods of Sec. 5.2, the precision of laser sampling is combined with the precision of the analysis method (e.g., ICP
or atomic-absorption spectrometry). Although there are as yet no
definitive data, the results appear to indicate that precision is probably limited by the reproducibility of laser sampling. That is, although
laser ablation offers a more direct method of solid-sample introduction
compared to reducing the material to a solution, the price paid is a
reduction in analysis precision. Not only must equal amounts of
material be removed on replicate laser shots, but the composition of
the material must also be the same for maximum reproducibility. For
example, in atomic absorption analysis, the same percentage of
material must be vaporized on each shot. Because of the strong dependence of the laser-surface interaction on the properties of the
laser pulse as well as the surface condition, attainment of this goal
would appear difficult, although normalizing the analyte line to a
reference line may help. Direct analysis of the laser spark plume
would exhibit the same sampling requirements as the ablation methods
with auxiliary analysis, with the added requirement that the excitation
characteristics of the laser plume remain constant from shot-to-shot.

The dynamic range of the auxiliary techniques discussed in Sec.
5.2 typically extends over several orders of magnitude in concentration. The ICP is particularly notable in this respect. However, the
dynamic range (and other capabilities) of these techniques may be
reduced when they are combined with laser sampling for at least two
reasons. First, laser ablation may exhibit matrix effects such that
the composition of the ejected material is influenced by large changes
in the composition of the matrix material. As discussed in Sec. 5.2.2,
there are experimental data that indicate the composition of ablated
particles differs from that of the bulk sample. Second, the outstanding analytical performance of many atomic emission and atomic

absorption techniques has been established using nebulized aqueous
solutions. The capabilities of these techniques may be degraded when
the samples are ablated solid particles. For example, large solid
ablated particles may not be completely vaporized by a plasma source
such as the ICP. The dynamic range of direct spark analysis is somewhat less than that of the auxiliary techniques. Calibration curves
usually show a loss of sensitivity at high analyte concentrations due
to self absorption because of the high density of material in the laser
plume. The dynamic range can be extended by carefully selecting
the analyte lines to use within a certain concentration range. The
stronger analyte lines, which exhibit self-absorption at high concentrations, would be used to calibrate the lower concentrations and the
weaker lines would be used to calibrate the higher concentration
ranges.

Accuracy refers to the degree of correspondence between the
measured and actual sample composition. Generally, all of the
auxiliary methods listed in Sec. 5.2 give accurate results when used
with conventional sample introduction techniques. However, because
in some cases laser ablation removes material having a composition
not representative of the bulk sample this accuracy would be degraded.
The exact nature of these effects and whether they can be calibrated
out of the analysis remains to be demonstrated rigorously. In direct
spark analysis, the vaporized material in the laser plume may be more
representative of the bulk material than the ablated particles because
insufficient time has passed for segregation phenomena to occur.

5.6 SUMMARY

As discussed in this chapter, laser plasmas have been investigated for
chemical analysis in a variety of ways. Only the techniques of laser
ablation for mass spectrometry analysis and the laser microprobe
with/without cross excitation have been incorporated into commercial
instruments. There are several reasons for this, but perhaps the
most important one is that the analytical capabilities of laser plasma
methods are generally poorer than conventional methods. Also, the
cost of a laser is sometimes equal to and often much higher than the
price of a conventional excitation source, although there is presently
much commercial activity to produce small, reliable, and economically
priced laser systems. However, it is doubtful that in the near future
a laser plasma technique will replace any analytical method currently
in routine laboratory use. One possible exception is laser ablation
into the ICP, but this would replace sample preparation rather than
an analysis method. It appears that laser plasma techniques are
best suited to applications where the use of standard laboratory instrumentation is not feasible and some reduction in analytical capabilities can be tolerated. For laser ablation methods, this includes

situations in which the removed material can be transported from a
harsh environment or inaccessible area to a remote location for analysis by a conventional auxiliary technique. For direct spark analysis
this includes situations requiring (1) in situ analysis, since only
optical access to the sampled medium is required; (2) a rapid analysis, since the laser spark prepares and excites the sample in one
step; or (3) analysis in harsh environments, where the simplicity
of the technique minimizes the extent of equipment "hardening" and
maximizes reliability. Particularly promising areas of application for
laser plasma techniques include environmental monitoring and industrial process control.

BIBLIOGRAPHY

Andersen, C. A., ed. (1973). *Microprobe Analysis*, Wiley, New York.
Chapters 12 to 15 are devoted to discussions of laser microprobe
analysis. Topics include laser microprobe instrumentation and
applications of the method to geology, biology, and analysis of
metals.

Hughes, T. P. (1975). *Plasmas and Laser Light*, Wiley, New York.
Contains detailed discussions of the theoretical aspects of laser
plasma formation and related phenomena as well as numerous
citations from the literature describing experimental results.
Plasmas formed in gases and on solids are treated in depth.

Laqua, K. (1979). Analytical spectroscopy using laser atomizers,
in *Analytical Laser Spectroscopy* (N. Omenetto, ed.), Wiley,
New York, pp. 47-118. Presents a comprehensive review of
laser plasmas used to atomize material for subsequent analysis.
Emphasis is on the analytical aspects of laser atomization. Only
minor consideration is given to direct monitoring of the light
from the laser spark.

Moenke, H., and Moenke-Blankenburg, L. (1973). *Laser Micro-Spec-
trochemical Analysis* (transl.), Crane, Russak, New York. A
very complete monograph describing the "nuts and bolts" of laser
microprobe analysis using cross excitation. Extensive references
indicate the many uses to which the laser microprobe has been
applied. Some discussion is made of direct spark analysis.

Petukh, M. L., and Yankovskii, A. A. (1978). Atomic emission
spectral analysis using lasers, *Zh. Prikl. Spektrosk.*, 29; 1109 (*J.
Appl. Spectrosc. USSR*, 29; 1527). Provides a general outline of
laser microprobe techniques with and without cross excitation and
provides an update to the references listed in Moenke and Moenke-
Blankenburg (1973). Numerous references are given to a wide

range of applications of the laser microprobe, although many are
listed in Soviet journals not readily available to many persons.

Raizer, Y. P. (1977). *Laser-Induced Discharge Phenomena* (transl.),
Consultants Bureau, New York. Offers a more detailed and
comprehensive examination of laser discharges than Raizer's
review paper "Optical Discharges." There is some discussion of
conventional types of discharges, and the accent is on the
physics rather than analytical applications of discharges.

Raizer, Y. P. (1980). Optical discharges, *Usp. Fiz. Nauk*, 132; 549.
(Sov. Phys. Usp., *23;* 789). A review paper describing the
physics of laser sparks and continuous optical discharges. Some
experimental results are presented, but no data concerning analytical results with laser plasmas are given.

Ready, J. F. (1971). *Effects of High-Power Laser Radiation*, Academic Press, New York. Provides an informative overview of
the variety of phenomena induced by high-power laser pulses.
Topics include optically-induced gas breakdown, effects of laser
light on biological systems, and the interaction between laser
radiation and surfaces, including optical damage of materials.

Ready, J. F. (1978). *Industrial Applications of Lasers*, Academic
Press, New York. Chapters 13 to 16 present a good introduction
to the interactions between laser radiation and solids from the
standpoint of materials processing. Some applications of lasers
in other areas, including holography, chemistry, fiber optics,
and laser measurement, are discussed in other chapters.

Scott, R. H., and Strasheim, A. (1978). Laser emission excitation
and spectroscopy, in *Applied Atomic Spectroscopy* (E. L. Grove,
ed.), Plenum Press, New York, pp. 73-118. Outlines some of
the basics of laser microprobe analysis with and without cross
excitation. Analytical figures of merit are presented for selected
topics.

REFERENCES

Abercrombie, F. N., Solvester, M. D., Murray, A. D., and
Barringer, A. R. (1978). Applications of inductively coupled
plasmas to emission spectroscopy, *1977 Eastern Anal. Symp.*,
Franklin Institute Press, Philadelphia.

Ahmad, N., Gale, B. C., and Key, M. H. (1969). Experimental
and theoretical studies of the time and space development of
plasma parameters in a laser induced spark in helium, *Proc.
R. Soc. London, A310;* 231.

Allemand, C. D. (1972). Spectroscopy of single-spike laser generated plasmas, *Spectrochim. Acta, B27*; 185.

Archbold, E., Harper, D. W., and Hughes, T. P. (1964). Time resolved spectroscopy of laser-generated microplasmas, *Br. J. Appl. Phys., 15*; 1321.

Baldwin, J. M. (1970). Q-switched laser sampling of copper-zinc alloys, *Appl. Spectrosc., 24*; 429.

Baldwin, J. M. (1973). Comment on "laser evaporation and elemental analysis," *J. Appl. Phys., 44*; 3362.

Bar-Isaac, C., and Korn, U. (1974). Moving heat source dynamics in laser drilling processes, *Appl. Phys., 3*; 45.

Barnes, R. M. (1978). Recent advances in emission spectroscopy: inductively coupled plasma discharges for spectrochemical analysis; in *CRC Crit. Rev. Anal. Chem.* Sept. 1978, P. 203.

Bingham, R. A., and Salter, P. L. (1976). Analysis of solid materials by laser probe mass spectrometry, *Anal. Chem., 48*; 1735.

Boitsov, A. A., and Zil'bershtein, K. I. (1981). Optical emission laser microprobe analysis of synthetic oxide monocrystals, *Spectrochim. Acta, B36*; 1201.

Bondybey, V. E., and English, J. H. (1981). Laser-induced fluorescence of metal clusters produced by laser vaporization: gas phase spectra of Pb_2, *J. Chem. Phys., 74*; 6978.

Bondybey, V. E., and English, J. H. (1982). Laser excitation spectra and lifetimes of Pb_2 and Sn_2 produced by YAG laser vaporization, *J. Chem. Phys., 76*; 2165.

Bondybey, V. E., Schwartz, G. P., and English, J. H. (1983). Laser vaporization of alloys: laser induced fluorescence of heteronuclear metal clusters, *J. Chem. Phys., 78*; 11.

Braerman, W. F., Stumpell, C. R., and Kunze, H. J. (1969). Spectroscopic studies of a laser-produced plasma in helium, *J. Appl. Phys., 40*; 2549.

Brech, F., and Cross, L. (1962). Optical microemission stimulated by a ruby laser, *Appl. Spectrosc., 16*; 59.

Carr, J. W., and Horlick, G. (1982). Laser vaporization of solid metal samples into an inductively coupled plasma, *Spectrochim. Acta, B37*; 1.

Cotter, R. J. (1984). Lasers and mass spectrometry, *Anal. Chem., 56*; 485A.

Cremers, D. A., and Radziemski, L. J. (1983). Detection of chlorine and fluorine in air by laser-induced breakdown spectrometry, *Anal. Chem.*, *55*; 1252.

Cremers, D. A. and Radziemski, L. J. (1985). Direct detection of beryllium on filters using the laser spark, *Appl. Spectrosc.*, *39*; 57.

Cremers, D. A., Radziemski, L. J., and Loree, T. R. (1984). Spectrochemical analysis of liquids using the laser spark, *Appl. Spectrosc.*, *38*; 721.

Cremers, D. A., Archuleta, F. L., and Martinez, R. J. (1985). Evaluation of the continuous optical discharge for spectrochemical analysis, *Spectrochim. Acta, B40*; 665.

Debras-Gudédon, J., and Liodec, N. (1963). De l'utilisation du faisceau d'un amplificateur a ondes lumineuses par émission induite de rayonnement (laser à rubis), comme source énergétique pour l' excitation des spectres d'émission des éléments, *C. R. Acad, Sci.*, *257*; 3336.

DeMichelis, C. (1969). Laser induced gas breakdown: a bibliographical review, *IEEE J. Quantum Electron.*, *QE-5*; 188.

Denoyer, E., Van Grieken, R., Adams, F., and Natusch, D. F. S. (1982). Laser microprobe mass spectrometry: 1. Basic principles and performance characteristics, *Anal. Chem.*, *54*; 26A.

Dutta, P. K., Rigan, D. C., Hofstader, R. A., Denoyer, E., Natusch, D. F. S., and Adams, F. C. (1984). Laser microprobe mass analysis of refinery source emissions and ambient samples, *Anal. Chem.*, *56*; 302.

Evtushenko, T. P., Zaidel, A. N., Ostrovskaya, G. V., and Chelidze, T. Y. (1966). Spectroscopic studies of a laser spark: I. Laser spark in helium, *Zh. Tekh. Fiz.*, *36*; 1506 (*Sov. Phys. Tech. Phys.*, *11*; 1126).

Felske, A., Hagenah, W.-D., and Laqua, K. (1972). Uber einige Erfahrungen bei der Makrospektralanalyse mit Laserlichtquellen: I. Durchschnittanalyse metallischer Proben, *Spectrochim. Acta, B27*; 1.

Franzen, D. L. (1973). Continuous laser-sustained plasmas, *J. Appl. Phys.*, *44*; 1727.

Generalov, N. A., Zimakov, V. P., Kozlov, G. I., Masyukov, V. A., and Raizer, Y. P. (1970). Continuous optical discharge, *Zh. Eksp. Teor. Fiz. Pisma Red.*, *11*; 447 (*JETP Lett.*, *11*; 302).

Generalov, N. A., Zimakov, V. P., Kozlov, G. I., Masyukov, V. A., and Raizer, Y. P. (1972). Experimental investigation of a continuous optical discharge, *Zh. Eksp. Teor. Fiz.*, *61*; 1434 (*Sov. Phys. JETP, 34*; 763).

George, E. V., Bekefi, G., and Ya'akobi, B. (1971). Structure of the plasma fireball produced by a CO_2 laser, *Phys. Fluids*, *14*; 2708.

Griem, H. R. (1964). *Plasma Spectroscopy*, McGraw-Hill, New York.

Hardin, E. D., Fan, T. P., Blakley, C. R., and Vestal, M. L. (1984). Laser desorption mass spectrometry with thermospray sample deposition for determination of nonvolatile biomolecules, *Anal. Chem.*, *56*; 2.

Harding-Barlow, I. (1974). Quantitative laser microprobe analysis, in *Laser Applications in Medicine and Biology* (M. L. Wolbarsht, ed.), Plenum Press, New York, p. 133.

Harding-Barlow, I., and Rosan, R. C. (1973). Application of the laser microprobe to the analysis of biological materials, in *Microprobe Analysis* (C. A. Andersen, ed.), Wiley, New York, p. 477.

Harding-Barlow, I., Snetsinger, K. G., and Keil, K. (1973). Laser microprobe instrumentation in *Microprobe Analysis* (C. A. Andersen, ed.), Wiley, New York, p. 423.

Hercules, D. M., Day, R. J., Balasanmugam, K., Dang, T. A., and Li, C. P. (1982). Laser microprobe mass spectrometry: 2. Applications to structural analysis, *Anal. Chem.*, *54*; 280A.

Hillenkamp. F., Unsold, E., Kaufmann, R., and Nitsche, R. (1975). A high-sensitivity laser microprobe mass analyzer, *Appl. Phys.*, *8*; 341.

Holm, R., Kampf, G., and Krichner, D. (1984). Laser microprobe mass analysis of condensed matter under atmospheric conditions, *Anal. Chem.*, *56*; 690.

Honig, R. E., and Woolston, R., Jr. (1963). Laser induced emission of electrons, ions, and neutral atoms from solid surfaces, *Appl. Phys. Lett.*, *2*; 138.

Hughes, T. P. (1975). *Plasmas and Laser Light*, Wiley, New York.

Ishizuka, T. (1973). Laser emission spectrography of rare earth elements, *Anal. Chem.*, *45*; 538.

Ishizuka, T., and Uwamino, Y. (1980). Atomic emission spectrometry of solid samples with laser vaporization-microwave induced plasma system, *Anal. Chem.*, *52*; 125.

Ishizuka, T., and Uwamino, Y. (1983). Inductively coupled plasma emission spectrometry of solid samples by laser ablation, *Spectrochim. Acta, B 38*; 519.

Ishizuka, T., Uwamino, Y., and Sunahara, H. (1977). Laser-vaporized atomic absorption spectrometry of solid samples, *Anal. Chem.*, *49*; 1339.

Jansen, J. A. J., and Witmer, A. W. (1982). Quantitative Inorganic Analysis by Q-switched laser mass spectroscopy, *Spectrochim. Acta, B 37*; 483.

Kagawa, K., and Yokoi, S. (1982). Application of the N_2 laser to laser microprobe spectrochemical analysis, *Spectrochim. Acta, B 37*; 789.

Ka'ntor, T., Polos, L., Fodor, P., and Pungor, E. (1976). Atomic-absorption spectrometry of laser-nebulized samples, *Talanta, 23*; 585.

Ka'ntor, T., Bezur, L., Pungor, E., Fodor, P., Nagy-Balogh, J., and Heincz, G. (1979). Determination of the thickness of silver, gold and nickel layers by a laser microprobe and flame atomic absorption technique, *Spectrochim. Acta, B 34*; 341.

Karyakin, A. V., and Kaigorodov, V. A. (1967). Effect of the base in spectrographic analysis with a laser, *Zh. Anal. Khim,,* *22*; 504 (*J. Anal. Chem. USSR, 22*; 444).

Karyakin, A. V., and Kaigorodov, V. A. (1968). The use of a pulsed laser in atomic absorption spectrographic analysis, *Zh. Anal. Khim.*, *23*; 930 (*J. Anal. Chem. USSR, 23*; 807).

Karyakin, A. V., Pchelintsev, A. M., Shidlovskii, A. I., Vul'fson, E. K., and Tsingarelli, M. N. (1973). Possibility of using lasers for the atomic-absorption analysis of geochemical objects, *Zh. Prikl. Spektrosk.*, *18*; 610 (*J. Appl. Spectrosc. USSR, 18*; 449).

Kawaguchi, H., Xu, J., Tanaka, T., and Mizuike, A. (1982). Inductively coupled plasma-emission spectrometry using direct vaporization of metal samples with a low-energy laser, *Bunseki Kagaku, 31*; E185.

Keefer, D. R., Henriksen, B. B., and Braerman, W. F. (1975). Experimental study of a stationary laser-sustained air plasma, *J. Appl. Phys.*, *46*; 1080.

Keil, K., and Snetsinger, K. G. (1973). Applications of the laser microprobe to geology, in *Microprobe Analysis* (C. A. Andersen, ed.), Wiley, New York, p. 457.

Keller, R. A., and Travis, J. C. (1979). Recent advances in analytical laser spectroscopy, in *Analytical Laser Spectroscopy* (N. Omenetto, ed.), Wiley, New York, p. 493.

Klocke, H. (1969). Utersuchungen zum Materialabbau durch Laserstrahlung, *Spectrochim. Acta, B24;* 263.

Klockenkamper, R., and Laqua, K. (1977). Uber die Nachweisgrenzen für absolute Mengen bei der Laser-Lokalanalyse mit Nachanregung durch Querfunken, *Spectrochim. Acta, B32;* 207.

Kozlov, G. I., Kuznetsov, V. A., and Masyukov, V. A. (1979). Sustained optical discharges in molecular gases, *Zh. Tekh. Fiz.,* 49; 2304 (*Sov. Phys. Tech. Phys.,* 49; 1283).

Krivchikova, E. P., and Demin, V. S. (1971). Use of a laser in atomic absorption analysis, *Zh. Prikl. Spektrosk.,* 14; 592 (*J. Appl. Spectrosc. USSR,* 14; 438).

Kwong, H. S., and Measures, R. M. (1979). Trace element laser microanalyzer with freedom from chemical matrix effect, *Anal. Chem.,* 51; 428.

Laqua, K. (1979). Analytical spectroscopy using laser atomizers, in *Analytical Laser Spectroscopy* (N. Omenetto, ed.), Wiley, New York, p. 47.

Leis, F., and Laqua, K. (1978). Emissionsspektralanalyse mit Anregung des durch Laserstrahlung erzeugten Dampfes fester Proben in einer Mikrowellenentladung: I. Grundlagen der Methode und experimentelle Verwirklichung, *Spectrochim. Acta, B33;* 727.

Leis, F., and Laqua, K. (1979). Emissionsspektralanalyse mit Anregung des durch Laserstrahlung erzeugten Damfes fester Proben in einer Mikrowellenentladung: II. Analytische Anwendungen, *Spectrochim. Acta, B34;* 307.

Lewis, A. L., II, and Piepmeier, E. H. (1983). Chemical and physical influences of the atmosphere upon the spatial and temporal characteristics of atomic fluorescence in a laser microprobe plume, *Appl. Spectrosc.,* 37; 523.

Lewis, A. L., II, Beenen, G. J., Hosch, J. W., and Piepmeier, E. H. (1983). A laser microprobe system for controlled atmosphere time and spatially resolved fluorescence studies of analytical laser plumes, *Appl. Spectrosc.,* 37; 263.

Litvak, M. M., and Edwards, D. F. (1966). Spectroscopic studies of
    laser-produced hydrogen plasma, *IEEE J. Quant. Electron.*,
    QE-2; 486.

Loree, T. R., and Radziemski, L. J. (1981a). Laser-induced break-
    down spectroscopy: detecting sodium and potassium in coal
    gasifiers, *Proc. 1981 Symp. Instr. Control Fossil Energy Prog.*,
    Argonne National Laboratory Press; Argonne Ill. 1982; ANL
    81-62/Conf-810607, p. 768.

Loree, T. R., and Radziemski, L. J. (1981b). Laser-induced break-
    down spectroscopy: time integrated applications, *J. Plasma Chem.
    Plasma Proc.*, *1*; 271.

Maker, P. D., Terhune, R. W., and Savage, C. M. (1964). Optical
    third harmonic generation, *Proc. 3rd Int. Conf. Quantum Electron.*,
    Paris, Columbia University Press, New York, Vol. 2, p. 1559.

Manabe, R. M., and Piepmeier, E. H. (1979). Time and spatially
    resolved atomic absorption measurements with a dye laser plume
    atomizer and pulsed hollow cathode lamps, *Anal. Chem.*, *51*; 2066.

Mandel'shtam, S. L., Pashinin, A. V., Prokhindeev, A. V., Prok-
    horov, A. M., and Sukhodrev, N. K. (1964). Study of the
    "spark" produced in air by focused laser radiation, *Zh. Eksp.
    Teor. Fiz.*, *47*; 2003 (*Sov. Phys. JETP, 20*; 1344).

Margoshes, M. (1973). Application of the laser microprobe to the
    analysis of metals, in *Microprobe Analysis* (C. A. Andersen, ed.),
    Wiley, New York, p. 489.

Marich, K. W., Carr, P. W., Treytl, W. J., and Glick, D. (1970).
    Effect of matrix material on laser-induced element spectral
    emission, *Anal. Chem.*, *42*; 1775.

Marich, K. W., Treytl, W. J., Hawley, J. G., Peppers, N. A., Myers,
    R. E., and Glick, D. (1974). Improved Q-switched ruby laser
    microprobe for emission spectroscopic element analysis, *J. Phys.*,
    E7; 830.

Matousek, J. P., and Orr, B. J. (1976). Atomic absorption studies
    of CO_2 laser-induced atomization of samples confined in a graphite
    furnace, *Spectrochim. Acta, B31*; 475.

Mayo, S., Lucatorto, T. B., and Luther, G. G. (1982). Laser
    ablation and resonance ionization spectrometry for trace analysis
    of solids, *Anal. Chem.*, *54*; 553.

Measures, R. M., and Kwong, H. S. (1979a). TABLASER—a new
    concept for a trace element laser microprobe, *Proc. Int. Conf.
    Lasers 1979*, p. 88.

Measures, R. M., and Kwong, H. S. (1979b). TABLASER: trace (element) analyzer based on laser ablation and selectively excited radiation, *Appl. Opt.*, *18*; 281.

Michiels, E., and Gijbels, R. (1983). Fingerprint spectra in laser microprobe mass analysis of titanium oxides of different stoichiometry, *Spectrochim, Acta, B38*; 1347.

Moenke, H., and Moenke-Blankenburg, L. (1973). *Laser Microspectrochemical analysis* (transl.), Crane, Russak, New York.

Montgolfier, P. de, Dumont, P., Mille, Y., and Villermaux, J. (1972). Laser-induced gas breakdown: spectroscopic and chemical studies, *J. Phys. Chem.*, 76; 31.

Mossotti, V. G., Laqua, K., and Hagenah, W.-D. (1967). Laser-microanalysis by atomic absorption, *Spectrochim. Acta, B23*; 197.

Nitsche, R., Kaufmann, R., Hillenkamp, F., Unsold, E., Vogt, H., and Wechsung, R. (1978). Mass spectrometric analysis of laser induced microplasmas from organic samples, *Isr. J. Chem.*, *17*; 181.

Osten, D. E., and Piepmeier, E. H. (1973). Atomic absorption measurements in a Q-switched laser plume using pulsed hollow cathode lamps, *Appl. Spectrosc.*, *27*; 165.

Ozaki, T. T., Takahashi, T., Iwai, Y., Gunji, K., and Sudo, E. (1982a). Giant pulse laser spectrochemical analysis of C, Si, and Mn in solid steel, *Tetsu To Hagane*, *68*; 863.

Ozaki, T. T., Takahashi, T., Iwai, Y., Gunki, K., and Sudo, E. (1982b). Giant pulse laser spectrochemical analysis of C, Si, and Mn in liquid iron, *Tetsu To Hagane*, *68*; 872.

Panteleev, V. V., and Yankovskii, A. A. (1965). Possibility of using lasers for spectral analysis of copper-base alloys, *Zh. Prikl. Spektrosk.*, 3; 96 (*J. Appl. Spectrosc. USSR*, 3; 70).

Peppers, N. A., Scribner, E. J., Alterton, L. E., Honey, R. C., Beatrice, E. S., Harding-Barlow, I., Rosan, R. C., and Glick, D. (1968). Q-switched ruby laser for emission microspectroscopic elemental analysis, *Anal. Chem.*, *40*; 1178.

Petukh, M. L., and Yankovskii, A. A. (1978). Atomic emission spectral analysis using lasers, *Zh. Prikl. Spektrosk.*, *29*; 1109 (*J. Appl. Spectrosc. USSR, 29*; 1527.

Petukh, M. L., Satsunkevich, V. D., and Yankovskii, A. A. (1976a). Application of a laser plasma without additional electric discharges for spectral analysis, *Zh. Prikl. Spektrosk.*, *25*; 786 (*J. Appl. Spectrosc. USSR, 25*; 1353).

Petukh, M. L., Satsunkevich, V. D., and Yankovskii, A. A. (1976b).
Investigation of laser sampling of material for spectral analysis,
Zh. Prikl. Spektrosk., *25*; 33 (*J. Appl. Spectrosc. USSR*, *25*; 828).

Piepmeier, E. H., and Malmstadt, H. V. (1969). Q-switched laser
energy absorption in the plume of an aluminum alloy, *Anal. Chem.*,
41; 700.

Piepmeier, E. H., and Osten, D. E. (1971). Atmospheric influences
on Q-switched laser sampling and resulting plumes, *Appl. Spectrosc.*, *25*; 642.

Quentmeier, A., Laqua, K., and Hagenah, W.-D. (1979). Atomab-
sorptionsspektrometrie mit Verdampfung fester Proben durch
Laserstrahlung:I. Optimierung der experimentellen Parameter,
Spectrochim. Acta, *B34*; 117.

Quentmeier, A., Laqua. K., and Hagenah, W.-D. (1980). Atomab-
sorptionsspektrometrie mit Verdamfung fester Proben durch
Laserstrahlung:II. Analytische Anwendugen, *Spectrochim. Acta*,
B35; 139.

Radziemski, L. J., and Loree, T. R. (1981). Laser-induced breakdown spectroscopy: time-resolved spectrochemical applications,
J. Plasma Chem. Plasma Proc., *1*; 281.

Radziemski, L. J., Loree, T. R., Cremers, D. A., and Hoffman, N. M.
(1983a). Time-resolved laser-induced breakdown spectrometry
of aerosols, *Anal. Chem.*, *55*; 1246.

Radziemski, L. J., Cremers, D. A., and Loree, T. R. (1983b). Detection of beryllium by laser-induced breakdown spectroscopy,
Spectrochim. Acta, *B38*; 349.

Radziemski, L. J., Cremers, D. A., and Niemczyk, T. M. (1985).
Measurement of the properties of a CO_2 laser-induced air-plasma
by double floating probe and spectroscopic techniques, *Spectrochim, Acta*, *B40*; 517.

Raizer, Y. P. (1977). *Laser-induced Discharge Phenomena* (transl.),
Consultants Bureau, New York.

Raizer, Y. P. (1980). Optical Discharges, *Usp. Fiz. Nauk*, *132*; 549
(*Sov. Phys. Usp.*, *23*; 789).

Raizer, Y. P. (1981). A CW optical discharge in helium can be an extremely hot steady-state source of ultraviolet light, *Pisma Zh.
Tekh. Fiz.*, *7*; 938 (*Sov. Tech. Phys. Lett.*, *7*; 404).

Rasberry, S. D., Scribner, B. F., and Margoshes, M. (1964). Characteristics of the laser probe for spectrochemical analysis, *Proc.
XII CSI*.

Rasberry, S. D., Scribner, B. F., and Margoshes, M. (1967a). Laser probe excitation in spectrochemical analysis: I. Characteristics of the source, *Appl. Opt.*, *6*; 81.

Rasberry, S. D., Scribner, B. F., and Margoshes, M. (1967b). Laser probe excitation in spectrochemical analysis: II. Investigation of quantitative aspects, *Appl. Opt.*, *6*; 87.

Ready, J. F. (1965). Effects due to absorption of laser radiation, *J. Appl. Phys.*, *36*; 462.

Ready, J. F. (1971). *Effects of High-Power Laser Radiation*, Academic Press, New York.

Ready, J. F. (1978). *Industrial Applications of Lasers*, Academic Press, New York.

Rudnevsky, N. K., Tumanova, A. N., and Maximova, E. V. (1984). Preliminary laser sampling in the microspectral analysis of metals and alloys, *Spectrochim. Acta, B39*; 5.

Runge, E. F., Minck, R. W., and Bryan, F. R. (1964). Spectrochemical analysis using a pulsed laser source, *Spectrochim. Acta, 20*; 733.

Runge, E. F., Bonfiglio, S., and Bryan, F. R. (1966). Spectrochemical analysis of molten metal using a pulsed laser source, *Spectrochim. Acta, 22*; 1678.

Schmeider, R. W. (1981). Combustion applications of laser-induced breakdown spectroscopy, *13th Ann. Electro-Opt./Laser Conf. 1981*, Anaheim, Calif.

Schmeider, R. W., and Kerstein, A. (1980). Imaging a conserved scalar in gas mixing by means of a linear spark, *Appl. Opt.*, *19*; 4210.

Schroeder, W. W., van Niekerk, J. J., Dicks, L., Strasheim, A., and Piepen, H. V. D. (1971). A new electronic time resolution system for direct reading spectrometers and some applications in the diagnosis of spark and laser radiations, *Spectrochim. Acta, B26*; 331.

Schron, W., Bombach, G., and Beuge, P. (1983). Schnellverfahren zur flammenlosen AAS-Bestimmung von Spuren-elementen in geologischen Proben, *Spectrochim. Acta, B38*; 1269.

Scott, R. H., and Strasheim, A. (1970). Laser induced plasmas for analytical spectroscopy, *Spectrochim. Acta, B25*; 311.

Scott, R. H., and Strasheim, A. (1978). Laser emission excitation and spectroscopy, in *Applied Atomic Spectroscopy* (E. L. Grove, ed.), Plenum Press, New York, p. 73.

414 Cremers and Radziemski

Shankai, Z., Conzemius, R. J., and Svec, H. J. (1984). Determination of carbon, nitrogen, and oxygen in solids by laser mass spectrometry,

Smith, D. C., and Fowler, M. C. (1973). Ignition and maintenance of a CW plasma in atmospheric-pressure air with CO_2 laser radiation, *Appl. Phys. Lett.*, *22*; 500.

Stevenson, R. W. (1975). Spectroscopic examination of carbon dioxide laser-produced gas breakdown, thesis, Naval Postgraduate School, Monterey, Calif.

Stricker, J., and Parker, J. G. (1982). Experimental investigation of electrical breakdown in nitrogen and oxygen induced by focused laser radiation at 1.064 μm, *J. Appl. Phys.*, *53*; 851.

Sukhov, L. T., Zolotukhin, G. E., and Zyabkina, S. M. (1976). Atomic-absorption method of analysis for biological samples with laser atomization, *Zh. Prikl. Spektrosk.*, *25*; 199 (*J. Appl. Spectrosc. USSR*, *25*; 942).

Sumino, K., Yamamoto, R., Hatayama, F., Kitamura, S., and Itoh, H. (1980). Laser atomic absorption spectrometry for histochemistry, *Anal. Chem.*, *52*; 1064.

Thompson, M., Goulter, J. E., and Sieper, F. (1981). Laser ablation for the introduction of solid samples into an inductively coupled plasma for atomic-emission spectrometry, *Analyst*, *106*; 32.

Torok, T., Mika, J., and Gegus, E. (1978). *Emission Spectrochemical Analysis*, Crane, Russak, New York.

Treytl, W. J., Orenberg, J. B., Marich, K. W., and Glick, D. (1971a) Photoelectric time differentiation in laser microprobe optical emission spectroscopy, *Appl. Spectrosc.*, *25*; 376.

Treytl, W. J., Marich, K. W., Orenberg, J. B., Carr, P. W. Miller, D. C., and Glick, D. (1971b). Effect of atmosphere on spectral emission from plasmas generated by the laser microprobe, *Anal. Chem.*, *43*; 1452.

Treytl, W. J., Orenberg, J. B., Marich, K. W., Saffir, A. J., and Glick, D. (1972). Detection limits in analysis of metals in biological materials by laser microprobe optical emission spectrometry, *Anal. Chem.*, *44*; 1903.

Treytl, W. J., Marich, K. W., and Glick, D. (1975). Spatial differentiation of optical emission in Q-switched laser-induced plasmas and effects on spectral line analytical sensitivity, *Anal. Chem.*, *47*; 1275.

Van Deijck, W., Balke, J., and Maessen, F. J. M. J. (1979). An assessment of the laser microprobe analyzer as a tool for quantitative analysis in atomic emission spectrometry, *Spectrochim. Acta, B34*; 359.

Vogt, H., Heinen, H. J., Meier, S., and Wechsung R. (1981). LAMMA 500—principles and technical description of the instrument, *Fesenius Z. Anal. Chim.*, *308*; 195.

Vul'fson, E. K., Karyakin, A. V., and Shidlovskii, A. I. (1973). Analytical possibilities of atomic-absorption analysis using a laser atomizer, *Zh. Anal. Khim.*, *28*; 1253 (*J. Anal. Chem. USSR, 28*; 1115).

Vul'fson, E. K., Dvorkin, V. I., Karyakin, A. V., and Khomyak, A. S. (1983). Possibilities of lowering the determination of elements in a laser flame from a graphite-containing target using an intraresonator method for recording atomic absorption, *Zh. Prikl. Spektrosk.*, *38*; 537 (*J. Appl. Spectrosc. USSR, 38*; 382).

Wennrich, R., and Dittrich, K. (1982). Simultaneous determination of traces in solid samples with laser-AAS, *Spectrochim. Acta, B37*; 913.

Young, M., and Hercher, M. (1967). Dynamics of laser-induced breakdown in gases, *J. Appl. Phys.*, *38*; 4393.

Young, M., Hercher, M., and Wu, C.-Y. (1966). Some characteristics of laser-induced air sparks, *J. Appl. Phys.*, *37*; 4938.

Zil'bershtein, K. I. (1980). Modern light sources for analysis of optical emission spectra, *Zavod. Lab.*, *12*; 1095 (*Ind. Lab. USSR, 12*; 1234).

6

Laser-Induced Molecular Dissociation
Applications in Isotope Separation and Related Processes

JOHN L. LYMAN *Los Alamos National Laboratory Los Alamos, New Mexico*

6.1 INTRODUCTION

Over about the last 10 years many research groups have investigated possible applications of laser photochemistry. The large majority of this research has been on laser isotope separation (LIS). Other related areas of research include chemical synthesis and gas purification by laser-induced chemical reactions. The purpose of this chapter is to review the research and development in these areas. The emphasis in the chapter is on *applications* of laser-induced chemical reactions of molecular species, not on the purely scientific aspects of laser photochemistry. Much of the data reviewed here will be useful for evaluating the economic feasibility of some of the photochemical processes; however, the chapter does not contain any specific economic studies. The chapter makes only a few references to photochemical processes in condensed phases. Processes based on photoionization of atomic species are covered in Chapter 3.

The literature contains many references to laser isotope separation. Individual sections of the chapter review the LIS research for hydrogen (both deuterium and tritium), carbon, sulfur, and uranium. An additional section reviews the work on enrichment of isotopes of other elements. A final section reviews purification and synthesis applications.

6.1.1 Isotopic Selectivity

In an attempt to allow for comparisons of the different methods of isotope separation, I have used the isotopic selectivity of the photochemical process as much as possible. The isotopic selectivity is the ratio of reaction probability of two isotopic forms of a molecule. It is not necessarily the most important feature of a given enrichment process, but it is essential. Also, most published papers give sufficient information to obtain a good estimate of the selectivity.

The nomenclature in the literature for isotopic selectivity is not uniform. The symbol used here is β, with the definition

$$\beta = \frac{X/(1 - X)}{X_0/(1 - X_0)}$$

where X is the fraction of a given isotope in the reaction products, and the zero subscript indicates the starting material. For low levels of photolysis of a minor isotope, this expression is very nearly

$$\beta = \frac{X}{X_0}$$

Many references use this definition. I have either used the selectivity stated in the original reference or evaluated it from the data given in the paper. This quantity is essentially equivalent to the parameter α used in many earlier papers.

One can also obtain the selectivity from the enrichment in the reactant species (β_2) and the fraction of the original reaction remaining,

$$\beta = 1 + \frac{\ln(\beta_2)}{\ln(f)}$$

where f is the fraction of the enriched isotope that remains unreacted.

The isotopic selectivity is an important parameter in determining the cost of an enrichment process. Its actual contribution to the process cost depends on the desired isotopic purity of the enriched product and on other process details. If one could achieve the desired purity in a single photochemical step, then substantially improving the selectivity would not result in a corresponding increase in process efficiency.

As an example, the natural abundance of deuterium is 1.5×10^{-4}. If one desires that the major fraction of the photoproduct be deuterium, an enrichment factor of 2×10^4 would give a reaction product of 75% deuterium.) For selective reaction of carbon-13 (natural abundance = 0.011) a selectivity of a few hundred is sufficient to obtain a product that is mostly carbon-13.

The photochemical process may involve the selective reaction of the undesired species. In this case the selectivity determines the time necessary to achieve the desired isotopic purity in the unreacted reagent and the amount of loss of the desired isotope to the photoproduct.

Several other parameters may contribute significantly to the process cost. These include the fraction of the process stream that undergoes photochemical reaction, the fraction of the laser energy that induces photochemical reaction, the cost of starting materials or feed preparation, the cost of removing the photoproduct from the process stream, and the efficiency of the process lasers.

6.1.2 Photolysis Methods: IRMPD,UV,IR-UV

Most of the photochemical applications in this chapter fall into three categories: infrared multiple-photon dissociation (IRMPD), ultraviolet photolysis (UV), and infrared-ultraviolet dissociation (IR-UV). Figure 6.1 gives a simple schematic representation of the three processes for a typical polyatomic molecule. The first of these, IRMPD, is molecular dissociation induced by absorption of many photons from an infrared laser pulse. The second, UV, is dissociation induced by excitation of the molecule to a dissociative or predissociative electronic state with a single ultraviolet photon. The third, IR-UV, is a combination of these two. Absorption of one or more photons produces some initial excitation, then the molecule dissociates by absorption of an ultraviolet photon.

The majority of the experiments, particularly the isotope separations, are with the technique of infrared multiple-photon dissociation (IRMPD). Study of this process has been extensive. Several review

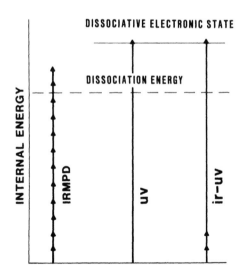

FIG. 6.1 Schematic representation of laser photolysis by three methods. The short arrows represent infrared photons and the long ones ultraviolet photons.

articles summarize the experimental [Letokhov, 1977, 1979, Robinson, 1983, 1977; Grunwald, 1978; Judd, 1979; Danen, 1981; Kompa, 1981; Francisco, 1982; Lyman, 1984], the theoretical [Quack, 1978; 1981; Stone, 1979; Hodgkinson, 1981; Galbraith, 1981], or both [Bloembergen, 1978; Cantrell, 1979; Harrison, 1980; Lyman, 1982; King, 1982] aspects of this processes.

The IRMPD process is the absorption of many infrared photons from an intense infrared laser beam by a single polyatomic molecule (Fig. 6.1). Most of the experiments have been with pulsed infrared lasers (generally CO_2) with low gas pressures. Absorption and dissociation do not require collisions [Brunner, 1977; Diebold, 1977; Sudbø, 1978, 1979], but they alter the processes [Quick, 1978; Lyman, 1984]. Collisions may alter the absorption process by changing the internal energy state of the molecule during the laser pulse. This generally increases the amount of laser energy absorbed per molecule. Collisions may also deactivate the highly excited molecules during or after the laser pulse, and thus decrease the yield of photolysis products. Higher pressures also inhibit heat transfer away from the irradiated region. This results in higher temperatures and more (nonselective thermal reaction.

The isotopic selectivity of an infrared photolysis reaction is determined mainly by the spectroscopic characteristics of the unexcited molecules. The isotopic selectivity can be very high if the infrared isotope shift is large, and if the laser frequency is appropriate.

The excitation processes are generally successive absorption of laser photons; however, true multiphoton absorption processes can and do occur at low internal energy. The anharmonic nature of molecular vibrational modes also influences the excitation process at low internal energy. An important feature of molecules that absorb intense infrared radiation is the increasing density of vibrational states with increasing excitation energy. Diatomic molecules do not dissociate by IRMPD, and the few triatomic molecules that dissociate generally requires collisions during the laser pulse. At a high vibrational state density mixing of the vibrational states occurs, which allows for interaction of the laser radiation with a much larger number of states than those in the pumped vibrational mode. This region of high state density is called the vibrational quasi-continuum in many of the published papers [Bloembergen, 1975; Heller, 1980]. In this energy region the normal-mode description of the molecule becomes less correct, the anharmonic frequency shift becomes less important, and the molecule continues to absorb the infrared photons.

When the molecule has absorbed more energy than the dissociation energy, it may dissociate in a manner similar to a thermal dissociation reaction. The reaction is generally from high vibrational levels of the ground electronic state. The RRKM unimolecular reaction-rate theory [Robinson, 1972] has proven to be a good method for describing the dissociation process.

The review articles cited above give extensive details about the experimentally observed properties of IRMPD. The energy absorbed generally increases linearly with laser fluence (laser energy per unit area) up to the point of some degree of saturation of the molecules in absorbing states. Above that level the energy absorbed increases with about the two-thirds power of the fluence [Judd, 1979]. The point of transition between these two regions is generally at very low fluence for small, rigid molecules and at very high fluence for large, nonrigid species.

Typically, the probability for dissociation of an irradiated molecule increases with some high power (3.4) of the laser fluence at the onset of dissociation. (This power dependence is weaker for larger species. For very large molecules the dissociation probability is very nearly a linear function of fluence [Cox, 1979].) Many authors report the threshold fluence for dissociation. The measurement of threshold fluence depends to some extent on the sensitivity of the detection technique. Because of the rapid increase of the dissociation probability with fluence, the threshold fluence does not depend strongly on the measurement technique. The references to threshold in this chapter are either the threshold that the authors report or the fluence necessary to dissociate 1% of the molecules.

While most IRMPD experiments support the general characteristics mentioned above, many research groups have noted exceptions for extreme properties of the laser, the molecule, or the experimental conditions. For example, the absorption characteristics and the isotopic selectivity sometimes vary considerably with the addition of a second laser frequency.

Another extension of the IRMPD technique is photolysis with a CW infrared laser. This is the limit of long pulse and low power. The important considerations for CW photolysis are collisional transfer of internal energy and conductive transport of thermal energy. Applications with CW infrared lasers can be useful when large departures from thermodynamic equilibrium are not necessary. Several research groups have demonstrated isotope enrichment, impurity removal, and chemical synthesis with CW lasers.

The UV method (Fig. 6.1) is just conventional ultraviolet photolysis. It is useful for laser photochemical applications (separations and purifications) for species with sharp or steep absorption features. This type of absorption feature may also be useful for synthesis applications in complex mixtures, such as vitamin D synthesis. The main application that we discuss in the chapter is isotope separation. This method is particularly important for hydrogen enrichment by UV predissociation of formaldehyde [Gelbart, 1980; Moore, 1983], but it also has applications with other species. The technique is quite simple, and it is successful when the UV spectra of the isotopic species differ sufficiently and when the dissociation quantum yield is high.

A variant of the UV method is to induce selective reactions by excitation of a molecule to a bound state that has a lifetime sufficiently long for reaction with other species in the gas mixture. Most of the demonstrated isotope separations and purifications have been with halogen diatomics such as Cl_2 or ICl.

The third method in Fig. 6.1 (IR-UV) is infrared irradiation of a sample followed by ultraviolet dissociation. Potential applications of this method are almost entirely isotope separation. One isotopic form of a molecular sample selectively absorbs one or more infrared photons. The fluence requirements for this step are usually much less than for IRMPD, and with the milder excitation the absorption selectivity can be much greater. One chooses the frequency of the ultraviolet laser such that the low level of vibrational excitation from the infrared laser enhances the absorption probability. The infrared laser provides the isotopic selectivity and the ultraviolet laser provides most of the energy for dissociation. One can sometimes increase the efficiency of this process by adding another infrared laser that further excites the vibrationally excited molecules.

Early research on the IR-UV method for isotope separation preceded the extensive research on the IRMPD method. Researchers at the institute of spectroscopy in Moscow published several papers [Ambartzumian, 1972a, 1972b, 1973] on isotope separation by this method with the molecular species HCl and NH_3. The early investigations of isotope separation at Los Alamos were also by this method. The investigations were mainly directed toward uranium enrichment [Jensen, 1976], with some attention to enrichment of isotopes of hydrogen [Jensen, 1974] and boron [Rockwood, 1974]. The method has been particularly useful for species with small infrared isotope shifts (UF_6). The section on uranium enrichment contains a more detailed discussion of the method.

6.2 LASER ENRICHMENT OF HYDROGEN ISOTOPES

The large isotope shifts in the absorption spectra of hydrogen-containing species suggest that hydrogen isotopes (deuterium and tritium) should be among the least difficult to enrich by laser methods. The experiments have demonstrated that this is so.

The most promising methods for deuterium enrichment are predissociation of HDCO with near-UV laser radiation and infrared photolysis of methane derivatives (like CDF_3) by IRMPD. In the opinion of Bigeleisen et al. [Bigeleisen, 1983] IRMPD of CDF_3 is the most promising, and probably, the only method with economic potential. On the other hand, Vanderleeden [Vanderleeden, 1977, 1978, 1980] thinks that the formaldehyde predissociation has considerable promise. He has published several papers on optimal optical design and economic analysis for this process.

Several groups have demonstrated hydrogen isotope separation by other methods. These include photochemical reactions induced by CW infrared lasers and IRMPD of other molecular species. We discuss these below.

An enrichment method based on water would have the advantage of a readily available feed source. The infrared-ultraviolet dissociation of water vapor has received some attention [Jensen, 1974], but no one has reported a successful demonstration.

The principal potential market for deuterium enrichment is for heavy-water nuclear reactors. The projected annual consumption of heavy water is between 100 and 10000 tons [Bigeleisen, 1983]. The reactor water needs to have a high deuterium purity for this application.

The economical considerations for tritium enrichment differ somewhat from deuterium enrichment. The main reason for interest in tritium is for development of a method for decontamination of water (or heavy water) that has been in contact with a nuclear reactor. A method for this purpose need not be capable of producing high-purity tritium, but it must be capable of removing a large fraction of the tritium from the contaminated water. Methods for both tritium-hydrogen separation and tritium-deuterium separation are of interest.

The tritium methods that have received the most attention are all IRMPD of hydrogen-containing species. These are similar in many respects to the methods for deuterium enrichment.

Because the isotopic selectivity can be so high for the hydrogen techniques, other steps in the enrichment process have the greatest impact on process costs. These include costs for handling the feed materials and the processed stream, and in the case of deuterium, the cost of supplying the feed material. The deuterium demands for a large-scale enrichment plant are so great that only a few materials could meet the demand. These include water (preferably seawater because of the higher deuterium content), hydrocarbons (oil and natural gas), and ammonia. Because no efficient enrichment methods have been developed for these materials, a laser enrichment method must rely on chemical exchange of the hydrogen isotopes and recycling of the processed material. The Bigeleisen paper [Bigeleisen, 1983] reports that fluoroform (CHF_3) undergoes rapid exchange of hydrogen isotopes in the presence of the conjugate bases of ammonia and several ammonia derivatives. This base-catalyzed exchange with several possible species (water, ammonia, methane, ethane, and hydrogen) as the deuterium source, is sufficiently rapid at ambient temperature to regenerate the deuterium-depleted fluoroform. These researchers concluded that hydrogen was the best deuterium source. This, however, would probably require an increase in the production rate of hydrogen. The estimate D_2O requirement is 100 to 1000 tons per year [Bigeliesen, 1983]. Current sources are sufficient for 100 tons per year.

One isotope-exchange possibility for formaldehyde feedstock is the catalytic exchange with water at an elevated temperature (~125°C). Some preliminary evaluations [Vanderleeden, 1977] suggest that one could use a platinum catalyst for isotope exchange. However, this process has one potential problem: the formaldehyde can decompose to molecular hydrogen and carbon monoxide on the catalyst. Consequently, the additional formaldehyde feedstock would increase the production cost.

Even laser methods may be applicable to the isotopic exchange problem. Parthasarathy et al. [Parthasarathy, 1982] have demonstrated exchange of hydrogen isotopes between fluoroform and ammonia. The exchange occurs when both species absorb the radiation from a pulsed CO_2 laser. This type of process may be too costly for a commercial application, but the authors suggest that it may have some potential in an enrichment process.

6.2.1 Deuterium Enrichment by Ultraviolet Photolysis

One of the early demonstration of laser isotope separation was the selective dissociation of D_2CO in the presence of H_2CO with a frequency-doubled ruby laser [Yeung, 1972]. Letokhov [Letokhov, 1972] suggested the possibility about the same time. Laser technology has developed dramatically since that demonstration, but UV laser photolysis of formaldehyde is still one of the better methods for hydrogen isotope separation. Table 6.1 summarizes some of the important experiments in the development of this technique. Marling and coworkers at Lawrence Livermore National Laboratory have done much of the recent research on this technique. The column labeled "λ" is the wavelength of the ultraviolet laser that gave the best enrichment. The next column is the highest value of the isotopic selectivity. The "comment" column gives additional information about the experiment.

Ultraviolet excitation of formaldehyde may result in one of two possible reactions [Yeung, 1972, 1973; Marling, 1977; Gelbart, 1980; Moore, 1983]: one that gives molecular products.

$$H_2CO \rightarrow H_2 + CO$$

and one that gives radical products,

TABLE 6.1 Deuterium LIS by UV Photolysis of Formaldehyde

Species	Laser	λ (nm)	Best β	Comment	Reference
D_2CO	Ruby (doubled)	347.2	5.6	Early demonstration of laser enrichment from H_2CO/D_2CO mixture	[Yeung, 1972]
D_2CO	Hg lamp	—	2.45	No laser, H_2CO/D_2CO mixture irradiated through H_2CO filter	[Bazhin, 1974]
HDCO	CW HeCd	325.03	> 14	1.4 to 3.4-torr sample	[Marling, 1975]
D_2CO	Nd:glass	353	9	H_2CO/D_2CO mixture	[Ambartzumian, 1975a]
HDCO	CW Xe ion	345.426	60	Xe III laser, complete formaldehyde study	[Marling, 1977]
D_2CO	CW Ne ion	332.377	180	Ne II laser	[Marling, 1977]
HDCO	Dye	345.6	254	N_2 pumped dye laser, dissociation of 4 torr of natural H_2CO	[Mannik, 1977]

$$H_2CO \rightarrow H + HCO$$

The molecular reaction is much more desirable for isotope separation because the reaction products are stable against isotope exchange. The radical reaction, on the other hand, leads to isotopic scrambling in reactions like

$$D + H_2CO \rightarrow HD + HCO$$

and

$$HCO + DCO \rightarrow HDCO + CO$$

The absorption in the region 300 to 355 nm contains much vibronic structure. The molecular reaction occurs below about 355 nm, and the radical reaction below 325 nm [Marling, 1977; Gelbart, 1980].

   The isotopic selectivity arises entirely from the spectral shifts in the near-ultraviolet region. The observed enrichments agree quite well with the isotopic differences in the spectrum, at least for photolysis in region of only molecular products. All of the atomic species of formaldehyde have resolvable isotopic shifts; this makes enrichment possible for hydrogen, oxygen, and carbon isotopes [Marling, 1977].

   The entries in Table 6.1 fall in three categories: an initial demonstration of the possibility of enrichment [Yeung, 1972], demonstrations of enrichment with existing (hopefully efficient) ultraviolet sources [Bazhin, 1974; Marling, 1975; Ambartzumian, 1975a], and selection of an optimum frequency for high selectivity [Marling, 1977; Mannik, 1979].

   The experiments that used H_2CO/D_2CO mixtures have been useful in understanding the photochemical processes, but they would not be practical for an enrichment application. The last two entires for HDCO have the most promise for an enrichment plant.

   The predissociation of formaldehyde is apparently sufficiently rapid to prevent selectivity and yield losses by collisional quenching and electronic excitation transfer. Several other processes, however, prevent enrichment at high pressures. One of these is polymerization. Near room temperature formaldehyde polymerizes to paraformaldehyde at pressures much above the 4 torr of the Mannik experiment [Mannik, 1979]. Furthermore, at higher pressures spectral broadening begins to reduce the absorption selectivity.

   In the Mannik experiment [Mannik, 1979] the expected selectivity from spectroscopic measurements was on the order of 10,000. They observed a considerably lower selectivity. The authors suggested two additional possibilities for the lower value. One of these was the difficulty of obtaining high-purity, spectrally narrow laser radiation. The other was the presence in the sample of formaldehyde containing minor carbon and oxygen isotopes. These species may well have strong absorptions at the laser frequency. This would produce small, but significant, amounts of H_2, and thus reduce the apparent selectivity for these conditions.

6.2.2 Deuterium Enrichment by IRMPD

Most of the research on laser enrichment of deuterium has been on infrared photolysis, the IRMPD method. Many of the deuterium experiments played an important role in understanding the IRMPD process. These include one of the earliest demonstrations of isotope separation by infrared photolysis [Yogev, 1975], several papers that clarify the dominant role of collisions for small molecules like formaldehyde [Koren, 1976; Orr, 1981] and fluoroformaldehyde [Abzianidze, 1981], and an experiment that demonstrated the deleterious effects of radical reactions [Gandini, 1977].

The major objective for much of the recent research has been to develop an enrichment process with high isotopic selectivity and high efficiency. Table 6.2 summarizes the demonstrations of highly selective enrichment of deuterium. The entries in this table have several features in common. All of the experiments involved dissociation of the deuterium-containing species, and each of these species has only one hydrogen atom. The species are all simple compounds. All but one are methane derivatives. The simpler species have simpler infrared spectra; this reduces the probability of absorption by (and dissociation of) the unwanted isotopic form. The simpler species tend to have larger isotope shifts. All experiments to date have used the highly efficient carbon dioxide laser.

For all entries in Table 6.2 the listed selectivities are only lower limits. The selectivities were so high that they were difficult to measure. This was especially true for experiments with enriched starting material as in the Abzianidze experiment [Abzianidze, 1981], where the initial fluoroform was 50% deuterated. For the highest measured selectivity [Herman, 1980] the authors used a clever technique involving ^{13}C labeling. They prepared a mixture of $^{12}CDF_3$ and $^{13}CHF_3$ after showing that the carbon isotope effect was small at the frequency of their experiment. The lower limit of 20,000 for the selectivity followed directly from the carbon isotope ratio in the C_2F_4 product.

As we discussed above, one of the important considerations for deuterium enrichment is replenishing of the photolysed material with deuterium from another source. The species in Table 6.2 do undergo isotopic-exchange reactions with species like hydrogen (CHF_3) [Bigeleisen, 1983] or water (CF_3CHCl_2) [Marling, 1979]. A major advantage of the ethane derivative is the rapid, base-catalyzed exchange of hydrogen isotopes with this species.

TABLE 6.2 Highly Selective Enrichment of Deuterium by IRMPD with CO_2 Laser Pulses

Species	ν (cm^{-1})	Best β	Comment	Reference
CDF_3	971.9	>5000	4-torr sample, natural D/H, focused laser	[Tuccio, 1979]
CF_3CDCl_2	938.8	>1400	10 J/cm^2, threshold = 1.5 J/cm^2, selectivity decreases above 1 torr	[Marling, 1979]
CDF_3	966.3 970.5 975.9	>20,000	0.2-torr CHF_3, 20-torr Ar, 1000-fold absorption selectivity, ^{13}C labeling used to measure	[Herman, 1980]
CDF_3	969.1	>100	0.5-torr CDF_3, 0.5-torr CHF_3, 400-torr Ar, diluent improves dissociation > 100 times	[Adbzianidze, 1981]
$CFCl_2D$	944.2	>10,000	4 torr optimum pressure, 10% dissociation at 6 J/cm^2	[Hason, 1982]
CDF_3	979.7	>2000	10-20 J/cm^2, 6-ns pulses, product collected as D_2, thorough study	[Evans, 1982]
CF_2DCl	980.9	>10,000	Pressure up to 15 torr with high selectivity, 6 J/cm^2, 30-ns pulse, optimum enrichment and efficiency	[Moser, 1983]

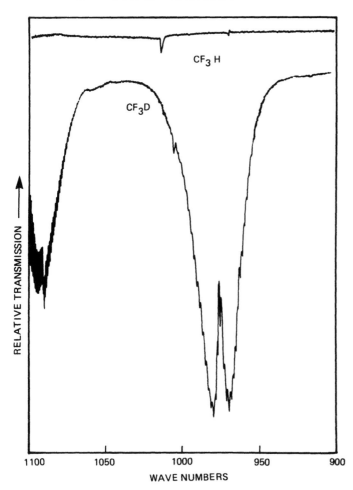

FIG. 6.2 Infrared spectra of CF_3H and CF_3D in the region of the CO_2 laser lines [Tuccio, 1979].

The highest selectivity in Table 6.2 is for photolysis of deuterated fluoroform. Figure 6.2 shows the infrared spectra of both the deuterated and normal fluoroform in the region of the ν_5 absorption band of CDF_3. The only infrared absorption feature in this region for CHF_3 is a weak combination band at 1015 cm^{-1}. This feature results in weak absorption by the undeuterated species. (The paper by Herman and Marling [Herman, 1980] has a more detailed spectrum of this feature.) However, because the natural concentration of deuterium is very low, absorption by the undeuterated species (even without dissociation) can significantly impact the efficiency of an enrichment

plant. Herman and Marling [Herman, 1980] measured the absorption
selectivity in the region of the ν_5 absorption feature of CDF_3 (Fig. 6.2
and Table 6.2). The absorption selectivity was 1000 compared to
>20,000 for the dissociation selectivity.

All of the experiments with CDF_3 listed in Table 6.2 were at frequencies near the center of the absorption band (Fig. 6.2). Again,
the best isotopic selectivities were achieved for excitation on the low-
frequency side of the band, but still in the intense absorption region.
The dissociation probability is maximum near the peak of the R branch
(971 cm^{-1}), and the molecule does not appear to exhibit a strong anharmonic shift to lower frequencies [Herman, 1983a].

The dominant photolysis reaction of fluoroform is elimination of HF

$$CHF_3 \xrightarrow{nh\nu} CF_2 + HF$$

The CF_2 radicals then recombine:

$$2CF_2 \rightarrow C_2F_4$$

Therefore, the hydrogen fluoride product contains the enriched deuterium. This species adsorbs quickly on the walls of the container.
If glass is present, the HF reacts to give SiF_4 and water. Evans
et al. [Evans, 1982] determined the enrichment by reacting the water
formed in this manner with a ribbon of uranium metal. The molecular
hydrogen formed in this reaction contained the enriched deuterium.
The selectivity they demonstrated (Table 6.2) is particularly impressive because of this added complication.

One disadvantage of the fluoroform photolysis is the high fluence
required to dissociate the molecule. At low pressure the threshold
for dissociation is about 22 J/cm^2. The other species in Table 6.2
all have lower thresholds. The threshold is 1.5 J/cm^2 for CF_3CDCl_2
and 2 J/cm^2 for CF_2DCl. This difference makes the CDF_3 somewhat
less attractive for an industrial processes. Another species that may
be a candidate for deuterium enrichment is $CDCl_3$ [Herman, 1983b].
This species would also have a lower threshold for dissociation.

One way to improve the enrichment efficiency with fluoroform is
to use two laser frequencies [Herman, 1983a], one near the band
center, and the other at 944 cm^{-1}. Although this procedure does not
reduce the total fluence requirements as it does in some other species,
it does allow one to use the second frequency for most of the laser energy. The CO_2 laser is considerably more efficient at that frequency.

Evans et al. [Evans, 1982] used shorter laser pulses (2 or 6 ns)
for some absorption and dissociation experiments with CDF_3. They
observed that the energy absorbed per molecule increased more rapid-

ly with laser fluence for these short pulses than for longer pulses with this and other species. For most small species [Judd, 1979; Lyman, 1984a] the energy absorbed per molecule increases with about the two-thirds power of the fluence. For CDF_3 with short pulses above 1 J/cm^2, the absorption increased with the 1.02 power of the fluence. This observation suggests that the shorter pulses dissociate the molecules more efficiently. The higher intensity of the short pulses could improve the efficiency of excitation to high (quasi-continuum) vibrational states by enhancing multiphoton absorption and by increasing the fraction of molecules absorbing laser energy (power broadening).

The species $CFCl_2D$ and CF_2DCl both have very high isotopic selectivities. They both have an advantage over CDF_3 of lower fluences required for dissociation. Good isotopic-exchange techniques are available for both of these species. Hydrogen exchange occurs rapidly for $CFCl_2H$ with NaOH/DMSO catalyst at 100°C [Hason, 1982]. For CF_2HCl hydrogen exchange is rapid, with high-temperature HCl in the gas phase [Moser, 1983].

The high selectivities for CF_2HCl at pressures up to 15 torr [Moser, 1983] makes deuterium enrichment with this species even more attractive. The higher pressure makes design of an efficient enrichment facility easier. The CF_2HCl molecule is also an excellent candidate for carbon enrichment. Kutschke et al. [Kutschke, 1983] suggested that one could enrich both species in a single facility; however, the isotopic selectivity they measured in the region of 1025 cm^{-1} was disappointingly low ($\beta = 35$). The much higher selectivity that Moser et al. measured at the lower frequency makes this synergistic idea much more feasible.

All of our discussion of deuterium enrichment by IRMPD has been concerned with reaction induced by carbon dioxide lasers. The HF or DF chemical lasers are also efficient infrared lasers. The frequencies of these lasers are near the hydrogen-stretching frequencies of many species. Species like HCOOH [Evans, 1978] and H_2CO [Evans, 1979] do dissociate when irradiated with these lasers, but the isotopic selectivities are low ($\beta < 50$). The research results to date favor the CO_2-laser methods.

6.2.3 Deuterium Enrichment with CW Lasers

The first demonstration of laser isotope separation was with the CW HF chemical laser [Mayer, 1970]. In this demonstration the authors irradiated a sample of Br_2 and methanol. The methanol was 50% D_3COD and 50% H_3COH. After the experiment an infrared analysis of the remaining gas indicated that 95% of the methanol was the deuterated form. Many people viewed this experiment with some skepticism, and for many years no good explanation of the result appeared. In 1976, Willis et al. [Willis, 1976] attempted to repeat the experiment with a pulsed CO_2 laser and undeuterated methanol. They concluded from their experiments that the quantum yield for reaction of vibrationally excited (O-H stretch) with bromine was very low (<0.001).

Another demonstration of deuterium enrichement with CW lasers was with another perdeuterated species. Chien and Bauer [Chien, 1976] irradiated a sample of D_3BPF_3 and H_3BPF_3 at very low pressure with a CW CO_2 laser. They measured an isotopic selectivity of about 4 for reaction of the deuterated species. This species has a very weak B-P bond. Only three infrared photons are sufficient to dissociate these molecules. The authors observed a cubic dependence on laser power. This was, apparently, a true photolytic reaction, but because of the nature of the starting materials it has never been seriously considered for an industrial process.

Hsu and Manuccia [Hsu, 1978] have studied the reaction of chlorine atoms with CH_2D_2. They induced the reaction by irradiation with a CW CO_2 laser. The deuterated methane reacted about 72% faster than the undeuterated form. These authors proposed a reaction mechanism based on competition between vibrational-vibrational and vibrational-translational energy transfer. They also suggested that the experiment of Mayer et al. [Mayer, 1970] may have a similar explanation.

The CW process may compete economically with other enrichment methods because of higher throughput and lower photon costs. In their evaluation of the economic feasibility of this process for deuterium enrichment, this group [Manuccia, 1979] estimated the energy requirement for deuterium enrichment. For the reaction of chlorine atoms with CH_3D induced by a CW CO_2 laser, that estimate was 2.3 keV per deuterium atom enriched from natural abundance to >99%. They noted that this energy consumption is modest when compared to costs for similar types of chemical processing.

One of the negative features of the enrichment with a methane (CH_3D) absorber is that the absorption cross sections are low for the CO_2 laser frequencies. This same group [Hsu, 1980a] made two changes in the method that improved the isotopic selectivity. The first of these was to use CO_2 in the gas mixture as an absorber of the CO_2 laser radiation. The isotopic selectivity relies on vibrational energy transfer, and transfer from the vibrationally excited CO_2 is sufficiently selective for the deuterium enrichment. The second change was to induce the reaction with Br atoms. The reaction with bromine is endothermic and thus more sensitive to the vibrational energy of the methane reactant. These changes gave a selectivity of 4.0 for CH_2D_2 and 1.83 for CH_3D.

6.2.4 Tritium Enrichment by IRMPD

The techniques for enrichment of tritium are similar in many respects to those for deuterium enrichment. The experiments are hampered by the radioactivity of the tritium. The tritium is generally a very small fraction of the hydrogen content of a given sample, and this adds complexity. Fortunately, radioactive counting techniques make quantitative isotopic assay possible.

The experiments published to date consider tritium-hydrogen separation, tritium-deuterium separation, and even separation of deuterium from a tritium-deuterium-hydrogen mixture. Table 6.3 summarizes tritium separation experiments. While the ultraviolet predissociation method would probably separate tritium, only the IRMPD method has received serious attention. The groups at the University of Tokyo (Makide et al.), the Institute of Physical and Chemical Research, Saitama, Japan (Takeuchi et al.), and the Department of Physics, Lawrence Livermore Laboratory (Herman et al.) have performed most of the research on this problem.

As in the case of deuterium enrichment, most of the experiments have been with fluoroform. The first demonstration [Makide, 1980] was with this species with focused CO_2 laser pulses. The analysis technique was to separate the components of the irradiated sample by gas chromatography. A thermal-conductivity detector gave the amount of each species and a proportional counter gave the tritium content in each species. The authors left a systematic optimization of selectivity and yield to later work.

In later experiments this group obtained much higher isotopic selectivity and better reaction yield. They found that an argon diluent increased the dissociation probability for CTF_3 by about a factor of 3 between 1 torr total pressure and 10 torr. The rate of increase was even greater above 10 torr. The dissociation probability for CHF_3 did not increase with argon dilution up to 10 torr, but it did increase above 10 torr [Makide, 1981a]. The higher pressure increased the selectivity as well as the reaction yield. Another improvement was the use of a longer-focal-length lens for focusing the laser radiation. This change reduced the laser fluence at the focus and raised the pressure at which dielectric breakdown occurred.

The presence of deuterium in the sample reduces the isotopic selectivity. However, Makide et al. [Makide, 1981b] demonstrated that they could selectively remove the deuterium from a mixture of all three hydrogen isotopes by irradiation at 970.5 cm^{-1} (Table 6.3).

As we noted in the discussion of deuterium enrichment, fluoroform requires high fluence for dissociation. The high diluent pressure greatly improves this situation. A pressure of 100 torr of argon reduces the critical fluence for dissociation of CTF_3 from 136 J/cm^2 to 34 J/cm^2 [Takeuchi, 1982]. (The critical fluence is the fluence ne-

TABLE 6.3 Tritium LIS by IRMPD

Species	Laser	ν (cm^{-1})	Best β	Comment	Reference
CTF$_3$	CO$_2$	1074.6	7.9	First tritium LIS	[Makide, 1980]
CTF$_3$	CO$_2$	1074.6	>500	5-torr CHF$_3$, 100-torr Ar, 0.2 ppm CTF$_3$, focused beam	[Makide, 1981a]
CTF$_3$	CO$_2$	1074.6	>1000	0.2 ppm CTF$_3$, 65 torr at 195 K	[Takeuchi, 1981]
CDF$_3$	CO$_2$	970.5	>10	Selective dissociation in mixture of H, D, T isotopes 43% D, 0.2 ppm T, 5 torr	[Makide, 1981b]
CTF$_3$	CO$_2$	1074.6	580	0.2 ppm CTF$_3$, 5-torr CHF$_3$, 100-torr AR, argon decreases critical, fluence (136 to 34 J/cm^2) and increases selectivity	[Takeuchi, 1982]

CTF_3	CO_2	1057	>10,000	70 torr at 195 K, selectivity the same but yield about 25% at 1052 cm^{-1}	[Makide, 1982a]
C_2TF_5	CO_2	944.2	>500	0.1-1.0 torr, critical fluence = 19 J/cm^2	[Makide, 1982b]
CF_3CTClF	CO_2	973	200	Critical fluence ~8 J/cm^2, 2-torr sample	[Kurihara, 1983]
C_2TF_5	CO_2	931.0	>3000	10 torr at 195 K, lower selectivity at higher pressure	[Makide, 1983]
$CTCl_3$	NH_3	828	>15,000	0.2-torr $CDCl_3$, 200 ppm $CTCl_3$, low fluence prepulse enhances yield	[Magnotta, 1982; Herman, 1984]

cessary to dissociate all of the compound. It is substantially greater than the threshold fluence, the fluence necessary to dissociate some of the compound.)

Reducing the temperature of the CHF_3 sample greatly improved both the isotopic selectivity and the pressure at which selective reaction occurs [Takeuchi, 1981; Makide, 1982a]. Figure 6.3 shows that reduction of the temperature from room temperature to dry-ice temperature produced an improvement in isotopic selectivity of about a factor of 50. It also increased the maximum pressure for selective reaction by about a factor of 5. A lower temperature narrows the absorption band by forcing the molecules into lower-energy vibra-

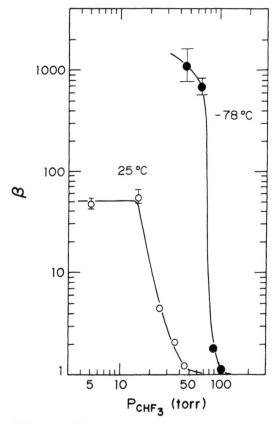

FIG. 6.3 Effect of pressure on the isotopic selectivity for dissociation of CTF_3 at two temperatures in CHF_3 [Takeuchi, 1981].

tional and rotational states. The paper by Takeuchi et al. [Takeuchi, 1981] contains a good summary of tritium LIS up to that date.

The group further improved the selectivity by about another factor of 10 [Makide, 1982a]. They did this by lowering the photolysis frequency to the low-frequency extreme of the CTF_3 absorption band ($1057 \ cm^{-1}$). The resulting isotopic selectivity was in excess of 10,000. The reaction yield at the lower frequency was about 60% of the maximum yield at $1072 \ cm^{-1}$. A further reduction in the laser frequency to $1072 \ cm^{-1}$ gave the same high isotopic selectivity, but it reduced the reaction yield by about 75%.

These near-perfect isotopic selectivities and high operating pressures make the enrichment of tritium by infrared photolysis of CTF_3 a very attractive process for commercial application.

The other entries in Table 6.3 have the advantage that they require a lower fluence for dissociation. In the case of C_2TF_5 [Makide, 1982b, 1983] we see that a lower temperature and a lower frequency again improve the isotopic selectivity.

The other ethane derivative, CF_3CTClF, has a low critical fluence. Consequently, one could use a still lower fluence for enrichment of tritium with this compound. The selectivity, however, is not as high as with the methane derivatives.

The last entry in Table 6.3 is a summary of experiments by the Livermore group on tritium enrichment by infrared photolysis of $CTCl_3$. They used an ammonia laser pumped with a CO_2 laser to photolysis this species in the presence of $CDCl_3$. They obtained extremely high isotopic selectivities (>15,000).

An added diluent increases the photolysis yield of $CTCl_3$. Addition of 10 torr of argon gave a dissociation probability of 80% at 16 J/cm^2 compared to less than 30% without the diluent. A major function of the diluent is to replenish the molecules in absorbing states by rotational relaxation.

The initial photochemical reaction is

$$CTCl_3 \rightarrow CCl_2 + TCl$$

Subsequent reactions of the CCl_2 radical with other constituents of the sample give a variety of carbon and chlorine products. The product distribution is a strong function of the gas pressure.

This group did some experiments with two-frequency photolysis of $CTCl_3$. They found that a prepulse of $2 \ J/cm^2$ enhanced the dissociation yield by a modest amount (about 30%). Both of the pulses were in the region of the ν_4 absorption band of $CTCl_3$ ($836 \ cm^{-1}$) from 5 to $20 \ cm^{-1}$ below the peak. The second pulse had a fluence of $10 \ J/cm^2$.

Enrichment of tritium by infrared photolysis of CTCl$_3$ also appears
to be attractive for commercial application. It has the advantage of
requiring lower fluence for dissociation, but it does have the complication of needing the ammonia laser.

6.3 LASER ENRICHMENT OF CARBON ISOTOPES

Research on carbon LIS has been intensive in recent years. Several
groups have developed enrichment techniques well beyond the state
of laboratory demonstrations. This is particularly true for the IRMPD
technique with several methane derivatives. For several of these
techniques research data are sufficient to make some meaningful economic evaluations. An improved, steady market for carbon isotopes
would allow some of these techniques to move quickly into industrial
production. A laser method would, however, compete with an existing
facility (the ICON plant at Los Alamos National Laboratory) for enriching carbon, nitrogen, and oxygen isotopes by cryogenic distillation of carbon monoxide and nitric oxide. The total world production
of ^{13}C was 5 kg in 1980 [Gauthier, 1982b].

Potential markets for isotopic carbon include tracers for environmental studies and isotopes for disease diagnosis by breath analysis.
The environmental studies with methane-21 (^{13}CD$_4$) is a current market. The breath tests are for diagnosis of such conditions as liver
dysfunction, fat metabolism, and bacterial overgrowth in the large in-
testine [Hackett, 1984]. These techniques are currently in a research
stage. Their widespread use could provide a large market for ^{13}C.

Most of the enrichment methods demonstrated to date fall into the
three major categories mentioned in the introduction: IRMPD, UV, and
IR-UV. The first of these, IRMPD, is the most promising for carbon
enrichment.

Another method of carbon enrichment is the use of a CW infrared
laser to induce isotopically selective reactions. We will not discuss
this method in detail. Section 6.2.3 contains a brief discussion of
this method for that isotope. Hsu and Manuccia [Hsu, 1980b] gave an
account of their research on carbon enrichment. They induced the
reaction of CH$_3$F with bromine atoms with CW CO$_2$ laser radiation. The
isotopic selectivity of the reaction was 2.0.

6.3.1 Carbon Enrichment by IRMPD

Most of the experiments on carbon enrichment by the IRMPD method
have been with methane derivatives such as CF$_3$I and CF$_2$HCl. These
two species both look promising for a carbon-enrichment process, but
much of the early work on the IRMPD process was with other carbon-
containing species.

The first demonstrations of carbon enrichment were with CF_2Cl_2 [Lyman, 1976] and CCl_4 [Ambartzumian, 1976a]. Neither of these have been given serious consideration for a commercial process. The two strong absorption bands of CF_2Cl_2 (1101 and 927 cm^{-1}) that one might use for an enrichment process are on the two extremes of the spectral region of the CO_2 laser. Similarly, the CCl_4 molecule has no strong absorption bands in the spectral region of the CO_2 laser. Selective dissociation of that species requires either irradiation in the region of weak combination bands [Ambartzumian, 1976a] or the use of an ammonia laser (780.5 cm^{-1}) [Ambartzumian, 1978]. The chemical reactions of the reaction products of these two species are also considerably more complex than those of some of the other methane derivatives.

Chou and Grant [Chou, 1981] demonstrated an anomolous temperature effect in some experiments with CF_2Cl_2. Most species show a decrease in isotopic selectivity with increasing temperature. The experiments with CF_2Cl_2 showed the opposite effect for one set of conditions. On the low-frequency side of the band at 1101 cm^{-1} an increase in the gas temperature destroys the isotopic selectivity. For irradiation on the high-frequency side of the band at 922 cm^{-1} increasing temperature improves the isotopic selectivity. The major effect of the temperature increase is to thermally populate vibrationally excited states. This, apparently, favorably alters the absorption process for excitation on the high-frequency side of the absorption band.

Tables 6.4 to 6.7 summarize the enrichment information on the methane derivatives that currently have the highest promise for a carbon-enrichment process. The first column in each table is the carbon isotope with the highest probability of dissociation. For all experiments the infrared source is a pulsed infrared CO_2 laser. The column labeled "ν" is the frequency of the infrared laser that gave the best enrichment. Many of the papers report additional experiments at other frequencies.

The high isotopic selectivity for CF_3I dissociation (Table 6.4) makes it very attractive for a carbon-enrichment process. Another big advantage of this species is the low laser fluence necessary to induce dissociation. The first detailed study [Bittenson, 1977] of IRMPD of this species showed that 0.25 J/cm^2 was the threshold fluence for dissociation of CF_3I (at 1074.6 cm^{-1}). The threshold fluence for dissociation of other small carbon-containing species is substantially greater than this. Figure 6.4 gives a comparison of the threshold fluences for some of these species.

The lower fluence greatly reduces the difficulty of designing an enrichment facility. A fluence of 1.3 J/cm^2 gives a dissociation probability of 10%. This is a good operating range for an enrichment

TABLE 6.4 Carbon LIS by IRMPD of CF_3I with CO_2 Laser Pulses

Species	$\nu\ (cm^{-1})$	Best β	Comment	Reference
^{12}C	1074.6	>25	0.1 torr, 193 K (low temperature improves selectivity)	[Bittenson, 1977]
^{12}C	1074.6	32	0.1 torr, selectivity drops rapidly with pressure, $\beta = 48$ at 228 K	[Bagratashvili, 1978, 1979a, 1979b]
^{12}C	1073.3	>100	0.3 torr, only 100 photons/molecule, dissociated (reflective cell used)	[Fuss, 1979]
^{12}C	1041.3	21.3	1-torr CF_3I and 20-torr NO,	[Abushelishvili,

Isotope	Frequency	Selectivity	Comments	Reference
			236 K-NO or Br_2 improve selectivity	1981]
^{13}C	1041.3	25	Low temperature improves selectivity, remains high up to 1 torr	[Abdushelishvili, 1982]
^{12}C	1076.7	50	Selectivity drops rapidly below 0.1 torr	[Abdushelishvili, 1982]
^{12}C	1078.6	4.5	0.2-torr CF_3I and 1.8-torr O_2	Kojima, 1983]
^{12}C	1074.6	40.8	0.1 torr, demonstrated 99.93% ^{13}C in reactant in two steps, 40% of original ^{13}C in sample	[Cauchetier, 1983]

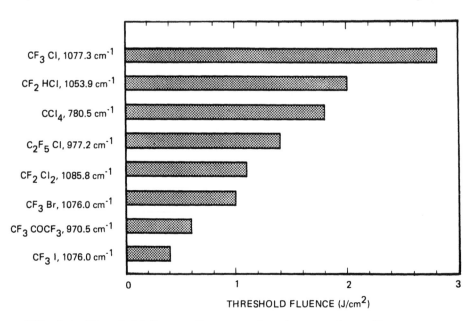

FIG. 6.4 Threshold fluence for several carbon species. (Data come from references in Table 6.12 and Lyman et al. [Lyman, 1984].)

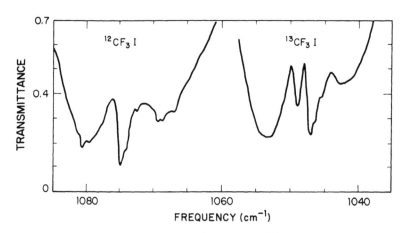

FIG. 6.5 Infrared spectra of $^{12}CF_3I$ and $^{13}CF_3I$ in the region of the CO_2 laser lines [Fuss, 1982].

TABLE 6.5 Carbon LIS by IRMPD of CF_3Br with CO_2 Laser Pulses

Species	ν (cm^{-1})	Best β	Comment	Reference
^{13}C	1035.5	<100	Selectivity increases with pressure up to 10 torr, selectivity lower at high frequency	[Doljikov, 1981]
^{13}C	1041.3	60-62	0.25-torr CF_3Br	[Adlzianidze, 1981]
^{13}C	1041.3	10-12	0.25-torr CF_3Br and 15-torr Ar, yield improved 20 to 1000 times	[Adlzianidze, 1981]
^{13}C	1035.5	150	0.3 torr (N_2 increases selectivity up to 500 torr)	[Abdushelishvili, 1982]
^{12}C	1077.3	High	0.5 torr 1.9 J/cm^2, $CF_3Br/O_2 = 1/2$; products = $^{12}COF_2$, $^{12}CO_2$, Br_2	[Borsella, 1983]

plant. One enrichment demonstration [Cauchetier, 1983] that produced macroscopic quantities of ^{13}C was with a fluence of slightly greater than 1 J/cm^2.

One feature of this species that is typical of the substituted methanes is the large carbon isotope shift in the infrared spectrum. Figure 6.5 from [Fuss, 1982] shows spectra of the ν_1 band of two isotopic forms of CF_3I. The bands are well separated; the isotope shift is 28 cm^{-1}. The spectral regions for both isotopes are accessible with the CO_2 laser. Table 6.4 lists demonstrations in both spectral regions. The isotopic selectivity is generally higher for dissociation of the lighter species; however, dissociation of the heavier species also has some advantages.

When irradiating the ^{12}C species, the isotopic selectivity decreases rapidly with increasing pressure above about 0.1 torr. On the other hand, the selectivity remains high up to about 1 torr for disso-

TABLE 6.6 Carbon LIS by IRMPD of CF_3Cl with CO_2 Laser Pulses

Species	ν (cm^{-1})	Best β	Comment	Reference
^{13}C	1067.5	45	0.5 torr, focused beam, yield increases with E^2 and with P^2	[Gauthier, 1980]
^{13}C	1071.9	>25	2 torr, focused beam 90-ns pulse lower selectivity for longer	[Neve de Mevergnies, 1981]
^{13}C	1073.3	9.25	1 torr, 0.5 J/cm^2, selectivity decreases with pressure	[Sarkar, 1981]
^{13}C	1048.7	65	10 torr, 3 J/cm^2 off resonance, selectivity increases with pressure	[Sarkar, 1981]

TABLE 6.7 Carbon LIS by IRMPD of CF_2HCl with CO_2 Laser Pulses

Species	ν (cm^{-1})	Best β	Comment	Reference
^{13}C	1046.9	-100	100 torr, 3.5 J/cm^2	[Gauthier, 1982a]
^{13}C	1041.3	-100	100 torr, 50-W laser, production rate = 100 g/yr	[Gauthier, 1982a]
^{12}C	1085.8	5.2	0.2-torr CF_2HCl and 1.8-torr O_2	[Kohima, 1983]
^{13}C	1031.5	-200	-60 torr, 50% of 1980 world production of ^{13}C	[Hackett, 1984]

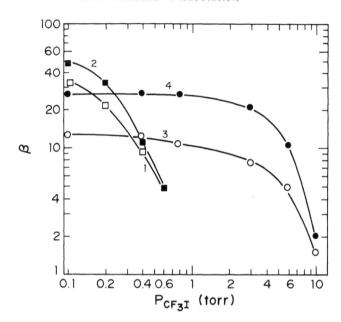

FIG. 6.6 Isotopic selectivity versus CF$_3$I pressure [Bagratashvili, 1978]. Numbers 1 and 2 are for dissociation of the ^{12}C species (1074.6 cm^{-1}), and numbers 3 and 4 are for dissociation of the ^{13}C species (1041.3 cm^{-1}). Numbers 1 and 3 are for 20°C and 2 and 4 are for −45°C.

ciation of the ^{13}C species. Figure 6.6 shows the difference for these two procedures. Several entries in Table 6.4 [Bagratashvili, 1978, 1979a, 1979b; Abdushelishvili, 1982] report this effect. A careful economic analysis would give the best approach for an enrichment plant.

Several of the research groups (Table 6.4 and 6.6) found that lowering the temperature improves the isotopic selectivity. The lower temperature narrows the absorption spectra by reducing the population of higher vibrational and rotational states.

The photochemical reaction for this species is

$$CF_3I \rightarrow CF_3 + I$$

Efficient enrichment requires chemical scavenging of at least one of the reaction products. Several possible approaches are discussed below.

The method of Cauchetier et al. [Cauchetier, 1983] was to have a metallic silver grid in the photolysis chamber. The iodine atoms produced in the photolysis reaction then diffused to the grid and reacted with the silver surface. This procedure prevented recombination of the photolysis products to CF_3I. The CF_3 radicals then recombined,

$$CF_3 + CF_3 \rightarrow C_2F_6$$

and the only gaseous product of the photolysis reaction was ^{13}C-depleted C_2F_6. This method of scavenging the iodine atoms was very successful.

This experiment also provided a demonstration of efficient, macroscopic enrichment of ^{13}C. These authors used a 15-W laser (1.5 J/pulse, 10 Hz) to produce 0.5 mg/h of CF_3I enriched to 90% ^{13}C in a 4.2-m-long cell (3-liter volume). In one experiment they performed a two-step enrichment to 99.93% ^{13}C. Their method was to photolyze the $^{12}CF_3I$ in the sample until the remaining CF_3I was 90% ^{13}C. They then removed the C_2F_6 and repeated the procedure to give the 130,000-fold enrichment in ^{13}C. Forty percent of the ^{13}C in the initial sample remained in the CF_3I.

Another successful scavenging procedure [Baklanov, 1980; Albushelishvili, 1981, 1982; Cauchetier, 1983] was to use either Br_2 or NO to react with the CF_3 radicals. These scavengers produced either CF_3Br or CF_3NO. These species improve isotopic selectivity as well as prevent recombination.

One variation of the enrichment with NO scavenger allows for staging the process to obtain higher enrichement. One can reformulate the CF_3I quite simply by heating the CF_3NO product in the presence of I_2. This allows another enrichment cycle with CF_3I. (The procedure also works with Br_2 to give CF_3Br.)

Another, perhaps simpler method for staging the enrichment process is to use Br_2 as the scavenger. One can then perform a second enrichment step with that species. Table 6.5 summarizes enrichment experiments with CF_3Br.

Abdushelishvili et al. [Abdushelishvili, 1982] made an observation that is probably general for photolysis by IRMPD at high pulse rates. To maintain high isotopic selectivity, the irradiated gas sample must clear the photolysis region between laser pulses. Otherwise, the resulting high temperature will destroy the isotopic selectivity of the absorption process.

The ν_1 absorption band for $^{12}CF_3Br$ is at a somewhat higher frequency (1085 cm^{-1}) than for CF_3I (1075 cm^{-1}). This makes the ^{12}C species less accessible with the CO_2 laser. Therefore, we see (Table

6.5) that most of the published experiments involve dissociation of the ^{13}C species. The selectivities for this process are high, but so is the dissociation threshold (Fig. 6.4).

As with CF_3I, dissociation of the minor isotopic species makes higher pressure more favorable. At the low-frequency extreme of the band the pressure of maximum selectivity is 10 torr [Doljikov, 1981]. For a diluent pressure (N_2) the maximum is much higher, 500 torr [Abdushelishvili, 1981]. The higher pressure also enhances the dissociation yield. The higher rate of rotational relaxation enables more molecules to absorb the laser radiation. In the experiment of Abzianidze et al. [Abzianidze, 1981] at a somewhat higher frequency the argon diluent decreased the selectivity, but it greatly enhanced the dissociation yield.

For CF_3Cl the ν_1 absorption band shifts to a still higher frequency (Table 6.6). The threshold for dissociation is higher than for the other CF_3X species (Fig. 6.4), and again we see that at the low-frequency extreme of the ^{13}C band an increase of pressure improves the selectivity [Sarkar, 1981]. Nearer the band center the selectivity decreases with pressure. The higher selectivity with shorter pulses [Neve de Mevergnies, 1981] is an indication in addition to the pressure dependence that vibrational energy transfer during the laser pulse leads to dissociation of the unwanted isotopic species.

Gauthier et al. [Gauthier, 1980] observed that the reaction yield increased with the square of both the pulse energy and the pressure. The pressure effect is probably mostly due to rotational relaxation during the pulse. The pulse energy effect has little meaning with focused laser pulses.

Table 6.7 summarizes the experiments with CF_2HCl. Carbon isotope separation on an industrial scale looks very promising for this species. The isotopic selectivity is so high that it is difficult to measure. This species has many of the advantages of the CF_3X species. For example, the selectivity increases with pressure up to 100 torr [Gauthier, 1982a] for photolysis of the minor isotopic species. The frequencies for the three experiments with high selectivity in Table 6.7 are all on the low-frequency extreme of the ν_9 band (1083 cm^{-1}) of $^{13}CF_2HCl$. Figure 6.7 shows the dependence of isotopic selectivity on pressure for several laser frequencies. Photolysis at the low-frequency extreme of the absorption band clearly improves the selectivity. At this band position only a small fraction of the absorbing species initially absorb the laser radiation. However, of those that do absorb, the probability of continuing to absorb is high because of the molecular anharmonicity. This reduces the chance for isotopic scrambling by collisional exchange of vibrational energy.

The efficiency of inducing the chemical reaction is also very high (50%). This is a very efficient use of the available laser energy, especially for photolysis of a minor isotope.

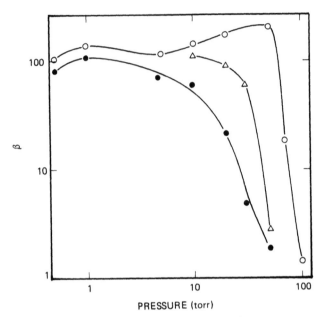

FIG. 6.7 Isotopic selectivity versus CF_2HCl pressure [Hackett, 1984]. The ^{13}C and ^{12}C species were in equal amounts. The laser frequencies, beginning at the top, were 1031.5, 1035.5, and 1041.3 cm^{-1}. The $^{13}CF_2HCl$ band center is at 1083 cm^{-1}.

The rate of production of ^{13}C is very high for photolysis at high pressure with a high-power pulsed laser [Gauthier, 1982b; Hackett, 1984]. Hackett's paper reports the best demonstrated rate to date for carbon LIS. It is about 50% of the world production rate of ^{13}C for the year 1980. And this was with a laboratory scale laser (10 J, 10 Hz).

The photolysis reaction for this species is HCl elimination,

$$CF_2HCl \rightarrow CF_2 + HCl$$

One can use a scavenger like O_2 for the CF_2 radical, but it is not necessary. In the absence of a scavenger the radicals form C_2F_4.

For the purpose of further enriching the carbon isotopes, one can reconstitute the CF_2HCl by heating the C_2F_4 in an excess of HCl [Moser, 1983; Hackett, 1984]. The C_2F_4 is also a good starting material for synthesis of other isotopically labeled compounds.

Several groups [Baldwin, 1981; Stephenson, 1978, 1979; Duperrex, 1981] have made careful studies of the absorption and dissociation processes in CF_2HCl. These include direct measurements of the rate of dissociation of vibrationally excited species.

One promising procedure for an industrial process that is not photolysis of a methane derivative is the two-frequency dissociation of perfluoroacetone [Hackett, 1981, 1981b]. This species dissociates with very high selectivity ($\beta \sim 700$) when irradiated with two frequencies of a CO_2 laser. The measured selectivity is for the central carbon atom. One can easily separate this atom from the reaction products. The reaction product that contains the central carbon is carbon monoxide. The other two carbon atoms remain bonded to the fluorine atoms. The overall photolysis reaction is

$$CF_3{}^{13}COCF_3 \rightarrow {}^{13}CO + C_2F_6$$

This enrichment process requires no scavenger to maintain the yield and selectivity.

The group at Frascati, Italy [Borsella, 1980, 1981], has performed several experiments with the substituted ethane, C_2F_5Cl. One advantage of this and other larger species is that they absorb the laser radiation more easily than most of the substituted methanes. Consequently, the fluence threshold (Fig. 6.4) is lower. The C_2F_5Cl reaction requires a scavenger for the reaction products. Both O_2 and Br_2 are good scavengers.

A two-frequency dissociation improves the reaction yield, and probably, the selectivity. In agreement with the usual pattern for this type of process, the best second frequency is on the low-frequency extreme of the infrared absorption band.

6.3.2 Carbon LIS by Ultraviolet Photolysis

Of the several demonstrations of carbon enrichment by ultraviolet photolysis (Table 6.8), only one of them, the photolysis of formaldehyde, has received serious consideration. Most of the research on formaldehyde photolysis has been directed toward hydrogen enrichment. We treated the topic more completely in that section.

In their experiment, Clark et al. [Clark, 1975] carefully studied the ultraviolet spectrum of H_2CO and selected the best wavelength for carbon enrichment. On the other hand, the purpose of the Marling experiment [Marling, 1977] was to demonstrate that an existing CW laser, the neon-ion laser, would induce an isotopically selective reaction. Marling did, however, require neon-22 in the laser to get high carbon selectivity.

Loree et al. [Loree, 1979] also used a readily accessible laser, the ArF excimer laser, for photolysis of CS_2. They found some selectivity with a broadband laser, but the selectivity improved with a frequency-narrowed, tunable laser. They used two prisms in the cavity to nar-

TABLE 6.8 Carbon LIS by Ultraviolet Photolysis

Species	Laser	λ (nm)	Best β	Comment	Reference
$H_2{}^{12}CO$	Pulsed dye	303.2	81	NO enhances enrichment, initial sample 90% ^{13}C	[Clark, 1975]
S-Tetrazine, ^{12}C	Pulsed dye	551.5	—	$\beta = 6$ for nitrogen	[Karl, 1975]
$H_2{}^{13}CO$	CW Ne-22	332.374	33	24% isotopic scrambling from radical products	[Marling, 1977]
$^{12}CS_2$	ArF	193.6	1.7	Prism-narrowed laser, 1-torr mixture, initial sample 70% $^{13}CS_2$	[Loree, 1979]
$^{13}CH_3NC$	CW dye	739.2	1.7	Vibrational isomerization to CH_3CN by direct excitation of fourth CH stretch overtone	[Reddy, 1980]

TABLE 6.9 Carbon LIS by an Infrared-Ultraviolet Method

Species	IR laser UV laser	ν (cm^{-1}) λ (nm)	Best β	Comment	Reference
CF$_3$I, ^{12}C	CO$_2$ XeF	1074.6 351	>48	IR fluence = 0.24 J/cm^2	[Knyazev, 1978, 1979]
CF$_3$I, ^{12}C	CO$_2$ XeCl	1074.6 308	2.3	IR fluence = 0.24 J/cm^2	[Knyazev, 1978, 1979]
OCS, ^{12}C	CO$_2$ KrF	1045.0 249	3.4	Scavengers; C$_2$F$_4$-Xe NO$_2$-Xe, or NO$_2$-O$_2$	[Zittel, 1983]
OCS, ^{13}C	CO$_2$ KrF	1033.5 249	1.7	Scavengers: C$_2$F$_4$-Xe or NO$_2$-Xe	[Zittel, 1983]

row the laser and to tune it over the ArF gain band. The isotopic
composition of the reaction products (both sulfur and carbon) depended strongly on the laser frequency.

For the s-tetrazine the best selectivity was obtained for nitrogen.
The authors [Karl, 1975] did, however, demonstrate carbon and hydrogen enrichment. The photolysis products were N_2 and HCN.

The final entry in Table 6.8 is not really UV photolysis. High
overtones of the hydrogen stretching vibrations absorb the visible
(dye-laser) radiation. This absorbed energy induces an isotopically
selective isomerization. In this example CH_3NC isomerizes to CH_3CN.
The low selectivity for this process makes it less attractive than
other methods.

6.3.3 Carbon Enrichment by IR-UV Dissociation

The most extensive study of LIS by IR-UV dissociation was for uranium. We discuss the technique in greater detail in Sec. 6.5. The
method involves a selective vibrational excitation with an infrared laser followed by dissociation of the vibrationally excited species with
an ultraviolet laser.

Table 6.9 summarizes some uses of the method for carbon enrichment. The first two entries show the strong dependence on the wavelength of the ultraviolet laser. At 351 nm the selectivity is comparable to infrared photolysis. At 308 nm it is much lower. The infrared
excitation tends to shift the ultraviolet spectrum to longer wavelengths
[Padrick, 1980]; therefore, the selectivity is best for regions of the
UV spectrum that have the greatest change of absorption cross section with wavelength.

One objective of the OCS expeirments [Zittel, 1983] was to demonstrate a laser photolysis that was selective for isotopes of all three
elements. The authors achieved that objective. The process does
require scavengers. The species C_2F_4, NO_2, and O_2 all prevent recombination of the products CO and S. They used xenon as a diluent
to enhance absorption by accelerating rotational relaxation.

6.4 LASER ENRICHMENT OF SULFUR ISOTOPES

The laser process that has received the most attention in the last 10
years is probably the laser enrichment of sulfur isotopes. One reason for this is the strong infrared absorption band of SF_6 (ν_3) in the
region of efficient operation of the CO_2 laser [Nowak, 1975]. The
demonstrations in 1975 [Ambartzumian, 1975b; Lyman, 1975] of very
high isotopic selectivities by IRMPD initiated the broad interest in
multiple-photon processes. The experiments with SF_6 laid the ground-
work for the other IRMPD enrichment processes discussed in this
chapter.

Presently, the major potential market for sulfur isotopes is in atmospheric studies of the causes of acid rain. Long-ranged (~1000 km) tracking of the distribution of emissions of sulfur dioxide could require quantities of the minor isotopes of sulfur. The current world production rate is in the range of 100 g per year. Another potential market for ^{33}S is for NMR studies of biological functions and other chemical reactions.

Sulfur has four stable isotopes: ^{32}S (95.0%), ^{33}S (0.076%), ^{34}S (4.22%), and ^{36}S (0.014%). Most of the experimental emphasis has been on separating the two most abundant isotopes, ^{34}S and ^{32}S, and most of these experiments have involved dissociation of compounds containing the most abundant isotope, ^{32}S. Several groups, however, have demonstrated selective reaction of the minor isotopic species.

By far the majority of published papers has been on the infrared dissociation of sulfur compounds (mostly SF_6). Other approaches include at least one published example for each of the following methods: dissociation sensitized by another absorber, ultraviolet photolysis, infrared-ultraviolet dissociation, and CW laser irradiation.

6.4.1 Sulfur Enrichment by IRMPD

The early demonstrations [Ambartzumian, 1975b; Lyman, 1975] of sulfur LIS were by infrared photolysis of SF_6. These experiments used focused CO_2 laser pulses (~1 J) and low pressures (~1 torr). The isotopic selectivity was much higher and the intensity (or fluence) required was much less than expected. These and subsequent experiments by many groups have given a good characterization of the observable processes that occur during infrared photolysis of SF_6. Theoretical investigations of the process have also been extensive. The review articles cited in the introduction to this chapter adquately evaluate that work. This section reviews some of the important experimental results that apply to understanding and implementing an enrichment process. As we discuss these experiments, note that the isotopic selectivity is lower ($\beta \sim 50$ [Lyman, 1977b]) than for similar experiments with hydrogen and carbon species. This is a direct consequence of the greater overlap of the infrared absorption band (ν_3) of SF_6 for the different isotopes. The isotope shift for this band is 8.5 cm^{-1} per mass unit.

The amount of energy absorbed per SF_6 molecule increases with the two-thirds power of the laser fluence from about 10^{-7} J/cm^2 to above 10 J/cm^2. This dependence holds for frequencies in the spectral region of the ν_3 absorption band. The review article [Lyman, 1984] gives a complete summary of absorption measurements under a broad range of conditions for the period up to 1980. It makes some reference to experiments since that time.

The threshold for dissociation of low-pressure SF$_6$ is about 1 J/cm^2. Figure 6.8 shows the dependence of reaction probability on laser fluence. Many groups have made this type of measurement. This figure includes measurements by three of those groups.

At higher pressures rotational relaxation feeds the absorbing rotational states during the laser pulse and enhances the absorption. The isotopic selectivity of the dissociation reaction generally decreases with pressure. However, a desirable experiment would be to measure the isotopic selectivity for higher pressures at frequencies near the low-frequency extreme of absorption bands of the minor isotopes. We noted several small species in the sections on hydrogen and carbon enrichment that showed a dramatic increase in selectivity for those conditions. Cooling of the gas may be necessary for SF$_6$ to show this effect.

Temperature has a very large effect on both the absorption and dissociation. At higher temperatures the absorption spectrum at high intensity tends to shift to lower frequencies and to broaden [Tsay, 1979]. The low-intensity absorption shows a similar, but perhaps not

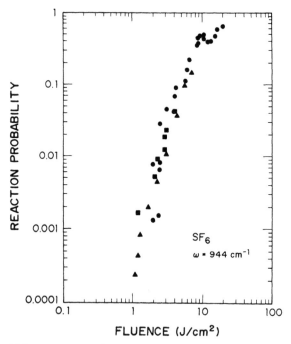

FIG. 6.8 Reaction probability of SF$_6$ versus fluence measured by three different laboratories, $\omega = 944$ cm^{-1} [Cantrell, 1979].

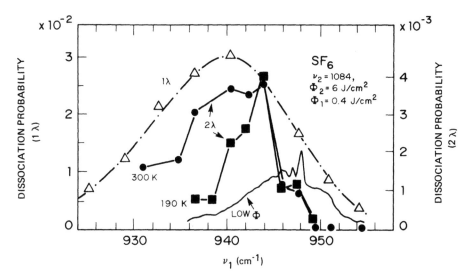

FIG. 6.9 Reaction probability versus frequency for SF_6 with one and two infrared lasers [Ambartzumian, 1976b].

so dramatic effect [Nowak, 1975]. The narrowing of the high-fluence absorption spectrum with decreasing temperature enhances the isotopic selectivity. This is particularly useful for selective dissociation of $^{33}SF_6$. Baranov et al. [Baranov, 1979a] demonstrated reasonably high isotopic selectivity ($\beta = 5$) for this species by cooling the sample to 170 to 190 K. They achieved the highest selectivity at 934.9 cm^{-1}.

The amount of energy absorbed by SF_6 may depend strongly on frequency, temperature, and so on, but a strong correlation exists between the energy absorbed and the dissociation probability. The amount of dissociation is very close to what one would expect from the measured dissociation rate [Lyman, 1977a] and a thermal distribution of the available energy among the vibrational states of the molecule [Lyman, 1984].

One can substantially reduce the total fluence necessary for dissociation by using two infrared frequencies. Two frequencies also reduce the width of the dissociation spectrum, or the plot of dissociation probability versus laser frequency. Ambartzumian et al. [Ambartzumian, 1976b] measured the dissociation spectrum for SF_6 with both one frequency and two frequencies. Figure 6.9 shows the results of these measurements. The two-frequency curve is about half the with. The two-frequency curve at a reduced temperature is still narrower. The figure also shows the low-fluence spectrum. In these ex-

TABLE 6.10 High-Volume Sulfur Enrichment: Production Rates and Laser Efficiencies

Dissociation rate for $^{32}SF_6$ (mg/h)	Laser power (W)	Photon Utilization (%)	Rate per watt (mg/h-W)	Reference
940	300	1.2	3.1	[Baranov, 1979b]
240	100	0.9	2.4	[McInteer, 1982]
20	5	1.6	4.0	[Fuss, 1979]

periments the first frequency is at a fluence of 0.4 J/cm^2. The second frequency is 1084 cm^{-1} at a fluence of 6 J/cm^2. For the single-frequency experiment the fluence was about 6 J/cm^2. In the two-frequency experiments the SF_6 did not dissociate with either laser by itself. The narrower dissociation spectrum improves the isotopic selectivity of the dissociation reaction. A later experiment [Alimpiev, 1982] shows a large reduction in the fluence necessary for dissociation when the second laser is at a frequency (888 cm^{-1}) well below the absorption band.

The isotopic selectivity is a function of the time delay between pulses. The decrease of selectivity with the decay time is a measure of the time for vibrational transfer among the SF_6 molecules. That time is about 1 μs-torr [Ambartzumian, 1976b].

Three groups have performed sulfur-enrichment demonstrations for macroscopic amounts of product (greater than 1 mg). Table 6.10 summarizes these experiments. The approach for all of these demonstrations was to selectively dissociate the $^{32}SF_6$, and the remaining SF_6 became enriched in the minor isotopes. With the reaction of the $^{32}SF_6$ one can optimize the enrichment process by either continuously adding SF_6 to the gas stream or by working with the SF_6 in batches.

The first column in Table 6.10 is the reported rate of reaction in the experiment in milligrams of ^{32}S per hour. The second column is the average laser power. (The lasers were all pulsed CO_2 lasers.) The third column is the photon efficiency, or the percent of the incident laser radiation that was actually used to dissociate the SF_6. Cleavage of a S-F bond in SF_6 requires 34 infrared photons—about 40 for rapid dissociation. The fourth column is the dissociation rate per watt of laser power. The latter rates are all similar. These measurements show the achievable enrichment rate for dissociation of the ^{32}S species. By carefully designing the irradiation chamber one could probably improve this rate. Photon efficiencies in the range of 20% should be possible. This would give a corresponding improvement in the dissociation rate per watt.

One reason for dissociating the abundant species ($^{32}SF_6$) is that the CO_2 laser operates best at frequencies that dissociate that species. Another is that the isotopic selectivity is best for the lighter species. Further research on dissociation of the minor species could be very fruitful.

These larger-scale processes do require scavengers for the fluorine atoms. Hydrogen is the most commonly used. After the initial photolysis reaction,

$$SF_6 \rightarrow SF_5 + F$$

additional photolytic and thermal reaction of the SF_5 occurs,

$$SF_5 \rightarrow SF_4 + F$$

The scavenging reactions rapidly consume the atomic fluorine,

$$F + H_2 \rightarrow HF + H$$

The experiments of McInteer et al. [McInteer, 1982] suggest that as SF_4 accumulates, the foregoing dissociation reactions of SF_6 and SF_5 may reverse and decrease the reaction yield. Furthermore, the HF product is a more effective quencher of vibrationally excited SF_6, which also decreases the reaction yield. The solution to these problems was to add water to the hydrogen scavenger, which induced the reaction

$$SF_4 + H_2O \rightarrow SOF_2 + 2HF$$

and to remove the HF by pumping the process gas through a filter containing NaF. The SOF_2 is stable with respect to the reaction products; furthermore, it is easier to separate from the SF_6 than is SF_4.

One method of separation of SF_6 from SOF_2 is distillation [McInteer, 1982]. The vapor pressures of these species differ by about a factor of 20. Baranov et al. [Baranov, 1979b] removed both SF_4 and HF by reaction with barium oxide,

$$SF_4 + BaO \rightarrow 2BaF_2 + 2BaSO_3$$

$$2HF + 2BaO \rightarrow BaF_2 + Ba(OH)_2$$

Sulfur-containing species other than SF_6 also undergo isotopically selective reactions [Leary, 1978; Lyman, 1977b, 1978, 1979]. A series of experiments with SF_5X molecules, where X is F, Cl, NF_2, and SF_5, show the effect of increasing molecular size. All of these species absorb CO_2 laser radiation by excitation of the same S-F stretching vibration, and the absorption cross sections are similar at very low fluence. The most notable feature of this set of experiments was the effect of molecular size on the fluence required for dissociation. The fluence required to dissociate 1% of the molecules ($\Phi_{1\%}$) increases by more than 100 times between S_2F_{10} (0.018% J/cm^2) and SF_6 (2.8 J/cm^2). The other two molecules in the series required intermediate fluences for 1% dissociation. The larger molecules absorb the intense laser radiation much more easily, but the isotopic selectivity does suffer somewhat in the larger species. The selectivity for SF_6 can be as high as 50, but the measured values for the other species are in the range 2 to 5. Higher costs and toxicity also make the larger species less attractive for an enrichment process.

6.4.2 Sulfur Enrichment by Other Methods

Several other methods for sulfur enrichment have received some preliminary attention, but no one has investigated them in detail. We mention them here for completeness.

Section 6.2.3 mentions the important role of vibrational-vibrational energy transfer in inducing isotopically selective reactions. That section contains a reference to enrichment by selective energy transfer. Cauchetier et al. [Cauchetier, 1982] published an account of sulfur enrichment by what they believe to be selective vibrational energy transfer. They irradiated mixtures of SF_5Cl and SF_6 with CO_2 laser pulses and observed isotopically selective dissociation of $^{32}SF_5Cl$ with a selectivity as high as 1.57. Under the conditions of the experiment neither of the species dissociates when the other is not present. The SF_6 strongly absorbs the laser radiation; the SF_5Cl does not. The proposed mechanism for this selective reaction is selective transfer of vibrational energy from the highly excited SF_6 to SF_5Cl, which then dissociates. The chemical bonds in SF_6 are considerably stronger (93 kcal/mol) than the S-Cl bond in SF_5Cl (61 kcal/mol).

We discussed the UV photolysis of CS_2 with the ArF excimer laser [Loree, 1979] in Sec. 6.3.2. This reaction also enriches sulfur. The best isotopic selectivity for this experiment was about 1.2. This is somewhat lower than the carbon selectivity by this method.

Zittel et al. [Zittel, 1983] have enriched sulfur by IR-UV dissociation of OCS. We discussed this experiment in Sec. 6.3.3. The authors showed that selective dissociation is possible for any of the three major isotopes of sulfur by changing the infrared frequency. The selectivities were in the range of 1.9 to 3.1. These photochemical reactions required scavenging with NO_2 or O_2.

The final example of sulfur is not really a photolytic process. Zellweger et al. [Zellweger, 1984] used a CW CO_2 laser to selectively excite one sulfur isotope in an expanding jet of SF_6 in argon diluent. As gas mixture expands the SF_6 component tends to condense, but the vibrational energy deposited by laser absorption prevents condensation of the SF_6. selected. The lighter, uncondensed SF_6 molecules diffuse more rapidly away from the center of the jet, while the larger clusters tend to remain near the center of the jet. With skimmers one can then separate the two fractions of the jet. This group demonstrated selectivities as high as 3.5 by this process. The highest dilution gave the best enrichment. Apparently, at high SF_6 mole fraction the collisional transfer of vibrational energy scrambled the intiial selective absorption.

6.5 LASER ENRICHMENT OF URANIUM ISOTOPES

Of all of the applications of laser photochemistry the enrichment of uranium isotopes probably has the greatest economic potential. The main reason for investigating uranium isotope separation is the promise of an economical method for producing bulk quantities of enriched uranium for reactor fuel. The light-water nuclear reactor requires 3.2% uranium-235. Natural uranium is only 0.7%. The current method for enriching uranium is gaseous diffusion. New diffusion facilities would be very expensive, and they consume large amounts of electricity. Other methods under investigation include gas centrifuge and laser ionization of atomic uranium. This section is confined to uranium enrichment by laser dissociation of molecular species.

Most of research on molecular processes has been on two methods: infrared-ultraviolet dissociation of UF_6, and infrared dissociation of UF_6 and a few other molecular species (IRMPD). The program for molecular laser isotope separation (MLIS) at Los Alamos Laboratory had as its major goal the development of an IR-UV process. The USDOE terminated that program in 1982. The group at Saclay, France [Alexandre, 1983], has also investigated this technique. A complete review of this research is not possible because of governmental and industrial restrictions on information transfer.

Another possible method for uranium enrichment is to induce isotopically selective bimolecular reactions of UF_6 by vibrational excitation with infrared lasers. This method has received some discussion and one claim of success [Eerkens, 1976]. The technique apparently involves the reaction of HCl with UF_6 under the influence of a CW CO_2

laser. However, the details of that claim have never been published in a refereed journal, and the published spectroscopy is not consistent with a selective chemical reaction.

Several authors have published reviews of laser enrichment of uranium. These include some that were mainly dedicated to the Los Alamos MLIS process [Jensen, 1976, 1980a, 1982; Robinson, 1977] and some that are more general [Becker, 1981, 1982; Martellucci, 1981; de Silvestri, 1981; Meyer-Kretschmer, 1983]. For the most part these papers give brief descriptions of the enrichment processes, but some discuss specific areas, such as laser availability [Martellucci, 1981] or photochemistry [de Silvestri, 1981].

6.5.1 Uranium Enrichment by IR-UV Photolysis of UF$_6$

The majority of the research on the IR-UV method for uranium enrichment has been at Los Alamos National Laboratory. This section will mainly be a description of that research. The two published reviews [Jensen, 1976, 1982] give some additional details.

The initiation of this research (1971) preceded the extensive research on multiple-photon excitation. The lasers for the process were not available until well into the research program.

The method of enrichment required several steps as outlined in Fig. 6.10. The first step was to cool a gas mixture containing UF$_6$ by expansion in a supersonic nozzle. While the gas was cold, but before the UF$_6$ condensed, one or more infrared lasers selectively excited the ^{235}UF$_6$. This excitation produced vibrationally excited UF$_6$. An ultraviolet laser then dissociated the excited molecules to UF$_5$ and F. The UF$_5$ product of this reaction rapidly condensed to give solid particles enriched in ^{235}U.

The reason for expansion cooling of the gas sample is to bring a large fraction of the UF$_6$ molecules to the ground vibrational state. Static cooling condenses the UF$_6$. At room temperature only 1 molecule in 2500 is in the ground vibrational state and the average molecule has a vibrational energy content of nearly 2000 cm^{-1}.

The high level of vibrational excitation results in broad, structureless infrared absorption bands [Jensen, 1982; Takami, 1984] contains sharp, well-defined absorption features of the ground-state molecules of both isotopic species (see the high-intensity spectrum, Fig. 6.11). These well-separated absorption features allow for selective excitation of the ^{235}UF$_6$ with an infrared laser (step 2, Fig. 6.10). Chapter 1 in this volume gives further details on the infrared spectroscopy of UF$_6$.

We have seen in our discussion of infrared excitation of other species that the resulting vibrational excitation shifts the absorption resonance to lower frequencies. This reduces the apparent absorp-

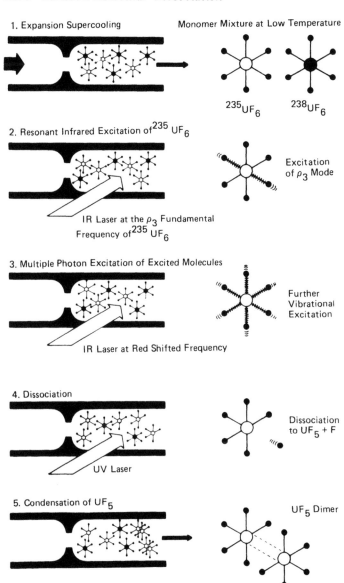

1. Expansion Supercooling   Monomer Mixture at Low Temperature

$^{235}UF_6$ $^{238}UF_6$

2. Resonant Infrared Excitation of $^{235}UF_6$

Excitation of ρ_3 Mode

IR Laser at the ρ_3 Fundamental Frequency of $^{235}UF_6$

3. Multiple Photon Excitation of Excited Molecules

Further Vibrational Excitation

IR Laser at Red Shifted Frequency

4. Dissociation

Dissociation to UF_5 + F

UV Laser

5. Condensation of UF_5

UF_5 Dimer

FIG. 6.10 Schematic representation of the Los Alamos method for uranium enrichment [Jensen, 1982]. The expansion cools the flowing gas, the two infrared lasers selectively excite the $^{235}UF_6$ molecules, and the ultraviolet laser dissociates them. The UF_5 molecules eventually grow to collectible particulates.

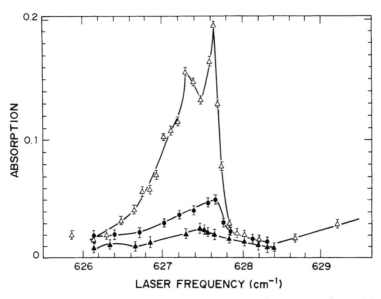

FIG. 6.11 Laser energy absorbed versus frequency for cold UF_6 in the region of the ground-state Q branch of the ^{238}U species. N_2 – UF_6, (0.5%); T = 105 K; ▲, 20 mJ/cm^2; ●, 4 mJ/cm^2; △, 0.4 mJ/cm^2.

tion cross section at the frequencies within the band and increases the cross section at lower frequencies. Figure 6.11 shows the effect of increasing the infrared fluence in the region of the ground-state Q branch of low-temperature $^{235}UF_6$. This decreasing cross section with increasing fluence results in a lowering of the selectivity of the infrared excitation. However, greater amounts of infrared excitation give greater discrimination in the UV dissociation step. One can increase the amount of vibrational excitation with a minimum loss of selectivity by using a second infrared laser (step 3, Fig. 6.10). This frequency is lower than the first. It is in the region where the initial excitation enhances the absorption cross section.

The fourth step in the enrichment process (Fig. 6.10) is dissociation of the excited UF_6 with the ultraviolet laser. The ultraviolet absorption cross section increases with decreasing wavelength below about 330 nm (see Fig. 6.12). Ultraviolet excitation in this region results in a direct dissociation reaction [Lyman, 1985]

$$UF_6 \rightarrow UF_5 + F$$

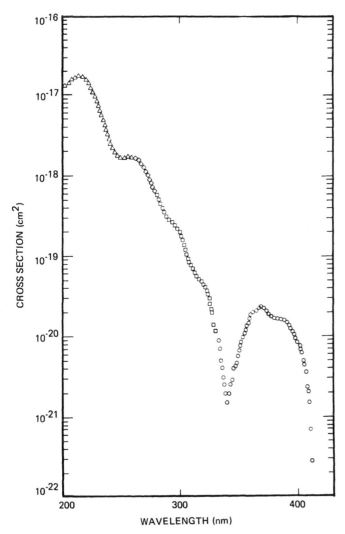

FIG. 6.12 Ultraviolet spectrum of UF_6. △, 233.7 K, 0.37 torr; ◇, 263.1 K, 7.26 torr; □, 272.5 K, 16.32 torr; ○, 296.1 K, 91.50 torr.

Vibrational excitation in the UF_6 tends to shift the ultraviolet spectrum to longer wavelengths. This property allows for preservation of the initial infrared selectivity in the dissociation process.

Because the expansion cooling takes the gas mixture below the
freezing point of the UF_6 in the flow, all of these laser processes must
occur in the short time before the supercooled gas condenses. Condensation of the UF_6 destroys the selectivity of the initial infrared excitation.

The last step in Fig. 6.10 is the dimerization (and subsequent
polymerization) of the UF_5 product. Fortunately, the dimerization
reaction

$$UF_5 + UF_5 \rightarrow (UF_5)_2$$

is much faster ($k = 1 \times 10^{-11}$ cm^3 mol^{-1} s^{-1} [Lyman, 1985]) than the
competing recombination reaction,

$$UF_5 + F \rightarrow UF_6$$

($k < 2 \times 10^{-12}$ cm^3 mol^{-1} s^{-1} [Lyman, 1985]). The dimerization reaction is just the first step in the growth of solid UF_5 particles. The
rate of polymerization tends to increase somewhat as the particles
grow. The fluorine-atom product may recombine to F_2, react with UF_5
or its higher polymers, or react with some scavenger species. A scavenger may be necessary to prevent some loss of enriched UF_5.

Following the steps outlined above and in Fig. 6.10, the gas flows
through a collector stage in the flow loop. Impact or filter collectors
then remove the enriched UF_5 particles from the gas stream.

The Los Alamos National Laboratory announced in 1976 that the
process does produce enriched uranium, but details of the enrichment
performance and economic analysis are not available.

The lasers for this process are the noble-gas/halogen lasers in
the ultraviolet and the CO_2 laser shifted to 16 μm by stimulated-Ra-
man scattering in parahydrogen. The development of both the UV
[Brau, 1975] and IR [Byer, 1978; Rabinowitz, 1978a; Kurnit, 1980]
lasers occurred during the research on this project.

6.5.2 Uranium Enrichment by IRMPD

The preceding section discussed some of the problems associated with
isotopically selective excitation of UF_6 with infrared lasers. These
problems are even more severe for infrared dissociation of UF_6. The
small infrared isotope shift, the high level of vibration excitation at
room temperature, and the loss of selectivity with increasing fluence
all tend to limit the selectivity of an infrared reaction. Several groups
[Rabinowitz, 1978b; Averin, 1979; Ambartzumian, 1979; Subbiah,
1980; Horsley, 1980a; Vasil'ev, 1980; Koren, 1981] have reported demonstrations of dissociation with an infrared laser at a single frequen-

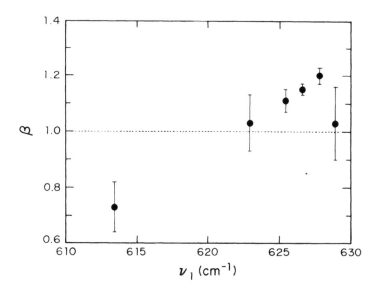

FIG. 6.13 Uranium isotopic selectivity with two infrared lasers versus the frequency of the first laser [Rabinowitz, 1982a, 1982b]. The data points are the average values for a range of fluences of the first laser. The frequency of the second laser was 598.5 cm^{-1} for the far-right point. $\nu_2 = 596.8$ cm^{-1}; $\phi_1 = 0.4$ to 3 J/cm^2; $\phi_2 = 2.4$ J/cm^2.

cy. None of these papers report an isotopically selective dissociation. In some of these papers the authors did not attempt, or at least did not report an attempt, to separate isotopes. In others the experiments indicated no isotopic selectivity in the reaction. It appears that the problems we outlined above prevent a selective dissociation.

One way around this difficulty is to use two infrared frequencies. Other groups with other species have found that two frequencies often improve the selectivity, but for UF_6 two frequencies appear to be essential for any selectivity. Table 6.11 summarizes some of the published attempts to enrich uranium by infrared dissociation of UF_6 with two laser frequencies.

For all of the experiments that showed selective reaction the species dissociation was $^{235}UF_6$. For the last entry [Rabinowitz, 1982a, 1982b], changing the frequency of the first laser frequency changed and reversed the selectivity. Figure 6.13 shows the dependence of isotopic selectivity on the first frequency. These experiments were with a molecular beam of UF_6 produced effusively. The second frequency in each case was 596.8 or 598.5 cm^{-1}. The infrared laser for both frequencies was the CO_2 laser shifted by stimulated Raman scat-

TABLE 6.11 Uranium-Enrichment Attempts by Dissociation of UF_6 with Two Infrared Frequencies

First Laser Second Laser	ν (cm^{-1}) ν (cm^{-1})	Best β	Comment	Reference
CF_4 CO_2	615 1077	—	No enrichment measurement, demonstrated large enhancement, first laser fluence -1 J/cm^2 second laser fluence -100 J/cm^2	[Tiee, 1978a]
CF_4 CO_2	615 900–1100	1.00 ± 0.3	No selectivity, observed, first laser fluence 0.1 J/cm^2, second laser fluence -100 J/cm^2	[Alimpiev, 1979]
CF_4 CO_2	620.6 1084.6	(1.18)	Selectivity estimated from luminescence and isotope shift, first laser 0.15 J/cm^2	[Alimpiev, 1981]
CF_4 CO_2	615.2 1073.3	1.05 ± 0.2	Assay by gamma spectroscopy 0.47 torr UF_6, first laser fluence 0.02 J/cm^2, second laser fluence 1 J/cm^2	[Koren, 1982]
CO_2-H_2 Raman CO_2-H_2 Raman	615.2 1073.3	$1.20 \pm .03$	Molecular beam, 50/50 isotope mix, both beams focused	[Rabinowitz,

TABLE 6.12 Uranium LIS by IRMPD with CO_2 Laser Pulses

Species	ν (cm^{-1})	Best β	Comment	Reference
U(OCH$_3$)$_6$	927.0	1.0315	Low-pressure photolysis, vapor pressure, 0.0001 torr at room temperature	[Miller, 1978]
UO$_2$(hfacac)$_2$·THF[a]	957.8	1.25	Molecular-beam enrichment, CW laser also enriches uranium, temperature ~390 K	[Cox, 1979]
Dimer[b]	938.7	1.53	Molecular-beam enrichment with CW CO_2 laser (1.8 kw/cm^2)	[Cox, 1980]
UO$_2$(hfacac)$_2$·THF	952.9	1.91	Molecular-beam enrichment at 333 K	[Horsley, 1980b]
U(OCH$_3$)$_6$	927.0	1.03	Enrichment peaks near 3.2 J/cm^2, no enrichment at 3 and 5.5 J/cm^2	[Cuellar, 1983]

[a]See Fig. 6.14.
[b][UO$_2$(hfacac)$_2$]$_2$.

tering in parahydrogen. This laser system gives a broader frequency range than the CF_4 laser used by the other groups in the earlier experiments (Table 6.11).

The first four entries in Table 6.11 used the CF_4 laser as the first laser in the frequency region of the strong ν_3 absorption band of UF_6. The low-frequency side of that band. Because of the weakness of the combination band and the large separation from the band center, dissociation with the CO_2 laser requires a high fluence. (The fluence of the second laser in the Koren paper [Koren, 1982] seems to be much too low for dissociation of UF_6.) It appears that having the second frequency in the 16-μm region is a superior technique. The selectivity listed in the third entry [Alimpiev, 1981] is not a measured quantity, but an estimate from the effect of the first laser frequency on the dissociation probability.

Most of the isotopic assays in these experiments were with a mass spectrometer, but Koren et al. [Koren, 1982] used the novel technique of gamma spectroscopy for the assay. The precesion of this technique was about 2%.

The papers summarized in Table 6.11 do not necessarily represent the optimum experimental conditions for uranium enrichment. Cooling the gas by methods such as the flow cooling discussed in the preceding section should enhance the isotopic selectivity.

The UF_6 molecule has the highest vapor pressure of any uranium compound. Several other species, however, have sufficient vapor pressure for enrichment experiments in the gas phase. Table 6.12 summarizes experiments with some of those species.

The species $U(OCH_3)_6$ has a structure similar to that of UF_6. The OCH_3 group replaces the F atom in that species. The methoxy compound has the advantage over UF_6 that it absorbs CO_2 laser radiation. However, the experiments [Miller, 1978; Cuellar, 1983] suggest that the selectivity of the reaction is extremely sensitive to the experimental conditions. For example, the reaction is selective only over a narrow fluence range [Cuellar, 1983]. Experiments with this species would probably benefit from two laser frequencies.

The Exxon group [Cox, 1979, 1980; Horsley, 1980b] have demonstrated fairly high enrichment factors with some very large uranium compounds. They have donse most of their research with uranyl hexafluoroacetylacetonate tetrahydrofuran [UO_2(hfacac)$_2$·THF]. Figure 6.14 shows the structure of that species. The dimer in Table 6.12 consists of two of those species, minus the THF group.

The infrared spectra of these species each have narrow absorption bands in the frequency region of the CO_2 laser. The infrared absorption in this region is very nearly linear. This is probably due to the very large size of the molecules. These species will even absorb sufficient energy for dissociation from CW lasers.

Because of the large size of these molecules, the lifetime for dissocation is very long, in the range of 1 ms. A higher laser intensity shortens the dissociation lifetime. The observed enrichment closely follows the selectivity estimated from the low-intensity infrared spectrum. Lowering the temperature [Cox, 1979; Horsley, 1980b] narrows the spectral features and thus improves the selectivity of the infrared absorption. The measured selectivities at 390 and 333 K show this effect (Table 6.12).

All of the experiments with these large molecules were with effusive molecular beams. The measurement of enrichment occurred before collisional processes could scramble the enrichment produced by photolysis. The problems of low volatility, expensive chemical synthesis, and collisional scrambling will probably prevent a process based on these species from developing into a high-volume enrichment plant.

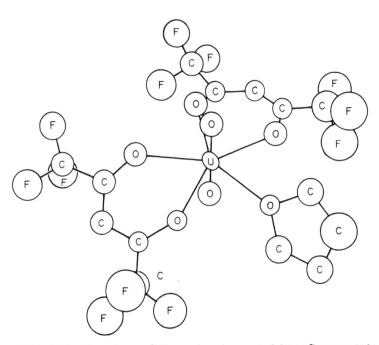

FIG. 6.14 Structure of the molecule uranyl hexafluoracetylacetonate tetrahydrofuran [Cox, 1979].

TABLE 6.13 Boron LIS by IRMPD with CO_2 Laser Pulses

Species	ν (cm^{-1})	Best β	Comment	Reference
$^{10}BCl_3$	978.5	~2	Early IRMPD demonstration,	[Ambartzumian,
$^{11}BCl_3$	938.7	~2	dissociation in O_2,	1974]
			BO emission shows selectivity	
$^{10}BCl_3$	978.5	~10	4-torr BCl_3, 20-torr air,	[Ambartzumian,
			BO emission diagnostic	
$^{10}BCl_3$	979.7	5.5	0.44-torr BCl_3, 11-torr air,	[Ambartzumian,
$^{11}BCl_3$	933.0	4.3	HBr better scavenger,	1976c]

Molecule				
$^{11}BCl_3$	944.2	3.2	0.6-torr BCl_3, 2.4-torr H_2, products degrade selectivity	[Lyman, 1976]
$^{10}BCl_3$	944.2	4	0.25-torr BCl_3, 4-torr air, wide frequency range studied	[Kolomiiskii, 1978]
$^{11}BCl_3$	944.2	3.3	0.5-torr BCl_3, 10-torr air	[Aslanidi, 1978]
$HClC = CH^{11}BCl_2$	935	1.67	3 torr, 0.6 J/cm^2,	[Jensen, 1980b]
$HClC = CH^{10}BCl_2$	986.6	1.76	$BCl_3 + H_2C_2$ products	
$HClC = CH^{10}BCl_2$	986.6	50	1 torr, 2 J/cm^2, selectivity lower at higher pressure	[Abzianidze, 1982]

6.6 LASER ENRICHMENT OF OTHER ISOTOPES

So far in this chapter we have discussed laser enrichment of isotopes of four elements: hydrogen, carbon, sulfur, and uranium. Most of the research effort on development of a commercial enrichment process has been with these elements. However, the literature contains reports of enrichment of many other isotopes. The main direction of this research has been to develop understanding of the basic photochemical processes, not to design an enrichment facility. We shall discuss some of the results of this research in this section.

Several programs are now in progress or in the planning stage that do have the goal of development of an enrichment facility. The researchers in these programs have not yet issued reports on their work. As examples of this type of activity, the Los Alamos National Laboratory has research programs on enrichment of zirconium, plutonium, and molybdenum. The different nuclear properties of isotopes of these elements provide the motivation for enrichment research. Information about the first two is restricted and the latter is in the planning stage.

6.6.1 Boron Enrichment

Much of the early work in multiple-photon excitation was with boron trichloride (BCl_3). This species has strong absorption bands in the frequency range of the CO_2 laser, and the bands for the two isotopes of boron are well separated. The bonds in this compound are strong (105 kcal/mol), and it is a small compound. Consequently, the laser fluence required for dissociation is high (29 J/cm^2) for 1% dissociation [Lyman, 1984a].

Table 6.13 summarizes some of the enrichment research with this and another boron compound, (*trans*-2-chloroethyenyl)dichloroborane (HClC = CHBCl$_2$). All of these experiments were with CO_2 laser pulses. All of the BCl_3 experiments used focused pulses. The larger compound requires a much lower fluence for dissociation, and consequently focusing was not necessary.

The first entry was the first demonstration of isotope separation by IRMPD. In that experiment the authors irradiated a sample containing a BCl_3 (50% for each boron isotope) and O_2 (as dry air) and measured the emission spectrum of the BO molecule formed int he resulting chemical reaction. The spectrum showed more intense emission from ^{11}BO when the laser frequency was in the region of the ^{11}BCl$_3$ band. When the frequency was in the region of the ^{10}BCl$_3$ band, the emission of ^{10}BO was stronger. At the time of the experiment the mechanism was not clear. The authors suggested that the infrared laser dissociated with BCl_3 and that subsequent reactions produced the electronically excited BO. Later experiments clarified the reaction mechanism and improved the selectivity.

The initial photochemical reaction is a simple bond cleavage.

$$BCl_3 \rightarrow BCl_2 + Cl$$

In the absence of a chemical scavenger for at least one of the products, they recombine to form the BCl_3. Most of the Russian experiments used dry air as a scavenger of the reaction products. Oxygen does prevent recombination of the reaction product to BCl_3, probably by reaction with the BCl_2 radical.

One of the experiments [Lyman, 1976] used hydrogen as a scavenger. This species is not completely satisfactory. The selectivity tends to decrease with the accumulation of reaction products. This indicates the participation of some product species in the reactions that scramble the boron isotopes. One of the major products of the reaction in the presence of hydrogen is $BHCl_2$. The CO_2 laser may also dissociate this species.

The isotope selectivities for BCl_3 are not as high as those for other species with similar infrared isotope shifts (carbon). Reactions that produce nonselective products and isotopic exchange ractions may be reasons for the lower selectivity. Another possibility is that most of these experiments were carried out before 1978, and techniques like those that were successful for carbon have not yet been applied to boron. The use of two infrared frequencies would probably be quite successful for BCl_3.

The last two entries in Table 6.13 suggest that boron enrichment with $HClC = CBCl_2$ may have promise for an industrial process. The fluence required for dissociation is low [Jensen, 1980b], and the selectivity is high at low pressure [Abzianidze, 1982]. Synthesis of this species is by a catalytic reaction of boron trichloride with acetylene at 150°C. The photolysis products are the starting materials, BCl_3 and C_2H_2.

Several groups have investigated boron enrichment by methods other than IRMPD. These, however, have not been very successful. These include enrichment by vibrationally enhanced chemical reactions and IR-UV photolysis. Freund and Ritter [Freund, 1975] studied boron enrichment by inducing reactions of BCl_3 and H_2S with CO_2 lasers. The reaction yield and selectivity were both very low in these experiments. The authors made some suggestions about the reaction mechanism, but it was never clarified. It is very likely that the reaction induced was dissociation (IRMPD), not a bimolecular reaction. Another early demonstration of born enrichment [Rockwood, 1974] was attributed to infrared-ultraviolet dissociation of BCl_3. Here again, the demonstrated enrichment was probably by IRMPD.

Schramm and Dize [Schramm, 1979] published an account of boron enrichment by inducing the reaction of BCl_3 with CH_4 with a CW CO_2 laser. The isotopic selectivity was low in these experiments.

6.6.2 Nitrogen, Oxygen, Silicon, and Selenium Enrichment

With the appropriate economic motivation one could probably develop
efficient enrichment methods for isotopes of other light elements. The
lighter elements generally form volatile compounds with large infrared
isotope shifts. Several groups have published accounts of enrichment
of some of these elements.

We mentioned in Sec. 6.3 the isotopically selective dissociation of
s-tetrazine with a dye laser [Karl, 1975]. Dissociation of this species
gives HCN and N_2 as products. The reported selectivity for nitrogen
was high (β = 6).

Other examples of nitrogen enrichment were by the IRMPD method
[Aslanidi, 1978; Chekalin, 1977] or a variant of that method, infrared
isomerization [Hartford, 1979]. The isotopic selectivity was low (β
< 2) for all of these experiments.

Because oxygen is a light atom, many volatile oxygen-containing
compounds would probably undergo isotopically selective photochemical
reactions. However, oxygen LIS has received very little attention.
One reason for this may be the efficient method of oxygen enrichment
by cryogenic distillation.

The carbon and sulfur sections discussed the IR-UV dissociation
of OCS [Zittel, 1983]. That technique enriches oxygen as well with
selectivities in the range 1.5 to 3.1.

High-fluence CO_2 laser pulses selectively dissociate SiF_4 [Lyman,
1976]. The Si-F bonds in this compound are very strong (140 kcal/
mol); consequently, high fluence (~9 J/cm^2) is necessary, and the
isotopic selectivity is low (β = 1.17).

Selenium has five stable isotopes with sufficient abundance for
easy detection. Tiee and Wittig [Tiee, 1978b, 1978c] have used two
methods for inducing selective reactions of these isotopes. The first
is the infrared dissociation of SeF_6 with the NH_3 laser, and the second
is the dissociation of the same species with two lasers, NH_3 and CO_2.

The SeF_6 molecule has a strong absorption band (ν_3) centered at
780 cm^{-1}. The best frequency for the NH_3 laser is 780.5 cm^{-1}. The
isotope shift for the ν_3 band is about 1.6 cm^{-1}/amu, and the absorption band has a width of about 20 cm^{-1}. One would therefore not expect the isotopic selectivity to be extremely high for this species.
The best selectivity was 1.34 for alternate masses for selenium (78 to
80) [Tiee, 1978b]. These experiments were with 0.1 to 0.5 torr of
SeF_6. The low output of the NH_3 laser (75 mJ) required focusing of
the laser beam. The use of the second laser (CO_2) enhanced the reaction yield by as much as a factor of 50. It did not, however, improve the isotopic selectivity.

6.6.3 Chlorine and Bromine

Several research groups have reported demonstrations of chlorine and
bromine enrichment by the IRMPD or CW infrared photolysis methods.

These, however, have all had very low isotopic selectivities. Considerably more development would be necessary for these to be considered for a commercial process.

The best method for enrichment of chlorine isotopes is probably one that we have not yet discussed. This method consists of laser excitation of electronic states of chlorine-containing diatomic molecules (Cl_2, ICl). These excited species then react with olefins or aromatic compounds. Table 6.14 summarizes these experiments. The excitation of ICl is superior because this species has only one chlorine atom. This improves the isotopic selectivity.

A final method for chlorine enrichment is vibrationally enhanced bimolecular reactions. Arnoldi et al. [Arnoldi, 1975] demonstrated enhancement of the rate of the reaction

$$Br + HCl \rightarrow HBr + Cl$$

by a factor of 10^{11} by vibrational excitation of HCl. These authors used an HCl chemical laser to produce vibrationally excited $H^{35}Cl$. The resulting selectivity in discharge flow reactor was 4.5.

6.6.4 Molybdenum and Osmium

The one example of molybdenum enrichment is the infrared dissociation of MoF_6 with CO_2 laser [Freund, 1978]. The molecule dissociates with very low reaction probability and low selectivity when irradiated in the region of the weak $\nu_3 + \nu_5$ combination band. Experiments at several frequencies showed preferential dissociation of the lighter isotopic species on the high-frequency side of the band and preferential dissociation of the heavier isotopic species on the low-frequency side of the band. The selectivity was in the range 1 to 2% per mass number. Both the selectivity and reaction yield would probably be much better for irradiation with a laser in the region of the strong ν_3 absorption band, or with two-frequency excitation.

Ambartzumian et al. [Ambartzumian, 1977a, 1977b] demonstrated dissociation (IRMPD) of OsO_4 with CO_2 lasers. With one infrared laser at several frequencies they observed no isotopic selectivity, but two lasers gave selectivities as high as 24% per mass number. The first frequency was within the ν_3 absorption band and the second was on the low-frequency side of the band. The pressures in these experiments were low (0.3 torr). Some experiments used chemical scavengers (OCS, C_2H_4, and NO), but they were not necessary for enrichment.

6.7 LASER SYNTHESIS AND PURIFICATION

As mentioned in the introduction, the major emphasis of this chapter is laser isotope separation. However, in this section we discuss the broader areas of laser synthesis and laser purification. The literature

TABLE 6.14 Chlorine and Bromine Enrichment by Photochemical Reaction of Diatomic Molecules

Diatomic	Reactant	Laser	Best β	Comments	Reference
ICl	trans-ClHC = CHCl	Dye (605.3 nm)	1.10	Photoexchange product is cis-ClHC = CHCl enriched in ^{37}Cl	[Liu, 1975]
ICl	BrHC = CHBr	Dye (605.3 nm)	1.33	Major photoexchange product is C_2H_2BrCl enriched in ^{37}Cl	[Liu, 1975]
Br_2	HI	Dye	1.02	Br_2 predissociation and subsequent reaction to HBr, subtracting dark reaction increases β to 1.15	[Suzuki, 1977]
ICl	C_6H_5Br	Dye (600-610 nm)	~1.5	Photoexchange product is chlorobenzene, enrichment of either chlorine isotope is possible	[Brenner, 1977]
Cl_2	C_2Cl_4	Ar ion (488 nm)	1.055	Photoaddition product is C_2Cl_6, residual Cl_2 enriched in ^{35}Cl	[Suzuki, 1978]
ICl	C_2H_2	Dye	100	Photoaddition product is C_2H_2ICl enriched in ^{37}Cl, high enrichment of ^{35}Cl also possible	[Stuke, 1978] [Stuke, 1977]

coverage in this section is not as complete as were the previous sections on isotope separation. We also omit several areas of laser photochemistry tha could possibly fit in this section, such as photochemical application in condensed phases and at surfaces. Our discussion is principally confined to examples of laser synthesis and purification that could possibly develop into industrial processes. The research and development in this area is not as extensive as that in isotope-separation research.

A recent review of laser applications [Hall, 1982] discusses some of the processes, in addition to isotope separation, that could develop into industrial processes. Hall's opinion is that none are as promising at the current time as isotope separation, but he also points out that none of them have had similar research emphasis.

Hall also mentions some of the economic constraints in these areas. The cost and complexity of any laser process make replacement of current, efficient chemical processing difficult. Industrial laser applications will come in areas that take full advantage of the laser's unique properties.

6.7.1 Laser Purification

The high costs of lasers and laser radiation become less important in several purification applications. The photochemical removal of a minor impurity keeps the purification costs low. One of the best examples of this application is the removal of the impurities arsine, phosphine, and diborane from silane by ultraviolet photolysis [Clark, 1978; Hartford, 1980a]. These species are present in most silane. They are particularly troublesome because they result in n-type impurities in semiconductor devices made from the silane.

All of these species (including silane) have high photodissociation quantum yields (<0.25) at the wavelength of the ArF excimer laser (193 nm). However, the absorption cross sections for the impurities are much greater than for silane. Table 6.15 lists the measured data on absorption cross sections for these species at the 193 nm [Hartford, 1980a; Clark, 1978]. The table lists the silane cross section and the ratio of the impurity cross section to the silane cross section. The lower temperature considerably enhances the ratio of absorption cross sections. The cross-section ratio exceeds 100 for diborane and is in the range of 10^4 for phosphine and arsine. This allows for efficient removal of these impurities, provided that the subsequent chemical reactions do not scramble the initial photochemical separation.

Fortunately, the photochemical processes are very favorable for removal of these three impurities. We shall discuss those processes for phosphine impurity in silane, but the other impurities behave similarly.

With irradiation, silane and the impurity both dissociate:

TABLE 6.15 Absorption Cross Sections for Silane, Arsine, Phosphine, and Diborane

Species	T (K)	σ (cm^2)	$\sigma/\sigma(SiH_4)$
AsH_3	295	$1.42 \pm 0.12 \times 10^{-17}$	9,500
PH_3	295	$1.0 \pm 0.12 \times 10^{-17}$	6,700
B_2H_6	295	$2.16 \pm 0.30 \times 10^{-19}$	180
SiH_4	295	$1.50 \pm 0.24 \times 10^{-21}$	1
AsH_3	198	$1.65 \pm 0.07 \times 10^{-17}$	17,000
PH_3	198	$1.08 \pm 0.10 \times 10^{-17}$	11,000
SiH_4	198	$9.66 \pm 0.73 \times 10^{-22}$	1

Sources: Clark and Anderson [Clark, 1978] and Hartford et al. [Hartford, 1980a].

$$SiH_4 \xrightarrow{193 \text{ nm}} SiH_2 + H_2$$

$$PH_3 \xrightarrow{193 \text{ nm}} PH + H_2$$

The reactive species (SiH_2, PH) react rapidly with SiH_4,

$$SiH_4 + SiH_2 \rightarrow Si_2H_6$$

$$SiH_4 + PH \rightarrow SiH_3PH_2$$

Reactions with the impurity species are much less frequent because of the lower impurity concentration. The product species then react further to form solid particles that contain photolyzed impurities. Scrambling reactions are not important.

For low impurity levels the major part of the photolysis product is still silicon, but impurity level is substantially higher than in the unphotolyzed gas. For the mechanism outlined above and for low impurity levels one can obtain approximate [Hartford, 1980a] expressions for the amount of silane and impurity remaining in the gas phase as a function of the amount of laser photolysis of the sample. The number of remaining impurity molecules is

$$n_A = n_{AO} \, \exp - \left(\frac{\phi_A \sigma_A e}{n_{SO} \sigma_S} \right)$$

and the number of silane molecules is

$$n_S = n_{SO} - 2\phi e$$

In these expressions the zero indicates initial conditions, the S subscript is silane, and the A subscript is the impurity. The symbols σ, ϕ, and e represent cross section, quantum yield, and total laser energy absorbed (photons), respectively. The available experimental data suggest that these expressions accurately represent the results of a purification with the ArF laser.

Purification of some gases is also possible by infrared photolysis of impurities. One such example is the purification of $AsCl_3$ by CO_2 laser photolysis of the impurities CCl_4 and $C_2H_4Cl_2$ [Ambartzumian, 1977c]. Photolysis of CCl_4 produced C_2Cl_4 and C_2Cl_6. The $C_2H_4Cl_2$ dissociation products were HCl, C_2H_2, and C_2H_3Cl. This type of purification lacks some of the favorable features of the UV laser purification of SiH_4. The utilization of laser energy was only about 1%, and photoproducts were gases that still required separation.

Another example of a purification technique with infrared lasers is silane purification by dielectric breakdown in silane-impurity mixtures [Freund, 1979]. The laser pulse produces very high temperatures, which in turn lead to decomposition of the molecular species (SiH_4 and B_2H_6). Because of the greater thermal stability of the silane, the diborane has a greater probability for irreversible decomposition to solid products. In experiments with 20 torr of SiH_4 and 1 torr of B_2H_6, five laser pulses removed 50% of the B_2H_6 and 12% of the SiH_4. (External heating of a sample cell to 200°C gave similar results.) From the experimental data to date, the UV photolysis technique is superior to this technique in selectivity and efficiency.

The removal of $COCl_2$ from BCl_3 by irradiation with a CW CO_2 laser is also effective [Merritt, 1977]. The mechanism for this reaction is probably also a thermal reaction.

The final example of a laser purification (or separation) process is similar to some of the isotope separation techniques we discussed above (Table 6.14) for chlorine and bromine by electronic excitation of diatomic halogen species. This process is the removal of *ortho*-I_2 molecules from an ortho-para mixture by laser-induced reaction with 2-hexane [Letokhov, 1975]. The argon laser (514.5 nm) selectivity excites *ortho*-I_2 molecules to an electronically excited state. These species then react with the olefin by predominantly a direct addition of the excited I_2 to the double bond in 2-hexane. This process is highly selective. The authors observed no decrease in the amount of *para*-I_2 in the sample. This technique would also be useful for iodine isotope separation.

6.7.2 Laser Synthesis

The literature contains many examples of experiments that could possibly lead to commercial laser processes for synthesis of chemical compounds. Improved product purity is a major advantage of many of the laser processes. Reasons for this improvement include the avoidance of hot walls in the reaction chamber for high-temperature synthesis and the termination of chemical reactions with the depletion of the absorbing reagent. The most promising of these are the laser production of vitamin D [Malatesta, 1981a] and laser synthesis of vinyl chloride [Wolfrum, 1981]. Both of these processes require electronic excitation of the absorbing species.

Malatesta et al. [Malatesta, 1981a] demonstrated high yield (80%) of previtamin D from 7-dehydrocholesterol in a two-step process. They used a KrF laser (248 nm) followed by a nitrogen laser (331 nm) for this synthesis. Single lasers or other combinations of frequencies gave lower yields.

The synthesis of vinyl chloride is a high-volume industry [Hall, 1982]. Laser initiation of this process could improve the product purity and the efficiency of synthesis of ths important industrial compound. The synthesis involves initiating a chain reaction in the starting material, $C_2H_4Cl_2$, by photolysis with a KrF laser,

$$C_2H_4Cl_2 + h\nu \rightarrow C_2H_4Cl + Cl$$

Subsequent reaction of these products produce C_2H_3Cl by a chain mechanism. The quantum yield for production of the vinyl chloride is about 20,000 at 500°C (the conventional synthesis temperature) and about 5000 at 300°C. The selectivity for production of vinyl chloride versus other species is actually better at the lower temperature (99% versus 85%). The product yield at the lower temperature is about the same as the conventional process (60%) but is somewhat higher at the higher temperature (85%).

Another synthesis process that is similar to this one is the laser polymerization of such species as styrene, N-vinylpyrrolidone, vinyl acetate, and acrylonitrile [Kuhlmann, 1977]. These processes involve initiation of polymerization chain reactions by either UV photolysis of the monomer species or by excitation of sensitizers to triplet states [Kuhlmann, 1977]. The sensitizers (benzophenone and related compounds) generally improved the polymerization efficiency.

One interesting chemical synthesis involves the laser-induced initiation of a heterogeneous reaction [Betteridge, 1979]. This reaction is KrF laser (248 nm) initiation of the reaction of ethylene with solid sulfur. The product of this reaction is ethylene episulfide (C_2H_4S). The laser irradiates the solid sulfur surface, and the yield increases with the 3.5 power of the laser intensity. The ethylene pressure for maximum yield is about 150 torr. The yield is lower at both higher and lower pressures. The suggested reaction mechanism is that the laser pulse evaporates the sulfur as S_8. It then dissociates those species to S_2 and excites the S_2 to electronic states. These excited S_2 molecules then dissociate unless quenched by sollisional processes. Irradiation with visible lasers (ruby) or infrared lasers (CO_2) do not give any of the episulfide product. This process is a synthesis of a chemically important compound from very inexpensive starting materials.

A number of groups have published accounts of chemical synthesis with infrared lasers, but we are not aware of any process that is near industrial implementation. Some of these reported syntheses use IRMPD, as in the examples we discussed earlier for isotope separation. Others use less selective processes, such as laser-induced breakdown or laser heating with CW or pulsed sources. Table 6.16 summarizes some of these published syntheses.

The synthesis of $BHCl_2$ shows several examples of advantages of the laser process. The reaction of BCl_3 with H_2 will occur without laser excitation at high temperature, but the reaction yields are generally lower because the product (BHCl2) is less stable than the reactant (BCl3). The lasers (CW or pulsed) provide local heating or, in the case of pulsed laser, photolysis of the BCl_3. The reaction can then proceed without hot cell walls. The heated gas can then cool rapidly and thus preserve the product.

Another example of this type of synthesis is the production of cylcobutene from cyclobutyl acetate (Nguyen, 1981]. The thermal reaction of this species produces butadiene and no cyclobutene. This is probably a true photochemical processes. The reaction of the laser-excited reactant must occur in an enrvironment that does not lead to ring opening of the cyclobutene. This ring opening occurs at a lower temperature than the thermal reaction of the cyclobutyl acetate.

TABLE 6.16 Laser Synthesis of Chemical Compounds with Infrared Sources

Compound	Reactants	Laser	Results	Reference
cis-$C_2H_2Cl_2$	trans-$C_2H_2Cl_2$	CO_2 pulsed	Isomerization best with no diluent, some dissociation occurs	[Ambartzumian, 1976d]
$BHCl_2$	BCl_3/H_2	CO_2 pulsed	Product purity ~100%	[Rockwood, 1975]
$BHCl_2$	BCl_3/H_2	CO_2 CW	He reduces yield, Xe improves yield	[Sazonov, 1983]
$HB(SH)_2$	B_2H_6/H_2S	CO_2 CW	Products do not form thermally	[Bachmann, 1976]

COF_2	CO_2/SF_6	CO_2 pulsed	Product purity ~100% by breakdown at 10–50 torr	[Malatesta, 1981b]
$B_{10}H_{14}$	B_2H_6/B_5H_9	CO_2 pulsed	65% yield, 1400 photons/molecule	[Hartford, 1980b]
Cyclobutene	Cyclobutyl acetate	CO_2 pulsed	85% yield with CF_4, synthesis impossible by thermal methods	[Nguyen, 1981]
Cyclopentadiene	Cyclopentene	CO_2 pulsed	Yield exceeds thermodynamic equilibrium	[Carr, 1982]
$(CF_3)_3CI$	$(CF_3)_3CBr/I_2$	CO_2 pulsed	70% yield, CF_3I is a major product	[Bagratashvili, 1983]

Excitation with an intense source produces a rapid reaction, and a diluent (CF_4 in this case) keeps the temperature low after the laser
pulse. The synthesis of the mercaptoboranes [Bachmann, 1976] is a
similar process.

The laser synthesis of the large borane compound ($B_{10}H_{14}$) occurs
with a much higher yield than the thermal synthesis, and the synthesis is reasonably efficient when one considers that each $B_{10}H_{14}$ molecule requires many reactant molecules. One advantage of the laser
synthesis is that it is generally possible to select a laser frequency
that excites the reactant species and not the product species. Thus,
when the reaction is complete the gas sample no longer reaches a
high temperature.

The synthesis of cyclopentadiene from cyclopentene [Carr, 1982]
shows that one can drive a chemical reaction beyond thermodynamic
equilibrium by producing high levels of vibrational excitation at a
low translational temperature.

The synthesis of $(CF_3)_3CI$ occurs by IRMPD of the bromine compound in the presence of I_2 [Bagratashvili, 1983]. The reaction
mechanism is complex, and the optimum synthesis depends on many
experimental conditions.

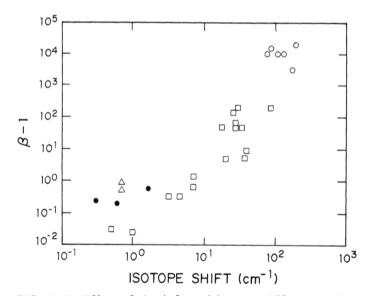

FIG. 6.15 Effect of the infrared isotope shift on the isotopic selectivity ($\beta = 1$ is no selectivity). The data points are from experiments
with many species. ○, lower limits; □, one frequency; ●, two frequencies; △ molecular beam.

TABLE 6.17 Summary of LIS Methods with Industrial Potential

Isotope	Method	Reference
Deuterium	HDCO UV predissociation	[Mannik, 1979]
Deuterium	CDF_3 IRMPD	[Herman, 1980; Moser, 1983]
Tritium	CTF_3 IRMPD	[Takeuchi, 1981; Makide, 1982a]
Tritium	$CTCl_3$ IRMPD	[Magnotta, 1982; Herman, 1984]
Carbon-13	CF_3I IRMPD	[Abdushelishvili, 1982; Cauchetier, 1983]
Carbon-13	CF_2HCl IRMPD	[Gauthier, 1982a, 1982b; Hackett, 1984]
Sulfur-33, 34, 36	SF_6 IRMPD	[Baranov, 1979b; McInteer, 1982; Fuss, 1979]
Uranium-235	UF_6 IR-UV dissociation	[Jensen, 1982]

The COF_2 synthesis [Malatesta, 1981b] from CO_2 and SF_4 mixture is an example of a high-purity synthesis with a very nonselective (laser-induced breakdown) excitation. The laser-induced breakdown does not require resonant excitation. It creates a very high temperature plasma, after which the atomic and radical fragments recombine to stable product. The only observed species that contained carbon was the desired product, COF_2. The thermal synthesis of this species requires high temperatures of SF_4/CO_2 mixtures. and the reaction gives only 10% COF_2. The major thermal product is CF_4.

6.8 SUMMARY

To conclude this chapter Table 6.17 contains a list of laser processes for isotope separation that appear to have sufficient promise for industrial application. These processes would require a stable market, and in some cases, some laser development. This list is somewhat subjective, and others could be added with some additional research and interest.

Other processes besides isotope separation that have economic promise include the laser purification of silane, the synthesis of vitamin D, and the laser production of vinyl chloride.

The many papers on isotope separation by IRMPD suggest that this is a very general technique. The maximum possible selectivity by this processes depends strongly on the infrared isotope shift. Figure 6.15 shows this dependence. It shows the isotopic selectivity from experiments summarized above versus the infrared isotope shift. The figure also shows that a second laser frequency can often improve the selectivity, especially when the shift is small.

ACKNOWLEDGMENTS

Work performed at the University of California, Los Alamos National Laboratory, Los Alamos, N. M., under the auspices of the U. S. Department of Energy under Contract W-7405-ENG-36.

The generous and prompt response from many authors for reprints and prepublication copies of papers and research reports is greatly appreciated.

REFERENCES

[Abdushelishvili, 1981] G. I. Abdushelishvili, O. N. Avatkov, A. B. Bakhtadze, V. M. Vetsko, G. I. Tkeshelashvili, V. I. Tomilina, V. N. Fedoseev, and Y. R. Kolomiiskii. Selective dissociation of CF_3I in a CO_2-laser infrared field in the presence of acceptors, Sov. J. Quantum Electron., 11, 326.

[Abdushelishvili, 1982] G. I. Abdushelishvili, O. N. Avatkov, V. N. Bagratashvili, V. Y. Baranov, A. B. Bakhtadze, E. P. Velikhov, V. M. Vetsko, I. G. Gverdtsiteli, V. S. Dolzhikov, G. G. Esadze, S. A. Kazakov, Y. R. Kolomiiskii, V. S. Letokhov, S. V. Pigul'skii, V. D. Pis'mennyi, E. A. Ryabov, and G. I. Tkeshelaskvili. Isotope separation by multiphoton dissociation of molecules using high-power CO_2 laser radiation. Scaling of the process for carbon isotopes, Sov. J. Quantum Electron., 12, 459 (1982).

[Abzianidze, 1982] T. G. Abzianidze, G. I. Abdushelishvili, A. B. Bakhtadze, I. G., Gverdtsiteli, A. V. Kaminski, A. G. Kudziev, G. I. Tkeshelashvili, and T. B. Tsinadze. Isotope effects in many-photon dissociation of dichloroborane, Sov. Tech. Phys. Lett., 8, 530 (1982).

[Abzianidze, 1981]   T. G. Abzianidze, A. S. Egiazarov, A. K. Petrov, and Y. N. Samsonov.   Isotopically selective dissociation of small molecules exposed to pulsed CO_2 laser radiation, *Sov. J. Quantum Electron.*, *11*, 343 (1981).

[Alexandre, 1983]   M. Alexandre, M. Clerc, R. Gagnon, M. Gilbert, P. Isnard, P. Nectoux, P. Rigny, and J.-M. Weulersse.   Dissociation D'UF_6 par un rayonnement UV en presence d'une excitation vibrationnelle, *J. Chim. Phys.*, *80*, 331 (1983).

[Alimpiev, 1979]   S. S. Alimpiev, A. P. Babichev, G. S. Baronov, N. V. Karlov, A. I. Karchevskii, S. Y. Kulikov, V. L. Martsynk'yan, S. S. Naviev, S. M. Nikiforov, A. M. Prokhorov, B. G. Sartakov, E. P. Skvortsova, and E. M. Khokhlov.   Dissociation of uranium hexafluoride molecules in a two-frequency infrared laser field, *Sov. J. Quantum Electron.*, *9*, 1263 (1979).

[Alimpiev, 1981]   S. S. Alimpiev, N. V. Karlov, S. S. Naviev, S. M. Nikiforov, A. M. Prokhorov, and B. G. Sartakov.   Investigation of the spectral dependences of the excitation of high vibrational levels of the UF_6 molecule, *Sov. J. Quantum Electron.*, *11*, 375 (1981).

[Alimpiev, 1982]   S. S. Alimpiev, B. O. Zikrin, B. G. Sartakov, and E. M. Khokhlov.   Excitation and dissociation of SF_6 molecules in a two-frequency infrared laser field, *Sov. Phys. JETP*, *56*, 943 (1982).

[Ambartzumian, 1972a]   R. V. Ambartzumian, and V. S. Letokhov.   Photoionization of atoms and photodissociation of molecules by laser radiation, *Appl. Optics*, *11*, 354 (1972).

[Ambartzumian, 1972b]   R. V. Ambartzumian, V. S. Letokhov, G. N. Makarov, and A. A. Puretsky.   "Two-Step photoionization of NH_3 molecules by laser radiation," *JETP Lett.*, *15*, 501.

[Ambartzumian, 1973]   R. V. Ambartzumian, V. S. Letokhov, G. N. Makarov, and A. A. Puretsky.   Laser separation of nitrogen isotopes, *JETP Lett.*, *17*, 63 (1973).

[Ambartzumian, 1974]   R. V. Ambartzumian, V. S. Letokhov, E. A. Ryabov, and N. V. Chekalin.   Isotopic selective chemical reaction of BCl_3 molecules in a strong infrared laser field, *JETP Lett.*, *20*, 273 (1974).

[Ambartzumian, 1975a]   R. V. Ambartzumian, U. M. Apatin, V. S. Letokhov, and V. I. Mishin.   High-power tunable narrow-band ultraviolet, visible, and infrared neodymium-glass laser for use in selective photochemistry, *Sov. J. Quantum Electron.*, *5*, 191 (1975).

[Ambartzumian, 1975b] R. V. Ambartzumian, Y. A. Gorokhov, V. S. Letokhov, and G. N. Makarov. Separation of sulfur isotopes with enrichment coefficient $<10^3$ through action of CO_2 laser radiation on SF_6 molecules, *JETP Lett.*, *21*, 171 (1975).

[Ambartzumian, 1975c] R. V. Ambartzumian, V. S. Dolzhikov, V. S. Letokhov, E. A. Ryabov, and N. V. Chekalin. Investigation of the dissociation of BCl_3 molecules in the field of an intense CO_2 laser pulse, *Sov. Phys. JETP*, *42*, 36 (1975).

[Ambartzumian, 1976a] R. V. Ambartzumian, Y. A. Gorokhov, V. S. Letokhov, G. N. Makarov, and A. A. Puretsky. Isotope-selective dissociation of CCl_4 molecule by excitation of composite vibrational bands by intense infrared fields, *Phys. Lett.*, *A56*, 183 (1976).

[Ambartzumian, 1976b] R. V. Ambartzumian, N. P. Furzikov, Y. A. Gorokhov, V. S. Letokhov, G. N. Makarov, and A. A. Puretsky. Selective dissociation of SF_6 molecules in a two-frequency infrared laser field, *Opt. Commun.*, *18*, 517 (1976).

[Ambartzumian, 1976c] R. V. Ambartzumian, Y. A. Gorokhov, V. S. Letokhov, G. N. Makarov, E. A. Ryabov, and N. V. Chekalin. Separation of $B10$ and $B11$ isotopes in a strong infrared CO_2 laser radiation field, *Sov. J. Quantum Electron.*, *5*, 1196 (1976).

[Ambartzumian, 1976d] R. V. Ambartzumian, N. V. Chekalin, V. S. Doljikov, V. S. Letokhov, and V. N. Lokhman. Selective trans-cis isomerization of the $C_2H_2Cl_2$ molecule in an intense infrared laser field, *Opt. Commun.*, *18*, 400 (1976).

[Ambartzumian, 1977a] R. V. Ambartzumian, Y. A. Gorokhov, G. N. Makarov, A. A. Puretsky, and N. P. Furzikov. Separation of osmium isotopes by dissociation of the OsO_4 in a two-frequency laser field, *Sov. J. Quantum Electron.*, *7*, 904 (1977).

[Ambartzumian, 1977b] R. V. Ambartzumian, N. P. Furzikov, Y. A. Gorokhov, V. S. Letokhov, G. N. Makarov, and A. A. Puretsky. Isotope-selective dissociation of the OsO_4 molecule by two pulses of infrared radiation at different frequencies, *Opt. Lett.*, *1*, 22 (1977).

[Ambartzumian, 1977c] R. V. Ambartzumian, Y. A. Gorokhov, S. L. Grigorovich, V. S. Letokhov, G. G. Makarov, Y. A. Malinin, A. Puretsky, E. P. Filippov, and N. P. Furzikov. Purification of materials in the gaseous phase by infrared laser radiation, *Sov. J. Quantum Electron.*, *7*, 96 (1977).

[Ambartzumian, 1978] R. V. Ambartzumian, N. P. Furzikov, V. S. Letokhov, A. P. Dyad'kin, A. Z. Grasyuk, and B. I. Vasil'yev.

Isotopically selective dissociation of CCl_4 molecules by NH_3 laser radiation, *Appl. Phys.*, *15*, 27 (1978).

[Ambartzumian, 1979] R. V. Ambartzumian, V. M. Apatin, N. G. Basov, A. Z. Grasyuk, A. P. Dyad'kin, and N. P. Furzikov. Dissociation of uranium hexafluoride by laser radiation, *Sov. J. Quantum Electron.*, *9*, 1546 (1979).

[Arnoldi, 1975] D. Arnoldi, K. Kaufmann, and J. Wolfrum. Chemical-laser-induced isotopically selective reaction of HCl, *Phys. Rev. Lett.*, *34*, 1597 (1975).

[Aslanidi, 1978] A. S. Aslanidi, A. B. Baktadze, K. V. Baiadze, B. I. Zainullin, M. N. Kerner, and Y. S. Turishchev. Separation of isotopes of carbon, boron, and nitrogen in a strong infrared laser field, *Soobshch. Akad. Nauk Gruz, SSR*, *90*, 573 (1978).

[Averin, 1979] V. G. Averin, A. P. Babichev, G. S. Baronov, A. I. Karchevskii, N. S. Krasnikov, A. Y. Kulikov, A. V. Merzlyakov, A. V. Morozov, A. I. Pisanko, and E. P. Skvortsova. Investigation of the dissociation of uranium hexafluoride gas at room temperature in pulsed infrared and ultraviolet laser fields, *Sov. J. Quantum Electron.*, *9*, 1565 (1979).

[Bachmann, 1976] H.-R. Bachmann, F. Bachmann, K. L. Kompa, H. Noth, and R. Rinck. Mercaptoborane aus der Reaktion von Diboran mit Schwefelwasserstoff, *Chem. Ber.*, *109*, 331 (1976).

[Bagratashvili, 1978] V. N. Bagratashvili, V. S. Dolzhikov, V. S. Letokhov, and E. A. Ryabov. Selective isotope dissociation of high-pressure CF_3I with a pulsed CO_2 laser, *Sov. Tech. Phys. Lett.*, *4*, 475 (1978).

[Bagratashvili, 1979a] V. N. Bagratashvili, V. S. Doljikov, V. S. Letokhov, A. A. Makarov, E. A. Ryabov, and V. V. Tyakht. Multiphoton IR excitation and dissociation of CF_3I: the experimental model, *Zh. ksp. Teor. Fiz.*, *77*, 2238 (1979).

[Bagratashvili, 1979b] V. N. Bagratashvili, V. S. Doljikov, V. S. Letokhov, and E. Ryabov. Isotopic selectivity of IR laser photodissociation of CF_3I molecules, *Appl. Phys.*, *20*, 231 (1979).

[Bagratashvili, 1983] V. N. Bagratashvili, N. Burimov, M. V. Kuzmin, V. S. Letokhov, and A. P. Sviridov. Multiple Photon IR dissociation of $(CF_3)_3CBr$ and synthesis of $(CF_3)_3CCl$ in laser-radical chemical reactions, *Laser Chem.*, *1*, 133 (1983).

[Baklanov, 1980] T. G. Baklanov, A. S. Egiazarov, A. K. Petrov, and Y. N. Samsonov. Chemical cascading and desired isotope production by laser separation of carbon isotopes, *Soobsch. Akad. Nauk. Gruz. SSR*, *98*, 934 (1980).

[Baldwin, 1981] A. C. Baldwin, and H. van den Bergh. Collisional energy transfer from excited polyatomic molecules produced by infrared multiple photon absorption, *J. Chem. Phys.*, *74*, 1012 (1981).

[Baranov, 1979a] V. Y. Baranov, E. P. Velikhov, Y. R. Kolomiiskii, V. S. Letokhov, V. G. Niz'ev, V. D. Pis'mennyi, and E. A. Ryabov. Isotope separation by multiphoton dissociation of molecules with high-power CO_2 laser radiation: IV. Enrichment with ^{33}S by irradiation of cooled SF_6 gas, *Sov. J. Quantum Electron.*, *9*, 621 (1979).

[Baranov, 1979b] V. Y. Baranov, E. P. Velikhov, S. A. Kazakov, Y. R. Kolomiiskii, V. S. Letokhov, V. D. Pis'mennyi, E. A. Ryabov, and A. I. Starodubtsev, Isotope separation by mutliphoton dissociation of molecules with high-power CO_2 laser radiation: III. Investigation of the process for sulfur isotopes and SF_6 molecules, *Sov. J. Quantum Electron.*, *9*, 486 (1979).

[Bazhin, 1974] H. M. Bazhin, G. I. Skubnevskaya, N. I. Sorokin, and Y. N. Molin. Photochemical separation of the isotopes H and D in an H_2CO-D_2CO mixture by the isotopic-filtration method, *JETP Lett.*, *20*, 18 (1974).

[Becker, 1981] F. S. Becker. Contribution to the clarification of the future prospects of uranium isotope separation using lasers, *Max-Planck-Inst. Quantenopt.*, *MPQ 1981*, 44 (1981).

[Becker, 1982] F. S. Becker, and K. L. Kompa. The practical and physical aspect of uranium isotope separation with lasers, *Nucl. Technol*, *58*, 329 (1982).

[Betteridge, 1979] D. R. Betteridge, and J. T. Yardley. Production of ethylene episulfide from laser irradiation of sulfur surfaces in the presence of ethylene, *Chem. Phys. Lett.*, *62*, 570 (1979).

[Bigeisen, 1983] J. Bigeleisen, W. B. Hammond, and S. Tuccio. Feed cycles for the industrial production of heavy water through multiphoton dissociation of fluoroform, *Nucl. Sci. Eng.*, *83*, 473 (1983).

[Bittenson, 1977] S. Bittenson, and P. L. Houston. Carbon isotope separation by multiphoton dissociation of CF_3I, *J. Chem. Phys.*, *67*, 4819 (1977).

[Bloembergen, 1975] N. Bloembergen. Comments on the dissociaton of polyatomic molecules by intense 10.6 µm radiation, *Opt. Commun.* *15*, 416 (1975).

[Bloembergen, 1978] N. Bloembergen and E. Yablonovitch. Infrared-laser-induced unimolecular reactions, *Phys. Today, 33*(5), 23 (1978).

[Borsella, 1980] E. Borsella, R. Fantoni, A. Giardini-Guidoini, and G. Sanna. Effects of wavelength, pressure and scavengers on the isotopically selective multiphoton dissociation of C_2F_5Cl, *Chem. Phys. Lett., 72*, 25 (1980).

[Borsella, 1981] E. Borsella, R. Fantoni, and A. Giardini-Guidoni. Enhancement of multiple-photon dissociation irradiating C_2F_5Cl by two IR lasers, *Chem. Phys. Lett., 84*, 313 (1981).

[Borsella, 1983] E. Borsella, C. Clementi, R. Fantoni, A. Giardini-Guidoni, and A. Palucci. Isotopic enrichment in laser-induced multiphoton dissociation. ^{13}C separation from natural CF_3Br in a flow system, *Nuovo Cimento, 73A*, 364 (1983).

[Brau, 1975] C. A. Brau and J. J. Ewing. 354 nm laser action on XeF, *Appl. Phys. Lett., 27*, 435 (1975).

[Brenner, 1977] D. M. Brenner, S. Datta, and R. N. Zare. Laser isotope separation. Photochemical scavenging of chlorine-37 by bromobenzene, *J. Am. Chem. Soc., 99*, 4544 (1977).

[Brunner, 1977] F. Brunner, T. P. Cotter, K. L. Kompa, and D. Prooch. Collisionless CO_2 laser-induced photodissociation of sulfur hexafluoride in a molecular beam experiment, *J. Chem. Phys., 67*, 1547 (1977).

[Byer, 1978] R. L. Byer and W. R. Trutna. 16-μm generation by CO_2-pumped rotational Raman scattering in H_2, *Opt. Lett., 3*, 144 (1978).

[Cantrell, 1979] C. D. Cantrell, S. M. Freund, and J. L. Lyman. Laser-induced chemical reactions and isotope separation in *Laser Handbook, Vol. 3* (M. L. Stich, ed.), North-Holland, Amsterdam, 1979, p. 485.

[Carr, 1982] R. J. Carr, Jr., and J. O. Shoemaker. Enhancement of chemical reaction yields by laser induced multiphoton processes beyond chemical equilibrium, *Chem. Eng. Commun., 19*, 91 (1982).

[Cauchetier, 1982] M. Cauchetier, M. Luce, and C. Angelie. Isotopic selectivity in the sensitized dissociation of SF_5Cl and CF_3I by multiphoton excitation of SF_6, *Chem. Phys. Lett., 88*, 146 (1982).

[Cauchetier, 1983] M. Cauchetier, O. Croix, M. Luce, and S. Tistchenko. Preparation d'enchantillons tres enriches en ^{13}C par dis-

sociation multiphotonique de $^{12}CF_3I$ au moyen d'un laser CO_2
TEA, *note CEA-N-2348*, Centre D'Etudes Nucleaires De Saclay,
May 1983.

[Chekalin, 1977] N. V. Chekalin, V. S. Dolzhikov, Y. R. Kolomisky,
V. S. Letokhov, V. N. Lokhman, and E. A. Ryabov. Experimental selection of molecules for isotope separation by multiple-photon
dissociation in an intense IR field, *Appl. Phys.*, *13*, 311 (1977).

[Chien, 1976] K.-R. Chien and S. H. Bauer. Laser augmented decomposition: 2. D_3BPF_3, *J. Phys. Chem.*, *80*, 1405 (1976).

[Chau, 1981] J.-S. J. Chou and E. R. Grant. Enhanced isotope
separation CF_2Cl_2 by infrared multiphoton dissociation at elevated
temperatures, *J. Chem. Phys.*, *74*, 5679 (1981).

[Clark, 1975] J. H. Clark, Y. Haas, P. L. Houston, and C. B.
Moore. Carbon isotope separation by tunable-laser predissociation of formaldehyde, *Chem. Phys. Lett.*, *35*, 82 (1975).

[Clark, 1978] J. H. Clark and R. G. Anderson. Silane purification
via laser-induced chemistry, *Appl. Phys. Lett.*, *32*, 46 (1978).

[Cox, 1979] D. M. Cox, R. B. Hall, J. A. Horsley, G. M. Kramer,
P. Rabinowitz, and A. Kaldor. Isotope selectivity of infrared la-
ser-driven unimolecular dissociation of a volatile uranyl compound,
Science, *205*, 390 (1979).

[Cox, 1980] D. M. Cox and E. T. Maas Jr. Isotope selective infrared
laser-induced unimolecular dissociation of dimeric bis(1,1,1,5,5,5-
hexafluoropentane-2,4-dionato)dioxouranium(VI), *Chem. Phys.
Lett.*, *71*, 330 (1980).

[Cuellar, 1982] E. A. Cuellar, S. S. Miller, T. J. Marks, and
E. Weitz. Chemistry, spectroscopy, and isotope-selective infrared photochemistry of a volatile uranium compound tailored for 10-
μm absorption: $U(OCH_3)_6$, *J. Am. Chem. Soc.*, *105*, 4580 (1983).

[Danen, 1981] W. C. Danen and J. C. Jang. Multiphoton infrared
excitation and reaction of organic compounds, *in Laser-Induced
Chemical Processes* (J. I. Steinfeld, ed.), Plenum Press, New
York, 1981, p. 45.

[DePoorter, 1975] G. L. DePoorter and C. K. Rofer-DePoorter. The
absorption spectrum of UF_6 from 2000 to 4200 Å, *Spectrosc. Lett.*,
8, 521 (1975).

[DeSilvestri, 1981] S. De Silvestri, O. Svelto, and F. Zaraga. Pho-
tophysical and photochemical properties of gaseous UF_6, *in Proc.
Int. School Phys. "Enrico Fermi,"* Course LXXIV, Developments
in High-Power Lasers and Their Applications, (C. Pellegrini, ed.),
North-Holland, Amsterdam, 1981 p. 310.

[Diebold, 1977] G. J. Diebold, F. Engelke, D. M. Lubman, J. C. Whitehead, and R. N. Zare. Infrared multiphoton dissociation of SF_6 in a molecular beam: observation of F atoms by chemi-ionization detection, *J. Chem. Phys.*, 67, 5407 (1977).

[Doljikov, 1981] V. S. Doljikov, Y. R. Kolomisky, and E. A. Ryabov. Effect of rotational relaxation on isotopic selectivity of IR multiphoton dissociation, *Chem. Phys. Lett.*, 80, 433 (1981).

[Duperrex, 1981] R. Duperrex and H. van den Bergh. Infrared multiphoton energy deposition in CF_2HCl, *J. Chem. Phys.*, 75, 3371 (1981).

[Eerkens, 1976] J. W. Eerkens. Spectral considerations in the laser isotope separation of uranium hexafluoride, *Appl. Phys.*, 10, 15 (1976).

[Evans, 1979] D. K. Evans, R. D. McAlpine, and F. K. McClusky. Laser isotope separation and the multiphoton decomposition of formaldehyde using a focused DF laser: the effect of single- or multi-line irradiation, *Chem. Phys. Lett.*, 65, 226 (1979).

[Evans, 1978] D. K. Evans, D. McAlpine, and F. K. McClusky. Laser isotope separation and the multiphoton dissociation of formic acid using a pulsed HF laser, *Chem. Phys.*, 32, 81 (1978).

[Evans, 1982] D. K. Evans, R. D. McAlpine, and H. M. Adams. The multiphoton absorption and decomposition of fluoroform-d: laser isotope separation of deuterium, *J. Chem. Phys.*, 77, 3551 (1982).

[Francisco, 1982] J. S. Francisco, W. D. Lawrence, J. I. Steinfeld, and R. G. Gilbert. Infrared multiphoton decompositon and energy-dependent absorption cross section of chloroethane, *J. Phys. Chem.*, 86, 724 (1982).

[Freund, 1975] S. M. Freund and J. J. Ritter. CO_2 TEA laser-induced photochemical enrichment of boron isotopes, *Chem. Phys. Lett.*, 32, 255 (1975).

[Freund, 1978] S. M. Freund and J. L. Lyman. Multiple-photon isotope separation in MoF_6, *Chem. Phys. Lett.*, 55, 435 (1978).

[Freund, 1979] S. M. Freund and W. C. Danen. Removal of diborane from silane by infrared laser induced dielectric breakdown and external heating, *Inorg. Nucl. Chem. Lett.*, 15, 45 (1979).

[Fuss, 1979] W. Fuss and W. E. Schmid. Praeparative und analytische Anwendung der isotopenselectiven Vielfotondissoziation, *Ber. Bunsenges Phys. Chem.*, 83, 1148 (1979).

[Fuss, 1982] W. Fuss. Fundamental and overtone infrared spectra of CF$_3$I, *Spectochim. Acta.*, *A38*, 829 (1982).

[Galbraith, 1981] H. W. Galbraith and J. R. Ackerhalt. Vibrational excitation in polyatomic molecules, in *Laser-Induced Chemical Processes* (J. I. Steinfeld, ed.), Plenum Press, New York, 1981 p. 1.

[Gandini, 1977] A. Gandini, C. Willis, and R. A. Back. Hydrogen/Deuterium selectivity in the infrared laser photolysis of chloroethylene, *Can. J. Chem.*, *55*, 4156 (1977).

[Gauthier, 1980] M. Gauthier, C. Willis, and P. A. Hackett. Etude paramétrique de l'enrichissement in carbone-13 dans la décomposition polyphotonique de CF$_3$Cl, *Can. J. Chem.*, *58*, 913 (1980).

[Gauthier, 1982a] M. Gauthier, C. G. Cureton, P. A. Hackett, and C. Willis. Efficient production of ^{13}C$_2$F$_4$ in the infrared laser photolysis of CHClF$_2$, *Appl. Phys.*, *B28*, 43 (1982).

[Gauthier, 1982b] M. Gauthier, P. A. Hackett, and C. Willis. Approaching world production rate in the CO$_2$ laser separation of ^{13}C, *Synth. Appl. Isot. Labeled Compounds*, Proc. Int. Conf., (W. F. Duncan and A. B. Susan, eds.), Kansas City, Mo, June 6-11, 1982, p. 413.

[Gelbart, 1980] W. M. Gelbart, M. L. Elbert, and D. F. Heller. Photodissociation of the formaldehyde molecule: does it or doesn't it? *Chem. Rev.*, *80*, 403 (1980).

[Grunwald, 1978] E. Grunwald, D. F. Dever, and P. M. Keehn. *Megawatt Infrared Laser Chemistry*, Wiley, New York, 1978, p. 107.

[Hackett, 1981a] P. A. Hackett, M. Gauthier, W. S. Nip, and C. Willis. Kinetic study of infrared multiphoton dissociation. Two frequency irradiation of CF$_3$^{13}COCF$_3$ molecules at natural abundance, *J. Phys. Chem.*, *85*, 1147 (1981).

[Hackett, 1981b] P. A. Hackett, V. Malatesta, W. S. Nip, C. Willis, and P. B. Corkum. Intensity and pressure effects in infrared multiphoton dissociation. Photolysis of hexafluoroacetone and trifluoromethyl bromide with 2-ns laser pulses, *J. Phys. Chem.*, *85*, 1152 (1981).

[Hackett, 1984] P. A. Hackett, M. Gauthier, and A. J. Alcock. Viable commercial ventures involving laser chemistry productions: two medium scale processes, *Proc. SPIE 458*, 65 (1984).

[Hall, 1982] R. B. Hall. Lasers in industrial chemical synthesis, *Laser Focus*, *18*(9), 57 (1982).

[Harrison, 1980] R. G. Harrison and S. R. Butcher. Multiple photon infrared processes in polyatomic molecules, *Contemp. Phys.*, *21*, 19 (1980).

[Hartford, 1979] A. Hartford, Jr., and S. A. Tuccio. Nitrogen isotope enrichment via infrared laser induced isomerization of methyl and ethyl isocyanide, *Chem. Phys. Lett.*, *60*, 431 (1979).

[Hartford, 1980a] A. Hartford, Jr., E. J. Huber, J. L. Lyman, and J. H. Clark. Laser purification of silane: impurity reduction to the sub-part-per-million level, *J. Appl. Phys.*, *51*, 4471 (1980).

[Hartford, 1980b] A. Hartford, Jr., and J. H. Atencio. CO_2-laser-driven reactions of B_2H_6 and B_2H_6/B_5H_9 mixtures, *Inorg. Chem.*, *19*, 3060 (1980).

[Hason, 1982] A. Hason, P. Gozel, and H. van den Bergh. Deuterium isotope separation in the infrared multiphoton dissociation of dichlorofluoromethane, *Helv. Phys. Acta*, *55*, 187 (1982).

[Heller, 1980] D. F. Heller and G. A. West. Electronically induced IR multiphoton absorption: a direct probe of the quasicontinuum, *Chem. Phys. Lett.*, *69*, 419 (1980).

[Herman, 1980] I. P. Herman and J. B. Marling. Ultrahigh single-step deuterium enrichment in CO_2 laser photolysis of trifluoromethane as measured by carbon-isotope labeling, *J. Chem. Phys.*, *72*, 516 (1980).

[Herman, 1983a] I. P. Herman. Two-frequency CO_2 laser multiple-photon dissociation and dynamics of excited state absorption in CDF_3, *Chem. Phys*, *75*, 121 (1983).

[Herman, 1983b] I. P. Herman, F. Magnotta, R. J. Buss, and Y. T. Lee. Infrared laser multiple-photon dissociation of $CDCl_3$ in a molecular beam, *J. Chem. Phys.*, *79*, 1789 (1983).

[Herman, 1984] I. P. Herman, F. Magnotta, and F. T. Aldridge. The status of the photochemistry and photophysics of tritium-from-deuterium isotope separation by infrared laser multiple-photon dissociation of chloroform, *Israel J. Chem.*, *24*, 192 (1984).

[Hodgkinson, 1981] D. P. Hodgkinson, A. J. Taylor, and A. G. Robiette. Multiphoton excitation of vibration-rotation states in the ν_3 mode of SF_6, *J. Phys.*, *B14*, 1803 (1981).

[Horsley, 1980a] J. A. Horsley, P. Rabinowitz, A. Stein, D. M. Cox, R. O. Brickman, and A. Kaldor. Laser chemistry experiments with UF_6, *IEEE J. Quantum Electron*, *QE-16*, 412 (1980).

[Horsley, 1980b] J. A. Horsley, D. M. Cox, R. B. Hall, A. Kaldor, E. T. Maas, Jr., E. B. Priestley, and G. M. Kramer. Isotopic selectivity in the laser induced dissociation of molecules with overlapping absorption bands, *J. Chem. Phys.*, *73*, 3660 (1980).

[Hsu, 1978] D. S. Y. Hsu and T. J. Manuccia. Deuterium enrichment by CW CO_2 laser-induced reaction of methane, *Appl. Phys.*, *Lett.*, *33*, 915 (1978).

[Hsu, 1980a] D. S. Y. Hsu and T. J. Manuccia. Enrichment of deuterium using vibrationally sensitized reaction of methane, *Appl. Phys. Lett.*, *36*, 715 (1980).

[Hsu, 1980b] D. S. Y. Hsu and T. J. Manuccia. Carbon isotope enrichment by retention of isotopic excitation selectivity in VV ladder climbing collisions, *Chem. Phys. Lett.*, *76*, 16 (1980).

[Jensen, 1974] R. J. Jensen and J. L. Lyman. Laser-induced recovery of deuterium and tritium from water, *2nd Proc. Eur. Electro-Opt. Mark. Technol. Conf.*, *1974*, Mack Brooks Exhib. Ltd., St. Albans, England, 1974, p. 106.

[Jensen, 1976] R. J. Jensen, J. G. Marinuzzi, C. P. Robinson, and S. D. Rockwood. Prospects for uranium enrichment, *Laser, Focus*, May 1976, 51 (1976).

[Jensen, 1980a] R. J. Jensen, J. A. Sullivan, and F. T. Finch. Laser isotope separation, *Sep. Sci. Technol.*, *15*, 509 (1980).

[Jensen, 1980b] R. J. Jensen, J. K. Hayes, C. L. Cluff, and J. M. Thorne. Isotopically selective infrared photodissociation of (trans-2-chloroethenyl) dichloroborane, *IEEE J. Quantum Electron.*, *QE-16*, 1352 (1980).

[Jensen, 1982] R. J. Jensen, O. P. Judd, and J. A. Sullivan. Separating isotopes with lasers, *Los Alamos Sci.*, *3*(1), 2 (1982).

[Judd, 1979] O. P. Judd. A quantitative comparison of multiple-photon absorption in polyatomic molecules, *J. Chem. Phys.*, *71*, 4515 (1979).

[Karl, 1975] R. R. Karl and K. K. Innes. Dye-laser-induced separation of nitrogen and carbon isotopes, *Chem. Phys. Lett.*, *36*, 275 (1975).

[King, 1982] D. S. King. Infrared multiphoton excitation and dissociation, in *Dynamics of the Excited State* (K. P. Lawley, ed.), Wiley, New York, 1982, p. 105.

[Knyazev, 1978] I. N. Knyazev, Y. A. Kudriavtzev, N. P. Kuz'mina, V. S. Letokhov, and A. A. Sarkisian. Laser isotope separation of carbon by multiple IR photon and subsequent UV excitation of CF_3I molecules, *Appl. Phys.*, *17*, 427 (1978).

[Knyazev, 1979] I. N. Knyazev, Y. A. Kudryavtsev, N. P. Kuz'mina, and V. S. Letokhov. Isotope-selective photodissociation of CF_3I molecules in multi-photon vibrational excitation followed by electron excitation by laser radiation, *Sov. Phys. JETP*, *49*, 650 (1979).

[Kojima, 1983] H. Kojima, K. Fukumi, S. Nakajima, Y. Maruyama, Y. Maruyama, and K. Kosasa. Laser isotope separation of ^{13}C by an elimination method, *Chem. Phys. Lett.*, *95*, 614 (1983).

[Kolomiiskii, 1978] Y. R. Kolomiiskii, and E. A. Ryabov. Frequency characteristics of isotopically selective dissociation of BCl_3 in a strong infrared CO_2 laser field, *Sov. J. Quant. Electron.*, *8*, 375 (1978).

[Kompa, 1981] K. L. Kompa. High-power tunable lasers and their applications to photochemistry and isotope separation, in *Proc. Int. School Phys. "Enrico Fermi,"* Course LXXIV, Developments in High-Power Lasers and Their Applications, (C. Pellegrini, ed.), North-Holland, Amsterdam, 1981, p. 274.

[Koren, 1981] B. Koren, M. Dahan, and U. P. Oppenheim. Multiple-photon absorption of a CO_2-pumped CF_4 laser radiation in UF_6, *Opt. Commun.*, *38*, 265 (1981).

[Koren, 1976] G. Koren, U. P. Oppenheim, D. Tal, M. Okon, and R. Weil. Deuterium separation in formaldehyde by an intense pulsed CO_2 laser, *Appl. Phys. Lett.*, *29*, 40 (1976).

[Koren, 1982] G. Koren, Y. Gertner, and U. Shreter. Isotope separation experiments in natural UF_6 by CF_4 and CO_2 lasers, analyzed by gamma-ray spectrometry, *Appl. Phys. Lett.*, *41*, 397 (1982).

[Kuhlmann, 1977] R. Kuhlmann and W. Schnabel. On the primary processes of sensitized photopolymerization of vinylmonomers. Laser flash photolysis studies and stationary polymerization experiments, *Angew, Makromol. Chem.*, *57*, 195 (1977).

[Kurihara, 1983] O. Kurihara, K. Takeuchi, S. Satooka, and Y. Makide. Tritium isotope separation by multiphoton dissociation of CF_3CTClF, *J. Nucl. Sci. Technol.*, *20*, 617 (1983).

[Kurnit, 1980] N. A. Kurnit, G. P. Arnold, L. W. Sherman, W. H. Watson, and R. G. Wenzel. CO_2-pumped p-H_2 rotational raman amplification in a hollow dielectric waveguide, CLEOS/ICF 1980.

[Kutschke, 1983] K. O. Kutschke, M. Gauthier, and P. A. Hackett. Separation of deuterium by IR multiphoton decomposition of chlorodifluoromethane. IR multiphoton absorption by and decomposition of a CF_2DCl/CF_2HCl mixture, *Chem. Phys.* 78, 323 (1983).

[Leary, 1978] K. M. Leary, J. L. Lyman, L. B. Asprey, and S. M. Freund. CO_2 laser induced reactions of SF_5Cl, *J. Chem. Phys.*, 68, 1671 (1978).

[Letokhov, 1972] V. S. Letokhov. Selective laser photochemical reactions by means of photopredissociation of molecules, *Chem. Phys. Lett.*, 15, 221 (1972).

[Letokhov, 1975] V. S. Letokhov and V. A. Semchishen. Selective reaction ortho-I_2 molecules on laser excitation, *Sov. Phys. Dokl.*, 20, 423 (1975).

[Letokhov, 1977] V. S. Letokhov. Laser separation of isotopes, *Ann. Rev. Phys. Chem.*, 28, 133 (1977).

[Letokhov, 1979] V. S. Letokhov. Laser isotope separation, *Nature*, 277, 605 (1979).

[Letokhov, 1983] V. S. Letokhov. *Nonlinear Laser Chemistry, Multiple-Photon Excitation*, Springer-Verlag, Berlin, 1983.

[Liu, 1975] D. D.-S. Liu, S. Datta, and R. N. Zare. Laser separation of chlorine isotopes. The photochemical reaction of electronically excited iodine monochloride with halogenated olefins, *J. Am. Chem. Soc.*, 97, 2557 (1975).

[Loree, 1979] T. R. Loree, J. H. Clark, K. B. Butterfield, J. L. Lyman, and R. Engleman. Carbon and sulfur isotope separation by ArF laser irradiation of CS_2, *J. Photochem.*, 10, 359 (1979).

[Lyman, 1975] J. L. Lyman, R. J. Jensen, J. Rink, C. P. Robinson, and S. D. Rockwood. Isotopic enrichment of SF_6 in [34]S by multiple absorption of CO_2 laser radiation, *Appl. Phys. Lett.*, 27, 87 (1975).

[Lyman, 1976] J. L. Lyman and S. D. Rockwood. Enrichment of Boron, carbon, and silicon isotopes by multiple-photon absorption of 10.6 μm laser radiation, *J. App. Phys.*, 47, 595 (1976).

[Lyman, 1977a] J. L. Lyman. A model for unimolecular reaction of sulfur hexafluoride, *J. Chem. Phys.*, 67, 1868 (1977).

[Lyman, 1977b] J. L. Lyman, S. D. Rockwood, and S. M. Freund. Multiple-photon isotope separation in SF_6: effect of laser pulse shape and energy, pressure, and irradiation geometry, *J. Chem. Phys.*, *67*, 4545 (1977).

[Lyman, 1978] J. L. Lyman and K. M. Leary. Absorption of infrared radiation by a large polyatomic molecule: CO_2 laser irradiation of disulfur decafluoride, *J. Chem. Phys.*, *69*, 1858 (1978).

[Lyman, 1979] J. L. Lyman, W. C. Danen, A. C. Nilsson, and A. V. Nowak. Multiple-photon excitation of difluoroamino sulfur penta-fluoride: a study of absorption and dissociation. *J. Chem. Phys.*, *71*, 1206 (1979).

[Lyman, 1982] J. L. Lyman, H. G. Galbraith, and J. R. Ackerhalt. Multiple-photon excitation, *Los Alamos Sci.*, *3*(1), 66 (1982).

[Lyman, 1984] J. L. Lyman, H. G. Galbraith, and J. R. Ackerhalt. Single-infrared-frequency studies of excitation and dissociation, in *Multiple-Photon Excitation and Dissociation* (C. D. Cantrell, ed.), Springer-Verlag, Heidelberg (in press).

[Lyman, 1985] J. L. Lyman, G. Laguna, and N. R. Greiner. Reactions of uranium hexafluoride photolysis products, *J. Chem. Phys.*, *82*, 175 (1985).

[Magnotta, 1982] F. Magnotta, I. P. Herman, and F. T. Aldridge. Highly selective tritium-from-deuterium isotope separation by pulsed NH_3 laser multiple-photon dissociation of chloroform, *Chem. Phys. Lett.*, *92*, 600 (1982).

[Makide, 1980] Y. Makide, S. Hagiwara, O. Kurihara, K. Takeuchi, Y. Ishikawa, S. Arai, T. Tominaga, I. Inoue, and R. Nakane. Tritium separation by CO_2 laser multiphoton dissociation of tri-fluoromethane, *J. Nucl. Sci. Technol.*, *17*, 645 (1980).

[Makide, 1981a] A. Makide, S. Hagiwara, T. Tominaga, K. Takeuchi, and R. Nakane. Tritium isotope separation by CO_2 laser-induced multiphoton dissociation of CTF_3, *Chem. Phys. Lett.*, *82*, 18 (1981).

[Makide, 1981b] Y. Makide, S. Hagiwara, T. Tominaga, O. Kurihara, and R. Makane. CO_2 laser isotope separation in ternary mixture of H/D/T compounds, *Int. J. Appl. Radiat. Iso. 32*, 881 (1981).

[Makide, 1982a] Y. Makide, T. Tominaga, K. Takeuchi, O. Kurihara, and R. Nakane. Tritium isotope separations with extremely large separation factors by CO_2 laser multiphoton dissociation of tri-fluoromethane-t, *ACS National Meeting. Symposium Laser Isotope Separation*, Las Vegas, Nev., Mar. 31, 1982.

[Makide, 1982b] Y. Makide, S. Kato, T. Tominaga, and K. Takeuchi.
CO_2-laser isotope separation of tritium with pentafluoroethane-t
(C_2TF_5), Appl. Phys., 28, 341 (1982).

[Makide, 1983] Y. Makide, S. Kato, T. Tominaga, and K. Takeuchi.
Laser isotope separation of tritium from deuterium: CO_2 laser induced multiphoton dissociation of C_2DF_5, Appl. Phys., B32,
33 (1983).

[Malatesta, 1981a] V. Malatesta, C. Willis, and P. A. Hackett. Laser
photochemical production of vitamin D, J. Am. Chem. Soc., 103,
6781 (1981).

[Malatesta, 1981b] V. Malatesta, P. A. Hackett, and C. Willis. Gas
phase fluorination of CO_2 by SF_6. J. Chem. Soc. Chem. Commun.,
1981 , 247 (1981).

[Mannik, 1979] L. Mannik, G. M. Keyser, and K. B. Woodall. Deuterium isotope separation by tunable-laser predissociation of formaldehyde, Chem. Phys. Lett., 65, 231 (1979).

[Manuccia, 1979] T. J. Manuccia and D. S. Y. Hsu. Deuterium enrichment by CW vibrational photochemistry of methane-economic
considerations, Laser-Induced Processes in Molecules. Physics
and Chemistry (Proc. Euro. Phys. Soc. Div. Conf., Heriot-Watt
Univ., Edinburgh, Scotland, Sept. 20-22, 1978) (K. L. Kompa
and S. D. Smith, eds.), Springer-Verlag, Berlin, 1979, p. 270.

[Marling, 1975] J. B. Marling. Laser isotope separation of deuterium, Chem. Phys. Lett., 34, 84 (1975).

[Marling, 1977] J. Marling. Isotope separation of oxygen-17, oxy-
gen-18, carbon-13, and deuterium by ion laser induced formaldehyde photopredissociation, J. Chem. Phys., 66, 4200 (1977).

[Marling, 1979] J. B. Marling and I. P. Herman. Deuterium separation with 1400-fold single-step isotopic enrichment and high yield
by CO_2-laser multiple-photon dissociation of 2,2-dichloro-1,1,1-
trifluoroethane, Appl. Phys. Lett., 34, 439 (1979).

[Martellucci, 1981] S. Martellucci and S. Somimeno. High-power 16
micrometre lasers for uranium isotope separation, in Proc. Int.
Phys. "Enrico Fermi," Course LXXIV, Developments in High-Power
Lasers and Their Applications, (C. Pellegrini, ed.), North-Holland, Amsterdam, 1981, p. 337.

[Mayer, 1970] S. W. Mayer, M. A. Kwok, R. W. F. Gross, and D. J.
Spencer. Isotope separation with the CW hydrogen fluoride laser, Appl. Phys. Lett., 17, 516 (1970).

[McInteer, 1982] B. B. McInteer, J. L. Lyman, G. P. Quigley, and A. C. Nilsson. A practical enrichment technique for 33-sulfur (34-sulfur), *Proc. Int. Conf. Synthesis and Applications of Isotopically Labeled Compounds* (W. F. Duncan and A. B. Susan, eds.), Kansas City, Mo., June 6-11, 1982, p. 397.

[Merritt, 1977] J. A. Merritt and L. C. Robertson. Removal of Phosgene impurity from boron trichloride by laser radiation, *J. Chem. Phys.*, *67*, 3545 (1977).

[Meyer-Kretschmer, 1983] G. Meyer-Kretschmer and H. Jetter. Uran-Anricherung mit Lasern, *Naturwissenschaften*, *70*, 7 (1983).

[Miller, 1978] S. S. Miller, D. D. Deford, T. J. Marks, and E. Weitz. Infrared photochemistry of a volatile uranium compound with 10-µm absorption, *J. Am. Chem. Soc.*, *101*, 1036 (1978).

[Moore, 1983] C. B. Moore and J. C. Weisshaar. Formaldehyde photochemistry, *Annu. Rev. Phys. Chem.*, *34*, 525 (1983).

[Moser, 1983] J. Moser, P. Morand, R. Duperrex, and H. Van Den Bergh. Deuterium separation and infrared photochemistry in Freon 22, *Chem. Phys.*, *79*, 277 (1983).

[Neve de Mevergnies, 1981] M. Neve de Mevergnies. Effect of gas pressure and pulse length on the isotopically selective multiphoton dissociation of CF_3Cl under CO_2 laser pulses, *Appl. Phys.*, *25*, 275 (1981).

[Nguyen, 1981] H. H. Nguyen and W. C. Danen. Pulsed infrared laser-induced reaction of cyclybutyl acetate. Laser synthesis of a thermally labile compound by a rapid heating-quenching process, *J. Am. Chem. Soc.*, *103*, 6253 (1981).

[Nowak, 1975] A. V. Nowak and J. L. Lyman. The temperature-dependent absorption spectrum of the v_3 band of SF_6 at 10.6 µm, *J. Quant. Spectrosc. Radiat. Transfer*, *15*, 945 (1975).

[Orr, 1981] B. J. Orr and J. G. Haub. Selective infrared excitation in D_2CO: rotational assignments and the role of collisions, *Opt. Lett.*, *6*, 236 (1981).

[Padrick, 1980] T. D. Padrick, A. K. Hays, and M. A. Parmer. Broadening of the CF_3I UV absorption spectrum by CO_2-laser-induced vibrational excitation, *Chem. Phys. Lett.*, *70*, 63.

[Parthasarathy, 1982] V. Parthasarathy, S. K. Sarkar, V. P. Singhal, A. Pandey, K. V. S. Rama Rao, and J. P. Mittal. TEA CO_2 laser-induced deuterium exchange in the ammonia-fluoroform-d system, *Chem. Phys. Lett.*, *86*, 259 (1982).

[Quack, 1978] M. Quack. Theory of unimolecular reactions induced by monochromatic infrared radiation, *J. Chem. Phys.*, *69*, 1282 (1978).

[Quack, 1981] M. Quack. Photochemistry with infrared radiation, *Chimia*, *35*, 463 (1981).

[Quick, 1978] C. R. Quick and C. Wittig. Infrared photodissociation of fluorinated ethanes and ethylenes. Collisional effects in the multiple photon absorption process, *J. Chem. Phys.*, *69*, 4201 (1978).

[Rabinowitz, 1978b] P. Rabinowitz, A. Stein, and A. Kaldor. Infrared multiphoton dissociation of UF_6, *Opt. Commun.*, *27*, 381 (1978).

[Rabinowitz, 1982a] P. Rabinowitz, A. Kaldor, A. Gnauck, R. L. Woodin, and J. S. Gethner. Two-color infrared isotopically selective decomposition of UF_6, *Opt. Lett.*, *7*, 212 (1982).

[Rabinowitz, 1982b] P. Rabinowitz, A. Kaldor, A. Gnauck. Two color infrared isotopically selective decomposition of UF_6, *Appl. Phys.*, *B28*, 187 (1982).

[Rabinowitz, 1978a] P. Rabinowitz, A. Stein, R. Brickman, and A. Kaldor. Stimulated rotational Raman scattering from *para*-H_2 pumped by a CO_2 TEA laser, *Opt. Lett.*, *3*, 147 (1978).

[Reddy, 1980] K. V. Reddy and M. J. Berry. Laser isotope separation by one-photon vibrational photochemistry: carbon isotope enrichment in methyl isocyanide photoisomerization, *Chem. Phys. Lett.*, *72*, 29 (1980).

[Robinson, 1972] P. J. Robinson and K. A. Holbrook. *Unimolecular Reactions*, Wiley, New York, 1972.

[Robinson, 1977] C. P. Robinson, R. J. Jensen, and C. D. Cantrell. Laser isotope separation, in *Lasers in Chemistry* (M. A. West, ed.), Elsevier, Amsterdam, 1977, p. 184.

[Rockwood, 1974] S. D. Rockwood and S. W. Rabideau. The status of laser separation of boron isotopes—June 1974, *LASL Rep. LA-5761-SR*, Nov. 1974.

[Sarkar, 1981] S. K. Sarkar, V. Parthasarathy, A. Pandey, K. V. S. Rama Rao, and J. P. Mittal. Isotope selective IR-laser multiphoton dissociation of CF_3Cl and selectivity enhancement with substrate pressure, *Chem. Phys. Lett.*, *78*, 479 (1981).

[Sazonov, 1983] V. N. Sazonov and G. V. Shmerling. Influence of rare gases on laser-chemical reactions: II. Experiments, *Sov. J. Quantum Electron*, *13*, 662 (1983).

[Schramm, 1979] B. Schramm and A. Diez. Laserangergte chemische reactonen mit bortrichlorid, *Ber. Bunsenges, Phys. Chem.*, *83*, 1279 (1979).

[Stephenson, 1978] J. C. Stephenson and D. S. King. Energy partitioning in the collision-free multiphoton dissociation of molecules: energy of $\tilde{X}$ CF_2 from CF_2HCl, CF_2Br_2, and CF_2Cl_2, *J. Chem. Phys.*, *69*, 1485 (1978).

[Stephenson, 1979] J. C. Stephenson, D. S. King, M. F. Goodman, and J. Stone. Experiment and theory for CO_2 laser-induced CF_2HCl decomposition rate dependence on pressure and intensity, *J. Chem. Phys.*, *70*, 4496 (1979).

[Stone, 1979] J. Stone and M. F. Goodman. A re-examination of the use of rate equations to account for fluence dependence, intramolecular relaxation, and unimolecular decay in laser-driven polyatomic molecules, *J. Chem. Phys.*, *71*, 408 (1979).

[Stuke, 1977] M. Stuke and F. P. Schafer. Enrichment of chlorine isotopes by selective photoaddition of iodine chloride to acetylene, in *Lasers in Chemistry* (M. A. West, ed.), Elsevier, Amsterdam, 1977 p. 195.

[Stuke, 1978] M. Stuke and E. E. Marinero. On-line computer controlled CW dye laser spectrometer for laser isotope separation, *Appl. Phys.*, *16*, 303 (1978).

[Subbiah, 1980] J. Subbiah, S. K. Sarkar, K. V. S. Rama Rao, and J. P. Mittal. Laser photochemistry of UF_6, *Indian J. Phys.*, *B54*, 121 (1980).

[Sudbø, 1978] A. S. Sudbø, P. A. Schulz, E. R. Grant, Y. R. Shen, and Y. T. Lee. Multiphoton dissociation products from halogenated hydrocarbons, *J. Chem. Phys.*, *68*, 1306 (1978).

[Sudbø, 1979] A. S. Sudbø, P. A. Schulz, E. R. Grant, Y. R. Shen, and Y. T. Lee. Simple bond rupture reactions in multiphoton dissociation of molecules, *J. Chem. Phys.*, *70*, 912 (1979).

[Suzuki, 1977] K. Suzuki, P. H. Kim, K. Taki, and S. Namba. Absorption spectrum and laser isotope separation of bromine, *Oyo Butsuri*, *45*, 932 (1977).

[Suzuki, 1978] K. Suzuki, P. H. Kim, and S. Namba. Chlorine isotope enrichment by the photochemical reaction of chlorine molecules using the Ar ion laser, *App. Phys. Lett.*, *33*, 52 (1978).

[Takami, 1984] M. Takami, T. Oyama, T. Watanabe, S. Namba, and R. Nakane. Cold jet infrared absorption spectroscopy. The ν_3 band of UF_6, *Jpn. J. Appl. Phys.*, Part 2, *23*, 88 (1984).

[Takeuchi, 1981] K. Takeuchi, O. Kurihara, S. Satooka, Y. Makide, I. Inoue, and R. Nakane. Tritium isotope separation by CO_2 laser irradiation at dry ice temperature, *J. Nucl. Sci. Technol.*, *18*, 68 (1981).

[Takeuchi, 1982] K. Takeuchi, I. Inoue, R. Nakane, Y. Makide, S. Kato, and T. Tominaga. "CO_2 laser tritium isotope separation: collisional effects in multiphoton dissociation of trifluoromethane, *J. Chem. Phys.*, *76*, 398 (1982).

[Tiee, 1978a] J. J. Tiee and C. Wittig. The photodissociation of UF_6 using infrared lasers, *Opt. Commun.*, *27*, 377 (1978).

[Tiee, 1978b] J. J. Tiee and C. Wittig. IR photolysis of SeF_6: separation and dissociation enhancement using NH_3 and CO_2 lasers, *J. Chem. Phys.*, *69*, 4756 (1978).

[Tiee, 1978c] J. J. Tiee and C. Wittig. Isotopically selective IR photo-dissociation of SeF_6, *App. Phys. Lett.*, *32*, 236 (1978).

[Tsay, 1979] W.-S. Tsay, C. Riley, and D. O. Ham. Thermal enhancement of multiple photon absorption by SF_6, *J. Chem. Phys.*, *70*, 3558 (1979).

[Tuccio, 1979] S. A. Tuccio and A. Hartford, Jr. Deuterium enrichment via selective dissociation of fluoroform-D with a pulsed CO_2 laser, *Chem. Phys. Lett.*, *65*, 235 (1979).

[Vanderleeden, 1977] J. C. Vanderleeden. Laser separation of deuterium, *Laser Focus*, June 1977, 51 (1977).

[Vanderleeden, 1978] J. C. Vanderleeden. Depletion optimization and photon efficiency in laser isotope separation of deuterium, *Appl. Opt.*, *17*, 785 (1978).

[Vanderleeden, 1980] J. C. Vanderleeden. Generalized concepts in large-scale laser isotope separation, with application to deuterium, *J. Appl. Phys.*, *51*, 1273 (1980).

[Vasil'ev, 1980] B. I. Vasil'ev, A. P. Dyad'kin, and A. N. Sukhanov. Dissociation of uranium hexafluoride at a composite frequency by NH_3 laser radiation, *Sov. Tech. Phys. Lett.*, *6*, 135 (1980).

[Willis, 1976] C. Willis, R. A. Back, R. Corcum, R. D. McAlpine, and F. K. McClusky. Infrared induced reactions in methanol-bromine mixtures: a negative result, *Chem. Phys. Lett.*, *38*, 336 (1976).

[Wolfrum, 1981] J. Wolfrum, M. Kneber, and P. N. Clough. Pulsed laser dehydrohalogenation of saturated halohydrocarbons, West German patent 293,853.

[Yeung, 1972] E. S. Yeung and C. B. Moore. Isotopic separation by photo-predissociation, *Appl. Phys. Lett.*, *21*, 109 (1972).

[Yeung, 1973] E. S. Yeung and C. B. Moore. Photochemistry of single vibronic levels of formaldehyde, *J. Chem. Phys.*, *58*, 3988 (1973).

[Yogev, 1975] A. Yogev and R. M. J. Benmair. Photochemistry in the electronic ground state: III. Isotope selective decomposition of methylene chloride by pulsed carbon dioxide laser, *J. Am. Chem. Soc.*, *97*, 4430 (1975).

[Zellweger, 1984] J.-M. Zellweger, J.-M. Philippoz, P. Melinon, R. Monot, and H. van den Bergh. Isotopically selective condensation and single photon infrared laser isotope separation, *Phys. Rev. Lett.*, *52*, 522 (1984).

[Zittel, 1983] P. F. Zittel, L. A. Darnton, and D. D. Little. Separation of O, C, and S isotopes by two-step, laser photodissociation of OCS, *J. Chem. Phys.*, *79*, 5991 (1983).

7

Laser Raman Techniques

JAMES J. VALENTINI   *University of California, Irvine, California*

7.1 INTRODUCTION

The absorption of light of one frequency and emission of another frequency was a well-known phenomenon by the 1920s, and such fluorescence from atomic or molecular species could be observed in gases, liquids, and solids. However, near the end of that decade C. V. Raman and K. S. Krishnan (1928) reported the then remarkable observation that a frequency shift in the light could be observed even in the absence of absorption. They found that when monochromatic optical radiation was incident upon optically transparent, dust-free liquids and gases, they observed scattered light not only at the incident frequency (the process of Rayleigh scattering), but also a very small fraction of the incident light scattered with a change in frequency. Within three months Landsberg and Mandelstam (1928) observed the same phenomenon in quartz, indicating that optically transparent, perfect solids could also produce this effect. The flurry of activity that followed this discovery established that the frequency shifts observed in the scattering were the vibrational frequencies of the molecules in the sample material. That is, the spectrum of scattered radiation could be described by $\omega_S = \omega_L - \omega_M$ and $\omega_{AS} = \omega_L + \omega_M$, where ω_L is the incident light frequency and ω_M is a molecular vibrational frequency. Those lines or bands characterized by a shift in the light to lower frequency, ω_S, are referred to as Stokes bands, while those corresponding to shifts to higher frequencies, ω_{AS}, are termed anti-Stokes bands.

That the frequency shifts in what soon came to be called the Raman effect were characteristic of molecular vibrations made it immediately apparent that this light-scattering phenomenon would be an important approach to molecular spectroscopy. The Raman effect soon developed into Raman spectroscopy, not only because the frequency shifts revealed the vibrational frequencies of molecules, but also because the phenomenon was universal and the experimental technique simple. Since it was based on the scattering of light, due to an induced dipole moment, rather than the absorption of light, due to a permanent dipole moment, it could be used to probe vibrations without dipole moments (e.g., symmetric vibrations in polyatomic molecules and the vibration of homonuclear diatomics). Hence it was applicable to any and all molecules. Also, since the effect could be observed with any excitation frequency, experiments could be carried out conveniently in the visible and ultraviolet spectral regions, where simple sources of monochromatic light, such as the mercury arc, were available. It is not surprising, then, that Raman light scattering soon became an important technique for molecular spectroscopy, particularly as a complement to infrared absorption spectroscopy, which also probes molecular vibrations. Subsequent work on the Raman effect demonstrated that frequency shifts corresponding to molecular rotational frequencies, electronic transition frequencies, and phonon modes in solids could also be observed, indicating that this phenomenon could be the basis for a spectroscopic technique of unparalleled breadth and convenience.

But the Raman effect had one serious deficiency that long prevented the realization of its full potential: the extremely low efficiency of the inelastic light scattering. The cross section for Raman scattering is typically 10^{-12} of an infrared absorption cross section (Fenner et al., 1973), and the intensity of the scattered light is 10^{-7} to 10^{-15} of the intensity of the incident light. The low light fluxes available from mercury arc and related sources meant that getting good Raman spectra required exposure times of many minutes to many hours. Consequently, spectroscopy of fluorescent or luminous samples, real-time monitoring, and minor species detection were essentially impossible. So for the first 40 years of its history the Raman effect served principally as a spectroscopic technique for pure solids and liquids, and fairly high pressure gases.

The commercial development of the laser in the mid-1960s radically changed this situation and revolutionized the nature of Raman spectroscopy. The high luminosity of the laser output, its narrow line width, and the small size and divergence of the beam were all advantageous for Raman scattering. While the inefficiency of the inelastic light scattering remained an unavoidable constraint imposed by nature, the laser reduced the severity of this constraint on the application of

Raman spectroscopy in the physical and biological sciences. Unlike most optical spectroscopies, which require frequency-tunable light sources, Raman spectroscopy took immediate advantage of the first lasers, which were fixed-frequency sources. A few years later, when tunable dye lasers became available, the range of applicability of Raman spectroscopy further widened. Resonance Raman spectroscopy, in which the incident light frequency is resonant with an electronic absorption of the sample, was now generally possible. This can enhance the efficiency of the Raman effect by a factor of 100 or more, greatly increasing the sensitivity of Raman scattering for the observation of species with low-lying electronic excited states. Tunable laser sources also made possible the development of coherent Raman spectroscopy, in which lasers provided both the incident excitation frequency and the Stokes or anti-Stokes frequency, to coherently drive the inelastic light scattering. Coherent techniques, such as coherent anti-Stokes Raman scattering (CARS) and stimulated Raman scattering (SRS), enhanced the sensitivity of Raman spectroscopy even for those cases in which electronic resonance enhancement was not possible.

As the result of these developments of the past 20 years, Raman spectroscopy has changed dramatically. It is now unquestionably one of the most important spectroscopic techniques for chemical and physical analysis in basic and applied science. It is used extensively for structural and kinetic studies of proteins, for concentration and temperature measurement in flames, internal combustion, and jet engines, for characterization of temperature and flow velocity in nozzles, for microscopic surface analysis, for detection of photofragments and reaction products, for forensic analysis, for studies of shocked materials, for structural analysis of polymers, for plasma diagnostics, for characterization of catalysts, for monitoring chemical vapor deposition, and for many other applications. It remains today as important a spectroscopic tool for the determination of molecular structures as it was in its early years. Advances in laser Raman techniques have led to very highly time-resolved Raman spectroscopy, presently with picosecond time resolution. Sensitive multichannel detectors now allow complete spectra to be obtained in a single nanosecond- or even picosecond-duration laser pulse. High spatial resolution is also now a hallmark of Raman spectroscopy, with a resolution of 10^{-4} cm and even less having been demonstrated. Fundamental restrictions of the uncertainty principle present the only real limits to the spatial and temporal resolution now achievable.

Recent technological advances in Raman spectroscopy and the consequent rapid growth in the application of this important spectroscopic technique have resulted in an enormous body of Raman literature. In the past two years more than 5000 publications dealing with some aspect or application of the Raman effect have appeared. In addition to the expected texts dealing with the theory and practice of Raman spectroscopy, one encounters monographs covering but a single area of application, and lengthy review articles on some subset of the field are now common. In view of this, it is clearly impossible to present a comprehensive review of Raman spectroscopy in a single book chapter. Rather, the aim of this chapter is to present only a selective overview of recent work in the use of Raman spectroscopy for chemical and physical analysis, that is the determination of the chemical composition and physical state (density, temperature, etc.) of a sample, including measurement of the temporal and spatial variation in composition and state. The review will deal with applications in combustion, biophysics, surface science, catalysis, polymer science, kinetics, photochemistry, plasma diagnostics, and other areas. The use of Raman spectroscopy for purely spectroscopic studies of molecular structure will not be discussed, even though such work still represents a major application of the Raman effect.

An outline of the phenomenology of the Raman effect, necessary for an understanding of its application, will be given first. Then a short discussion of the experimental techniques of spontaneous and coherent Raman spectroscopy will be presented. Many examples of the applications of Raman scattering will be reviewed to illustrate its prominence in chemical and physical analysis today. Representative references will be cited to provide the interested reader with an introduction to the published literature on specific topics. I conclude by summarizing the current status of applied Raman spectroscopy and attempting to assess its future direction.

7.2 PHENOMENOLOGY OF RAMAN SCATTERING

7.2.1 General Considerations

The Raman effect is an inelastic light-scattering phenomenon in which the illuminated sample absorbs one photon while simultaneously emitting another phonon of a different frequency. When the emitted photon is of lower frequency than the absorbed photon, the process is termed Stokes scattering, while if the emitted photon is of higher frequency, the process is called anti-Stokes scattering. These two processes are schematically represented in the energy diagrams of Fig. 7.1. The diagrams illustrate that Raman scattering is a two-photon process which connects the molecular (or atomic) energy levels g and r. The dashed line indicates a virtual electronic state, v, an electronic level that is not a real, stationary eigenstate of the molecule or atom. States o and p are the lowest-lying excited electronic levels (i.e., electronic eigenstates) of the molecule (or atom). Raman scattering is very much like a two-photon absorption process taking the system from

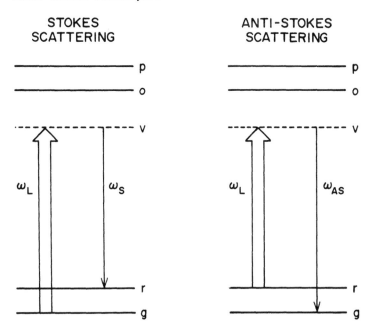

FIG. 7.1 Energy-level diagram for spontaneous Raman scattering. The bold arrows indicate the excitation (laser) photons at frequency ω_L, while the thin arrows represent the inelastically scattered photons, at frequency ω_S or ω_{AS}. States labeled g, r, o, and p are real molecular eigenstates, while v is a virtual electronic state.

state g to r (or r to g), except that instead of two photons being absorbed, one is absorbed and one emitted. Raman spectroscopy generally is carried out using visible or near-UV excitation frequencies. Since most molecular systems do not display absorption at these optical frequencies, the virtual level, v, lies well below the excited eigenstates, o and p. The levels g and r are most often molecular vibrational levels, for which the transition frequencies* are in the range 0 to 4000 cm^{-1}. These frequencies are much smaller than the optical excitation frequencies, typically, 20,000 to 40,000 cm^{-1}. The so-called resonance Raman effect arises when the excitation, Stokes, or anti-Stokes frequencies are equal to the frequencies of one-photon allowed transitions between states r or g and states o or p, as illustrated in Fig. 7.2. This electronic resonance enhancement can increase the efficiency of the Raman scattering by a factor of 100 or more. It is par-

---

*In this chapter we use the symbol ω to indicate a frequency in rad/s; however, we may sometimes refer to "frequencies" in wavenumbers (cm^{-1}), as is customary in spectroscopy.

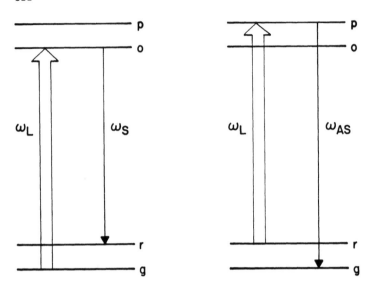

FIG. 7.2 Energy-level diagram for resonance Raman scattering. The labeling in the same as in Fig. 7.1. The intermediate state for the Raman transition from g to r (or r to g) is a real electronic eigenstate of the molecule, rather than a virtual state.

ticularly important in biophysical applications of Raman spectroscopy, since many proteins have chromophores with visible absorption bands.

The processes depicted in Figs. 7.1 and 7.2 are often referred to as spontaneous Raman scattering, in analogy to spontaneous emission phenomena, since the inelastically scattered photons are emitted incoherently. In contrast, the inelastic light scattering in processes such as coherent anti-Stokes Raman scattering (CARS) and stimulated Raman scattering (SRS) is marked by stimulated or coherent emission, as the names imply. In these coherent Raman spectroscopies, laser fields at both the excitation frequency and the Stokes-shifted frequency are incident on the sample, and they coherently drive the light scattering. In stimulated Raman scattering the Raman process coherently adds photons to the laser field at ω_S and coherently absorbs photons from the laser field at ω_L. The Raman signal is detected as a gain in the intensity at ω_S or as a loss in the intensity at ω_L. The former is the basis for stimulated Raman gain spectroscopy (SRGS), and the latter is the basis for inverse Raman spectroscopy (IRS). The laser field at which the gain or loss in intensity is measured is referred to as the "probe" laser, while the other is called the "pump" laser. Figure 7.3 depicts the IRS and SRGS phenomena. Coherent anti-Stokes Raman scattering also involves applied laser fields at ω_L and ω_S, but the signal is not detected as a gain or loss at these frequencies. Rather, the CARS process results in the coherent generation of an entirely new field at the anti-Stokes frequency, ω_{AS}, as shown in Fig. 7.4. In CARS the laser fields at ω_L and ω_S are mixed by a nonlinear Raman effect, which produces a coherent signal beam at higher frequency than either input beam. Two photons at ω_L are absorbed, resulting in emission of one photon at ω_S and one photon at ω_{AS}.

Since these coherent Raman processes result in stimulated light scattering, the signal intensities can be many orders of magnitude greater than those observed in incoherent, spontaneous Raman scattering. Furthermore, since the detected signals are laser beams, either heterodyned on the input beams in SRS or generated in CARS, signal collection efficiencies can be much higher than in spontaneous

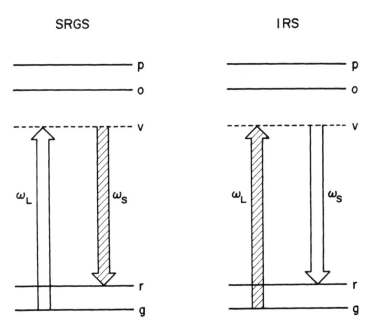

FIG. 7.3 Energy-level diagram for stimulated Raman scattering. The bold arrows indicate the excitation (ω_L) and Stokes (ω_S) laser photons. The inelastically scattered photons add at ω_S and subtract at ω_L. The crosshatching indicates the laser field whose intensity is monitored. In inverse Raman spectroscopy (IRS) the Raman scattering is detected as a loss in the intensity at ω_L. In stimulated Raman gain spectroscopy (SRGS) the Raman scattering is detected as a gain in intensity at ω_S.

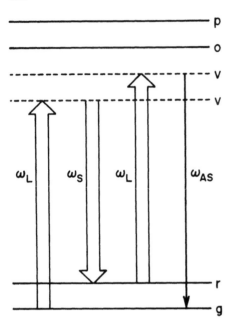

FIG. 7.4 Energy-level diagram for coherent anti-Stokes Raman scattering (CARS). The bold arrows indicate photons from the applied laser fields at ω_L and ω_S, while the thin arrow indicates the anti-Stokes signal photons.

Raman scattering, where the signal photons are emitted over 4π steradians. Also, in coherent Raman spectroscopy the signal actually increases with increasing spectral resolution, whereas in spontaneous Raman spectroscopy the opposite is true. These features have made coherent Raman spectroscopies popular, despite their much greater experimental complexity.

The phenomenon of the Raman effect arises because molecular vibrations modulate the frequency of the dipole induced in a molecule by an incident field. The electric dipole moment induced in a molecule by the electric field $\underline{E}$, can be expanded in a power series:

$$\underline{\mu} = \underline{\alpha}\underline{E} + \underline{\beta}E^2 + \underline{\gamma}E^3 + \cdots \tag{7.1}$$

The field $\underline{E}$, and induced dipole, $\underline{\mu}$, are vectors which generally are not oriented in the same direction, since α, β, and γ are tensor quantities. $\underline{\alpha}$ is the polarizability tensor of rank 2, while $\underline{\beta}$, $\underline{\gamma}$, etc., are the hyperpolarizabilities of rank 3, 4 etc. While the tensor properties of $\underline{\alpha}$, $\underline{\beta}$, and $\underline{\gamma}$ are an intrinsic feature of the Raman effect, they rarely need to be explicitly considered in applications of Raman spectroscopy in chemical and physical analysis. Thus, for simplicity we will neglect the tensor nature of the polarizability coefficients of Eq. (7.1) and treat them as purely scalar quantities. Thus we rewrite Eq. (7.1) as the scalar equation

$$\mu = \alpha E + \beta E^2 + \gamma E^3 + \cdots \tag{7.2}$$

Spontaneous Raman scattering arises due to the linear term in Eq. (7.2), while coherent Raman effects are connected with the term cubic in E.

7.2.2 Spontaneous Raman Scattering

While a quantum mechanical treatment is necessary for a complete explanation of Raman scattering, many aspects of the phenomenon can be described reasonably well by the classical electromagnetics of the induced dipole and the molecular vibration. Following the treatment of Placzek (1934) we expand the polarizability, α, in a Taylor series as a function of the vibrational normal coordinate, q:

$$\alpha = \alpha_0 + \left(\frac{\partial \alpha}{\partial q}\right)_0 q + \frac{1}{2}\left(\frac{\partial^2 \alpha}{\partial q^2}\right)_0 q^2 + \cdots \tag{7.3}$$

The subscript "0" indicates that the expansion is centered at the equilibrium molecular configuration. To understand the nature of the Raman effect it is only necessary to consider the first derivative term in Eq. (7.3), and we write more compactly:

$$\alpha = \alpha_0 + \alpha'q \tag{7.4}$$

where $\alpha' = (\partial \alpha / \partial q)_0$. Treating the molecular vibration as simple harmonic motion, we have

$$q = q_0 \cos(\omega_M t + \phi) \tag{7.5}$$

where q_0 is the vibrational amplitude, ϕ is a phase factor, and ω_M is the harmonic vibrational frequency. Substituting Eq. (7.5) in Eq. (7.4), we have

$$\alpha = \alpha_0 + \alpha'q_0 \cos(\omega_M t + \phi) \tag{7.6}$$

Substituting Eq. (7.6) in Eq. (7.2) and considering only the linear term, which is responsible for spontaneous Raman scattering, we see that the induced dipole moment is

$$\mu = \alpha E$$

$$= [\alpha_0 + \alpha' q_0 \cos(\omega_M t + \phi)] E_L \cos(\omega_L t)$$

$$= \alpha_0 E_L \cos(\omega_L t)$$

$$+ \frac{1}{2} \alpha' q_0 E_L \cos[(\omega_L - \omega_M)t + \phi] \qquad (7.7)$$

$$+ \frac{1}{2} \alpha' q_0 E_L \cos[(\omega_L + \omega_M)t + \phi]$$

Here we have taken the electric field to have amplitude E_L and harmonic frequency ω_L.

This oscillating dipole, μ, radiates energy not only at the incident frequency, ω_L (Rayleigh scattering), but also at the beat frequencies $\omega_L - \omega_M$ and $\omega_L + \omega_M$, the Stokes and anti-Stokes Raman scattered frequencies. The power radiated by the molecule at the Stokes frequency is given by the classical relation

$$P(\omega_S) = \frac{\omega_S^4}{3c^3} |\mu(\omega_S)|^2 \qquad (7.8)$$

or

$$P(\omega_S) = \frac{\omega_S^4}{3c^3} \left(\frac{1}{2} \alpha' q_0 E_L \right)^2 \qquad (7.9)$$

This expression is more commonly presented in terms of the phenomenological Raman differential cross section per unit solid angle, $d\sigma/d\Omega$, given by (Tolles et al., 1977)

$$\frac{d\sigma}{d\Omega} = (a' q_0)^2 \left(\frac{\omega_S}{c} \right)^4 \qquad (7.10)$$

and the relationship between the scattered light at ω_S and the incident light at ω_L is

$$P(\omega_S) = N \ell \frac{d\sigma}{d\Omega} \Delta \Omega \, P(\omega_L) \qquad (7.11)$$

N is the density of molecules in the molecular energy level responsible for the scattering, ℓ is the sample length, and $\Delta\Omega$ the solid angle over which the light is collected. Raman cross sections are typically only 10^{-30} cm^2 sr^{-1}, so for a sample at 1 torr pressure with a 1-cm sample length and a 1-sr collection solid angle we find $P(\omega_S)/P(\omega_L) = 3 \times 10^{-14}$. This illustrates the "feebleness of the effect" as C. V. Raman described it. Because the phase factors, ϕ, for the independently radiating dipoles are random, this scattered light is incoherent.

Classical electromagnetism also reveals the principal selection rule for Raman scattering. From Eq. (7.7) we see that for the Raman effect to be observed for a particular vibrational normal mode we must have $\alpha' \neq 0$ for that mode. That is, the derivative of the polarizability with respect to that normal coordinate, evaluated at the equilibrium position, must be nonzero if that vibration is to be Raman "active." For diatomic molecules this selection rule implies that the vibration of both homonuclear and heteronuclear molecules will be Raman active. Since the polarizability increases for positive displacement from the equilibrium bond length and decreases for negative displacement, α' is greater than zero. For polyatomic molecules, establishing whether α' is zero or nonzero for a specific vibration is more complicated. However, a general selection rule holds for vibrational transitions from the ground state to the first excited state. These transitions are Raman active only if at least one Cartesian component of α' is of the same symmetry species as the vibration itself. The significance of this selection rule is that many vibrations which are not infrared active, such as totally symmetric vibrations, are Raman active. This is why Raman spectroscopy is complementary to IR spectroscopy, and why the Raman effect has been so important to molecular spectroscopists.

All molecular gases and liquids have at least one vibrational mode that is Raman active. It is this fact that makes Raman spectroscopy singularly significant in applied spectroscopy, for it means that every molecular species can be detected by Raman spectroscopy. Since all molecular vibrations have frequencies $\omega_M/2\pi c$ between 0 and 4160 cm^{-1}, it is possible in principle to detect all molecular species in a range of no more than 2000 Å upon excitation at visible wavelengths. No other spectroscopic technique offers such generality and experimental convenience. This advantageous feature of Raman spectroscopy is more than simply one that we recognize as being important "in principle," for as will be discovered later in this chapter, it is being exploited in actual spectroscopic practice.

The classical electromagnetic picture of the Raman effect outlined above provides considerable insight into the nature of the phenomenon. However, a classical picture is not really adequate to describe this inelastic light-scattering process. A time-dependent quantum mechanical analysis is required, but such an analysis will not be presented here. An adequate description of the quantum mechanics is beyond the scope of this chapter, and furthermore is really not necessary for an appre-

ciation of the application of Raman spectroscopy, which is our principal interest here. The reader is referred to a very concise, yet
thorough quantum mechanical treatment of Raman scattering given by
Long (1977).

7.2.3 Coherent Raman Scattering

The term in Eq. (7.2) which is linear in E is responsible not only for
Raman scattering, but also other classical linear optical phenomena,
such as refraction, and as we have seen above, Rayleigh scattering.
The terms in Eq. (7.2) that involve higher powers of E are responsible for a nonlinear optical phenomenon. The quadratic term gives
rise to second harmonic generation, sum and difference frequency
generation, as well as what are called hyper-Rayleigh and hyper-Raman scattering. The terms cubic in E are the source of coherent Raman phenomena such as coherent anti-Stokes Raman scattering (CARS)
and stimulated Raman scattering (SRS), as well as other effects, such
as third-harmonic generation. Hyper-Raman scattering produces inelastically scattered light at frequencies $2\omega_L \pm \omega_M$, which is many orders of magnitude less intense than the already weak light produced
by the linear Raman effect. It is useful as a purely spectroscopic
technique, since the selection rules for it differ from those of linear
Raman spectroscopy, and it therefore provides additional information
on molecular structure. However, as a technique for applied spectroscopy it has little value, due to the weakness of the scattering. Were
it not for the coherent nature of the scattering, the usefulness for
chemical and physical analysis of techniques that depend on the third
power of the field also would be negligible. However, the coherent
properties of CARS and SRS make them highly desirable for applied
spectroscopy, largely because the coherence of the scattered light
makes its intensity much larger than that in linear Raman scattering
in many applications.

To analyze these third-order Raman effects we again invoke classical electromagnetics, and the resultant description is only slightly
more complicated than that of spontaneous Raman scattering. Because
these are coherent scattering processes, they are usually described
not in terms of dipoles induced in the individual molecules of the sample, but rather in terms of a bulk polarization of the sample. That is,
Eq. (7.1) is replaced by

$$\underline{P} = \underline{\chi}^{(1)}\underline{E} + \underline{\chi}^{(2)}\underline{E}^2 + \chi^{(3)}\underline{E}^2 + \cdots \tag{7.12}$$

where $\underline{\chi}^{(1)}$, $\underline{\chi}^{(2)}$, $\underline{\chi}^{(3)}$, ... are the first-order, second-order, third-
order,... dielectric susceptibility tensors. This polarization $\underline{P}$ is re-
lated to the induced dipole $\underline{\mu}$ by

$$P = N\zeta\underline{\mu} \tag{7.13}$$

where N is the particle density and ζ is a factor that relates the applied macroscopic field to the field acting on the individual particles (Armstrong, et al., 1962; Kittel, 1971; Pantell and Puthoff, 1969). For most materials the optical index of refraction is near unity and the macroscopic and microscopic fields are nearly identical, so

$$\underline{P} = N\underline{\mu} \tag{7.14}$$

and

$$\chi^{(1)} = N\underline{\alpha}$$
$$\chi^{(2)} = N\underline{\beta} \tag{7.15}$$
$$\underline{\chi}^{(3)} = N\underline{\gamma}$$

The term in Eq. (7.12) of interest to us now is the cubic one. As we did when considering spontaneous Raman scattering, we will ignore the tensor nature of $\chi^{(3)}$ and treat the sample as if its dielectric susceptibility were a scalar. We need consider then only the scalar properties of $\underline{P}$, $\chi^{(3)}$, and $\underline{E}$, and write

$$P^{(3)}(r,t) = \chi^{(3)}E^3(r,t) \tag{7.16}$$

where we have explicitly stated the spatial as well as the temporal dependence of $P^{(3)}$ and E, since the coherent Raman phenomena are integrative effects whose magnitude depends strongly on the position in the sample. In CARS and SRS two lasers, one at the excitation frequency and another at the Stokes frequency, are incident on the sample, so the electric field is made up of two harmonic components. For simplicity we will take these laser fields to be plane light waves, with parallel field vectors, propagating along the space-fixed z axis:

$$E(z,t) = E_L \cos(\omega_L t - k_L z) + E_S \cos(\omega_S t - k_S z) \tag{7.17}$$

where the wave vector $k_j = \omega_j n_j/c$, and n_j is the index of refraction at ω_j. Substituting Eq. (7.17) into Eq. (7.16) and expanding, we find that the nonlinear polarization $P^{(3)}(z,t)$ has harmonic components at many frequencies, including $2\omega_L - \omega_S$. The component at $2\omega_L - \omega_S$ is the CARS polarization, and by analogy to spontaneous scattering we have

$$2\omega_L - \omega_S = \omega_L + (\omega_L - \omega_S)$$

$$= \omega_L + \omega_M \qquad\qquad (7.18)$$

$$= \omega_{AS}$$

that is, the CARS polarization is at the anti-Stokes frequency. The CARS polarization is

$$P_{CARS}^{(3)}(z,t) = \frac{3}{4} \chi^{(3)}(-\omega_{AS}, \omega_L, -\omega_S, \omega_L)$$

$$\times E_L^2 E_S \cos[(2\omega_L - \omega_S)t - (2k_L - k_S)z] \qquad (7.19)$$

where $\chi^{(3)}(-\omega_{AS}, \omega_L, -\omega_S, \omega_L)$ is the third-order dielectric susceptibility for the process in which two photons at ω_L are absorbed,
while simultaneously a photon at ω_S and a photon at ω_{AS} are emitted.
We will abbreviate this as $\chi_{CARS}^{(3)}$.

This polarization acts as a source term in Maxwell's equations to
produce an electromagnetic wave at ω_{AS}. For plane-wave fields at
ω_L and ω_S we find (Regnier et al., 1974; Druet and Taran, 1981) that
the generated wave at ω_{AS} also is a plane wave,

$$E(z,t) = E_{AS}(z) \cos(\omega_{AS}t - k_{AS}z) \qquad\qquad (7.20)$$

and

$$E_{AS}(\ell) = \frac{3\pi\omega_{AS}}{2c} E_L^2 E_S \chi_{CARS}^{(3)} \ell \, \frac{\sin(\Delta k \, \ell/2)}{\Delta k \, \ell/2} \qquad\qquad (7.21)$$

for a sample that extends from $z = 0$ to $z = \ell$. The quantity $\Delta k = 2k_L - k_S - k_{AS}$ is the phase mismatch between the generated wave at ω_{AS}
and the driving field at $2\omega_L - \omega_S$. This yields a signal intensity at
the anti-Stokes frequency given by

$$I_{AS}(\ell) = \left(\frac{12\pi^2\omega_{AS}}{c^2}\right)^2 I_L^2 I_S \left|\chi_{CARS}^{(3)}\right|^2 \ell^2 \left[\frac{\sin(\Delta k \, \ell/2)}{\Delta k \, \ell/2}\right]^2 \qquad (7.22)$$

where I_j is the power per unit area at ω_j. When $\Delta k = 0$ the beams are
referred to as phase matched, and

$$I_{AS}(\ell) = \left(\frac{12\pi^2\omega_{AS}}{c^2}\right)^2 I_L^2 I_S \left|\chi_{CARS}^{(3)}\right|^2 \ell^2 \tag{7.23}$$

Equation (7.22) indicates that I_{AS} will vary sinusoidally with path length ℓ. We can define a coherence length, ℓ_c, as the path length over which the maximum signal is reached,

$$\ell_c = \frac{\pi}{\Delta k} \tag{7.24}$$

The coherence length is a function of the wavelength dispersion of the sample and is given by (Hauchecorne et al., 1971)

$$\ell_c = \frac{\pi c}{(\omega_L - \omega_S)^2} \left(2\frac{\partial n}{\partial \omega} + \omega_L \frac{\partial^2 n}{\partial \omega^2}\right)^{-1} \tag{7.25}$$

For most gases ℓ_c is about 100 cm at STP and is inversely proportional to density. Since the final term in Eq. (7.22) is nearly unity for $\ell < \ell_c/3$, Eq. (7.23) is a valid approximation for almost all gas samples. However, for condensed phases ℓ_c is of the order of millimeters, and the CARS signal is coherence-length limited. To increase ℓ_c the input beams at ω_L and ω_S are crossed at an angle θ, as shown in Fig. 7.5, so that $\Delta k = 0$. This angle, which is usually 1 to 3°, is known as the phase-matching angle. For crossed beams at ω_L and ω_S the signal becomes limited by the finite spatial overlap of the beams.

The four-wave mixing process that produces the CARS signal takes place in all materials. When $\omega_L - \omega_S$ is not equal to the frequen-

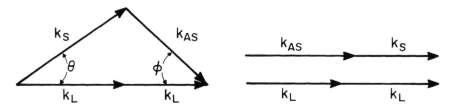

FIG. 7.5 Phase-matching diagrams for CARS. On the right is the collinear phase matching appropriate in dispersionless media (low-pressure gases). On the left is the crossed-beam phase matching necessary in media having finite wavelength dispersion (liquids, solids, high-pressure gases).

cy of some Raman active mode of the sample, the signal is referred to as nonresonant and arises because of the electronic susceptibility of the sample. However, when $\omega_L - \omega_M$ the CARS signal is greatly enhanced due to the existence of a two-photon Raman resonance in $\chi_{CARS}^{(3)}$. The magnitude of $\chi_{CARS}^{(3)}$ at such a resonance can be related to the Raman cross section, $d\sigma/d\Omega$ (Tolles et al., 1977):

$$\chi_{CARS}^{(3)} = \frac{\Delta N_{rg} c^4}{\hbar \omega_S^4} \frac{d\sigma}{d\Omega} \frac{\omega_M}{\omega_M^2 - (\omega_L - \omega_S)^2 - i\Gamma_M(\omega_L - \omega_S)} \tag{7.26}$$

where $\Delta N_{rg} = N_g - N_r$ is the population (number density) difference between the energy levels connected by the Raman transition at frequency ω_M, and Γ_M is the Raman transition line width. The complex quantity $\chi_{CARS}^{(3)}$ is usually written as

$$\chi_{CARS}^{(3)} = \chi' + i\chi'' + \chi^{NR} \tag{7.27}$$

where

$$\chi' = \frac{\Delta N_{rg} c^4}{\hbar \omega_S^4} \frac{d\sigma}{d\Omega} \frac{\omega_M [\omega_M^2 - (\omega_L - \omega_S)^2]}{[\omega_M^2 - (\omega_L - \omega_S)^2]^2 + \Gamma_M^2(\omega_L - \omega_S)^2} \tag{7.28}$$

and

$$\chi'' = \frac{\Delta N_{rg} c^4}{\hbar \omega_S^4} \frac{d\sigma}{d\Omega} \frac{\omega_M \Gamma_M(\omega_L - \omega_S)}{[\omega_M^2 - (\omega_L - \omega_S)^2]^2 + \Gamma_M^2(\omega_L - \omega_S)^2} \tag{7.29}$$

where

χ^{NR} = the nonresonant contribution to the dielectric susceptibility

χ^{NR} = real and independent of ω_L, ω_S, and ω_{AS} (i.e.,

χ^{NR} = constant)

The CARS signal at the anti-Stokes frequency is proportional to $|\chi_{CARS}^{(3)}|^2$, which is

$$\left|\chi_{CARS}^{(3)}\right|^2 = (\chi')^2 + (\chi'')^2 + 2\chi'\chi^{NR} + (\chi^{NR})^2 \tag{7.30}$$

When $\chi^{NR} \ll \chi', \chi''$, only the first two terms in Eq. (7.30) are important, and when $\chi^{NR} \gg \chi', \chi''$, only the last two are. In the former case $I_{AS} \propto [\Delta N(d\sigma/d\Omega)]^2$, while in the latter $I_{AS} \propto [\Delta N(d\sigma/d\Omega)]$ + constant. The CARS signal magnitude can depend either quadratically or linearly on the product of the population difference and the spontaneous Raman cross section. When the resonant and nonresonant contributions to $\chi_{CARS}^{(3)}$ are comparable, all terms in Eq. (7.30) are important and the dependence of the signal magnitude on species density is more complicated. The dependence of the CARS signal on both χ' and χ'' also complicates the signal line shape, since χ', χ'', $(\chi')^2$, and $(\chi'')^2$ have distinctly different line-shape functions, as illustrated in Fig. 7.6

The coherent Raman signals generated in stimulated Raman scattering (SRS) are simpler in both line shape and dependence on molecular density. Stimulated Raman scattering arises because $P^{(3)}(r,t)$ has frequency components at ω_S and ω_L. The polarization at ω_S, for plane light waves propogating along the space-fixed z axis, is

$$P_{SRGS}^{(3)}(z,t) = \frac{3}{2}\chi^{(3)}(-\omega_S,\omega_S,-\omega_L,\omega_L)E_L^2 E_S \cos(\omega_S t - k_S z)$$

$$\tag{7.31}$$

This polarization yields a signal intensity of (Eesley, 1981)

$$I_S(\ell) = I_S(0)\exp\left(\frac{24\pi^2\omega_S}{c^2}\chi_{SRGS}''\ell I_L\right) \tag{7.32}$$

where χ_{SRGS}'' is the imaginary part of $\chi^{(3)}(-\omega_S,\omega_S,-\omega_L,\omega_L)$. For IRS the signal is a loss at frequency ω_L and is given by

$$I_L(\ell) = I_L(0)\exp\left(-\frac{24\pi^2\omega_L}{c^2}\chi_{IRS}''\ell I_S\right) \tag{7.33}$$

where χ_{IRS}'' is the imaginary part of $\chi^{(3)}(\omega_L,-\omega_L,\omega_S,-\omega_S)$. Note that $\chi^{(3)}(\omega_L,-\omega_L,\omega_S,-\omega_S) \equiv \chi^{(3)}(-\omega_S,\omega_S,-\omega_L,\omega_L)$, so $\chi_{IRS}'' \equiv \chi_{SRGS}''$. The signal gain/loss observed in stimulated Raman scattering is usually quite small, 1% or less, so we may use the linear approximations to Eqs. (7.32) and (7.33).

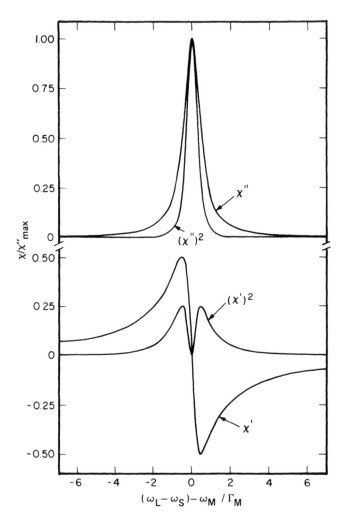

FIG. 7.6 Line-shape functions for the real (χ') and imaginary (χ'') parts of the third-order optical susceptibility.

$$\frac{I_S(\ell)}{I_S(0)} = 1 + \frac{24\pi^2 \omega_S}{c^2} \chi''_{SRGS} \ell I_L \tag{7.34}$$

and

$$\frac{I_L(0)}{I_L(0)} = 1 - \frac{24\pi^2 \omega_L}{c^2} \chi''_{IRS} \ell I_S \tag{7.35}$$

Comparison of these results with the corresponding relationship for the CARS signal intensity, Eq. (7.22), reveals some interesting differences. Unlike CARS the SRS signals depend only on the imaginary part of $\chi^{(3)}$ and have a linear rather than a quadratic dependence. SRS signals are thus simpler in line shape and always depend linearly on ΔN_{rg}, the population (number density) difference between the molecular levels connected by the Raman transition. Also, the SRS gain/loss signal is independent of χ^{NR}, the nonresonant contribution to the third-order susceptibility, so IRS and SRGS are background-free. Another difference is that there is no phase-matching condition that must be satisfied in SRS experiments. As a result, there are no special beam crossing geometries; the laser beams at ω_L and ω_S may be collinear or crossed at any angle.

7.2.4 Signal Magnitudes

The magnitude of the signal to be expected in a particular application of Raman spectroscopy is an important consideration in evaluating the feasibility of a Raman technique for that application. An estimate of the signal level expected from spontaneous Raman scattering, CARS, and SRS can be obtained from Eqs. (7.11), (7.22), (7.34), and (7.35). So many factors enter into these equations, however, that it is impossible to give a typical or representative example to illustrate such a feasibility calculation. There is no "typical" application of Raman scattering. Even a comparison of spontaneous versus coherent Raman signals is strongly influenced by the conditions to be encountered in a particular application. Factors that have to be considred in feasibility calculations include the total molecular density of the sample, the fraction of the sample which is composed of the species of interest, and the distribution of the species of interest over the electronic, vibrational, and rotational states available to it. The line width of the Raman transition and the resolution required are important. The level of luminescence or fluorescence from the sample and how close to the sample the collection optics can be located can dramatically affect the feasibility of a Raman measurement, particularly a spontaneous Raman measurement, and the optical transmission of the sample or surrounding media at the excitation frequency also can be critical. The power, line width, and stability of the lasers available for the measurement must be considered. Signal magnitude and signal-to-noise calculations are more complicated for coherent Raman spectroscopies than for spontaneous Raman spectroscopy, but several recent reviews deal extensively with such calculations for CARS, SRS, and other coherent Raman techniques (Harvey, 1981; Eesley, 1981; Levenson and Song, 1980).

Since the CARS signal depends on the intensity (power per unit area) of the pump and Stokes excitation lasers, it can be dramatically increased by focusing. Equation (7.23) shows that the CARS signal will increase in inverse proportion to the square of the focused-beam area, so even mild focusing can give large signal increase. However, Eq. (7.23) is derived from a plane-wave analysis of the CARS process and is not strictly valid for focused input beams. An appropriate analysis of the focused-beam case by Bjorklund (1975) and Shaub et al. (1977) indicates that the signal power does increase this dramatically with tighter focusing of the input beams, but only until the confocal parameter of the focus is about half the sample length.

Unlike spontaneous Raman scattering, for which signal saturation is unachievable, it is possible to saturate the Raman transition from state g to state r (see Fig. 7.4) using CARS. Hence Eq. (7.22) may not describe the CARS signal magnitude for high laser powers. It is possible to estimate the saturation power levels of the input laser beams (Tolles et al., 1977; Eckbreth and Schreiber, 1981), but the only really reliable check for saturation is an experimental determination of the signal power dependence. The CARS signal will scale at less than $P_L^2 P_S$ when saturation is near. If saturation does not limit the signal, optical breakdown of the sample or the dynamic stark effect (Rahn et al., 1980) can.

7.3 TECHNIQUES OF RAMAN SPECTROSCOPY

7.3.1 Spontaneous Raman Scattering

The apparatus necessary to carry out spontaneous Raman spectroscopy is quite simple. The essential elements, which are shown in Fig. 7.7, consist only of a fixed-frequency laser source, a monochromator to disperse the scattered light, and a detector. Since the excitation frequency, ω_L, can be any frequency, any laser can be used as the excitation source. However, most Raman experiments involve visible laser sources, partly for experimental convenience, but more because of the ready availability of both CW and pulsed lasers having visible output frequencies. Because of the dependence of the scattered light power on ω^4 [see Eqs. (7.8)–(7.11)], it is helpful to use as short a wavelength excitation source as practicable. However, it is generally necessary to avoid photochemical degradation of the sample, so UV excitation is often precluded. Since CW lasers, such as argon ion lasers, have higher average powers than pulsed lasers, CW excitation yields larger signals. However, CW excitation may not give the highest signal-to-noise ratio if the noise source is also continuous. The choice of CW or pulsed excitation depends on the specifics of the experiment, and both CW and pulsed lasers are routinely used for Raman spectroscopy. Many applications of Raman spectroscopy exploit its potential for highly time-resolved measurements, and these require pulsed-laser excitation sources. The commercial availability of nanosecond-and even picosecond-pulse-duration lasers is expanding the use of Raman scattering for highly time-resolved spectroscopy.

Due to the low intensity of the Raman scattered light compared to the intensity of the excitation source, $1:10^{14}$ typically for a gas at 1 torr pressure, it is necessary to use a double monochromator to disperse the scattered light in a Raman experiment. For visible excitation the Raman scattered light generally is shifted only 20 to 1000 Å from the excitation wavelength, so high stray light rejection in the monochromator is essential for a successful measurement. Raman spectra are most often obtained using a photomultiplier detector, with scanning of this monochromator. However, the use of multichannel detectors based on video cameras is growing. Multichannel detection

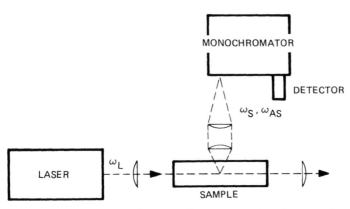

FIG. 7.7 Schematic diagram of an apparatus for spontaneous Raman spectroscopy.

allows simultaneous observation of all or at least a large part of the
Raman spectrum. This can significantly increase the sensitivity of
Raman scattering, by reducing the time necessary to acquire a spectrum. A recent analysis (Freeman et al., 1981) indicates that the improvement is sensitivity can be a factor of 1000 or more.

Figure 7.7 shows collection of the Raman scattered light in a direction orthogonal to the propagation direction of the excitation laser
beam. This orientation minimizes collection of the unwanted scattered
laser light, and generally is the geometry of choice when other constraints are not present. However, collection of the Raman scattered
light, which is distributed over 4π steradians, can be made in other
directions if necessary. For example, the use of Raman scattering
for remote atmospheric sensing in LIDAR experiments (Cooney, 1983;
Leonard, 1981) requires collection of directly backscattered light.

High spatial resolution can be achieved in Raman experiments by
some combination of tight focusing of the excitation beam, careful
aperturing of the collection optics, and precise spatial mapping of the
inelastically scattered light. The high spatial resolution capabilities
of Raman spectroscopy are well illustrated by the Raman microprobe
(Delhaye and Dhamelincourt, 1975), the so-called MOLE (molecular
optic laser examiner). The Raman microprobe couples Raman scattering with conventional optical microscope techniques to allow species-
selective optical microscopy. Spatial resolution of 10^{-4} cm has been
obtained with the Raman MOLE.

7.3.2 Coherent Anti-Stokes Raman Scattering

A schematic diagram of the important elements of a CARS apparatus is
given in Fig. 7.8. Both signal generation and signal collection differ
considerably from the techniques used in spontaneous Raman experiments. Excitation at both the pump frequency, ω_L, and the Stokes
frequency, ω_S, is required, necessitating the use of two laser
sources. Since Raman spectra are obtained by scanning the difference, $\omega_L - \omega_S$, one of the two lasers must have frequency-tunable
output. Since the CARS signal scales as $P_L^2 P_S$, high laser powers
are desirable and pulsed lasers are almost always used in CARS.
Fairly tight focusing (50 to 200-μm-diameter spot size) of the laser
beams also is generally used. As discussed in Sec. 7.2.4, saturation,
optical breakdown, and the dynamic Stark effect limit the magnitude
of the laser intensity at the focus. Nitrogen-pumped and excimer-
pumped dye-laser systems, as well as ruby lasers with ruby-laser-
pumped dye lasers can and have been used for CARS spectroscopy.
However, the most useful and most common setup uses a pulsed
Nd:YAG laser system. The second harmonic (532 μm) of the YAG
can provide the fixed-frequency CARS pump beam at ω_L, as well

as pump a dye laser to generate the tunable-frequency Stokes beam at ω_S. The fairly high repetition rate, 10 to 30 pulses per second, high peak powers, 1 MW or more at both ω_L and ω_S, and very high beam quality make the Nd:YAG/dye system well suited for CARS spectroscopy.

The angle θ at which the beams at ω_L and ω_S are crossed in the sample is chosen to satisfy the phase-matching condition, $\Delta k = 2k_L - k_S - k_{AS} = 0$, discussed in Sec. 7.2.3 and illustrated in Fig. 7.5. For condensed phase samples $\theta = 1$ to $3°$, while for gases, which are nearly dispersionless, $\theta = 0$. When the phase-matching angle is non-zero, the anti-Stokes signal beam is angularly separated from the input beams, facilitating signal detection. When $\theta = 0$, the input beams are collinearly focused into the sample, and the anti-Stokes signal beam propagates out of the cell collinearly with the input beams. When $\theta = 0$, dichroic mirrors usually are used to combine the ω_L and ω_S beams and to separate the signal beam from the input beams. Figure 7.9 shows a photograph of the beam combining optics used in our laboratory for collinear phase matching. The nonlinear nature of the CARS process and the tight focusing of the laser and Stokes beams necessitates very precise (a few micrometers) spatial overlap of the excitation beams.

A monochromator is necessary to separate the anti-Stokes signal beam from the input beams in the case of collinear phase matching and is often used even when $\theta \neq 0$. The monochromator functions only as a high-quality bandpass filter in these CARS experiments, since the spectral resolution derives from, and is determined by, the line widths of the beams at ω_L and ω_S. Since the anti-Stokes signal beam

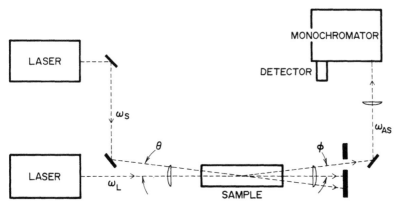

FIG. 7.8 Schematic diagram of a coherent anti-Stokes Raman spectroscopy (CARS) apparatus.

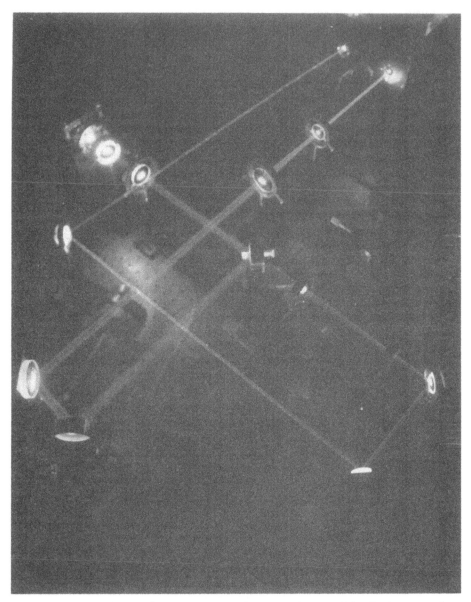

FIG. 7.9 Example of the beam combining optics used in CARS experiments.

is itself a laser beam of small divergence and high coherence, very little of the signal is lost in the process of spectrally and spatially separating the signal from the input beams. In our laboratory, we use a CARS system with collection optics having an effective f-number of about 100 and the equivalent of a double monchromator, with stray light rejection of 10^{10}, yet our signal collection efficiency is nearly 50%. That such efficient signal collection is possible with large f-number optics in CARS is one of the principal reasons this technique is so useful in highly luminescent media such as flames and plasmas. Because the signal levels in CARS experiments are often quite high, photodiodes are adequate detectors in many cases. However, because the dynamic range of photomultipliers is much larger than that of photodiodes, most CARS experiments employ photomultipliers as detectors.

As with spontaneous Raman spectroscopy, simultaneous observation of an entire spectrum using multichannel detectors is possible and advantageous in CARS spectroscopy. For such multiplex experiments the tunable dye laser which generates ω_S is operated broadband, by replacing the grating in the cavity with a total reflector. This will give usable output over a range of 100 cm^{-1} or more, sufficient to cover an entire spectral region of interest in many experiments. With a broadband source at ω_S and a fixed-frequency, single-line source at ω_L the monochromator becomes the important element for dispersing the broadband anti-Stokes signal spectrum and determines the spectral resolution. In these multiplex CARS techniques photodiode arrays and video camera multichannel detectors are used in place of photodiodes and photomultipliers. With multiplex CARS it is often possible to record an entire Raman spectrum in a single nanosecond- or picosecond-duration laser pulse. This capability is frequently exploited in application of CARS to combustion and kinetics studies. As an example of this, Fig. 7.10 shows single-pulse as well as multiple-pulse CARS spectra of nitrogen within the cylinder of a firing internal combustion engine (Klick et al., 1981).

Since the CARS signal is generated only where the ω_L and ω_S beams overlap, and these beams are very tightly focused, highly spatially resolved measurements are possible with CARS. The crossed-beam geometries necessary for phase matching in condensed media automatically restrict the sampled volumes. For example, a 2° crossing angle for input beams of 1 cm diameter focused by a 50-cm-focal-length lens will yield a sample volume about 2 mm long and less than 0.1 mm in diameter (Valentini, 1985). However, for collinear phase matching in gases the resolution along the beam propagation direction is quite poor, of the order of 1 cm or more (Valentini, 1985). In order to allow highly spatially resolved measurements in dispersionless media, special crossed-beam phase-matching geometries have been developed.

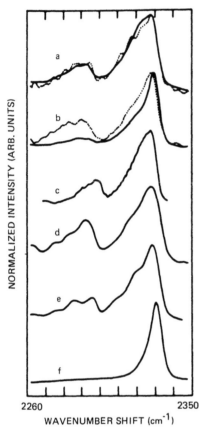

FIG. 7.10 Multiplex CARS spectra of nitrogen within the cylinder
of a firing internal combustion engine: (a) single-pulse (dotted
line) and three-pulse average (solid-line) taken 20 ms after firing
a propane-fueled engine; (b) single-pulse (dotted line) and 15-pulse
average (solid line) taken 10 ms after firing a propane-fueled engine (one or more misfires must have occurred during averaging):
(c) calculated spectrum with T = 2200 K; (d), (e), (f) three-pulse
average spectra taken 10, 20, and 30 ms, respectively, after firing
a methanol-fueled engine. (After Klick et al., 1981.)

The most common of these is a two-frequency, three-beam approach
known as BOXCARS, developed by Eckbreth (1978). In BOXCARS
the pump beam at frequency ω_L is split into two parts, which are
separately imaged into the sample at an angle 2γ, as shown in the

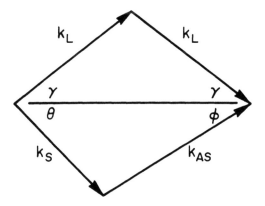

FIG. 7.11 Phase-matching

phase-matching diagram of Fig. 7.11. For a given value of the angle γ the angles θ and ϕ are uniquely determined by the phase-matching requirement. The beams can be phase-matched for any angle γ, and thus any spatial resolution can be obtained, down to the limit set by the focal spot size of the beams. A resolution of 0.1 mm is practical.

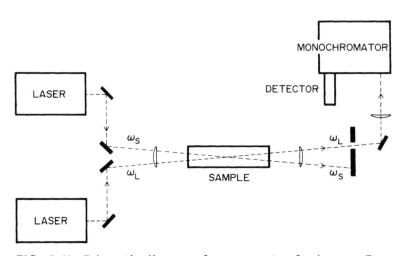

FIG. 7.12 Schematic diagram of an apparatus for inverse Raman spectroscopy.

7.3.3 Stimulated Raman Scattering

The apparatus necessary for carrying out stimulated Raman spectroscopy (SRS) is very similar to that used in CARS experiments. As
illustrated in Fig. 7.12, two laser sources are needed, one at the laser
frequency, ω_L, and a second at the Stokes frequency, ω_S. As with
CARS, Raman spectra are generated by scanning the frequency difference, $\omega_L - \omega_S$, over the Raman frequencies of interest, so one of
the two lasers must be frequency tunable. The two laser beams are
focused into the sample, usually crossed at some angle that makes selective detection of the beam of interest convenient. Phase matching
is achieved in stimulated Raman spectroscopy for any crossing angle
(see Sec. 7.2.3), so selection of the magnitude of this angle is based
principally on experimental convenience. However, since the probed
volume decreases as the crossing angle increases, the signal magnitude is also affected by the crossing angle.

In stimulated Raman gain spectroscopy (SRGS) the signal is detected as an increase in the intensity of the beam at the Stokes frequency, ω_S, when the laser beam at ω_L is present. In inverse Raman
spectroscopy (IRS), which is depicted in Fig. 7.12, the signal is a
decrease in the ω_L intensity. In stimulated Raman scattering the detected beam is referred to as the "probe" and the other beam is called
the "pump." The intensity of the probe beam is measured by a photo-
metric detector, after filtering through a monochromator to isolate it
from the pump.

As Eqs. (7.32) and (7.33) indicate, the stimulated Raman scattering signal depends exponentially on the intensity of the pump beam.
Thus, as in CARS, the use of high-power pulsed lasers for the pump
is desirable. For the probe beam a "quasi-CW" laser is often used
(Esherick and Owyoung, 1982). This is a CW laser whose output is
pulsed by a mechanical chopper synchronized with the repetition rate
of the pulsed pump laser, with a typical "pulse" duration of 100 µs.
These CW probe beams are employed in IRS and SRGS because CW
sources have very high amplitude stability, an important attribute for
these signal gain/loss measurements. The probe beam is pulsed because the gain or loss is effected only during the short time (a few
nanoseconds typically) during which the pump pulse is on. Since the
detector views the full probe-beam intensity, pulsing the probe avoids
detector saturation and allows higher probe powers to be used. Such
experiments use fast photodiodes to monitor the probe-beam signal
level.

It is also possible to effect stimulated Raman spectroscopy without
actually detecting an optical gain or loss. The stimulated Raman scattering populates an excited vibrational state of the sample molecules,
and subsequent vibrational-to-translational energy relaxation will in-

duce a pressure change in the sample. The pressure change can be detected by a simple microphone. This technique is the basis for photoacoustic Raman spectroscopy (PARS), which has been reviewed recently by West et al. (1983).

7.4 APPLICATIONS OF RAMAN SPECTROSCOPY

Since the commercial development of the laser some 20 years ago, the applications of Raman spectroscopy have grown in both number and scope. The high light intensity of the laser has mitigated the principal drawback of Raman scattering, the extreme inefficiency of the scattering, and had made feasible the application of Raman spectroscopy to many problems of chemical and physical analysis. The expansion of applied Raman spectroscopy continues today, driven largely by technological improvements in laser sources. The development of and continuous improvement in laser sources has made it possible to take advantage of the attractive features of Raman spectroscopy, which were clearly recognized early in its development. Among the attractive features is the species specificity of the scattering; that is, the Stokes or anti-Stokes shifts uniquely identify the molecule and its quantum state, so Raman spectroscopy is useful for chemical "fingerprinting" and temperature determination. Since all molecules have at least one Raman-active vibration, these capabilities are quite general. The spectral shift is independent of the wavelength of the excitation source. In contrast to fluorescence spectroscopy there are no quenching or lifetime effects to complicate the analysis of the signals. The instantaneous nature of the scattering allows very highly temporally resolved measurements, and three-dimensionally resolved measurements are also practical. These features give Raman spectroscopy great utility in a wide variety of diagnostic applications. Among the areas where this light-scattering technique is very valuable are combustion, chemical vapor deposition, and kinetics. Raman spectroscopy is also used extensively for analysis and characterization, of solids and solid surfaces, polymers, and thin films. It is used for studying molecular jets and beams, as well as shock waves. The biophysical applications include identification, structural characterization, and kinetic studies. Raman spectroscopy has been used for remote sensing of the atmosphere. Examples of these and other applications are discussed in detail in the sections that follow.

7.4.1 Combustion

The determination of species concentration and temperatures in combustors is one of the most important and exciting applications of Ra-

man spectroscopy. Both spontaneous Raman and coherent Raman
techniques have been used, although coherent Raman spectroscopies
are now more common, because of their lower sensitivity to sample
luminosity, laser-induced fluorescence, and laser-induced incande-
scence. Raman spectroscopy is particularly well suited for combustion applications. The combustion medium can be probed remotely
and nonintrusively, with high spatial and temporal resolution, to
yield information on temperature and species distributions. Common
combustion species such as N_2, O_2, CO, CO_2, H_2O, H_2, CH_4, and
so on, all can be detected by Raman scattering. Multichannel detection techniques allow several species to be detected simultaneously.
Also, coherent Raman techniques such as CARS display large discrimination against sample luminescence. Experiments have been carried out not only in laboratory-scale laminar and turbulent flames,
but also in practical combustion environments such as internal combustion engines (Allessandretti and Violino, 1983; Klick et al., 1981;
Stenhouse et al., 1979) and gas turbine combustors (Greenhalgh
et al., 1983; Druet and Taran, 1981; Eckbreth, 1980). Raman applications in combustion are so numerous that several articles reviewing
the field have been published. Two recent examples are the reviews
by Eckbreth et al. (1982) and by Mattern and Rahn (1980). Although not recent, an excellent discussion of Raman combustion diagnostics is provided in a monograph edited by Lapp and Penney
(1974).

The need for accurate, nonintrusive, spatially resolved temperature measurement in combustion environments has been the motivation
for much of the development of Raman spectroscopy in combustion
science. The Raman spectrum of molecular nitrogen serves as the
probe of temperature. Most reactors are air-fed, so nitrogen is usually the majority species in combustors. Raman temperature measurement is most often carried out by analysis of the band profile of the
nitrogen vibrational Q branch, for which the Raman transition is near
2300 cm^{-1}. However, temperature measurements also have been made
by analysis of rotational Raman spectra of N_2 at small (20 to 200 cm^{-1})
Stokes shifts, as for example in the recent work of Burlbaw and Armstrong (1983). Measurement of temperature with uncertainty of only
±50 K at 2000 K is possible. Temperature determination with spontaneous Raman is limited to flames of low luminosity, because of the
low signal intensities. However, coherent Raman techniques like CARS
allow temperature measurement even in bright, highly sooting flames.

The spatially and temporally resolved determination of temperature has become a hallmark of Raman spectroscopy in combustion applications. An example of the quality of spatially resolved temperature measurements is shown in Fig. 7.13, taken from the work of
Eckbreth and Hall (1979). Their measurements, taken in a laminar

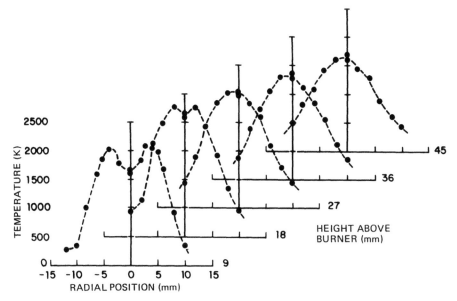

FIG. 7.13 Temperature profiles in a highly sooting, laminar propane diffusion flame. The temperatures were determined from the nitrogen CARS Q-branch rotational band profile. (Reprinted by permission of the publisher from CARS thermometry in a sooting flame, by A. C. Eckbreth and R. J. Hall, *Combust. Flame, 36*, 87; copyright 1979 by The Combustion Institute.)

propane diffusion flame, are believed to be accurate to ±50 K and have 1-mm spatial resolution. The temperatures were obtained by analysis of the nitrogen CARS vibrational Q-branch band contours. Using multiplex (i.e., broadband) CARS, as described in Sec. 7.3.2, it is possible to record the entire N_2 Q branch in a single laser pulse, and thus to determine the temperature in a single laser pulse. This makes highly time-resolved thermometry possible, as evidenced by the work of Klick et al. (1981), some of whose results are given in Fig. 7.10. These temperature measurements were made in a firing internal combustion engine. The spatial resolution is ~ 1 mm, and the temporal resolution is 10 ns for the single-pulse spectra. Such single-pulse temperature measurements are very important in non-steady-state combustion environments, such as jet engines, internal combustion engines, and turbulent flames.

While much combustion-related Raman work involves nonintrusive thermometry, considerable effort has also gone into developing species concentration measurements, particularly spatial mapping of species

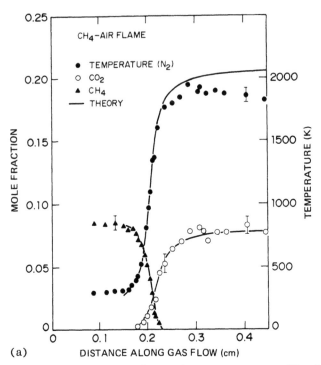

(a)

FIG. 7.14 Concentration and temperature profiles in an atmospheric-pressure, laminar methane-air flame. The points are measurements made by spontaneous Raman spectroscopy. Temperature and concentration profiles of CO_2 and CH_4 are in (a), temperature and concentration profiles for H_2O and O_2 are in (b), temperature and concentration profiles for CO and H_2 are in (c). All temperature measurements were made by analysis of the nitrogen Raman Q-branch band profile. (Reprinted by permission of the publisher from Atmospheric pressure pre-mixed hydrocarbon-air flames: theory and experiment, by J. H. Bechtel, R. J. Blint, C. J. Dasch, and D. A. Weinberger, *Combust. Flame, 42*, 197; copyright 1981 by the Combustion Institute.)

distributions. Long et al. (1983) have reported instantaneous two-dimensional mapping of fuel-gas concentrations in turbulent diffusion flames using spontaneous Raman spectroscopy. Determination of concentration profiles for H_2, O_2, CO, CO_2, H_2O, CH_4, and C_3H_8 in pre-mixed hydrocarbon-air flames has been reported by Bechtel et al. (1981), also by spontaneous Raman scattering. Figure 7.14 shows the

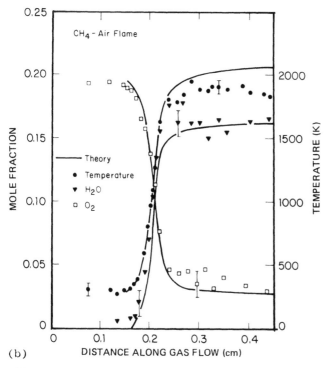

(b)

FIG. 7.14 (Continued)

concentration and temperature profiles they obtained in a methane-air flame. That a single Raman experiment like this can yield the temperature profile and concentration profiles for six different species with better than 1 mm spatial resolution is one of the reasons Raman scattering is so useful in combustion diagnostics. Detection of species other than stable molecules has been demonstrated in flames. Dasch and Bechtel (1981) detected oxygen atoms by spontaneous Raman scattering, and Teets and Bechtel (1981) detected the same species by CARS spectroscopy.

In related work, Taylor (1984) has been using CARS to determine species concentration and temperature in coal gasifiers. He has shown that nonintrusive, real-time measurements of both major (CO) and minor (H_2S) components of the gas stream can be made. The detectability is ~200 ppm. These measurements were made in full-scale coal gasifiers under full-operation conditions.

One problem associated with CARS spectroscopy in combustion applications is the difficulty of minor species detection. As explained

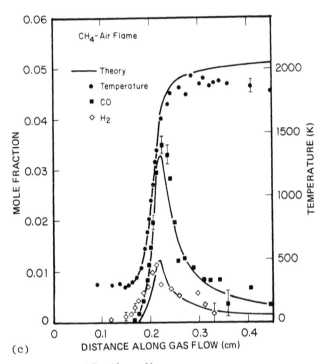

(c)

FIG. 7.14 (Continued)

in Sec. 7.2.3, the contribution to the CARS signal by a species at low relative concentration can be much smaller than the nonresonant background signal. Nonresonant background interference is significant enough that accurate concentration measurements in combustion media are difficult at concentrations less than 0.1 mol%. There are several special, or "advanced," CARS-like spectroscopies which reduce the nonresonant background interference, by separating the resonant signal from the nonresonant signal on the basis of the different electric field polarization direction dependence of the resonant and nonresonant third-order susceptibilities. Recall that $\underline{\chi}^{(3)}$ is a fourth-rank tensor and thus has $3^4 = 81$ elements. The elements of the resonant and nonresonant components of $\underline{\chi}^{(3)}$ are different and obey different symmetries. As a result, it is possible to cross the polarization direction of the incident fields at some angle, such that the resonant contribution to the signal ω_{AS} has an electric field vector component in a direction along which the nonresonant contribution is zero. Thus, by using a properly oriented polarization-sensitive detector one can detect the resonant signal and reject the nonresonant signal.

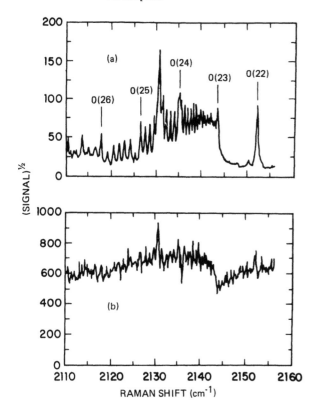

FIG. 7.15 Coherent Raman spectra of CO in a methane-air flame: (a) background-free CARE spectrum; (b) convential CARS spectrum. In (a) lines labeled O(number) are O-branch lines of N_2; all others are CO Q-branch transitions. (After Rahn et al., 1979.)

One such technique is called coherent anti-Stokes Raman ellipsometry (CARE). Using this technique, Pudar et al. (1979) were able to reduce the nonresonant background interference by more than 10^4, allowing them to detect weak resonant signals in dilute liquid mixtures of benzene in carbon tetrachloride. Rahn et al. (1979) have compared CARS to CARE for detection of minor species in flames. They observed an improvement in signal-to-noise ratio of more than a factor of 200 with CARE, with a resultant increase of more than 25 in the detectability of CO in a methane-air flame. Figure 7.15, taken from their publication, dramatically illustrates this improvement.

7.4.2 Chemical Vapor Deposition and Plasmas

The requirements of a good optical diagnostic for chemical vapor deposition (CVD) are very similar to those for combustion. Spatially and temporally resolved species concentration and temperature determination is needed, in a luminescent, non-steady-state environment, which must be probed remotely. In view of the similarities, it is not surprising that Raman scattering is becoming as important a tool for studying CVD as for studying combustion. Several recent reports indicate that both spontaneous and coherent Raman spectroscopies can be very useful diagnostic techniques in CVD. Hanabusa and Kikuchi (1983) used CARS to identify important chemical species in CO_2-laser-driven CVD of silane. Measurements of temperature and species distribution in boron CVD from a H_2-BCl_3 reactive mixture have been made by Bouix et al. (1982) using spontaneous Raman scattering. Breiland et al. (1983) used the Raman scattering induced by a pulsed UV laser to determine temperature as a function of height above the susceptor in chemical vapor deposition cells. They also measured concentration profiles for silane in silicon CVD, and tungsten hexafluoride in tungsten CVD. Figure 7.16, taken from their work, illustrates the sensitivity and spatial resolution possible with Raman spectroscopy.

Raman techniques also have been used to probe plasmas. For example, Tanaka et al. (1982) used stimulated Raman scattering to in-

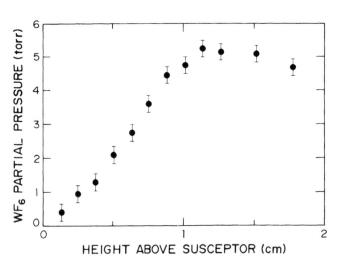

FIG. 7.16 Partial-pressure profile of tungsten hexafluoride in a chemical vapor deposition cell. (After Breiland et al., 1983).

vestigate UV-laser-produced plasmas. Pealat et al. (1981) measured H_2 rotational and vibrational population distributions by CARS in a low-pressure hydrogen plasma. They determined that the rotational distribution could be described by a temperature of 475 K, but a non-Boltzmann vibrational state distribution was evident.

7.4.3 Biology and Biophysics

Raman scattering is well established as an important tool in the life sciences, particularly for elucidating the structure and dynamics of macromolecules. Raman spectroscopy, both spontaneous and coherent, is used to determine molecular structure and establish functionality. Its value for detection and identification of proteins and even simple organisms such as algae has been demonstrated. Raman spectroscopy has become singularly important in identification and characterization of photolytic and reactive intermediates in protein kinetics. Resonance Raman spectroscopy plays a more significant role in these applications of Raman scattering than in any other. Many proteins and other biological macromolecules contain visible chromophores, making resonance Raman effects generally important. Since electronic resonance enhancement can increase Raman cross sections by a factor of 100 or more (see Sec. 7.2.1), the utility of Raman spectroscopy in biology and biophysics is greatly enhanced because of the widespread existence of such chromophores. The significance of biological and biophysical applications of Raman spectroscopy is documented by the large number of recent publications reviewing these applications. Tu (1982) and Carey (1982) have written books dealing exclusively with the application of Raman spectroscopy in biology and biochemistry. Hilinski and Rentzepis (1983) have reviewed the biological applications of picosecond Raman spectroscopy. Carey (1983) has discussed the use of Raman spectroscopy for analysis of biomolecules.

Of the many examples of the use of Raman spectroscopy in biology and biophysics we will cite only a few to illustrate the range of problems that can be addressed. The interested reader is referred to the reviews mentioned above for a more extensive discussion of examples of these applications. Wallach and Verma (1980) carried out Raman spectroscopic studies characterize changes in erythrocyte membrane connected with Duchense muscular dystrophy. Yu et al. (1982) obtained Raman spectra in situ of the eye lens of an anesthetized rabbit. Raman spectroscopy has been used to detect and identify algae in water (Brahma et al., 1983) and to detect hemoproteins in the eluate from high-performance liquid chromatrograph (Iriyama et al., 1983).

Time-resolved Raman studies are particularly important in biophysics. Among the many examples of this application is the study of carboxyhemoglobin photolysis by Friedman and Lyons (1980), who demonstrated, using nanosecond time-resolved Raman scattering, that photolysis at 532 nm leads to deoxyhemoglobin. They also measured the kientics of recombination to reform the carboxyhemoglobin. Such time-resolved Raman studies are now being carried out with picosecond time resolution, as evidenced, for example, by the work of Terner et al. (1980). They obtained detailed structural information on a car-boxyhemoglobin photointermedate, which was probed by resonance Raman spectroscopy within 30 ps of its formation. Some of the Raman spectra they obtained are shown in Fig. 7.17.

7.4.4 Photochemistry and Kinetics

The utility of Raman scattering for time-resolved spectroscopy is illu-strated by the application of spontaneous and coherent Raman spec-troscopy to quantum-state-resolved detection of photofragments and reaction products in the gas phase, to nanosecond- and picosecond-resolution studies of vibrational relaxation in solids, liquids, and gases, and to measurements of rate constants in solids and aqueous solutions.

Pealat et al. (1983) have detected vibrationally excited H_2 from UV photolysis of formaldehyde using CARS, while Valentini and coworkers (Valentini et al. 1981; Valentini, 1983) have extracted nascent rota-tional and vibrational quantum state distributions from collision-free CARS spectra of O_2 formed in the visible and UV photolysis of ozone. Gerrity and Valentini (1984) have extended this use of CARS to the study of chemical reactions under single-collision conditions. They photolyzed HI to make H atoms in an HI/D_2 gas mixture using a nano-second laser pulse. After a time delay that allowed only about one collision of H with D_2, they used a CARS laser pulse of nanosecond duration to record the spectrum of the HD product of the $H + D_2$ re-action. One of the HD CARS spectra they recorded is shown in Fig. 7.18. From such CARS spectra the HD nascent quantum state distributions were obtained.

Several groups have used Raman spectroscopy to investigate the dynamics of vibrational relaxation in gases, liquids, and solids. Lau-bereau (1980) has reviewed picosecond relaxation studies in liquids. The recent picosecond CARS work of Dlott et al. (1982) and that of Ho et al. (1983) illustrate the use of Raman spectroscopy to investi-gate vibrational dephasing in molecular crystals. Gladkov et al. (1983) have combined two-photon Raman pumping with CARS probing to study vibrational energy transfer in gases. They pumped selected vibrational levels of CO_2 or SF_6, and determined the rates of energy transfer to other vibrational levels.

There are several recent examples of the use of Raman scattering in determining rates and mechanisms of chemical reactions in aqueous

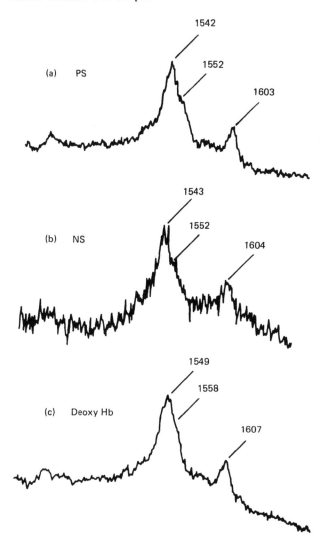

FIG. 7.17 Time-resolved resonance Raman spectrum of the photo-transient of carboxyhemoglobin: (a) with a 30-ps laser pulse, and (b) with a 20-ns laser pulse. (c) shows the steady-state resonance Raman spectrum of deoxyhemoglobin. The Raman spectral peaks are labeled by their transition frequencies. (Reprinted with permission from J. Terner, T. G. Spiro, M. Nagumo, M. F. Nicol, and M. A. El-Sayed, Resonance Raman spectroscopy in the picosecond time scale: the carboxyhemoglobin photointermediate, *J. Am. Chem. Soc.*, *102*, 3228; copyright 1980, American Chemical Society.)

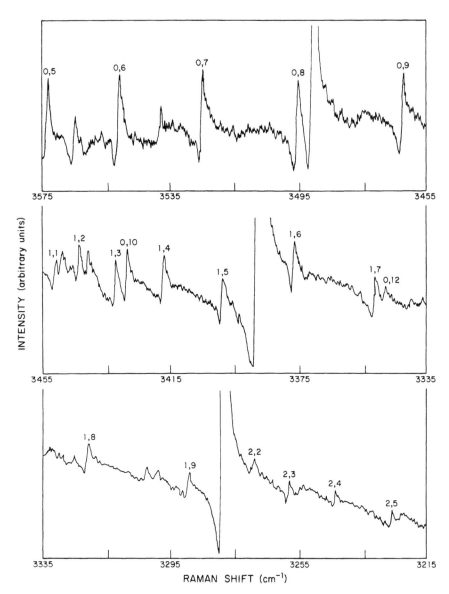

FIG. 7.18 Vibrational Q-branch CARS spectrum of HD formed in the
reaction $H + D_2 \rightarrow HD + D$ under single-collision conditions. The
spectrum was recorded 5 ns after a photolysis laser pulse dissociated
HI to produce H atoms in an HI/D_2 gas mixture. The peaks are labeled in v,J format. Unlabeled peaks are due to D_2. (After Gerrity
and Valentini, 1984).

solution and in solids. Prasad et al. (1982) have reviewed laser Raman studies of solid-state reactions. Nishimura and Tsuboi (1982) used time-resolved Raman to study deuteration kinetics of N-methylacetamide in a stopped-flow solution reactor. In a novel experiment, Rossetti et al. (1984) studied surface redox reactions in aqueous colloidal semiconductor solutions using transient Raman spectroscopy. One optical pulse above the semiconductor band gap was used to create a transient population of electrons and holes in the crystallite. After a fixed delay a second optical pulse below the band gap was used to generate Raman spectra of surface reaction products. CARS has been used to probe the decomposition kinetics of the propellant RDX by Aron and Harris (1984). They documented the transient species above the surface of solid RDX burning unconfined in air. Keiser et al. (1982) monitored the reaction products by Raman spectroscopy in situ during electrochemical reduction of rust films.

7.4.5 Solids, Polymers, and Surfaces

The applications of Raman spectroscopy to the characterization of solids, polymers, and surfaces are particularly diverse. The structure, reactivity, and phase transformation of catalysts, while in preparation or in use, have been studied by Raman techniques. The structure, orientation, and bonding in polymer films and fibers has been investigated by Raman spectroscopy. Raman has been used to characterize samples of coal, coke, and graphite. It has been used for forensic identification of particles in tissue samples. Temperatures of solids undergoing laser annealing can be accurately determined by Raman scattering from the surface. One very exciting recent development in Raman spectroscopy is the observation of enhancements in the Raman scattering efficiency of many orders of magnitude for molecules on surfaces. This effect may make Raman spectroscopy a very sensitive and important tool for analysis of molecular species on surfaces. However, at present the phenomenon is not well understood, and research in surface-enhanced Raman scattering rightly focuses on explaining the phenomenology rather than developing it as the basis for an applied spectroscopic technique. The interested reader is referred to recent reviews by Chang (1983) and Chang and Furtak (1982) for more discussion of the surface-enhanced Raman effect.

As with many other applications of Raman spectroscopy, the ability to perform highly spatially resolved species identification is important in Raman analysis of solids and surfaces. The Raman microprobe, or MOLE (molecular optic laser examiner), has been developed to facilitate spatially resolved Raman analysis of solids and surfaces. The Raman microprobe, developed about 10 years ago (Delhaye and Dhamelincourt, 1975), is the Raman analog of the optical microscope. A sam-

ple is viewed as it would be in a microscope, but it is the spatially resolved Raman scattered light from the sample that is detected. This
allows a spatial map of the distribution of a particular chemical species
in or on the surface of the sample. Duncan et al. (1982) recently demonstrated a CARS version of the Raman microprobe. These Raman
techniques can have spatial resolution of 10^{-4} cm, so they allow very
precise spatial imaging.

The Raman microprobe technique is used for analysis of many kinds
of solids and surfaces. For example, Couzi et al. (1983) have used it
to study structural and chemical inhomogeneities in silicon carbide
filaments deposited by CVD. It has been used to characterize the
structure of coals and cokes, and to identify minerals in those materials (Green et al., 1983). The Raman microprobe has also been used
in forensic and pathological investigations to identify particles in situ
in tissue samples (Abraham et al., 1980). Annealed spots in laser-an-
nealed implanted silicon have been analyzed by microprobe Raman spectroscopy (Nissim et al., 1983). The Raman microprobe is also important in Raman characterization of catalysts. Payen et al. (1982) studied solid-phase transformations in cobalt and nickel molybdates and
investigated the reactivity of these hydrosulfurization catalysts using
the microprobe.

Raman spectroscopy can be quite valuable in catalyst characterization; it can be used to study the preparation of the catalyst, from
the precursor solutions to the activated form, and it can be used to
detect and characterize adsorbed species and poisons. All these experiments can be done in situ, and even opaque pellets or powders
can be investigated. For example, $NiO/Cr_2O_3/MgSiO_3$ methanation
catalysts have been studied by Raman spectroscopy after being reduced, sulfided, and tested for methanation activity (Stencel et al.,
1980). The Raman spectra allowed determination of the structure of
the catalyst and the structural effects of reduction and sulfiding.
Schrader et al. (1981) used Raman spectroscopy to characterize bismuth molybdate selective oxidation catalysts in situ during preparation. Raman spectra of the catalyst as a function of time during preparation are shown in Fig. 7.19.

Polymer structure and orientation has been investigated using Raman spectroscopy. For example, Bower and Ward (1982) determined
the molecular orientation in 0.001-cm-diameter poly(ethylene terphtha-
late) fibers. Rabolt et al. (1980) were able to get Raman spectra of
polymer films as thin as 8×10^{-6} cm. Raman spectroscopy is valuable
in studying ion corrosion (Keiser et al., 1982). In combination with
ion bombardment it can be used to obtain chemical profiles for thin
oxide films. Hamilton et al. (1982) applied this combination of techniques to establish the chemical composition as a function of depth in

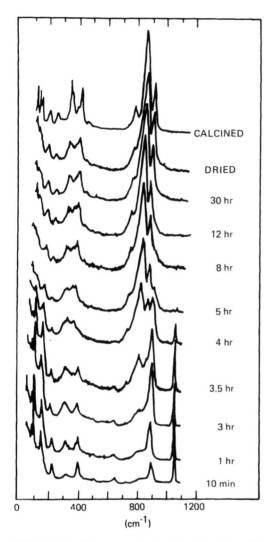

CALCINED

DRIED

30 hr

12 hr

8 hr

5 hr

4 hr

3.5 hr

3 hr

1 hr

10 min

0   400   800   1200

(cm^{-1})

FIG. 7.19 Raman spectra obtained in situ during the preparation of the bismuth molybdate selective oxidation catalyst γ-Bi_2MoO_6. (After Schrader et al., 1981).

films of iron oxide on iron foils and stainless steel. Lo and Campaan (1980) made time-resolved Raman measurements of the lattice temperature during pulsed laser annealing of silicon.

7.4.6 Molecular Jets and Shocks

Since Raman spectroscopy is nonintrusive and is capable of high spatial and temporal resolution, it is an especially good technique for probing flow fields in shocks and molecular jets. Identification of constituent species, determination of temperature, and even measurement of flow velocity, all with high temporal and spatial resolution, can be done with Raman techniques. For example, Herring et al. (1983) carried out inverse Raman spectroscopy to measure the flow velocity in a nitrogen supersonic flow. The Doppler shift of a Raman transition was used to determine the flow velocity with an uncertainty of less than 5%. Snow et al. (1983) were able to obtain spectrally dispersed CARS signals from 20 points along a line across a nitrogen

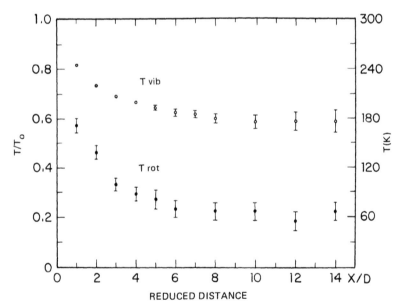

FIG. 7.20 Rotational and vibrational temperature profiles for SF_6 in a supersonic nozzle expansion, obtained from spontaneous Raman spectra of SF_6. T_0 is the nozzle temperature, 300 K, while x is the distance from the nozzle orifice, and D is the nozzle orifice diameter. (After Luijks et al., 1981.)

or oxygen nozzle flow, in a single laser pulse. They used a multiplex CARS crossed-laser-beam geometry which gave 0.1 mm resolution at each of the 20 sampled points. Very detailed characterization of molecular relaxation processes in supersonic jets can be effected with Raman spectroscopy, as evidenced by the work of Luijks et al. (1981). These workers carried out Raman spectroscopy in free-jet expansions of N_2, CO_2, CH_4, and SF_6, and made detailed measurements of molecular temperature as a function of distance from the nozzle orifice. Figure 7.20 presents their results for an SF_6 jet, demonstrating that there are different temperatures for vibrational and rotational degrees of freedom.

An example of the application of Raman spectroscopy to shock-tube diagnostics is the work of Glaser and Lederman (1983). They were able to analyze the temperature and the concentration of both molecu-

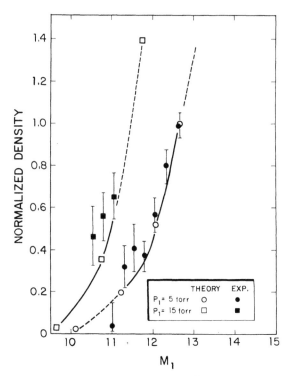

FIG. 7.21 Concentration of N_2^+ behind the reflected shock in a shock tube of N_2, as a function of Mach number, determined by Raman spectroscopy. (After Glaser and Lederman, 1983; copyright American Institute of Aeronautics and Astronautics.)

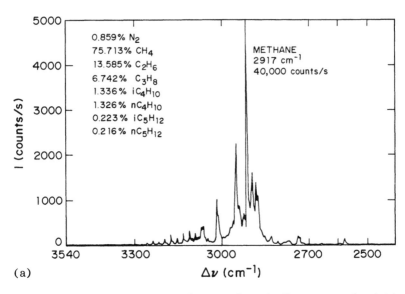

FIG. 7.22 Raman spectrum of a gravimetrically prepared, eight-component, natural gas type mixture. (After Diller and Chang, 1980.)

lar nitrogen and singly ionized molecular nitrogen behind the reflected shock produced in a shock tube filled with nitrogen gas. Figure 7.21 shows their measurement of N_2^+ density as a function of Mach number behind the reflected shock. Raman scattering has also been used to identify transient species in shocked explosives (Tailleur and Cherville, 1982) and to measure vibrational frequency shifts in shocked liquids (Schmidt et al., 1983).

7.4.7 Remote Sensing

Since the laser source and the scattered light collection optics can be located at an arbitrary distance from the sampled region, Raman spectroscopy can be used for remote sensing. This application of Raman scattering is described in detail in Chapter 8, so it will be discussed only briefly here. Cooney (1983) and Leonard (1981) have reviewed the technique and applications of remote Raman sensing. A powerful pulsed laser of nanosecond pulse duration serves as the excitation source, while an optical telescope located at the source collects the Raman-shifted light which is backscattered. Ranging information is provided by the time delay of the scattered light with respect to the excitation pulse. Arshinov et al. (1983) used remote Raman sensing to measure atmospheric temperatures at up to 1 km altitudes, analyzing rotational Raman spectra of N_2 and O_2 to get this information. Their

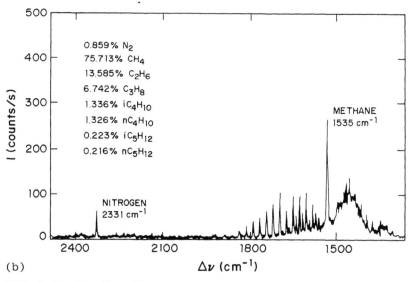

(b)

FIG. 7.22 (Continued)

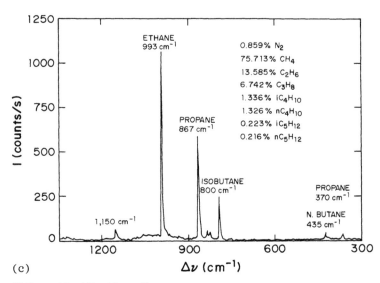

(c)

FIG. 7.22 (Continued)

measurement accuracy was ±0.8 K at altitudes less than 400 m, and
±1.5 K at 1 km. Renaut et al. (1980) obtained water vapor concentration profiles with 30-m resolution and 10% accuracy at altitudes up to
1 km.

7.4.8 Other Applications

There are several other examples of applications of Raman spectroscopy
which warrant mention but do not fit in any of the categories discussed
above. The capability of Raman spectroscopy to provide spectral
"fingerprinting" of compounds allows convenient analysis of mixtures.
Gaufres et al. (1981) studied the equilibrium between the monomer and
dimer of acetic acid in the gas phase by using Raman spectroscopy
to establish the relative amounts of the monomer and dimer. Multichannel Raman spectroscopy provided a means for real time in situ
analysis of the compounds in automobile exhaust (D'Orazio and Hirschberger, 1983). Determination of the composition of combustion gases
from different burning materials has been made with Raman scattering
(Alden et al., 1983). Diller and Chang (1980) demonstrated that the
composition of mixtures of natural gas components could be determined
with only 0.002 mole fraction uncertainty using Raman spectroscopy.
A Raman spectrum which they obtained for an eight-component mixture is shown in Fig. 7.22. The generality and sensitivity of applied
Raman spectroscopy for analyzing mixtures of very similar compounds
is illustrated by these results. The spatial resolution capabilities of
Raman spectroscopy have been put to use in novel ways. For example, Simpson et al. (1983) measured by Raman spectroscopy the
mixing times of crossed streams of microdroplets (130 μm diameter)
and found the times to be ~200 μs. Quantitative determination of
deuterium in laser fusion targets has been reported by Daigreault
et al. (1983), who used inverse Raman spectroscopy to probe these
hollow glass spheres, which have diameters of only about 0.01 cm.

7.5 SUMMARY

The applications of Raman spectroscopy discussed in Sec. 7.4 testify
to the impressive utility and extensive range of this technique for
chemical and physical analysis of gases, liquids, and solids. Development of new applications most certainly will continue in the future,
driven by scientific and technological progress, which presents new
and ever-changing demands for optical diagnostics. Improvements in
the capabilities of spontaneous and coherent Raman spectroscopy can
also be expected, as technological improvements in lasers and detectors take place. Greater sensitivity, improvements in simultaneous

multicomponent detection capabilities, and development of the ability to make instantaneous, three-dimensional concentration or temperature measurements can be anticipated. Other developments that would enhance the utility of Raman spectroscopy may also occur. For example, new or improved methods for suppressing nonresonant background interference in CARS experiments would extend the minor species detection capabilities of this already highly utilized technique. However, we are approaching fundamental limits on some of the measurement capabilities of Raman spectroscopy. For example, the currently achievable spatial and temporal resolution are close to limits imposed by the uncertainty principle.

ACKNOWLEDGMENTS

Work performed in part at the University of California, Los Alamos National Laboratory, Los Alamos, N. M.

The author gratefully acknowledges Martha Maes and Evelyn Cone for their expeditious and accurate typing of the manuscript.

BIBLIOGRAPHY

Bechtel, J. H., Blint, R. J. Dasch, C. J., and Weinberger, D. A. (1981). Atmospheric pressure pre-mixed hydrocarbon-air flames: theory and experiment, *Combust. Flame, 42,* 197. Describes the measurement of temperature and species concentration with 0.1-mm spatial resolution in many types of flames, using spontaneous Raman scattering. Concentration profiles of the fuel, O_2, CO, H_2, CO_2, and H_2O are determined and compared with the predictions of theoretical models of flame propagation and structure.

Breiland, W. G., Coltrin, M. E., and Ho, P. (1983). Laser-base studies of chemical vapor deposition, *Proc. Soc. Photo-Opt. Instrum. Eng., 385,* 146. The use of pulsed UV-laser-Raman spectroscopy for measurement of temperature and species concentration profiles in chemical vapor deposition cells is reported. Rotational Raman spectra of hydrogen and nitrogen were used to obtain temperature profiles at 0 to 2.5 cm above the susceptor with ~1 mm resolution. Concentration profiles of silane and tungsten hexafluoride with similar spatial resolution also are reported.

Carey, P. R. (1982). *Biochemical Applications of Raman and Resonance Raman Spectroscopies*, Academic Press, New York. Extensively reviews the use of Raman spectroscopy in biochemical research.

Delhaye, M., and Dhamelincourt, P. (1975). Raman microprobe and microscope with laser excitation, *J. Raman Spectrosc.*, *3*, 33. Describes the design and uses of a Raman microprobe.

Eesley, G. L. (1981). *Coherent Raman Spectroscopy*, Pergamon Press, Oxford. Presents a unified and consistent description of the phenomenology of all of the important coherent Raman spectroscopies. A brief summary of the applications of many of the techniques is also provided.

Gerrity, D. P., and Valentini, J. J. (1984). Experimental study of the dynamics of the H + D_2 → HD + D reaction at collision energies of 1.30 eV and 0.5 eV, *J. Chem. Phys.*, *81*, 1298. Illustrates the great power of CARS for highly time resolved, quantum-state specific detection.

Harvey, A. B., ed. (1981). *Chemical Applications of Nonlinear Raman Spectroscopy*, Academic Press, New York. This very useful book contains review chapters by several authors describing the applications of coherent Raman spectroscopies to combustion, high-resolution spectroscopy, and chemical analysis. A brief introduction to the theory of nonlinear optical effects is presented.

Leonard, D. A. (1981). Remote Raman measurement techniques, *Opt. Eng.*, *20*, 91. A succinct description of the use of Raman spectroscopy in remote sensing applications. Presents an analysis of the feasibility of such measurements, and examples in atmospheric and oceanographic monitoring are discussed.

Long, D. A. (1977). *Raman Spectroscopy*, McGraw-Hill, New York. An excellent textbook that describes the Raman effect and presents good classical electromagnetic and quantum mechanical treatments of inelastic light scattering. Some of the applications of Raman spectroscopy for molecular spectroscopy are reviewed.

Luijks, G. Stolte, S., and Reuss, J. (1981). Molecular beam diagnostics by Raman scattering, *Chem. Phys.*, *62*, 217. The authors describe spatially resolved measurement of rotational and vibrational temperatures in free-jet nozzle expansions. Beams of N_2, CO_2, CH_4, and SF_6 have been characterized.

Prasad, P. N., Swiatkiewicz, J., and Eisenhardt, G. (1982). Laser Raman investigation of solid state reactions, *Appl. Spectrosc. Rev.*, *18*, 59. Extensive review of the application of Raman spectroscopy to studies of solid-state kinetics.

Terner, J., Spiro, T. G., Nagumo, M., Nicol, M. F., and El-Sayed, M. A. (1980). Resonance Raman spectroscopy in the picosecond time scale: the carboxyhemoglobin photointermediate, *J. Am. Chem. Soc.*, *102*, 3228. Illustrates the use of time-resolved resonance Raman spectroscopy for the detection, identification, and characterization of photochemical and photobiological intermediates.

Tu, A. T. (1982). *Raman Spectroscopy in Biology: Principles and Applications*, Wiley, New York. A good overview of the applications of Raman spectroscopy in the biological sciences.
Valentini, J. J. (1985). *Spectrometric Techniques*, Vol. 4 (G. A. Vanasse, ed.), Academic Press, New York, Chap. 2. An introduction to the phenomenology of CARS, the most important coherent Raman technique, and an extensive discussion of the applications of CARS.

REFERENCES

Abraham, J. L., Andersen, M. E., and Muggli, R. Z. (1980). Raman microprobe identification of pathologically and forensically important particles in tissue, *Proc. Eighth Int. Conf. Raman Spectros*, Bordeaux, France, p. 809.

Alden, M., Blomqvist, J., Edner, H., and Lundberg, H. (1983). Raman spectroscopy in the analysis of fire gases, *Fire Mater.*, 7, 32.

Allessandretti, G. C., and Violino, P. (1983). Thermometry by CARS in an automobile engine, *J. Phys.*, D16, 1583.

Armstrong, J. A., Bloembergen, N., Ducuing, J., and Pershan, P. S. (1962). Interactions between light waves in a nonlinear dielectric, *Phys. Rev.*, 127, 1918.

Aron, K., and Harris, L. E. (1984). CARS probe of RDX decomposition, *Chem. Phys. Lett.*, 103, 413.

Arshinov, Y. F., Bobrovnikov, S. M., Zuev, V. E., and Mitev, V. (1983). Atmospheric temperature measurements using a pure rotational Raman lidar, *Appl. Opt.*, 22, 2984.

Bechtel, J. H., Blint, R. J., Dasch, C. J., and Weinberger, D. A. (1981). Atmospheric pressure pre-mixed hydrocarbon-air flames: theory and experiment, *Combust. Flame*, 42, 197.

Bjorklund, G. C., (1975). Effects of focusing on third-order nonlinear processes in isotropic media, *IEEE J. Quantum Elec.*, QE-11, 287.

Bouix, J., Berthet, M. P., Boubehira, M., Dazord, J., and Vincent, H. (1982). Determination of temperature and partial pressure by Raman scattering inside a CVD reactor, *J. Electrochem. Soc.*, 129, 2338.

Bower, D. I., and Ward, I. M. (1982). Quantitative characterization of orientation in PET fibers by Raman spectroscopy, *Polymer*, 23, 645.

Brahma, S. K., Hargraves, P. E., Howard, W. F., Jr., and Nelson, W. H. (1983). A resonance Raman method for the rapid detection and identification of algae in water, *Appl. Spectrosc.*, 37, 55.

Breiland, W. G., Coltrin, M. E., and Ho, P. (1983). Laser-based studies of chemical vapor deposition, *Proc. Soc. Photo-Opt. Instrum. Eng.*, 385, 146.

Burlbaw, E. J., and Armstrong, R. L. (1983). Rotational Raman interferometer measurement of flame temperatures, *Appl. Opt.*, 22, 2860.

Carey, P. R. (1982). *Biochemical Applications of Raman and Resonance Raman Spectroscopies*, Academic Press, New York.

Carey P. (1983). Raman spectroscopy for the analysis of biomolecules, *Trends Anal. Chem. (Pers. Ed.)*, 2, 275.

Chang, R. K. (1983). Surface-enhanced Raman scattering, *J. Phys. Paris Colloq.*, C10, 283.

Chang R. K., and Furtak, T. E. (1982). *Surface Enhanced Raman Scattering*, Plenum Press, New York.

Cooney, J. A. (1983). Use of Raman scattering for remote sensing of atmospheric properties of meterological significance, *Opt. Eng.*, 22, 292.

Couzi, M., Cruege, F., Martineau, P., Mullet, C., and Pailler, R. (1983). Characterization of composite materials by means of the Raman microprobe, *Microbeam Anal.*, 18, 274.

Daigreault, G. R., Morris, M. D., and Schneggenburger, R. G. (1983). Quantitative determination of deuterium in laser fusion targets by inverse Raman spectroscopy, *Appl. Spectrosc.*, 37, 443.

Dasch, C. J., and Bechtel, J. H. (1981). Spontaneous Raman scattering by ground state oxyten atoms, *Opt. Lett.*, 6, 36.

Delhaye, M., and Dhamelincourt, P. (1975). Raman microprobe and microscope with laser excitation, *J. Raman Spectrosc.*, 3, 33.

Diller, D. E., and Chang, R. F. (1980). Composition of mixtures of natural gas components determined by Raman spectrometry, *Appl. Spectrosc.*, 34, 411.

Dlott, D. D., Schosser, C. L., and Chronister, E. L. (1982). Temperature-dependent vibrational dephasing in molecular crystals: a picosecond CARS study of napthalene, *Chem. Phys. Lett.*, *90*, 386.

D'Orazio, M., and Hirschberger, R. (1983). Multichannel Raman spectrometer for the study of dynamical processes in analytical chemistry, *Opt. Eng.*, *22*, 308.

Druet, S. A., and Taran, J.-P.E. (1981). CARS spectroscopy, *Prog. Quantum Electron.*, *7*, 1.

Duncan, M. D., Reintjes, J., and Manuccia, T. J. (1982). Scanning coherent anti-Stokes Raman microscope, *Opt. Lett.*, *7*, 350.

Eckbreth, A. C. (1978). BOXCARS: crossed-beam phase-matched CARS generation in gases, *Appl. Phys. Lett.*, *32*, 421.

Eckbreth, A. C. (1980). CARS thermometry in practical combustors, *Combust. Flame*, *39*, 133.

Eckbreth, A. C., and Hall, R. J. (1979). CARS thermometry in a sooting flame, *Combust. Flame*, *39*, 133.

Eckbreth, A. C., and Schreiber, P. (1981). *Chemical Applications of Nonlinear Raman Spectroscopy* (A. B. Harvey, ed.), Academic Press, New York, p. 27.

Eckbreth, A. C., Hall, R. J., Shirley, J. A., and Verdieck, J. F. (1982). Spectroscopic investigations of CARS for combustion applications, *Adv. Laser Spectrosc.*, *1*, 101.

Esherick, P., and Owyoung, A. (1982). High resolution stimulated Raman spectroscopy, *Adv. Infrared Raman Spectrosc.*, *9*, 130.

Fenner, W. R., Hyatt, H. A., Kellam, J. M., and Porto, S. P. S. (1973). Raman cross section of some simple gases, *J. Opt. Soc. Am. 63*, 73.

Freeman, J. J., Hendra, P. J., Prior, J., and Reid, E. S. (1981). Raman spectroscopy with high sensitivity, *Appl. Spectrosc.*, *35*, 196.

Friedman, J. M., and Lyons, K. B. (1980). Transient Raman study of CO-haemoprotein photolysis: origin of the quantum yield, *Nature*, *284*, 570.

Gaufres, R., Maillots, J., and Tabacik, V. (1981). Composition of a gaseous mixture in chemical equilibrium as studied by Raman spectroscopy: gas phase acetic acid, *J. Raman Spectrosc.*, *11*, 442.

Gerrity, D. P., and Valentini, J. J. (1984). Experimental study of the dynamics of the $H + D_2 \rightarrow HD + D$ reaction at collision energies of 1.30 eV and 0.55 eV, *J. Chem. Phys.*, *81*, 1298.

Gladkov, S. M., Karimov, M. G., and Korateev, N. I. (1983). Two-photon Raman excitation and coherent anti-Stokes Raman spectroscopy probing of population changes in polyatomic molecules: a novel non-linear optical technique for vibrational relaxation studies. *Opt. Lett.*, *8*, 298.

Glaser, J. W., and Lederman, S. (1983). Shock tube diagnostics utilizing laser Raman scattering, *AIAA J.*, *21*, 85.

Green, P. D., Johnson, C. A., and Thomas, K. M. (1983). Applications of laser Raman microprobe spectroscopy to the characterization of coals and cokes, *Fuel*, *62*, 1013.

Greenhaigh, D. A., Porter, F. M., and England, W. A. (1983). The application of coherent anti-Stokes Raman scattering to turbulent combustion thermometry, *Combust. Flame*, *49*, 171.

Hamilton, J. C., Mills, B. E., and Benner, R. E. (1982). Depth profiling of metal oxides using Raman spectroscopy with ion bombardment, *Appl. Phys. Lett.*, *40*, 499.

Hanabusa, M., and Kikuchi, H. (1983). Coherent anti-Stokes Raman spectroscopic study of CO_2 laser CVD of silane, *Jpn. J. Appl. Phys.*, *Part 2*,

Harvey, A. B., ed. (1981). *Chemical Applications of Nonlinear Raman Spectroscopy*, Academic Press, New York.

Hauchecorne, G. Kerherve, F., and Mayer, G. (1971). Mesure des interactions entre ondes lumineuses dans diverses substances, *J. Paris*, *32*, 47.

Herring, G. C., Lee, S. A., and She, C. Y. (1983). Measurements of a supersonic velocity in a nitrogen flow using inverse Raman spectroscopy, *Opt. Lett.*, *8*, 214.

Hilinski, E. F., and Rentzepis, P. M. (1983). Biological applications of picosecond spectroscopy, *Nature*, *302*, 481.

Ho, F., Tsay, W.-S., Trout, J., Velsko, S., and Hochstrasser, R. M. (1983). Picosecond time-resolved CARS in isotopically mixed crystals of benzene, *Chem. Phys. Lett.*, *97*, 141.

Iriyama, K., Ozaki, Y., Hibi, K., and Ikeda, T. (1983). Raman spectroscopic detection of hemoproteins in the eluate from high-performance liquid chromatography, *J. Chromatogr.*, *254*, 285.

Keiser, J. T., Brown, C. W., and Heidersbach, R. H. (1982). The electrochemical reduction of rust films on weathering stainless steel surfaces, *J. Electrochem. Soc.*, *129*, 2686.

Kittel, C. (1971). *Introduction to Solid State Physics*, Wiley, New York.

Klick, D., Marko, K. A., and Rimai, L. (1981). Broadband single-pulse CARS spectra in a fired internal combustion engine, *Appl. Opt.*, *20*, 1178.

Landsberg, G., and Mandelstam, L. (1928). Eine neue Erscheinung bei der Lichtzerstreuung in Krystallen, *Naturwissenschaften*, *16*, 557.

Lapp, M., and Penney, C. M., eds. (1974). *Laser Raman Gas Diagnostics*, Plenum Press, New York.

Laubereau, A. (1980). Ultrafast vibrational relaxation processes of polyatomic molecules, investigated by picosecond light pulses, in *Relaxation of Elementary Excitations* (R. Kubo and E. Hanamura, eds.), Springer-Verlag, Berlin, p. 101.

Leonard, D. A. (1981). Remote Raman measurement techniques, *Opt. Eng.*, *20*, 91.

Levenson, M. D., and Song, J. J. (1980). Coherent Raman spectroscopy, in *Coherent Nonlinear Optics* (M. S. Feld and V. S. Letokhov, eds), Springer-Verlag, Berlin, p. 293.

Lo, H. W., and Compaan, A. (1980). Raman measurement of lattice temperature during pulsed laser heating of silicon, *Phys. Rev. Lett.*, *44*, 1604.

Long, D. A. (1977). *Raman Spectroscopy*, McGraw-Hill, New York.

Long, M. B., Fourguette, D. C., and Escoda, M. C. (1983). Instantaneous Ramanography of a turbulent diffusion flame, *Opt. Lett.*, *8*, 244.

Luijks, G., Stolte, S., and Reuss, J. (1981). Molecular beam diagnostics by Raman scattering, *Chem. Phys.*, *62*, 217.

Mattern, P. L., and Rahn, L. A. (1980). Coherent Raman combustion diagnostics, *Prepr. Pap. Nat. Meet. Div. Pet. Chem. Am. Chem. Soc.*, *25*, 105.

Nishimura, Y., and Tsuboi, M. (1982). Deuteration kinetics of *N*-methylacetamide by means of stopped-flow multi-detector Raman spectrophotometry, *J. Raman Spectrosc.*, *12*, 138.

Nissim, Y. I., Sapriel, J., and Oudar, J. L. (1983). Microprobe Raman analysis of picosecond laser annealed implanted silicon, *Appl. Phys. Lett.*, *42*, 504.

Oudar, J.-L., Smith, R. W., and Shen, Y. R. (1979). Polarization-sensitive coherent anti-Stokes Raman spectroscopy, *Appl. Phys. Lett.*, *34*, 758.

Pantell, R., and Puthoff, H. (1969). *Fundamentals of Quantum Electronics*, Wiley, New York.

Payen, E., Dhamelincourt, M. C., Dhamelincourt, P., Grimblat, J., and Bonnelle, J. P. (1982). Study of Co (or Ni)-Mo oxide phase transformation and hydrodesulfurization catalysts by Raman microprobe equipped with new cells, *Appl. Spectrosc.*, *36*, 30.

Pealat, M., Taran, J. P. E., Taillet, J., Bacal, M., and Bruneteau, A. M. (1981). Measurement of vibrational populations in low-pressure hydrogen plasma by coherent anti-Stokes Raman scattering, *J. Appl. Phys.*, *52*, 2687.

Pealat, M., Debarre, D., Marie, J.-M., Taran, J.-P. E., Tramer, A., and Moore, C. B. (1983). CARS study of vibrationally excited H_2 produced in formaldehyde dissociation, *Chem. Phys. Lett.*, *98*, 299.

Placzek, G. (1934). *Hanbuch der Radiologie*, Akademische Verlagsgesellschaft, Wiesbaden, West Germany.

Prasad, P. N., Swiatkiewicz, J., and Eisenhardt, G. (1982). Laser Raman investigation of solid state reactions, *Appl. Spectrosc. Rev.*, *18*, 59.

Rabolt, J. F., Santo, R., and Swalen, J. P. (1980). Measurements on thin polymer films and organic monolayers, *Appl. Spectrosc.*, *34*, 517.

Rahn, L. A., Zych, L. J., and Mattern, P. L. (1979). Background-free CARS studies of carbon monoxide in a flame, *Opt. Commun.*, *30*, 249.

Rahn, L. A., Farrow, R. L., Koszykowski, M. L., and Mattern, P. L. (1980). Observation of an optical Stark effect on vibrational and rotational transitions, *Phys. Rev. Lett.*, *45*, 620.

Raman, C. V., and Krishnan, K. S. (1928). A new type of secondary radiation, *Nature*, *121*, 501.

Regnier, P. R., Moya, F., and Taran, J. P. E. (1974). Gas concentration measurement by coherent Raman anti-Stokes scattering, *AIAA J.* *12*, 826.

Renaut, D., Pourny, J. C., and Capitini, R. (1980). Daytime Raman-LIDAR measurements of water vapor, *Opt. Lett.*, *5*, 223.

Rossetti, R., Beck, S. M., and Brus, L. E. (1984). Direct observation of charge-transfer reactions across semiconductor: aqueous solution interfaces using transient Raman spectroscopy, *J. Am. Chem. Soc.*, *106*, 980.

Schmidt, S. C., Moore, D. S., Schiferl, D., and Shaner, J. W. (1983). Backward stimulated Raman scattering in shock-compressed benzene, *Phys. Rev. Lett.*, *50*, 661.

Schrader, G. L., Basista, M. S., and Bergman, C. B. (1981). Laser Raman spectroscopy of precipitates: applications in catalysis, *Chem. Eng. Commun.*, *12*, 121.

Shaub, W. M., Harvey, A. B., and Bjorklund, G. C. (1977). Power generation in coherent anti-Stokes Raman spectroscopy with focused laser beams, *J. Chem. Phys.*, *67*, 2547.

Simpson, S. F., Kincaid, J. R., and Haller, F. J. (1983). Microdroplet mixing for rapid reaction kinetics with Raman spectrometric detection, *Anal. Chem.*, *55*, 1420.

Snow, J. B., Zheng, J.-B., and Chang, R. K. (1983). Spatially and spectrally resolved multipoint coherent anti-Stokes Raman scattering from N_2 and O_2 flows, *Opt. Lett.*, *8*, 599.

Stencel, J. M., Bradley, E. B., and Brown, F. R. (1980). Infrared and Raman spectra of a sulfur-resistant methanatioan catalyst, *Appl. Spectrosc.*, *34*, 319.

Stenhouse, I. A., Williams, D. R., Cole, J. B., and Swords, M. D. (1979). CARS measurements in an internal combustion engine, *Appl. Opt.*, *22*, 3819.

Tailleur, M. H., and Cherville, J. (1982). Study of PETN initiation conditions by ultrafast Raman spectrometry, *Propellants Explos. Pyrotech.*, *7*, 22.

Tanaka, K., Goldman, L. M., Seka, W., Richardson, M. C., Sources, J. M., and Williams, E. A. (1982). Stimulated Raman scattering from UV-laser-produced plasmas, *Phys. Rev. Lett.*, *48*, 1179.

Taylor, D. J. (1984). Unpublished.

Teets, R. E., and Bechtel, J. H. (1981). Coherent anti-Stokes Raman spectra of oxygen atoms in flames, *Opt. Lett.*, *6*, 458.

Terner, J., Spiro, T. G., Nagumo, M., Nicol, M. F., and El-Sayed, M. A. (1980). Resonance Raman spectroscopy in the picosecond time scale: the carboxyhemoglobin photointermediate, *J. Am. Chem. Soc.*, *102*, 3228.

Tolles, W. M., Nibler, J. W., McDonald, J. R., and Harvey, A. B. (1977). A review of the theory and application of coherent anti-Stokes Raman spectroscopy (CARS), *Appl. Spectrosc.*, *31*, 253.

Tu, A. T. (1982). *Raman Spectroscopy in Biology: Principles and Applications*, Wiley, New York.

Valentini, J. J. (1983). Anomalous rotational state distribution for the O_2 photofragment in the UV photodissociation of ozone, *Chem. Phys. Lett.*, *96*, 395.

Valentini, J. J. (1985). *Spectrometric Techniques*, Vol. 4 (G. A. Vanasse, ed.), Academic Press, New York, Chap. 2.

Valentini, J. J., Moore, D. S., and Bomse, D. S. (1981). Collision-free coherent anti-Stokes Raman spectroscopy (CARS) of molecular photofragments, *Chem. Phys. Lett.*, *83*, 217.

Wallach, D. F. H., and Verma, S. P. (1980). Some Biomedical Applications of Raman Spectroscopy, *Proc. Eighth Int. Conf. Raman Spectrosc*, Bordeaux, France, p. 801.

West, G. A., Barrett, J. J., Siebert, D. R., and Reddy, K. V. (1983). Photoacoustic spectroscopy, *Rev. Sci. Instrum.*, *54*, 797.

Yu, N. T., Kuck, J. F. R., and Askren, C. C. (1982). Laser Raman spectroscopy of the lens in situ, measured in an anesthetized rabbit, *Curr. Eye Res.*, *1*, 615.

8

Laser Remote Sensing Techniques

WILLIAM B. GRANT   *Jet Propulsion Laboratory, California Institute of Technology, Pasadena, California*

8.1 INTRODUCTION

"Laser remote sensing techniques" is used in this chapter to refer to any of a number of techniques that use lasers to measure properties of a medium or its constituents at some distance from a laser system. This definition includes remote measurement of atmospheric trace species (molecules, atoms, ions), aerosols, meteorological parameters, chemical composition, and other properties of the earth's surface, bodies of water, chemicals in solution, and atmospheres of other planets. In many cases the laser is used as an "active" probe of the medium, interacting directly with it. For example, a laser beam can be propagated into the atmosphere and the resultant elastic scattering (Mie from aerosols, Rayleigh from molecules, or resonance fluorescence), inelastic scattering (Raman or fluorescence), or the absorption combined with backscattering can be measured. In other cases, the laser is merely used as a local oscillator for a "passive" heterodyne radiometer, providing the central frequency that mixes with radiation from inside or beyond the medium incident on the detector.

   This chapter introduces the reader to the field of laser remote sensing and refers to much of the recent literature in the field. Listed at the end of the chapter are a number of excellent review articles, books, and conference proceedings that develop the theory of laser remote sensing and discuss the state of the art of the field and its status at the time of their publication.

   Several acronyms often used in this field are defined here: lidar stands for "light detection and ranging," DIAL for "differential absorption lidar," and DISC for "differential scattering." Other acronyms sometimes encountered in the literature which are nearly synonymous with DIAL are DASE for "differential absorption and scattering of energy," and DAS for "differential absorption and scattering."

8.1.1 Laser Properties Useful for Remote Sensing

The properties that make lasers indispensible for remote sensing applications are their tunability, monochromaticity, high intensity or power, high spectral brightness, and coherence. In combination, these properties allow lasers to be used to generate probe beams that interact with specific properties of targets, yielding backscattered radiation which carries information about the concentrations and distributions of the components. Tunability permits operation in spectral regions where particular molecules, ions, or geologic features have characteristic absorptions or reflections. Monochromaticity allows different absorption or reflection features to be resolved. High intensities allow measurements to be made at large distances, giving the technique stand-off capability. High-resolution spectrographic techniques take advantage of the high spectral brightness (intensity divided by bandwidth). Coherence allows calculation of beam propagation according to the theory of Gaussian optics (Siegman, 1979). Highly collimated beams propagate energy efficiently and permit the use of narrow fields of view which reduce the impact of background radiation on the measurement.

8.1.2 Advantages of Laser Remote Sensing

The primary advantages of laser remote sensing include (McClenny, 1978):

1. The ability to probe regions inaccessible to point analysis. Examples include detection of plumes and hazardous gas leaks in the troposphere, molecules in the stratosphere, and metal ions and atoms in the thermosphere.
2. The ability to rapidly probe a large area or volume as in the measurement of SO_2 in the effluent of plumes from power plants (Hawley et al., 1983) or ozone in an air basin (Shumate et al., 1983).
3. The ability to measure concentrations and distributions *in situ*, when taking samples would result in sampling degradation.
4. The ability to probe "actively," removing time-of-day restrictions for remote sensing measurements that use the sun as the radiation source, [e.g., SO_2 and NO_2 measurements by correlation spectrometry (Hoff and Millan, 1981)].

Disadvantages of the method include the cost of the apparatus, possible eye-safety considerations, the sophisticated nature of the signal detection and processing, and the potential ambiguity in the interpretation. As we will see later, some model of the laser system and sampled region is usually needed to perform the data analysis.

8.1.3 History of Laser Remote Sensing

The earth's atmosphere was the first major target of investigation for
laser remote sensing. Early in the 1960s, pulsed ruby lasers were
used to measure aerosol density distributions in the stratosphere (Fiocco and Smullin, 1963). Soon thereafter, a temperature-tuned ruby
lidar was used to measure the vertical water vapor profile using the
DASE (DIAL) technique (Schotland, 1966). By the late 1960s, Raman
scattering techniques were being used to detect various atmospheric
trace gases (Leonard, 1967; Cooney, 1968), and fluorescence techniques to detect metal ions such as sodium in the lower thermosphere
(80 to 100 km) (Bowman et al., 1967). In the early 1970s, the DIAL
technique was developed using dye lasers to determine concentrations
of atmospheric pollutants such as NO_2 (Rothe et al., 1974), and SO_2
and O_3 (Grant and Hake, 1975). In the mid-1970s, the DIAL technique
was extended to the middle infrared spectral region, where many molecules have absorption bands, by the measurement of water vapor using
a CO_2 lidar (Murray et al., 1976; Murray, 1978).

Closely related to the DIAL technique is the laser long-path technique for the meausrement of path-averaged molecular gas concentrations. In this case, either a topographic target or a retroreflector is
used to define the path and reflect laser radiation back to a receiver
near the laser transmitter. The use of a CO_2 laser with topographic
targets for the measurement of a number of gases was suggested by
Hanst (1970), but the first laser long-path measurements were made
on ethylene (C_2H_4) in automobile exhaust by using a tunable diode laser and a retroreflector (Hinkley, 1972).

Meteorological properties of the atmosphere can also be studied using laser remote sensing techniques. The measurement of tropospheric
wind velocities using laser Doppler velocimetry was suggested by Jelalian and Huffaker (1967). In this approach, small Doppler frequency
shifts of aerosols as they are transported by the wind are measured
with homodyne or heterodyne detection techniques (i.e., radio-frequency detection is used for the signal at the difference frequency
between the local oscillator and the backscattered energy). In addition, atmospheric pressure measurements can be made using differential absorption by oxygen (Korb and Weng, 1983), and atmospheric
temperature can be measured using differential absorption by oxygen
(Mason, 1975) or CO_2 (Murray et al., 1979).

The laser can also be used to make passive measurements of atmospheric constituents using the laser heterodyne radiometer approach.
The laser is used as a local oscillator, defining the spectral region in
which measurements will be made. This approach has been applied
successfully to chlorine monoxide in the stratosphere (Menzies, 1983)
and CO_2 in the atmosphere of Mars (Deming et al., 1983).

In addition, the earth's surface has features that can be studied using active lidar systems: fluorescence techniques can be used to identify oils spilled on water surfaces (reviewed by Measures, 1984), or to locate deposits of uranium and many other ores (deNeufville et al., 1981; Kasdan et al., 1981). Also, differential spectral reflection in the thermal infrared can be used to identify locations of geological features (Kahle et al., 1984).

8.2 NONRESONANT TECHNIQUES

While many interesting laser remote sensing techniques use spectroscopic features, for many others the laser frequency is not tuned into a resonance and is of secondary importance. The explanation of nonresonant techniques serves as a background to understanding the resonant methods. Examples include measurement of aerosol distributions and measurement of wind velocities, both of which use Mie backscatter from aerosols. Since Mie backscatter is much less dependent on wavelength than is absorption by molecules and atoms, fixed-frequency lasers can be used. Commonly, the Nd:YAG at 1.064 μm (or frequency-doubled at 0.53 μm) is used for aerosol distributions, and the CO_2 laser in the region 9 to 11 μm is used for wind profiles through laser Doppler velocimetry. The theory and applications of nonresonant techniques are presented in this section.

8.2.1 The Lidar Equation and Elastic Backscatter

The lidar equation describing the dependence of the atmospheric elastic backscatter detected by a receiver coaxial with a laser transmitter is given in a general form, derived from the Beer-Lambert-Bouguer law for an absorbing medium, by

$$P_r = \frac{P_t A \eta \beta(R) f(R) c \tau \exp[-2(\alpha C + \sigma)R]}{2R^2} \tag{8.1}$$

where

P_r = received power (W)
P_t = transmitted power (W)
A = area of the receiver (m^2)
η = optical efficiency of the receiver and detector
β = volume backscatter coefficient of the atmosphere (m^{-1} sr^{-1})
R = distance to the atmospheric region doing the backscattering (m)
$f(R)$ = a function relating the overlap between the transmitted beam
        and the receiver field of view

c = speed of light (m s^{-1})

α = absorption due to molecules (atm^{-1} m^{-1})

C = concentration of the molecules (atm)

σ = scattering coefficient due to molecules and aerosols (m^{-1})

τ = width of the laser pulse (s)

The factor of 2 in the exponential function comes from the two-way transmission through the atmosphere to the target volume. The A/R^2 term is the solid angle of the collecting aperture, and the $c\tau/2$ term represents the instantaneous pulse path length. Note that the lidar equation has some terms related solely to the system and some solely to the atmosphere. Table 8.1 gives typical parameter values for various lidar systems.

The volume backscattering coefficient, β (m^{-1} sr^{-1}), is defined as the fractional amount of the incident energy scattered per steradian in the backward direction per unit atmospheric path length. It can be a sum of many scattering contributions:

$$\beta = n_1 \frac{d\sigma(\pi)_1}{d\Omega} + n_2 \frac{d\sigma(\pi)_2}{d\Omega} + \cdots = \sum_i n_i \frac{d\sigma(\pi)_i}{d\Omega} \qquad (8.2)$$

where n_i are species densities and $d\sigma(\pi)_i/d\Omega$ are differential scattering cross sections. A few calculations will illustrate the orders of magnitude involved. Cross sections and species densities are often given in units of centimeters, while the final result is often given in meters. Number densities of aerosols and dusts in the lower atmosphere can vary from 10^2 to 10^5 cm^{-3}, with Mie scattering cross sections up to 10^{-8} cm^2 sr^{-1}. Dominant molecular species densities can range from 10^{17} to 10^{19} cm^{-3} at sea level, with Raman cross sections of 5×10^{-32} cm^2 sr^{-1} at 1.06 μm. Hence β can vary from 10^{-1} m^{-1} sr^{-1} in the former case to 10^{-13} m^{-1} sr^{-1} in the latter.

Consider a 1-J, 10-ns outbound pulse ($\lambda = 1.06$ μm) with the following numerical values for the other parameters:

$A/R^2 = 10^{-7}$ sr

$\eta f(R) = 0.003$

electronics limit = $\dfrac{10c\tau}{2} = 15$ m

$\alpha C + \sigma = 0.11$ km^{-1}

R = 1 km

The number of outgoing photons per pulse is 1 J/hν = 5×10^{18}. The number of returning photons in the case of Mie scattering from aero-

TABLE 8.1 Typical LIDAR System Parameters

Laser	Dye	Doubled Dye	Ruby	Nd:YAG	Doubled Nd:YAG	CO_2
Spectral region (μm)	0.4-0.9	0.25-0.4	0.694	1.064	0.532	9-11
Pulse energy (J)	0.05-0.5	0.05	1	0.5	0.2	0.1-10
Pulse width (ns)	10-1000	10	10	10	10	100-1000
Pulse repetition frequency (Hz)	10	10	1	10	10	0.1-100
Receiver area (m^2)	0.1-0.2	0.1-0.2	0.1	0.1	0.1	0.02-0.2
Receiver/detector efficiency	0.02-0.1	0.1	0.03	0.003	0.1	0.2
Typical applications	NO_2, H_2O, metals	SO_2, O_3, metals	Aerosols, H_2O	Aerosols	Aerosols	Molecules, winds

sols would be 2×10^9, and from Raman scattering from dominant species -6×10^{-4}. The latter result indicates that Raman scattering is not useful at $1.06 \ \mu m$ at large distances for even major components. However, Raman scattering cross sections increase as the fourth power of laser frequency, and detectors are more efficient at shorter wavelengths, so an improvement of about three orders of magnitude could be expected by making measurements in the ultraviolet spectral region.

We have assumed in the analysis above for aerosol backscatter that the scattered wavelength was the same as the incident wavelength. Frequency shifts do exist due to the Doppler effect, originating in scattering from moving aerosols and molecules (Ansmann, 1985). These shifts are not important in most lidar systems that use bandwidths on the order of a few wavenumbers ($1 \ cm^{-1} = 30$ GHz). They are important to lidar systems that use heterodyne detection techniques with a bandwidth of typically several tenths of a megahertz for measurement of tropospheric wind velocities, as well as for lidar systems that use a narrow-bandpass Fabry-Perot filter to separate Rayleigh and Mie scattering (Shipley et al., 1983; Sroga et al., 1083), and for DIAL measurements of narrow molecular absorption features (Ansmann, 1985).

As noted above, there are other forms of scattering besides Rayleigh and Mie. When these give rise to return signals at frequencies other than the incident frequency, Eq. (8.1) must be modified. Details of the modifications and resulting equations can be found in Measures (1984, Chap. 7).

One of the primary requirements in using a lidar system for making reliable quantitative measurements is a proper calibration of the lidar response function. This calibration is especially important for a single-frequency lidar used to make measurements of atmospheric backscatter because some factors such as the transmitter/receiver overlap functions are difficult to calculate. A block diagram of the measurements and calculations necessary to calibrate an aerosol backscatter measuring lidar system is shown in Fig. 8.1. The first aspect of this procedure is to use a target with well-characterized reflection properties—in both wavelength and angle. If this target can be positioned at several locations, it can be used to measure the transmitter/receiver overlap function; if not, the overlap function will have to be estimated. Next, attenuation due to molecular gases and aerosols has to be determined for the laser frequencies being used. The calibration may be valid for only a short time due to changes in atmospheric properties and possible changes in lidar alignment. Once the lidar is calibrated, it can be used for making atmospheric measurements, but again, care must be exercised in determining other properties of the atmosphere that might influence the measurements of interest. In three journal articles (Kavaya et al., 1983; Menzies et al., 1984b; Ka-

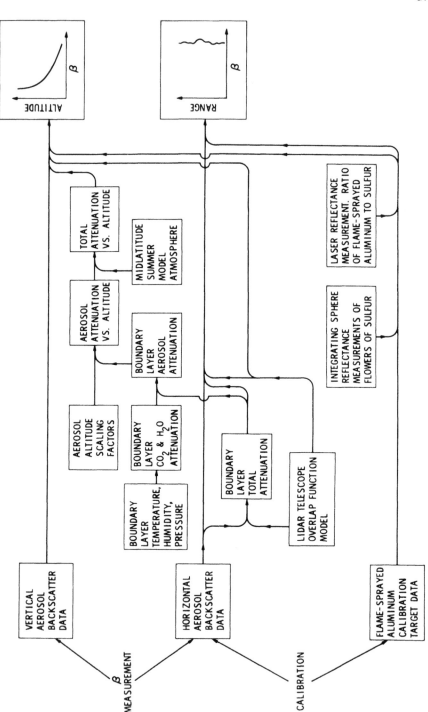

FIG. 8.1 Flow diagram depicting the main components in the lidar calibration and β profile computation (Kavaya et al., 1983).

vaya and Menzies, 1985), this point is discussed in considerable detail in reference to calibrating a CO_2 lidar, emphasizing the properties of the solid targets used for the reflectance calibration. As the authors demonstrate, rough surfaces cannot be necessarily assumed to give Lambertian scattering (angular dependence varies with cos θ, where θ is the angle of incidence), as had heretofore been assumed. Laser-beam polarization is also shown to be important in the scattering function. Under the best circumstances, only a ±50% measurement accuracy is expected, due to quantities, such as aerosol extinction and transmitter/receiver overlap function, which must be modeled.

8.2.2 Aerosol Measurements

Aerosols were historically the first atmospheric constituents to be measured using lidar and are closely connected with much of the subsequent work in laser remote sensing of the atmosphere. The parameters of aerosols usually studied with in situ monitors include their spatial distribution, size distribution, physical shape and state, and chemical composition. Lidar techniques are also being applied to measure these parameters.

8.2.2A. Spatial Distribution

The spatial distribution of aerosols is needed for studies of dynamics of the earth's boundary layer, plume dispersion and opacity, the distribution of water clouds, and fluctations in the density of the stratospheric aerosol layer (15 to 25 km). Lidar systems can be used from the earth's surface, from airplanes, or, in the future, from spaceborne platforms. For many aerosol measurements, an airborne platform allows coverage of large regions rapidly. For some measurements of the earth's boundary layer, a stationary or mobile surface-mounted platform is most appropriate. For example, the height of the temperature-inversion layer that traps urban pollution in a region called the "mixing layer" (so called because of the large degree of turbulence and mixing in this layer) can be monitored in a routine fashion from the ground. Airborne aerosol-measuring lidar systems can also be used, as described, for example, by Uthe (1983). In a 1979 study, a pair of flights over the Los Angeles Basin clearly show the boundary layer structure and some of its diurnal variation (see Fig. 8.2).

Aerosol plume dispersion is ideally studied by lidar, since a three-dimensional profile of the plume can be measured rapidly with a spatial resolution on the order of 15 to 30 m along the line of sight (limited by the laser pulse width and the electronics processing rate), and a few meters perpendicular to the line of sight, controlled by the scanning rate.

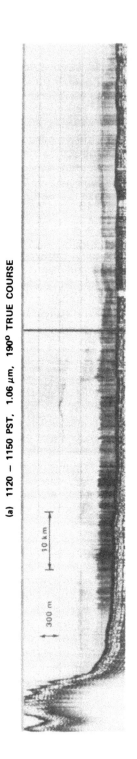

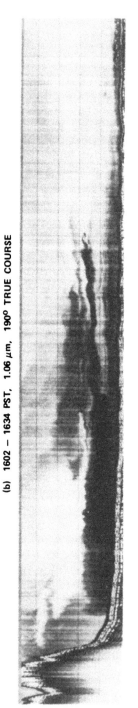

(a) 1120 – 1150 PST, 1.06 μm, 190° TRUE COURSE

(b) 1602 – 1634 PST, 1.06 μm, 190° TRUE COURSE

FIG. 8.2 Los Angeles boundary layer structure measured using an airborne Nd:YAG lidar, December 16, 1979 (Uthe, 1983).

574

Lidar has been used routinely to measure aerosol backscatter from the stratosphere at a number of lidar installations around the earth, primarily in the 30 to 55° north latitude regions. There is a fairly stable aerosol layer at an altitude of 15 to 25 km above mean sea level called the Junge layer (Junge and Manson, 1961). This layer is important because the aerosols in this layer can reflect solar radiation, thus reducing the average temperature of the earth's atmosphere and surface such as after a volcanic eruption [for a lidar determination of this effect, see Labitzke et al. (1983)]. Also, after volcanic eruptions, such aspects as the increase in aerosol mass (if the aerosol size distribution and composition are known), transport, and decay of the total mass can be determined by lidar (McCormick and Swissler, 1983). The transport is followed either by the informal international network of lidar stations (McCormick, 1982) or by a lidar system mounted in an aircraft (McCormick et al., 1984).

8.2.2B. Aerosol Size Determination

The key to aerosol size determination is the dependence of the Mie backscatter efficiency on the size parameter, b, defined as the ratio of particle circumference to incident radiation wavelength. For values of the size parameter less than about 2, the backscatter efficiency rapidly increases with the size parameter. Above a value of 2, the efficiency oscillates considerably; it will normally increase if the imaginary portion of the aerosol index of refraction, associated with absorption, is small, but decrease if it is large (see, e.g., Collis and Uthe, 1972; McCartney, 1976). Thus, in general, if the wavelength is much larger than the diameter of the aerosol, the scattering will be weak, but if it is equal to or smaller than the particle diameter, the backscatter will be quite large. An example of this is given in a measurement by Uthe et al. (1982) in which a "two-color" lidar (1.06 and 0.532 μm from a Nd:YAG laser) was used to meausre forest-fire plume backscatter from an airborne platform. The backscatter at 0.532 μm was three or four times more intense than that at 1.06 μm, indicating that the predominant aerosol diameter was about 0.1 μm. The presence of gases that absorb strongly at 1.06 μm but not at 0.532 μm could affect the measurement, but no such gases were expected. This result agreed with independent size estimates. Martonchik et al. (1984), calculated the optical properties of NH_3 ice for particle sizes from 1 to 100 μm to aid in understanding the clouds of Jupiter and Saturn. Their calculations clearly show the influences of index of refraction and particle diameter on the single-particle scattering function and on the extinction efficiency from the near ultraviolet to the far-infrared spectral regions.

8.2.2C. Aerosol Physical and Chemical Properties

In addition to the size parameter, the physical shape and state and chemical composition of the aerosol can also affect the wavelength dependence of the backscatter. The Mie theory was developed for uniform spherical droplets. Naturally occurring aerosols tend to be spherical when they have a high liquid content and are traveling with the surrounding air mass. Thus aerosols present during high humidity are likely to be larger and more nearly spherical than those present during low relative humidities. The shape affects the scattering because the radiation that travels around the surface is able to constructively or destructively interfere at different angles, depending on the shape, orientation, and size parameter of the particle. [For a discussion of the effect of aerosol shape on backscatter, see Hill et al. (1984).]

The chemical composition can affect the real and imaginary indices of refraction of the aerosol and thus the backscatter. This effect is superimposed on the size and shape effects discussed previously. Nonetheless, it gives a large enough "signature" that aerosol chemical species can be distinguished from each other. Mie theory was checked experimentally by Mudd et al. (1982) for ammonium sulfate and sulfuric acid aerosols in the spectral region 9 to 11 μm. Dramatic changes in aerosol backscatter occur in regions where the chemical has a strong absorption. However, Nevitt and Bohren (1984) have recalculated the backscatter for ammonium sulfate using randomly oriented anisotropic oscillators (i.e., Rayleigh ellipsoids) and have obtained a better fit between theory and experimental results. A recent paper by Volz (1983) shows the absorption by a number of atmospheric aerosols and indicates the spectral regions for which the DISC technique might be useful. The real and imaginary parts of the refractive index of ammonium sulfate [which agree moderately well with the work by Mudd et al.] are given.

8.2.3 Meteorological Parameters

Lidar systems are being developed and applied to the measurement of meteorological parameters of the earth's troposphere and stratosphere, such as wind velocities, temperature, pressure, density, and humidity. Lasers in principle allow the atmosphere to be probed in three dimensions from a remote location both day and night on a continuous basis. Point monitors often have to remain in a fixed location or be taken into the atmospheric region of interest, often on an infrequent basis. Laser systems also offer the promise of being used routinely from air- or space-borne platforms, which would allow coverage over large geographic regions. This section describes techniques for the measurement of wind velocities; the discussion of pressure, temperature, and humidity measurements is deferred to the section on resonant measurement techniques.

8.2.3A. Wind Velocities

The goal of global tropospheric wind velocity measurements from a space-borne free-flying platform is the driving force behind much of the research and development of laser remote measurement of wind velocity inferred from the Doppler shift via Mie backscatter from atmospheric aerosols. The Doppler frequency shift occurs for the component of velocity along the direction of beam propagation. Thus, in order to determine the horizontal wind vector field, at least two measurements at orthogonal directions are required at each location. The Doppler shift, $\Delta \nu$, is given approximately by

$$\Delta \nu = 2 \nu \frac{v}{c} \qquad (8.3)$$

where

ν = frequency of laser radiation
v = wind speed along the path of the beam
c = speed of light

For $v = 1$ m/s, for the $^{12}C^{18}O_2$ laser 9.11 μm (1098 cm^{-1}), $\Delta \nu$ = 3.6×10^{-5} cm^{-1} = 0.22 MHz; for the Nd:YAG laser at 1.064 μm, $\Delta \nu$ = 1.880 MHz. Since wind velocity measurements are desired with an accuracy of 1 m/s, these values give approximate values for the laser frequency stability, the laser pulse width, and the frequency measurement resolution required.

CO$_2$ laser technology for wind measurement has been under development since the late 1960s, and since then several pulsed CO$_2$ lidar systems have been developed for routine field use. One such system is flown in an aircraft and has been used in a number of field programs (Bilbro et al., 1984). During flights, it is alternately pointed 20° ahead and 20° behind the normal to the flight track in order to measure the horizontal two-dimensional vector wind fields. Out to distances of 10 km, the measurement accuracy has been limited to a few meters per second due to uncertainties in the aircraft drift angle. See, also, the recent paper by Hall et al. (1984).

Currently, much of the activity in this field is directed toward laser development and feasibility studies for placing a CO$_2$ laser system on a free-flying satellite in the mid-1990s. One aspect of this work is to develop a space-qualified CO$_2$ laser that emits 10 J/pulse at 8 Hz in a 3 to 5-μs pulse width with minimal frequency "chirp." Methods developed thus far to limit the frequency bandwidth to 0.2 MHz and which scale to the pulse energies required are: 1) to injection-lock the pulsed laser using an external CW CO$_2$ laser (Menzies et al., 1984a), and (2) to use a low-pressure oscillator and a high-pressure amplifier in a "master oscillator-power amplifier" or MOPA configuration. Another aspect of the satellite laser work is to develop long-life operation for the $^{12}C^{18}O_2$ laser because this rare isotope has the following advantages: (1) the atmospheric CO$_2$ attenuation is much less than for $^{12}C^{16}O_2$, and (2) Mie scattering is about 10 to 20% stronger at 9.11 μm than at 10.6 μm (Menzies et al., 1984b). Catalysts for gas regeneration are being developed for long-lived operation. A third area of investigation is to determine of the global distribution of tropospheric aerosol backscatter. Some data sets exist (Kent et al., 1983; Post, 1984), but they are usually from populated areas, representing only a small fraction of the earth's surface. A space-borne system should be able to monitor global tropospheric wind velocities in a 1-km vertical by less than 100-km horizontal grid using a scanning monitor with a 60° scan angle (see, e.g., Huffaker et al., 1984). From a polar orbit, the earth's surface would be covered once every 1 to 3 days.

Another approach suggested for measuring global wind velocities is to use a Nd:YAG laser either directly at 1.064 μm with heterodyne detection (Kane et al., 1984), or at the second harmonic at 0.532 μm, using a multiple Fabry-Perot filter to measure the Doppler frequency shift (Abreu, 1979; Hays et al., 1984). The primary advantages of the Nd:YAG laser approach are the stronger Mie backscatter at these wavelengths and the larger Doppler frequency shifts. The primary disadvantages are the lack of a suitable laser source, the large pulse energies required (10 to 100 J), and the eye-safety problem. The disadvantages are significant enough that the use of the Nd:YAG laser for measuring global wind velocities is being reconsidered (Menzies, 1985). However, this laser could be used in some other wind-measuring application, such as regional measurements from ground or aircraft, or planetary measurements from spacecraft, where use of a compact system would be advantageous.

8.3 ACTIVE RESONANT TECHNIQUES

Considerable progress has been made since 1973 in laser remote sensing systems for selective field measurements of gaseous atmospheric species [see, e.g., reviews by Grant and Menzies (1983) and Measures (1984) and conference proceedings edited by Killinger and Mooradian (1983)]. Systems employing the DIAL technique have been popular for tropospheric and stratospheric species, as have systems using long-path monitoring with differential absorption and a retroreflector or topographic target to define the path. DIAL or similar approaches have also been used for measurements of atmospheric humidity, temperature, and pressure, while fluorescence techniques have proven to be very useful for measuring metal ions and atoms in the thermosphere. Raman scattering has been used in a number of cases, but does not in general have the range capability or the sensitivity of the DIAL technique, although it is useful in certain applications. A discussion of each technique, together with examples of field measurements, follows.

8.3.1 DIAL Technique

The DIAL technique usually uses two laser frequencies to determine the spatial distribution of atmospheric trace gases. One frequency is tuned to a strong absorption in the gas of interest, while the other is chosen a small frequency interval away from where the gas has little, if any, absorption, the latter serving to correct the lidar return at the first frequency for such things as spatial variations in Mie and/or Rayleigh backscatter, attenuation by other atmospheric species, and for instrumental parameters. Two methods exist for returning the signal: backscatter from the atmosphere and reflection from a human-made or topographic target. In the latter, range resolution is lost and only column content is measured. For the former, the molecular concentration (C) is given by solving the equation

$$C = \frac{1}{2 \, \Delta \alpha \, \Delta R} \left[\ln \frac{P_1(R_i)}{P_1(R_j)} - \ln \frac{P_2(R_i)}{P_2(R_j)} \right] \tag{8.4}$$

where

$\Delta \alpha$ = differential absorption coefficient at the two laser frequencies ($atm^{-1} \, m^{-1}$)

ΔR = range interval ($R_i - R_j$) (m)

$P_{1(2)}$ = power received at the first (second) laser frequency (W)

$R_{i(j)}$ = ith (jth) target range (m)

Thus, four independent power measurements are required for each determination of concentration in a particular region in space. Each measurement can be made with an accuracy approaching 1 to 4% with the appropriate averaging of a number of lidar signals, giving a total measurement accuracy for the absorptance ($2C \, \Delta \alpha \, \Delta R$) usually in the range 2 to 10%. It should be noted that Eq. (8.4) is an idealized equation, which ignores many other factors that have to be considered in actual measurements, such as background radiation, detector noise, interference from other gases, special properties of the backscattering target, spectral properties of the atmospheric attenuation, and so on.

As the laser beam propagates through the atmosphere and back at a round-trip rate of 0.15 km/μs, the data are recorded in a range-gated fashion. If, for example, the data sampling rate is 10 MHz, the range gating occurs every 15 m. If the laser pulse width is 300 ns, a range resolution of about 45 m is possible using the lidar system.

When choosing a laser for DIAL measurements of one or more molecular species, three general factors have to be considered: (1) the spectral regions where the gas absorbs with sufficient strength

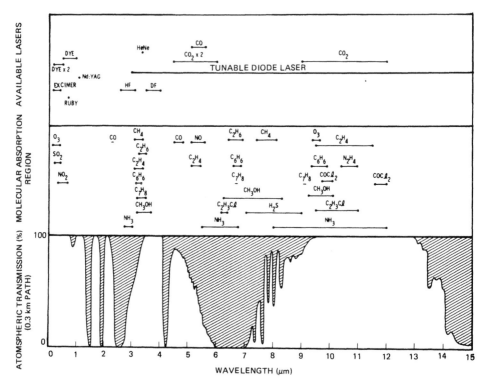

FIG. 8.3 Wavelength dependence of lasers, absorption regions for
several pollutant gases, and transmission of the atmosphere over a
0.3-km path near sea level from 0.2 to 15 μm. Hatched areas indicate
regions of low atmospheric transmission (Grant and Menzies, 1983).

that it can be measured at the distance, spatial resolution, and accuracy desired for the concentration path lengths likely to be present; (2) the transmission and backscattering properties of the atmosphere; and (3) the tunability, pulse energy (or power), pulse
length, size, and cost of the system. The primary cause of atmospheric extinction in the spectral region 1 to 12 μm is absorption
by water vapor and carbon dioxide, with additional absorption by
other gases in selected regions, and scattering and absorption by
aerosols. At shorter wavelengths, molecular (Rayleigh) scattering
also becomes important. Figure 8.3 shows aspects of these factors as
a function of wavelength, with the atmospheric transmission spectrum
taken in low resolution. There are some narrow atmospheric transmission windows in regions of generally poor transmission, allowing,
for example, carbon monoxide to be measured in the presence of

water vapor (Killinger et al., 1980). Note, too, the overlap between the absorption bands of the various gases, which may give rise to interference when trying to measure specific gases.

8.3.2 Molecular Absorption Coefficients

Essential parameters are the absorption coefficients of the molecules of interest for the laser frequencies, laser line widths, temperatures, and pressures of the atmosphere during the field measurement. The absorption coefficients are affected by a number of factors, including the atmospheric pressure (collisional broadening), the atmospheric temperature (Doppler broadening and thermal population of electronic, vibrational, and rotational energy levels), and the density of energy levels that have nearby transitions.

Absorption coefficients can be measured either with a scanning instrument, such as a grating or interference spectrometer, or with a tunable laser source. The scanning instruments have the advantage of being able to measure a large spectral region rapidly, but this advantage must be weighed against the disadvantage that the spectral resolution may not be the same as for the lidar system to be used in remote sensing. An example of a comparison of the two approaches is given by Molina and Grant (1984), in which a Fourier-transform infrared (FTIR) spectrometer was used with a resolution of 0.05 cm^{-1} (1.5 GHz) to measure the absorption coefficients of three hydrazines and four of their air-oxidation products, including methanol and ammonia, in the spectral region 9 to 12 μm. (Low-pressure CW CO_2 lasers have line widths of a few megahertz, but atmospheric-pressure pulsed lasers can lase in a 1-GHz envelope.) Absorption coefficients determined using an FTIR spectrometer were compared with the data in the literature obtained using $^{12}C^{16}O_2$ lasers. For most of the molecules, especially those which have broad absorption features, the agreement was good, in the range of ± 5 to $\pm 25\%$. The FTIR spectrometer allowed the easy determination of the absorption coefficients at rare-isotope CO_2 laser frequencies as well as in spectral regions where tunable diode lasers might be useful for long-path measurements (see Fig. 8.4).

It is important to measure the pressure dependence of absorption coefficients if field measurements will be made at pressures other than 760 torr, for example, if the laser system is not at sea level, or if vertical measurements are made. The pressure dependence of the absorption coefficients can also be useful for determining an altitude profile of a gas when a column-content measurement is made from an air-borne or space-borne platform. For example, Menzies (1976a) has shown that the pressure dependence of the ozone absorption coefficients at CO_2 laser wavelengths is sufficient to allow an altitude resolution of approximately 5 km of the ozone profile in the troposphere.

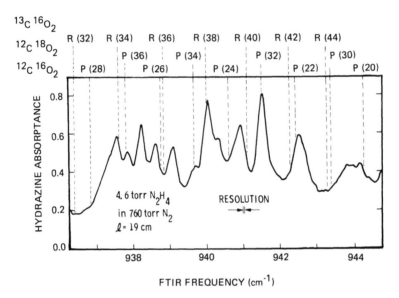

FIG. 8.4 Hydrazine absorptance in the spectral region 936 to 945 cm^{-1}, showing the coincidence with CO_2 laser lines (Molina and Grant, 1984).

The temperature dependence of the absorption coefficients can also be important. Usually, the measurements reported in the literature are determined at 760 torr and 290 to 300 K, as these are convenient laboratory operating parameters, but are not necessarily those pertaining to the field conditions encountered. Gases in exhaust plumes, for example, may be very hot near the stack exit, but decline in temperature away from the source. Some work has been done on the temperature dependence of the water vapor absorption coefficients at CO_2 laser frequencies. A very strong temperature dependence is caused by the formation of dimers at temperatures below about 350 K (see Loper et al., 1983). The temperature dependence of ammonia and ethylene absorption coefficients at CO_2 laser frequencies has also been studied (Persson et al., 1980).

The Air Force Geophysical Laboratories (AFGL) spectral data tapes (Rothman et al., 1983a,b) contain information about absorption coefficients, energy levels, line widths, and so on, that can be used to generate synthetic spectra. These spectra would be first approximations to the real atmosphere in predicting atmospheric transmission with both normally occurring and pollutant molecules.

8.3.3 DIAL Measurements

NO_2, SO_2, and O_3, the first pollutant molecules to be studied using the DIAL technique, were chosen for the following reasons: (1) they have strong, structured absorption bands in the visible or ultraviolet spectral regions, where dye and frequency-doubled dye lasers (available in the early 1970s) operate; (2) atmospheric backscatter is strong in these regions; (3) detector technology was well advanced in these regions; and (4) these gases are economically important pollutants. The early measurements of NO_2 showed that the DIAL technique worked generally as expected, and that urban or industrial concentrations could be usefully measured (Rothe et al., 1974). In the late 1970s, when Nd:YAG-pumped dye lasers became commercially available, lidar systems were constructed for dedicated field use.

On such ground-based UV-visible DIAL system for these molecules was assembled by SRI International under contract to the Electric Power Research Institute (EPRI) to study SO_2 plume dispersion from electric power plants (Hawley et al., 1983). The key feature of this system is a pair of Nd:YAG-pumped dye lasers which are frequency doubled to 299.38 and 300.05 nm for the weak and strong absorptions by SO_2, respectively, with a differential absorption coefficient of 25.4 atm^{-1} cm^{-1}. The pair of lasers allows the two laser frequencies to be transmitted within 100 μs of each other, with an accuracy that has been demonstrated three different ways. First, measurements of ambient NO_2 were made which were compared with point measurements from a local air pollution control board station a few kilometers away (Baumgartner, 1979). Good agreement was found during periods of low wire speed, but not during periods of high wind speed, when the two instruments were probably measuring different air masses. Second, the lidar system was first used to measure SO_2 in a sample chamber with quartz windows, and agreed within a few percent of those of *in situ* measurements. Third, the lidar-determined values for a power-plant plume were compared to those from point monitors on the ground under the lidar path with very good agreement. The usefulness of this lidar system has been demonstrated by the measurements made in a horizontal plane in the vicinity of a coal-burning power plant. The SO_2 concentrations from 50 to 300 ppb were mapped in a 3.5-km^2 area in only 10 min with a resolution of 50 m × 200 m. To achieve the same results with fixed-location point monitors would require 400 such instruments.

A similar lidar system for use from an airborne platform was constructed at NASA Langley Research Center (Browell, 1982; Browell et al., 1983), to measure concentrations of ozone in the troposphere up to the tropopause, and has been used to study transport of ozone over large regions by measuring both the ozone and the aerosols and following the air masses as they converge on

a particular point (Shipley et al., 1984). In addition, the system
has been used to measure water vapor in the lower 1 to 2 km.
Radiosondes were used to calibrate the system for water vapor, and
point monitors flown in another aircraft below the flight altitude of
the aircraft containing the lidar calibrated for ozone. An advanced
version of this lidar system, using a tunable solid-state alexandrite
laser is being designed for use on a high-flying aircraft, the ER-2,
operated by the NASA Ames Research Center.

It is also possible to use one laser to generate both laser frequencies or lidar measurements of pollutant gas concentrations by
switching the laser frequency between pulses. This saves considerable expense and is quite useful when the pollutant and aerosol concentrations change only gradually with time and position. A disadvantage with this approach is that the distribution of atmospheric
aerosols and gas of interest can change between wavelengths. Several
UV-visible DIAL systems taking this approach have been described
in the literature [see, e.g., Egebach et al., (1984) for SO_2 and
Fredriksson and Hertz (1984) for NO_2 (see Fig. 8.5)]. SO_2 measurements with a range resolution of 50 m, a concentration accuracy of
100 $\mu g/m^3$, and a distance of 1400 m were reported.

Ground-based XeCl excimer laser systems have been used to
measure tropospheric (0 to 15 km altitude) and stratospheric (15 to
45 km) ozone concentrations (Uchino et al., 1983a,b; Werner et al.,
1983, 1984; Megie et al., 1985). The XeCl excimer laser can generate
more energetic laser pulses than can frequency-doubled dye lasers.
The tuning range is not as great [XeCl lasing can be tuned from
307.5 to 308.5 nm (McDermid et al., 1983)], but additional frequencies can be generated by using a high-pressure gas cell to
Raman shift the excimer laser frequency. For O_3 measurements,
Werner et al. used a cell containing methane (2916 cm^{-1}) or hydrogen (4156 cm^{-1}) to shift the wavelength to 338 or 353 nm. Both
fundamental and Raman-shifted wavelengths are transmitted simultaneously, then sent to two detectors by means of a dichroic filter/
beamsplitter. This system was perched on the Zugspitze in southern
Germany at an altitude of 2964 m. The attenuation due to the atmospheric Rayleigh and Mie scattering was reduced by at least a
factor of 3 compared to systems located at mean sea level, thus enabling O_3 concentrations to be made to altitudes of 45 km, above
which there is little O_3 (see Fig. 8.6). The lidar measurements
have been compared with those made by ozonesondes, and the agreement was found to be within a few percent when the measurements
were taken during periods of calm winds allowing for measurements
of the same air mass. Lidar measurements of O_3 can be useful for
studying O_3 depletion by chlorofluorocarbons because they are expected to have the greatest impact above 35 to 40 km.

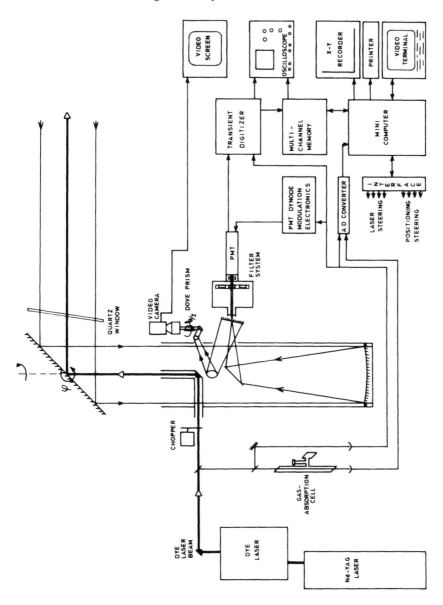

FIG. 8.5 Schematic diagram of the electronic and optical arrangements of a mobile DIAL system used for measuring NO_2 (Fredriksson and Hertz, 1984).

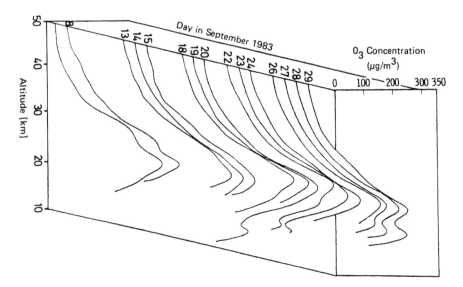

FIG. 8.6 Daily ozone profiles for September 1983 measured with a
XeCℓ laser on the Zugspitze, West Germany. The slant axis gives
the day of the month; the vertical axis gives the altitude in kilometers; the horizontal axis gives the ozone concentration (Werner et
al., 1984).

DIAL measurements have also been extended to the spectral region 9 to 11 μm, using carbon dioxide (CO_2) lasers, where a large
number of atmospheric trace and pollutant gases have useful differential absorption coefficients. This is a more difficult spectral region
in which to work than the UV or visible regions because molecular
backscatter is extremely weak, aerosol backscatter is much lower
than in middle infrared regions (by one to two orders of magnitude),
and detector plus background radiation noise can be significant.
Also, the CO_2 laser tunes discretely, rather than continuously, so
that DIAL measurements must rely on accidental coincidences between
laser frequencies and absorption features. Rare isotope CO_2 lasers
(e.g., $^{12}C^{18}O_2$, $^{13}C^{16}O_2$, and $^{14}C^{16}O_2$) can be used to extend the
tuning range to 12 μm and fill in some of the gaps of the $^{12}C^{16}O_2$
laser (Freed et al., 1980), although a high-pressure (10 atm) CO_2
laser that is continuously tunable could also be constructed. For
simultaneously attempting to detect several molecules with complex
spectra, the CO_2 laser should be rapidly line switched, for example,
by using a rotating cylinder with a number of gratings affixed to it,
and firing the laser synchronously with various positions of the
cylinder (Faxvog and Mocker, 1982).

In spite of these difficulties, useful DIAL measurements using the $^{12}C^{16}O_2$ laser have been made with of such trace gases as water vapor (Murray et al., 1976; Murray, 1978; Rothe, 1980), ozone (Asai et al., 1979), and ethylene (Rothe et al., 1983). When "direct" detection of the backscattered radiation is used, as it was in these measurements, DIAL measurements are limited to distances of about 1.5 km, since background radiation in both a 0.2- to 4-m bandwidth and laser backscatter are measured. However, with "heterodyne" detection, the measurement range can be extended to 5 to 10 km, since the thermal background component is restricted to a few megahertz bandwidth by the use of a local oscillator and narrow-bandpass electronics. In this approach, a second laser at a frequency shifted by a few megahertz from the transmitted frequency acts as a local oscillator for the detector. The second frequency is then mixed with the backscattered radiation. Signals arising at the difference frequency can be detected using narrow-bandwidth radio-frequency detection techniques (see, e.g., Menzies, 1976b). This procedure gives a reduction in background noise by two to three orders of magnitude in comparison with direct detention techniques.

While the use of CO_2 lasers with heterodyne detection for DIAL measurements of atmospheric gases was proposed years ago (Kobayasi and Inaba, 1975), experimental demonstrations of the feasibility of this approach have been reported only recently by Fukuda et al. (1984) and Hardesty (1984a,b) with a hybrid TEA CO_2 laser emitting 140 mJ per pulse at a 5-Hz rate, and transmitter and receiver aperatures at a diameter of 15 cm. Fukuda et al. demonstrated a measurement range of 5 km with a range resolution of 300 m. They found that multipulse integration could decrease the measurement error down to 10% for 50 pulses, but further reduction was limited by atmospheric composition variations at the 5-Hz repitition rate. Finally, they demonstrated the detection of a gas (Freon 12) in a sample chamber at a distance of 1100 m. Hardesty reported water vapor measurements out to 10 km. These results using heterodyne detection for hazardous and pollutant gases. However two CO_2 lasers should be used for nearly simultaneous transmission of two wave lengths in order to improve the measurement accuracy.

8.3.4 DIAL Measurement Error

DIAL measurements are limited in either minimum detectable concentrations or uncertanties in the measured concentrations by a number of error mechanisms. It is important to understand these mechanisms so that systems can be better designed and so that the impact of the error mechanisms on the measurements is minimized. The error sources are of six general types (Schotland, 1974; Menyuk and Killinger 1983, Grant; 1986): those arising from (1) knowledge of absorption coefficients, (2) differential spectral backscatter and attenuation, (3) interfering molecular species, (4) speckle, (5) atmospheric turbulence, and (6) the lidar system itself.

Imperfect knowledge of the absorption coefficients is an important measurement error. First, it is difficult to measure the absorption coefficients to better than 5 to 10% for standard pressure and room temperature (Menyuk and Killinger, 1983). Several parameters have to be measured to determine each absorption coefficient, and it is difficult to measure the ensemble of parameters with an accuracy better than a few percent. Comparisons of measurements from different laboratories usually show agreement of about 5 to 15% (see also Molina and Grant, 1984). Second, the pressures and temperatures encountered in field measurement situations are generally different from 760 torr and 293 K, which can strongly affect the absorption coefficients. Laboratory measurements of the temperature dependencies of water vapor, ammonia, and ethylene absorption coefficients were discussed in Sec. 8.3.2. In addition to line shape changes, there can also be shifts in the position of line center (see, e.g., Giver et al., 1982; Bosenberg, 1985, for the effect on water vapor lines). Third, the absorption coefficients may have a dependence on the partial pressure of the species of interest due to absorption coefficients (Cahen and Megie, 1981; Korb and Weng, 1982; Brassington et al., 1984).

The finite line width of the laser radiation can also affect the measurement accuracy due to the possibility of a departure from the Beer-Lambert-Bouger law for an absorbing medium: different frequencies across the laser bandwidth will experience different absorption coefficients (Cahen and Megie, 1981; Korb and Weng, 1982; Brassington et al., 1984).

Differential spectral backscatter and attenuation arise from the spectral dependence of the scattering by the atmospheric aerosols and molecules. While molecular (Rayleigh) scattering has a smooth wavelength dependence, aerosol (Mie) scattering is strongly affected by the chemical composition, and, to a lesser extent, the physical shape and size distribution of the aerosols. If the aerosol distribution is uniform in the region of the measurement, the spectral dependence will generally ratio out to a large extent using the DIAL equation, Eq. 8.4, especially if aerosols contribute primarily to backscatter but not extinction, as can be the case in the middle IR. However, when the distribution is nonuniform, large errors can enter. Browell et al. (1985), and Sasano et al. (1985), have treated this in the UV, while Petheram (1981) has treated it for the middle IR, showing that the errors arising from aerosol backscatter spectral variations can be equivalent to a few percent differential absorption for several gases of interest at CO_2 laser frequencies. The effect can be minimized by using a small frequency interval, or by using three laser frequencies for the measurement in order to measure the slowly varying spectral dependence of backscatter.

Spectral interference by molecular species other than those of interest can affect the measurements. Possible errors from this cause can be minimized by several approaches. The first is to choose laser frequencies for which the differential absorption of interfering species is minimized. The second is to measure the concentration of the interfering gases and correct the measurements accordingly. An example of interference is that caused by ozone when SO_2 is measured near 300 nm. The effect is minimized both by choosing laser frequencies that have a small separation (0.7 nm), and by measuring or estimating the ozone concentration. Another example is the interference of water vapor on measurements of ethylene, using a CO_2 laser. When long-path measurements of ethylene indicated that "negative" values of ethylene were present (Murray, 1978), the absorption by water vapor was checked using the AFGL spectral data tapes, (McClatchey et al., 1973; Rothman et al., 1983a,b). A tail of a water vapor line was found at the reference CO_2 laser frequency. Water vapor concentrations were measured using a point monitor, and by compensating for the water vapor, the ethylene concentrations quickly became "positive." Care must be exercised in relying on the AFGL spectral data tapes because they represent experimental and theoretical values which do not always agree with the real atmosphere, especially for weak lines that have not been studied experimentally (see, e.g., Giver et al., 1982).

Laser speckle can affect the measurement accuracy because a small, finite number of speckle lobes may be incident on the receiver/ detector, each of which has essentially the same intensity. The normalized uncertainty, σ/I, in a measurement due to a finite number of speckle lobes, M, collected by the receiver is given by:

$$\frac{\sigma}{I} = (M)^{-1/2} \tag{8.5}$$

M can be estimated using the expression (Goodman, 1975) for the speckle lobe radius, r:

$$r = \frac{\lambda R}{D_s} \tag{8.6}$$

where

λ = wavelength of the laser radiation (m)
R = distance to the target (m)
D_s = diameter of the laser beam on the target (m)

The area of each speckle lobe can be found from the radius, and M can be found by dividing the area of the receiver by the area of one speckle lobe. For atmospheric backscatter, the longitudinal coherence length is also important. It can vary from a few meters for a multimode laser to 300 m for a single longitudinal mode CO_2 laser (Flamant et al., 1984). In practice, speckle is an important consideration primarily in the middle IR spectral region (8 to 12 μm). Even there, for direct detection, proper system design, such as using a large transmitter divergence and receiver field of view, can reduce the impact. For heterodyne detection, however, the transmitter and receiver diameters and divergences are usually closely matched, so that the receiver collects one speckle lobe at a time, giving rise to a large measurement error with each individual measurement. Thus, to overcome the speckle problem, several pulse pairs must be averaged for a measurement.

Atmospheric turbulence can have several effects on DIAL measurements. By affecting the index of refraction, it can distort the laser beam and the backscattered radiation: the beam can be bent, focused, defocused, broken up, and lose phase coherence, etc. These effects have been discussed several places, to which the reader is referred (see, e.g., Hill et al., 1980; Holmes, 1983). In addition, turbulence can cause mixing and wandering of aerosols and gases, which can affect measurements made over a finite time interval. One advantage of turbulence is that it decorrelates aerosols in a few microseconds so that the speckle pattern on the receiver changes between laser pulses (Churnside and Yura, 1983), which guarantees statistical independence of the speckle patterns measured with different lidar pulses.

The laser system itself can contribute significantly to measurement error unless it has been carefully designed, assembled, and calibrated. For example, the lidar transmitter-receiver overlap function aften does not converge for several hundred meters or more, making it difficult to make short-range measurements. In DIAL systems, the two laser frequencies may have slightly different transmit/ receiver overlap functions, which can give rise to spurious range-dependent molecular concentrations. However, this effect can be checked or reduced by reversing the laser frequencies for some of the measurements (Lahman et al., 1985). Detector noise, together with shot noise and background noise, is also important (Measures, 1984). Signal digitization error can also affect the measurement. An 8-bit transient digitizer has 0.4% bit error at full scale, but 4% bit error at 10% of full scale. Sometimes the detector signal amplification can be increased stepwise with time (distance) so that the signal level can be maintained in the upper register for the digitizer. Finally, refractive optical elements can have a spectral dependence. Ahlberg et al. (1985) have discussed how beam splitters must either be wedged or thick in order to avoid having the reflected beams from the two sides of the element interfer and cause a spectral dependence of transmission or reflectance into monitors.

With these and more factors contributing to measurement error, the question naturally arises just how accurate can DIAL measurements of gas concentrations be made? A pessimestic point of view was taken in three recent papers (Menyuk et al., 1982; Menyuk and Killinger, 1983; Menyuk et al., 1985) based on their results with a lidar system aimed at a building 2.7 km away. While the prevailing wisdom expects that averaging over N measurements will reduce the single measurement error by a factor of $N^{-1/2}$ without limit, these papers reported experimental results and theory to show that a factor of 10 reduction in error from the single pulse pair measurement to the average of a large number of pulse pairs. However, close analysis of their experimental set up shows that this result is due primarily to measuring nearly the same speckle pattern for each pulse (Grant, 1986). Atmospheric turbulence causes each speckle pattern to vary, but with partial correlation, so that the set of measurements is not really an independent set of values. In order to make completely independent measurements, the target area should be scanned (Shapiro, 1985; Grant, 1986). Indeed, Grant et al. (1983) were able to make measurements down to the 4% differential absorptance level using the Laser Absorption Spectrometer, an airborne CW CO_2 laser system using heterodyne detection, which is strongly affected by speckle, by traversing a desert while measuring ozone. When a pair of lasers is used with atmospheric backscatter, the $N^{-1/2}$ rule should apply, since the atmospheric aerosols decorrelate (Churnside and Yura, 1983). Indeed, Staehr et al. (1985) reported that they measured an $N^{-0.42}$ dependence for a dual dye laser system operating at 450 nm, and Fukuda et al. (1984) measured nearly $N^{-1/2}$ dependence for aerosol backscatter near 10 µm. In practice, differential absorptances of around 2 to 5% can be made with DIAL systems using atmospheric backscatter.

8.3.5 Column-Content Measurements

Column-content measurements can be made using differential absorption and topographic targets to provide the backscattered radiation. This approach allows the laser system to be used in a "single-ended" manner (i.e., the laser transmitter and receiver are co-located) so that the instrument can be aimed at any convenient target, but sacrifices the ability to determine the concentration-with-distance profile, yielding only the average column-content concentration of the molecule over the path length. However, by using targets at different distances along the same general line of sight it is possible to obtain some range-resolved information (Killinger et al., 1980). The

significant feature of this approach is that the lower-power laser
systems and simpler data systems required can be carried by a person or mounted on a small cart.

Two aircraft-mounted CO_2 laser long-path systems have been
used to measure column abundances of atmospheric trace gases
(Shumate et al., 1982, 1983; Englisch et al., 1983). For example, the
Laser Absorption Spectrometer (LAS) described by Shumate et al.
transmits low-power CW CO_2 laser radiation (100 mW) to the ground
at an angle of a few degrees ahead of nadir so that the backscattered
radiation is Doppler shifted 1 to 3 MHz for heterodyne detection.
The LAS underwent extensive calibration tests in Virginia, where
an aircraft with an in situ ozone monitor flew in a modified spiral
below the aircraft carrying the LAS in order to measure the vertical
profile of ozone. These measurements indicated that the LAS had a
measurement uncertainty of 30 to 40 ppb, which was large compared with the ozone present (50 to 100 ppb). Eventually, it was
realized that the differential spectral reflectance of the earth's surface was responsible for a large fraction of this measurement error
(Shumate et al., 1982; Grant, 1982a), due to a 7-cm^{-1} laser line
separation in the spectral region where quartz has a strong reflection peak. Subsequently, another pair of laser lines with a separation of 2 cm^{-1} was used, and the measurement error decreased to
15 to 20 ppb (Shumate et al., 1983). Later, the LAS was used to
measure geological features in Death Valley, California, with the reflectance at a pair of lines at 9.23 and 10.27 μm. Excellent agreement was found between these measurements and those of a thermal
IR sensor that measured emission in half-micrometer bands near 9.1
and 10.4 μm from the same area (Kahle et al., 1984). Note that
DIAL measurements using this approach can be improved by using
a four-laser system, as proposed by Wiesemann and Lehmann (1985).

An approach that may have important economic consequences
is the use of helium neon (HeNe) laser systems for remote sensing
of methane. The HeNe laser has a strong line at 3.3913 μm, where
methane has an absorption coefficient of about 8.3 $\text{atm}^{-1}\text{ cm}^{-1}$ (Kucerovsky et al., 1973). If a cell containing methane is inserted in the
cavity, lasing at this line is quenched, allowing one at 3.3903 μm to
lase (Moore, 1965), where the methane absorption coefficient is only
0.6 $\text{atm}^{-1}\text{ cm}^{-1}$, which can be used as the reference frequency.
The small (0.9 cm^{-1}) line separation is sufficient to minimize the
effects of differential spectral reflectance. Thus a HeNe laser system could be used to locate natural gas leaks from underground
pipelines of landfill sites using topographic targets at ranges to 100
m (Grant, 1982b, 1986). In addition, a HeNe laser system could be
used as a perimeter monitor for natural gas storage tanks.

Another approach that may also be useful in searching for leaks
of hazardous gases incorporates tunable diode lasers (TDLs) in por-

table systems. It has been demonstrated that TDLs can be used with topographic targets to measure gas concentrations down to the sub-ppm-m level at 10 m, and out to 30 m with lower sensitivity (Reid et al., 1985). Although TDLs have low power ($\leqslant 1$ mW), they can be tuned to strong absorption features of gases and can be frequency modulated over intervals of up to 0.5 cm^{-1} (15 GHz), allowing sensitive harmonic-detection techniques to be used. However, for TDLs to be routinely useful, technology must be improved to where TDLs can operate reliably above 77 K to allow for liquid-nitrogen cooling. Then the cumbersome helium refrigerator can be dispensed with.

With retroreflectors, measurements may be made over longer paths, with lower laser power, and at higher measurement accuracy, than generally possible using topographic targets, and in locations where topographic targets do not exist. These measurements might be more properly considered long-path *in situ* measurements, rather than remote measurements, especially if an average measurement is made of the ambient gas. The primary use of retroreflectors with laser long-path systems has been with TDLs. An early example includes the measurement of carbon monoxide (Ku et al., 1975). More recent work includes measurement of NO in the stratosphere using a balloon-borne TDL system with a 500-m separation between the TDL and the retroreflector. A value of 17 ± 5 ppmv was measured at 36 km (Webster and Menzies, 1984). The CW HF laser has been applied to monitoring HF concentrations in the vicinity of an aluminum processing plant (Tonnissen et al., 1979). Retroreflectors were used to define paths around the plant. Concentrations to 10 mg/m^3 were measured.

8.3.6 Fluorescence

Some atmospheric trace species are best determined in fluorescence rather than in absorption. These include atoms and ions in the thermosphere, which can be measured using resonance fluorescence, and the hydroxyl radical (OH) in the troposphere (McDermid et al., 1983) and stratosphere (Heaps and McGee, 1985), using nonresonance fluorescence. For the theory of fluorescence lidar see, for example, McIlrath (1980). The thermosphere (above 80 km) has a number of metal atoms and ions that arrive there either by the ablation of meteorites or by upwelling from the earth's surface. In fact, one of the interesting experiments that can be performed is to measure sodium atom concentrations before and after a meteor shower. One such measurement found a fourfold increase of sodium following the passage of the Geminids asteroid group (Hake et al., 1972). Related to this result is the measurement of the relative abundances of sodium and potassium to further distinguish between extraterrestrial and terrestrial origin of the metals in the thermosphere (Megie et al., 1978).

The hydroxyl radical is extremely important in atmospheric chemistry because it is involved in many photochemical reactions. It is thought to exist in concentrations of around 10^5 to 10^6 cm^{-3} in the troposphere. It is difficult to observe in absorption, although this has been accomplished over a 10-km path using a broadband light source and a UV spectrometer with a repititively scanned aperture (Hubler et al., 1982). OH has a couple of absorption bands in the UV that are accessible by laser radiation—one near 282 nm, the other near 308 nm. The band near 282 nm is difficult to use for reliable measurements in the troposphere because ozone absorbs there strongly too, releasing an $O(^1D)$ atom which can react rapidly with water vapor to produce two OH radicals. This problem can be avoided by working near 308 nm, where the ozone absorption coefficient is much lower, as well as the quantum yield of $O(^1D)$. The best laser to use is a xenon chloride (XeCl) excimer laser because of its short pulse width, its high pulse energy, and its tunability. From 307.6 to 308.4 nm, its output overlaps several OH absorption lines (Pacala et al., 1982; McDermid et al., 1983), with preferred detection at 309 nm using a very narrow bandpass filter. However, this is a difficult and expensive measurement and is not yet funded.

Fluorescence laser sensors can also be used to detect oil spills on water (e.g., O'Neil et al., 1980), chlorophyl in water (e.g., Bristow et al., 1981; Hoge and Swift, 1983), and uranium deposits on land (deNeufville et al., 1981; Kasdan et al., 1981).

8.3.7 Raman Scattering

Raman scattering, discussed in Chapter 7, was seen in the early 1970s as a potentially powerful laser remote sensing technique for tropospheric trace species. The primary advantage of the Raman technique is that one laser frequency can be used to measure a large variety of molecules; the secondary advantage is that the signal for each species is proportional only to the concentration of that species and the scattering coefficient of that species. During the 1960s and 1970s, a number of demonstration measurements were performed with ruby or N_2 lidars on such targets as automobile exhaust gases (Inaba and Kobayasi, 1972), water vapor (Melfi et al., 1969), SO_2 and kerosene (Hirschfeld, 1973), and SO_2 (Poultney, 1977). The field as of 1975 was well reviewed by Inaba and Kobayasi (1976). Generally, reasonable agreement was found between the Raman scattering measurements and those made using some other technique. However, the sensitivities were not sufficient for the technique to be generally applicable, except for strong sources or high concentrations. Several approaches were attempted to improve the sensitivity, such as using higher-frequency harmonics of the laser (Hirschfeld, 1973), or by using tunable lasers near the molecular absorption bands so the the scattering would be enhanced by "resonance Raman scattering" (Rosen et al., 1975). The sensitivities were still found to be limited by background radiation during the daytime, and interference by fluorescence of other materials in some cases (Hirschfeld, 1977). Mobile Raman scattering lidar systems can be useful for remote sensing of high concentrations of gases measured at short distances from the lidar (e.g., gases leaking or being emitted from an industrial plant). However, if a large receiver telescope can be used, vertical profiles of atmospheric constituents can be usefully measured (Melfi and Whiteman, 1985).

8.3.8 Meteorological Parameters

Meteorological parameters such as humidity, temperature, and pressure are measured using either absorption, fluorescence, Rayleigh scattering, or Raman scattering by various atmospheric molecules or atoms. Atmospheric temperature can be measured by absorption by water vapor or carbon dioxide or by Raman scattering by nitrogen in the troposphere, and by sodium atom fluorescence in the thermosphere. Atmospheric pressure can be measured by absorption by oxygen in the troposphere, or by Rayleigh scattering in the stratosphere. These and other approaches are discussed in this section.

8.3.8A. Water Vapor Concentration

Although water vapor is a molecular gas, it can properly be considered under meteorological parameters because of its strong connection with weather and climate. The primary advantage offered by lidar techniques is the ability to provide profiles of water vapor concentration. Historically, water vapor was the first molecule to be measured using the DIAL technique, for which a temperature-tuned ruby lidar was used (Schotland, 1966). Subsequently, the Raman technique was used (e.g., Melfi et al., 1969; Cooney, 1970; Renault et al., 1980). Other DIAL approaches were also demonstrated using the CO_2 laser (Murray, 1978) and a pair of Nd:YAG-pumped dye lasers operating near 720 nm (Cahen et al., 1982; Browell, 1982). The most appropriate technique for a given measurement scenario depends on such factors as the measurement range, expected water vapor concentration, and other measurements to be made with the same lidar.

Over a short range (to 1.5 km) and with moderately high relative humidities, it is difficult to choose between the DIAL and Raman techniques. Schwiesow (1983) prefers the Raman technique because the same lidar can also be used for temperature measurements. Cooney (1983) also prefers the Raman technique for its greater simplicity of apparatus and measurement strategy, and has demonstrated

its usefulness (Cooney et al., 1985), as have Melfi and Whiteman
(1985). For measurement of water vapor at intermediate distances
(up to 8 km altitude), either the Nd:YAG laser-pumped dye lidar
using the DIAL technique or a frequency-doubled or frequency-
tripled Nd:YAG laser or an excimer laser using Raman scattering
can be used if the receiver telescope area is large enough, such as
the 1.5-m diameter searchlight mirror used by Melfi and Whiteman
(1985). Cahen et al. (1982) have demonstrated the DIAL technique
to heights of 8 km in southern France. However, for space-borne
measurements, the DIAL technique would have to be employed.
Wilkerson and Schwemmer (1982) and Browell and Ismail (1984) have
presented the results of calculations which indicate that water vapor
from Space Shuttle altitudes from 0 to 10 km with less than a 10%
error, but with a 250-km ground track.

8.3.8B. Atmospheric Temperature

Atmospheric temperature can be determined by measuring the
strength of the absorption by a relatively abundant, thoroughly
mixed species which has an absorption line with a strong temperature dependence. In the troposphere, oxygen, carbon dioxide, and
water vapor have been proposed for such measurements. Mason
(1975) suggested a three-frequency differential absorption measurement for oxygen (two oxygen lines plus one reference frequency).
Murray et al. (1979) measured an average temperature measurement over 5 km [CO_2 lidar on the P(38) and P(20) lines in the
10-μm band], with an uncertainty of ± 1.1 K. This would be a difficult measurement to make in a range-resolved fashion due to the
low aerosol backscatter at 10 μm and the variation in backscatter
with frequency (Petheram, 1981) over the large interval required
for a measurable differential absorption.

Range-resolved atmospheric temperature measurements have a
better chance of success with the approach of Korb and Weng
(1982) with a two-wavelength lidar with the oxygen A band near
13,000 cm^{-1}. Measurement accuracies to better than ± 1 K are expected for a ground-based system if the laser bandwidth is less
than 0.01 cm^{-1} and if the atmospheric pressure can be determined.
The P_{P29} line is shown to have a temperature dependence of the
absorption coefficient at 1 km altitude that varies from 0.05 km^{-1}
at 200 K to 0.36 km^{-1} at 300 K. A long-path measurement over 1
km using this approach has been demonstrated by Kalshoven et al.
(1981). The authors calculate that this approach could be used from
a space-borne platform, although a long-lived pulsed laser with high
wall plug efficiency would have to be developed for this application.

The Raman scattering technique can also be used to measure
range-resolved temperature profiles over medium distances (1 to 3

km) in the troposphere. Cooney (1983) made measurements using the rotational envelope of the Q branch of nitrogen with an uncertainty of ±0.85 K and Arshinov et al. (1983) did so with an uncertainty of ±1.5 K at 1 km.

Another way to measure atmospheric temperature between the aerosol layer on clouds to about 50 km is to measure the atmospheric density profile, then analyze these measurements with the hydrostatic relation and the ideal gas law. Russell and Morley (1982) considered this approach for use from a space-borne platform with a Nd:YAG laser at the fundamental and second harmonic. They concluded that measurements could be made with RMS errors of ±1.2 to 2.5 K in the best cases, with an increase in error in the presence of volcanic aerosol. Hauchecorne and Chanin (1980) made an experimental demonstration of this approach from the ground, and demonstrated ±2 K measurement accuracy at 35 km, falling to ±15 K at 60 km.

Blamont et al. (1972) demonstrated that the Doppler line profile of sodium atoms can be used to measure the temperature in the thermosphere (80 to 100 km). They used a sodium-vapor cell at various temperatures to filter the backscattered fluorescence for a measurement accuracy of ±20 K.

8.3.8C. Atmospheric Pressure

Weng and Korb (1983) discussed the "trough" technique for lidar measurement of atmospheric pressure based on absorption by oxygen in the trough between two absorption lines near 13,150 cm^{-1} (760.5 nm) with a reference frequency 3 to 20 cm^{-1} away. They presented theory demonstrating that the differential absorption is proportional to the second power of the pressure. They have used an alexandrite (Cr:BeAl$_2$O$_4$) laser over a 1-km horizontal path to demonstrate the technique. An error analysis is presented which considers nine error sources. They concluded that surface pressure could be measured from the Space Shuttle with an accuracy of ±0.2%, and that the pressure profile could be measured from the Space Shuttle with a 5-km resolution and a reduced measurement accuracy but would require greater laser pulse energy. Such measurements from an unmanned space platform would require a laser with a much higher wall-plug efficiency and lifetime than is presently available.

8.4 OTHER TECHNIQUES

8.4.1 Laser Heterodyne Radiometers

Laser heterodyne radiometers employ heterodyne detection of naturally occurring emission or absorption to measure such properties of gases as total concentration, altitude variation of concentration, temperature, pressure, velocities, and distribution of molecules among their vibrational and rotational energy levels. A frequency-stable, narrow-bandwidth laser is used as the local oscillator. The heterodyne detection electronics comprises a set of RF filters and detection electronics tuned to observe intervals along the frequency region where a particular molecule is expected to have an emission or absorption line.

An example of such an instrument is the CO_2 laser heterodyne radiometer, which has been flown from a balloon to measure chlorine monoxide (ClO) in the stratosphere (Menzies et al., 1981; Menzies, 1983). ClO is an important intermediate species arising from chloro-fluorocarbons and is responsible for ozone destruction. In this instrument, a $^{14}C^{16}O_2$ laser operating on the P(14) line at 853.181 cm^{-1} is used as the local oscillator. Six RF filters in the region 1650 to 1850 MHz are used to measure ClO absorption at 1686 ± 6 MHz from the CO_2 laser line center while viewing the sun through a "limb" of the earth's atmosphere, allowing the ClO concentration to be measured from 20 to 40 km with a resolution of 2 to 3 km. Values on the order of 1 part per billion (volume) have been measured, which compares reasonably well with theory.

While the measurement of ClO relies on an "accidental" near coincidence between a CO_2 laser frequency and a ClO absorption line, it is possible to use a tunable diode laser (TDL) as the local oscillator (see, e.g., Glenar et al., 1982). While there are some problems with TDLs (low power, cryogenics requirements, difficulty in obtaining single-mode operation, etc.), their technology should develop sufficiently in the future for their use in this application (Allario et al., 1983).

Another interesting set of measurements using heterodyne radiometry is that of planetary atmospheres from ground-based observatories (see, e.g., Johnson et al., 1976; Deming et al., 1983; Kostiuk et al., 1983; Kostiuk and Mumma, 1983). The CO_2 laser is particularly useful in these experiments because its emission lines fall within the 10-μm atmospheric window and many molecules of astronomical interest exhibit rotational-vibrational bands within this spectral region. Thus, many laser lines overlap absorption lines of target molecules. The most interesting result to date is the finding that CO_2 on Mars exhibits naturally occurring laser radiation. By making measurements with 5-MHz resolution, the emission lines appear as sharp spikes inside a much broader absorption line. The emission lines have local kinetic temperatures of 135 ± 20 K and arise from population inversion of the CO_2 ν_3 relative to the $2\nu_2$ level at altitudes above 60 km on Mars (see Fig. 8.7). Kostiuk et al., also show how the pressure dependence of the ethane emission lines can be used to determine the vertical mixing of ethane on Jupiter, yielding a constant mixing ratio of 1.2×10^{-6}.

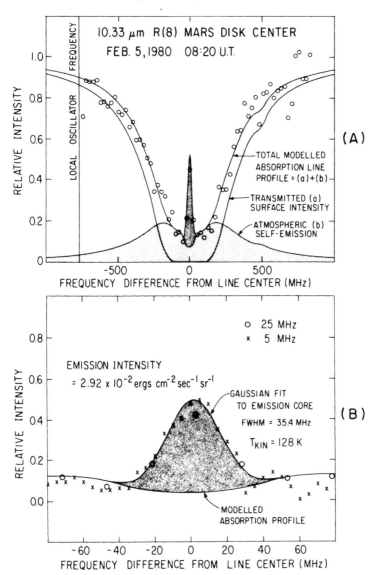

FIG. 8.7 Example showing observations and modeling of the 10.33-μm R(8) line of $^{12}C^{16}O_2$ at the center of the Martian disk. The top portion shows 25-MHz data and modeled profiles; the bottom portion includes the 5-MHz observations of the emission core (Deming et al., 1983).

8.4.2 Remote Sensing Using Fiber Optics

Although most laser remote sensing relies on atmospheric transmission of the laser radiation, there are a number of applications for remote sensing in which there is no atmospheric-path access to the region or material of interest. Fiber optics can sometimes play an invaluable role in such applications. It is possible to use fiber optics to direct laser radiation to remote locations and detect the presence of elevated gas concentrations via differential absorption, as has been demonstrated for NO_2 at 0.5 μm and CH_4 at 1.33 μm (Inaba, 1983). This technique could have important industrial applications but is limited to some extent because fibers work best in the spectral region 0.2 to 3.7 μm, where not all gases have useful absorption strengths. However, fibers are being developed that are useful out to 9 to 12 μm and would allow CO_2 lasers to be used for strong absorption lines of many gases (Harrington and Standlee, 1983; Miyagi et al., 1983).

Another aspect of this subfield is monitoring fluorescence using an "optrode" at the end of an optical fiber in the material of interest. The optrode may be a small sphere that is efficient at coupling laser radiation into the material, and then coupling fluorescent radiation back to the fiber, where it is returned to the spectrally filtered detector. This process works best when the material of interest fluoresces. When it does not, it may still be possible to get a fluorescent signal by coating the optrode with a substance that fluoresces in proportion to some parameter of the material, such as the pH. See Maugh (1982) and Hirschfeld (1983) for a more complete discussion of this technique, which has applications to industrial and nuclear chemical processing. It is also possible to detect Raman scattering from a bulk liquid (Newley et al., 1984).

8.5 FUTURE DIRECTIONS FOR LASER REMOTE SENSING

Despite the considerable progress in the field of laser remote sensing during the past 20 years, there remains considerable research and development to bring the field to full fruition. Currently, there are only a few laser remote sensing instruments being used in routine field measurements. The rest are either being studied in laboratories or used for demonstration and testing purposes in field situations; it should also be noted that advances in other fields can have an impact on this field. For example, improvements in laser technology will enable better and/or different measurements to be made. With this in mind, what are some of the trends expected for laser remote sensing?

8.5.1 Instruments

8.5.1A. Improved Laser Sources

Laser remote sensors depend on the availability of appropriate laser sources. For example, flash-lamp-pumped dye lasers allowed atmospheric NO_2, SO_2, O_3, and sodium to be measured in the visible and ultraviolet spectral regions in early mid-1970s, but it was not until Nd:YAG-pumped dye lasers were made commercially available in the late 1970s that these gases as well as metal atoms and ions in the thermosphere could be measured easily. These lasers have high reliability, a short (10-ns) pulse width, which improved the spatial resolution and increased the efficiency of generating the second harmonic, and a useful PRF (10+ Hz). It should be noted that the useful spectral region for chemical species measurements (approximately 0.2 to 12 μm, with gaps in the troposphere where atmospheric water vapor and carbon dioxide absorb strongly) is only partly covered: with dye lasers from 0.25 to 0.9 μm, with the alexandrite laser from 0.72 to 0.78 μm, and with CO_2 lasers from 9 to 12 μm. Other lasers, such as excimer and HF and DF lasers are useful in limited regions or in applications for Raman scattering or fluorescence measurements (Laudenslager et al., 1984). However, the usefulness of excimer lasers is being extended via Raman shifting of the laser frequency. Thus, additional laser types could be useful if they fill in some of the gaps.

One category of new laser types is the "all-solid-state" laser. This is generally taken to include such lasers as the Nd:YAG laser and other lasers or crystals it pumps, even though the Nd:YAG laser is usually pumped with a xenon flash lamp. However, GaAlAs diode lasers near 800 nm are being developed as pump sources for the Nd:YAG laser, so that the inclusion of Nd:YAG lasers in this category is appropriate. The Nd:YAG laser has been used to pump the $LiNbO_3$ optical parametric oscillator, which emits radiation from 1.4 to 4.0 μm, and the Nd:YAG laser at 1.3 μm is being used to pump the $Co:MgF_2$ laser, which lases from 1.5 to 2.3 μm. For further information on these and other solid-state lasers, see Rediker et al. (1984) and Byer et al. (1983).

Nonlinear crystals can be used both to generate harmonics of fundamental laser frequencies, and to "mix" two frequencies to give a "sum" frequency. For example, the third harmonic of CO_2 lasers can be generated by summing the second harmonic of one CO_2 laser with the first harmonic of another CO_2 laser using tandem $CdGeAs_2$ crystals (Menyuk and Iseler, 1979), giving laser radiation from 3.06 to 3.6 μm using $^{12}C^{16}O_2$ lasers, and from 3.0 to 4.0 μm using rare isotope CO_2 lasers.

Excimer lasers are being developed as sources for remote sensing measurements. They have the advantages of requiring no flash

lamps, of being scalable to high energies, and being down- or up-
shifted by Raman scattering. These properties should make excimer
lasers useful for measurements of wind velocities, aerosols, and
water vapor using Raman scattering (Laudenslager et al., 1984),
in addition to ozone already measured using the DIAL technique
(Werner et al., 1983a,b).

One advantage that would help laser long-path remote sensing
would be the development of continuously tunable lasers with sufficient power and tunability to allow them to be frequency modulated
at kilohertz rates so that derivative spectroscopy techniques could
be applied. This tunability allows tunable diode lasers to make
very sensitive measurements (Reid et al., 1985), but their low
power ($\sim$1 mW) restricts the measurement path length to 10 to 30 m
unless retroreflectors are employed. About an order-of-magnitude
improvement in sensitivity is possible with the use of two laser
frequencies, if the detected power is sufficient.

8.5.1B. Compact Remote Sensing Instruments

There are a number of applications where a compact, perhaps
personally portable laser remote sensor would be quite useful. One
of these is in searching for leaks of hazardous gases: in industrial
settings where personnel are exposed; where leaks represent economic losses; or where leaks can generate high enough concentrations that explosions can occur. The HeNe, CO_2, and tunable diode
lasers are leading candidates for this application now because of
their tunability in the spectral regions where many gases absorb,
and because they have the potential for miniaturization. They still
require some heavy support components, such as power supplies
and for the TDLs, refrigerators, but these can be separated from
the laser to allow the laser head and receiver to be moved easily,
and will be reduced in weight when TDLs operate reliably above
77 K.

8.5.1C. Large, Multispecies DIAL Systems

At the other end of the size spectrum are large lidar systems
with computer data systems for the remote measurement of some
chemical species. Several such systems have been or are being
constructed for the measurement of naturally occurring atmospheric
trace species, common pollutant gases, hazardous industrial gases,
and chemical warfare agents. These systems tend to be quite expensive and usually require that the lidar make unique measurements
of considerable importance in order to justify their existence. The
number of such systems should increase as the existing or planned
ones prove their usefulness, as technology advances, and as users
find ways to justify or share the expenses.

8.5.2 Meteorological Parameters

The potential of lidar instruments to measure meteorological parameters other than such items as mixing heights is slowly being
realized. Wind measurements from ground and air-borne and space-
borne platforms may have the greatest impact. For example, turbulence at airports could be monitored with a ground-based system,
helping to avoid turbulence-caused crashes during takeoff and
landing. Wind-sensing lidar mounted on aircraft might be useful in
helping to avoid clear-air turbulence, and perhaps in locating the
flight altitude with the most favorable wind velocity. Vertical temperature profiles would be useful in pollution episode forecasting.
Water vapor measurements would be useful for index-of-refraction
determinations, aiding radar for long-distance applications. The
proposal by Schwiesow (1983) for development of a lidar-based 1-km
meteorological tower might be implemented.

The use of lidar for measuring properties of aerosols (distribution, size, chemical composition, shape) has been well demonstrated
in some applications, yet requires further development in others.
Such areas as plume opacity and airport visibility are now being
approached with prototype lidar systems. As the physics becomes
better understood and the technology improves, there should be many
lidar systems in routine use for a variety of aerosol measurements.

8.5.3 Space-Borne Applications of Lidar

NASA has development and planning programs for placing lidar systems in space on either manned or unmanned platforms. A range of
measurement options for the Space Shuttle lidar was studied in the
late 1970s, and over two dozen potential measurements were identified where a space-borne platform would allow a global coverage
that would have valuable scientific or practical benefits (Browell,
1979). NASA, NOAA, AFGL, and others are working on developing
a CO_2 Doppler lidar system that could be used on a satellite to
measure the global tropospheric wind field (Huffaker et al., 1984;
Gurk et al., 1984). Such a laser system should greatly improve
weather forecasting as well as assist in determining appropriate
flight altitudes for minimal fuel consumption. Also, space-borne
lidar systems could be useful in measuring global stratospheric
aerosol distributions, especially after volcanic eruptions (McCormick
and Swissler, 1983), the planetary boundary layer (Uthe, 1983;
Melfi et al., 1985), and cloud-top heights (Spinhirne et al., 1982,
1983). Also, polar ice cap heights to 10-cm accuracy could be
measured (Bufton et al., 1983). DIAL measurements of water vapor
and ozone are also feasible, but further improvements in lasers and
lidar technology are required for their realization. For further
analyses on space-borne lidar applications, see, for example, Rems-

berg and Gordley (1978), Megie and Menzies (1980), Abreu (1980), Atlas and Korb (1981), Russell et al. (1982), Russel and Morley (1982), Megie et al. (1983), and Browell and Ismail (1983).

8.5.4 Better Understanding of Laser Remote Sensing

The history of laser remote sensing has seen an interplay of ideas, technology, laboratory demonstrations, and field measurements. Each area has contributed in a valuable way to the progress in the field. For example, use of the Laser Absorption Spectrometer (Shumate et al., 1983) in the field showed that differential spectral reflectance presented a problem when using topographic targets to provide the backscatter of the laser radiation, which led to a better general understanding of that problem and to the use of CO_2 laser systems to identify geologic features on the earth's surface (Kahle et al., 1984).

A publication by Menyuk and Killinger (1983) describes the current state of knowledge of the important mechanisms that affect DIAL measurements of atmospheric trace species. There is more work that can be done in elucidating these mechanisms and in finding ways to minimize their effects on the measurements. In general, if error mechanisms are understood, ways can be found to design improved systems or analyze the data better. When the mechanisms are not known, large measurement errors can arise.

In addition to understanding the error mechanism better, it may be possible to better understand the physical processes that give rise to new applications of laser remote sensing. The field is nearly 20 years old, yet a number of the ideas being developed today were suggested only in the past 2 to 4 years, for example, the random-modulation continuous-wave (RM-CW) lidar technique for getting range-resolved measurements of aerosols and gases using a CW laser (Takeuchi et al., 1986), the "trough" technique for measuring atmospheric pressure (Korb and Weng, 1983), and laser tomography for obtaining a two-dimensional map of atmospheric species from a series of line-of-sight measurements (Wolfe and Byer, 1982; Bennett et al., 1984) and extinction coefficients of atmospheric aerosols (Weinman, 1984). Thus it seems highly probable that new ideas in the field will continue to be generated and developed, allowing the field of laser remote sensing to mature usefully.

BIBLIOGRAPHY

Browell, E. V., ed. (1979). *Shuttle Atmospheric Lidar Research Program*, NASA SP-433. Identifies 26 lidar measurements that are considered feasible from Space Shuttle. Several of these are still being considered for the 1990s, while others will be postponed or dropped. The report is a good source of ideas for advanced lidar projects.

Browell, E. V., and Woods, P. T., eds. (1986). *Assessment of DIAL Data Collection and Analysis Techniques*, National Aeronautics and Space Administration, Washington, D. C. (to appear). This Nasa reference document is the joint work of forty plus workers in the field of DIAL measurement of atmospheric trace gases. The various chapters provide the collective wisdom on such topics as transmitters and atmospheric effects, while the appendix provides a snapshot of current DIAL research at over twently laboratories worldwide.

Byer, R. L. (1975). Remote air pollution measurement, *Opt. Quantum Electron.*, *7*, 147. The first major survey of progress in the field of laser remote sensing of atmospheric trace species using the DIAL technique.

Byer, R. L., and Garbuny, M. (1973). Pollutant detection by absorption using Mie scattering and topographic targets as retroreflectors, *Appl. Opt.*, *12*, 1496. Develops the theory for measurement of atmospheric trace gases using differential absorption with atmospheric or topographic backscatter, again showing the advantages of the DIAL technique over the Raman and fluorescence techniques.

Carswell, A. L. (1983). Lidar measurements of the atmosphere, *Can. J. Phys.*, *61*, 378. A good overview of laser remote sensing of the atmosphere for both chemical species and meteorological parameters. The author's specialty, multiple scattering in clouds, is covered in some detail.

Grant, W. B., and Menzies, R. T. (1983). A survey of laser and selected optical systems for the remote measurement of pollutant and gas concentrations, *J. Air Pollut. Control Assoc.*, *33*, 187. Reviews the status and outlook for DIAL, laser long-path, and passive optical systems for measurement of pollutant gas concentrations as of the end of 1982.

Hinkley, E. D., ed. (1976). *Laser Remote Sensing of the Atmosphere*, Springer-Verlag, Berlin. The standard reference in the field for many years. It has half a dozen chapters authored by the leaders in the field on such fields as heterodyne detection, Raman scattering, DIAL, long-path monitoring using tunable diode lasers, and aerosol distribution measurement.

Inaba, H., ed. (1972). Laser measurement of atmospheric pollution, *Opto-electronics 4*, 69-186. This issue of *Opto-electronics* contains ten articles by pioneers in the field of laser remote sensing of pollutants. Many of the authors are still active in the field today, and have used the ideas expressed in these articles to build very successful field lidar systems.

Killinger, D. K., and Mooradian, A., eds. (1983). *Optical Sensing of the Atmosphere*, Springer-Verlag, Berlin. Forty-seven papers presented at a conference held in Monterey, California, in February 1982. The goal of the conference was to review the state-of-the-art laser and optical remote sensing of atmospheric trace gases and aerosols, meteorological parameters, and earth surface features. Most of the leading groups active in the United States, Europe, and Japan at the time are represented.

Measures, R. M. (1984). *Laser Remote Sensing: Fundamentals and Applications*, Wiley, New York. This book, which starts from Maxwell's equations and progresses logically and thoroughly through the theory for laser remote sensing of atmospheric trace species, meteorological parameters, and fluorescent materials on the earth's surface to a review of field work described by the end of 1982, has become the new standard reference book in the field.

Measures, R. M., ed. (1987). *Laser Remote Chemical Analysis*, Wiley, New York (to appear). This book will also be a textbook with contributions from several leaders in the field, covering such topics as infrared absorption, ground-based pollution measurements, and the use of tunable diode lasers.

Uthe, E. E. (1983). Application of surface based and airborne lidar systems for environmental monitoring, *J. Air Pollut. Control Assoc.*, *33*, 1149. Reviews the author's extensive work in measurements of atmospheric aerosols, including boundary layer structure, stationary source emissions, plume transport, and aerosol size distribution.

These are several ways to keep abreast of the field: (1) read *Applied Optics*, as many important articles in the field, especially those describing new systems, and measurement and analysis techniques, are published there; (2) write to the authors in your subfields of interest and request reprints of their latest work; (3) attend conferences sponsored by the Optical Society of America or the American Meteorological Society; (4) look for citations of articles of interest in *Science Citation Index*, Institute for Scientific Information, Philadelphia, PA 19104. This index is published bimonthly and is a valuable source for rapidly locating new publications in this or any other field.

REFERENCES

Abreu, V. J. (1979). Wind measurements from an orbital platform using a lidar system with incoherent detection: an analysis, *Appl. Opt. 18*, 2992.

Abreu, V. J. (1980). Lidar from orbit, *Opt. Eng., 19*, 489.

Ahlberg, H., Lundqvist, S., Shumate, M. S., and Persson, U. (1985). Analysis of errors caused by optical interference effects in wavelength-diverse CO_2 laser long-path systems, *Appl. Opt., 24*, 3917.

Allario, F. S., Katzberg, J., and Hoell, J. M. (1983). Tunable laser heterodyne spectrometer measurements of atmospheric species, in *Optical and Laser Remote Sensing* (D. K. Killinger and A. Mooradian, eds.), Springer-Verlag, Berlin, p. 50.

Ansmann, A. (1985). Errors in ground-based water-vapor DIAL measurements due to Doppler-broadened Rayleigh backscattering, *Appl. Opt. 24*, 3476.

Arshinov, Y. F., Bobrovnikov, S. M., Zuev, V. E., and Mitev, V. M. (1983). Atmospheric temperature measurements using a pure rotational Raman lidar, *Appl. Opt., 22*, 2984.

Asai, K., Itabe, T., and Igarashi, T. (1979). Range-resolved measurements of atmospheric ozone using a differential absorption CO_2 laser radar, *Appl. Phys. Lett., 35*, 60.

Atlas, D., and Korb, C. L. (1981). Weather and climate needs for lidar observations from space and concepts for their realization, *Bull. Am. Meteorol. Soc., 62*, 1270.

Baumgartner, R. A. (1979). Comparison of lidar and air quality station NO_2 measurements, *J. Air Pollut. Control Assoc., 29*, 1162.

Bennett, K. W., Faris, G. W., and Byer, R. L. (1984). Experimental optical fan tomography, *Appl. Opt. 23*, 2678.

Bilbro, J., Fichtl, G., Fitzjarrald, D., Krause, M., and Lee, R. (1984). Airborne Doppler lidar wind field measurements, *Bull. Am. Meteorol. Soc., 65*, 348.

Brassington, D. J., Felton, R. C., Jolliffe, B. W., Marx, B. R., Mancrieff, J. T. M., Rowley, W. R. C., and Woods, P. T. (1984). Errors in spectroscopic measurements of SO_2 due to nonexponential absorption of laser radiation, with application to the remote monitoring of atmospheric pollutants, *Appl. Opt., 23*, 469.

Bristow, M., Nielsen, D., Bundy, D., and Furtek, R. (1981). Use of water Raman emission to correct airborne laser fluorosensor data for effects of water optical attenuation, *Appl. Opt. 20*, 2889.

Browell, E. V., ed. (1979). *Shuttle Atmospheric Lidar Research Program*, NASA SP-433.

Browell, E. V. (1982). Lidar measurements of tropospheric gases, *Opt. Eng., 21*, 128.

Browell, E. V., and Ismail, S. (1984). Spaceborne lidar investigations of the atmosphere, in *Space Laser Applications and Technology*, European Space Agency, 8-10 rue Mario-Nikis, 75738 Paris 15, France, p. 181.

Browell, E. V., Ismail, S., and Shipley, S. T. (1985). Ultraviolet DIAL measurements of O_3 profiles in regions of spatially inhomogeneous aerosols, *Appl. Opt. 24*, 2827.

Browell, E. V., Carter, A. F., Shipley, S. T., Allen, R. J., Butler, C. F., Mayo, M. N., Siviter, J. H., Jr., and Hall, W. M. (1983). NASA multipurpose airborne DIAL sustem and measurements of ozone and aerosol profiles, *Appl. Opt., 22*, 522.

Bufton, J. L., Robinson, J. E., Femiano, M. D., and Flatow, F. S. (1983). Satellite laser altimeter for measurement of ice sheet topography, *IEEE Trans. Geosci. Remote Sensing, GE-20*, 544.

Byer, R. L., Kane, T., Eggleston, J., and Long, S. Y. (1983). Solid laser sources for remote sensing, in *Optical and Laser Remote Sensing* (D. K. Killinger and A. Mooradian, eds.), Springer-Verlag, Berlin, p. 245.

Cahen, C., and Megie, G. (1981). A spectral limitation on the range resolved differential absorption lidar technique, *J. Quant. Spectrosc. Radiat. Transfer, 25*, 151.

Cahen, C., Megie, G., and Flamant, P. (1982). Lidar monitoring of the water vapor cycle in the troposphere, *J. Appl. Meterol., 21*, 1506.

Churnside, J. H., and Yura, H. T. (1983). Speckle statistics of atmospherically backscattered laser light, *Appl. Opt. 22*, 2559.

Collis, R. T. H., and Uthe, E. E. (1972). Mie scattering techniques for air pollution measurement with lasers, *Opto-electronics 4*, 87.

Cooney, J. A. (1983). Uses of Raman scattering for remote sensing of atmospheric properties of meteorological significance, *Opt. Eng., 22*, 292.

Cooney, J. A. (1968). Measurements of the Raman component of laser atmospheric backscatter, *Appl. Phys. Lett. 12,* 40.

Cooney, J., Petri, K., and Salik, A. (1985). Measurements of high resolution atmospheric water vapor profiles by use of a solar blind Raman lidar, *Appl. Opt. 24,* 104.

Deming, D., Espenak, F., Jennings, D., Kostiuk, T., Mumma, M., and Zipoy, D. (1983). Observations of the 10-μm natural laser emission from the mesospheres of Mars and Venus, *Icarus, 55,* 347.

deNeufville, J. P., Kasdan, A., and Chimenti, R. J. L. (1981). Selective detection of uranium by laser-induced fluorescence: a potential remote-sensing technique: 1. Optical characteristics of uranyl geologic targets, *Appl. Opt., 20,* 1279.

Egeback, A. L., Fredriksson, K. A., and Hertz, H. M. (1984). DIAL techniques for the control of sulfur dioxide emissions, *Appl. Opt., 23,* 722.

Englisch, W., Wiesemann, W., Boscher, J., and Rother, M. (1983). Laser remote sensing measurements of atmospheric species and natural target reflectivities, in *Optical and Laser Remote Sensing* (D. K. Killinger and A. Mooradian, eds.), Springer-Verlag, Berlin, p. 38.

Faxvog, F. R., and Mocker, H. W. (1982). Rapidly tunable CO_2 TEA laser, *Appl. Opt., 21,* 3986.

Fiocco, G., and Smullin, L. D. (1963). Detection of scattering layers in the upper atmosphere (60-140 km) by optical radar, *Nature, 199,* 1275.

Flamant, P. H., Menzies, R. T., and Kavaya, M. J. (1984). Evidence for speckle effects on pulsed CO_2 lidar signal returns from remote targets, *Appl. Opt. 23,* 1412.

Fredriksson, K. A., and Hertz, H. M. (1984). Evaluation of the DIAL technique for studies on NO_2 using mobile lidar system, *Appl. Opt., 23,* 1403.

Freed, C., Bradley, L. C., and O'Donnell, R. G. (1980). Absolute frequencies of lasing transitions in seven CO_2 isotopic species, *IEEE J. Quantum Electron., 16,* 1195.

Fukuda, T., Matsuura, Y., and Mori, T. (1984). Sensitivity of coherent range-resolved differential absorption lidar, *Appl. Opt., 23,* 2026.

Giver, L. P., Gentry, B., Schwemmer, G., and Wilkerson, T. D. (1982). Water absorption lines, 931-961 nm; selected intensities,

N_2-collision broadening coefficients and pressure shifts in air, *I. Quant. Spectrosc. Radiat. Transfer, 27,* 423.

Glenar, D., Kostiuk, T., Jennings, D. E., Buhl, D., and Mumma, M. J. (1982). Tunable diode-laser heterodyne spectrometer for remote observations near 8 μm, *Appl. Opt., 21,* 253.

Goodman, J. W. (1975). Statistical properties of laser speckle patterns in *Laser Speckle and Related Phenomena* (J. C. Dainty, ed.), Springer-Verlag, Berlin, p. 9.

Grant, W. B. (1982a). Effect of differential spectral reflectance on DIAL measurements using topographic targets, *Appl. Opt., 21,* 2390.

Grant, W. B. (1982b). Helium-neon laser remote measurement of methane, in *Resource Recovery from Solid Wastes* (S. Sengupta and K.-F. V. Wong, eds.), Pergamon Press, Elmsford, N.Y., p. 265.

Grant, W. B. (1986). He-Ne and CW CO_2 laser long-path systems for gas detection, *Appl. Opt., 25,* 709.

Grant, W. B., and Hake, R. D., Jr. (1975). Calibrated remote measurements of SO_2 and O_3 using atmospheric backscatter, *J. Appl. Phys., 46,* 3019.

Grant, W. B., and Menzies, R. T. (1983). A survey of laser and selected optical systems for remote measurement of pollutant gas concentrations, *J. Air Pollut. Control Assoc., 33,* 187.

Grant, W. B., Gary, B. L., and Shumate, M. S. (1983). Remote Measurements of Ozone, Water Vapor, and Liquid Water Content, and Vertical Profiles of Temperature in the Lower Troposphere, JPL Publication No. 83-80.

Gurk, H. M., Kaskiewicz, P. F., and Altman, W. P. (1984). Windsat free-flyer using the advanced Tiros-N satellite, *Appl. Opt., 23,* 2537.

Hake, R. D., Jr., Arnold, D. E., Jackson, D. W., Evans, W. E., Ficklin, B. P., and Long, R. A. (1972). Dye-laser observations of the nighttime atomic sodium layer, *J. Geophys. Res., 77,* 6389.

Hall, F. F. Jr., Huffaker, R. M., Hardesty, R. M., Jackson, M. E., Lawrence, T. R., Post, M. J., Richter, R. A., and Weber, B. F. (1984). Wind measurement accuracy of the NOAA pulsed infrared Doppler lidar, *Appl. Opt. 23,* 2503.

Hanst, P. L. (1970). Infrared spectroscopy and infrared lasers in air pollution research and monitoring, *Appl. Spectrosc., 24,* 161.

Hardesty, R. M. (1984a). Measurement of Range-Resolved Water Vapor Concentration by Coherent CO_2 Differential Absorption Lidar, *NOAA, Tech. Memo. ERL WPL-118*, Wave Propagation Laboratory, Boulder, Colo.

Hardesty, R. M. (1984b). Coherent DIAL measurement of range-resolved water vapor concentration, *Appl. Opt.*, *23*, 2545.

Harrington, J. A., and Standlee, A. G. (1983). Attenuation at 10.6 µm in loaded and unloaded polycrystalline KRS-5 fibers, *Appl. Opt.*, *22*, 3073.

Hauchecorne, A., and Chanin, M.-L. (1980). Density and temperature profiles obtained by lidar between 35 and 70 km, *Geophys. Res. Lett.*, *7*, 565.

Hawley, J. G., Fletcher, L. D., and Wallace, G. F. (1983). Ground-based ultraviolet differential absorption lidar (DIAL) system and measurements, in *Optical and Laser Remote Sensing* (D. K. Killinger and A. Mooradian, eds.), Springer-Verlag, Berlin, p. 128.

Hays, P. B., Abreu, V. J., Sroga, J., and Rosenberg, A. (1984). Analysis of a 0.5 micron spaceborne wind sensor, in *Preprints, Conf. Satellite/Remote Sensing and Applications*, Clearwater Beach, Fla., June 25-29, sponsored by the American Meteorological Society, New York, p. 266.

Heaps, W. S., and McGee, T. J. (1985). Progress in stratospheric hydroxyl measurement by balloon-borne lidar, *J. Geophys. Res.*, *90*, 7913.

Hill, R. J., Clifford, S. F., and Lawrence, R. S. (1980). Refractive-index and absorption fluctuations in the infrared caused by temperature, humidity, and pressure fluctuations, *J. Opt. Soc. Am.*, *70*, 1192.

Hill, S. C., Hill, A. C., and Barber, P. W. (1984). Light scattering by size/shape distributions of soil particles and spheroids, *Appl. Opt.*, *23*, 1025.

Hinkley, E. D. (1972). Tunable infrared lasers and their applications to air pollution measurements, *Opto-electronics*, *11*, 69.

Hirschfeld, T., (1977). On the nonexistence of nonfluorescent compounds and Raman spectroscopy, *Appl. Spectrosc.*, *31*, 328.

Hirschfeld, T., Schildkraut, E. R., Tannenbaum, H., and Tanenbaum, D. (1973). Remote spectroscopic analysis of ppm-level air pollutants by Raman spectroscopy, *Appl. Phys. Lett.*, *22*, 38.

Hirschfeld, T., Deaton, T., Milanovich, F., and Klainer, S. (1983).
Feasibility of using fiber optics for monitoring groundwater contaminants, *Opt. Eng.*, *22*, 527.

Hoff, R. M., and Millan, M. M. (1981). Remote SO_2 mass flux
measurements using COSPEC, *J. Air Pollut. Control Assoc.*, *31*,
381.

Hoge, F. E., and Swift, R. N. (1983). Airborne dual laser excitation and mapping of phytoplankton photopigments in a Gulf
Stream Warm Core Ring, *Appl. Opt. 22*, 2272.

Hubler, G., Perner, D., Platt, U., Tonnissen, A., and Enhalt, D.
H. (1982). Groundlevel OH radical concentration: new measurements by optical absorption, *Preprint Vol., Second Symp. Composition Nonurban Troposphere*, May 25-28, Williamsburg, Va.,
American Meteorological Society, Boston.

Huffaker, R. M., Lawrence, T. R., Post, M. J., Priestley, J. T.,
Hall, F. F. Jr., Richter, R. A., and Keeler, R. J. (1984).
Feasibility studies for a global wind measuring satellite system
(Windsat): Analysis of simulated performance, *Appl. Opt.*, *23*,
2523.

Inaba, H. (1983). Optical remote sensing of environmental pollution
and danger by molecular species using low-loss optical fiber
network system, in *Optical and Laser Remote Sensing* (D. K.
Killinger and A. Mooradian, eds.), Springer-Verlag, Berlin, p.
288.

Inaba, H., and Kobayasi, T. (1972). Laser-Raman radar, *Opto-
electronics*, *4*, 101.

Inaba, H., and Kobayasi, T. (1976). Detection of atoms and mole-
cules by Raman scattering and resonance fluorescence, in *Laser
Monitoring of the Atmosphere* (E. D. Hinkley, ed.), Springer-
Verlag, Berlin, p. 153.

Jelalian, A. V., and Huffaker, R. M. (1967). Laser Doppler techniques for remote wind velocity measurements, *Mol. Radiat.
Conf.*, Marshall Space Flight Center, Huntsville, Ala., Oct. 19.

Johnson, M. A., Betz, A. L., McLaren, R. A., Sutton, E. C., and
Townes, C. H. (1976). Non-thermal 10 μm CO_2 emission lines in
the atmospheres of Mars and Venus, *Astrophys. J.*, *208*, L145.

Junge, C. E., and Manson, J. E. (1961). Stratospheric aerosol
studies, *J. Geophys. Res.*, *66*, 2163.

Kahle, A., Shumate, M. S., and Nash, D. B. (1984). Active airborne infrared laser system for identification of surface rock
and minerals, *Geophys. Res. Lett.*, *11*, 1149.

Kalshoven, J. E., Jr., Korb, C. L., Schwemmer, G. K., and Dombrowski, M. (1981). Laser remote sensing of atmospheric temperature by observing resonant absorption of oxygen, *Appl. Opt.*, *20*, 1967.

Kane, T. J., Zhow, B., and Byer, R. L. (1984). Potential for coherent Doppler wind velocity lidar using neodymium lasers, *Appl. Opt.*, *23*, 2477.

Kasdan, A., Chimenti, R. J. L., and deNeufville, J. P. (1981). Selective detection of uranium by laser-induced fluorescence: a potential remote-sensing technique: 2. Experimental assessment of the remote sensing of uranyl geologic targets, *Appl. Opt.*, *20*, 1297.

Kavaya, M. J., and Menzies, R. T. (1985). Lidar aerosol backscatter measurements: Systematic, modeling, and calibration error considerations, *Appl. Opt.*, *24*, 3444.

Kavaya, M. J., Menzies, R. T., Haner, D. A., Oppenheim, U. P., and Flamant, P. H. (1983). Target reflectance measurements for calibration of lidar atmospheric backscatter data, *Appl. Opt.*, *22*, 2619.

Kent, G. S., Yue, G. K., Farrukh, U. O., and Deepak, A. (1983). Modeling atmospheric aerosol backscatter at CO_2 laser wavelengths: aerosol properties, modeling techniques, and associated problems, *Appl. Opt.*, *22*, 1655.

Killinger, D. K., and Mooradian, A., eds. (1983). *Optical Sensing of the Atmosphere*, Springer-Verlag, Berlin.

Killinger, D. K., Menyuk, N., and DeFeo, W. E. (1980). Remote sensing of CO using frequency-doubled CO_2 laser radiation, *Appl. Phys. Lett.*, *36*, 402.

Kobayasi, T. and Inaba, H. (1975). Infra-red heterodyne laser radar for remote sensing of air pollutants by range-resolved differential absorption, *Opt. and Quant. Elect.* *7*, 319.

Korb, C. L., and Weng, C. Y. (1982). A theoretical study of a two-wavelength lidar technique for the measurement of atmospheric temperature profiles, *J. Appl. Meteorol.*, *21*, 1346.

Korb, C. L., and Weng, C. Y. (1983). Differential absorption lidar technique for measurement of the atmospheric pressure profile, *Appl. Opt.*, *22*, 3754.

Kostiuk, T., and Mumma, M. J. (1983). Remote sensing by IR heterodyne spectroscopy, *Appl. Opt.*, *22*, 2644.

Kostiuk, T., Mumma, M. J., Expenak, F., Deming, D., Jennings, D. E., Maguire, W., and Zipoy, D. (1983). Measurements of stratospheric ethane in the Jovian south polar region from infrared heterodyne spectroscopy of the ν_9 band near 12 microns, *Astrophys. J., 265*, 564.

Ku, R. T., Hinkley, E. D., and Sample, O. J. (1975). Long-path monitoring of atmospheric carbon monoxide with a tunable diode laser system, *Appl. Opt., 14*, 854.

Kucerovsky, Z., Brannen, E., Paulekat, K. C., and Rumbold, D. G. (1973). Characteristics of a laser system for atmospheric absorption and air pollution experiments, *J. Appl. Meteorol., 12*, 1387.

Labitzke, K., Naujokat, B., and McCormick, M. P. (1983). Temperature effects on the stratosphere of the April 4, 1982, eruption of El Chichón, Mexico, *Geophys. Res. Lett., 10*, 24.

Lahman, W., Staehr, W., Baumgart, R., Breinig, A., Weitkamp, C., and Michaelis, W. (1985). Validation measurements of the inherent calibration of differential absorption lidar, *Technical Digest, Optical Remote Sensing of the Atmosphere*, Incline Village, Nev., Jan. 15-18, 1985, paper WC-12, sponsored by the Optical Society of America, Washington, D.C.

Laudenslager, J. B., Pacala, T. J., McDermid, I. S., and Rider, D. M. (1984). Applications of excimer lasers for atmospheric species measurements, *Proc. Soc. Photo-Opt. Instrum. Eng., 461*, 34.

Leonard, D. (1967). Observation of Raman scattering from the atmosphere using a pulsed nitrogen ultraviolet laser, *Nature, 216*, 142.

Loper, G. L., O'Neill, M. A., and Gelbwachs, J. A. (1983). Water-vapor continium CO_2 laser absorption spectra between 27°C and -10°C, *Appl. Opt., 22*, 3701.

Martonchik, J. V., Orton, G. S., and Appleby, J. F. (1984). Optical properties of NH_3 ice from the far IR to the near UV, *Appl. Opt., 23*, 541.

Mason, J. B. (1975). Lidar measurements of temperature: a new approach, *Appl. Opt., 14*, 76.

Maugh, T. H., II. (1982). Remote spectrometry with fiber optics, *Science, 218*, 875.

McCartney, E. J. (1976). *Optics of the Atmosphere*, Wiley, New York.

McClatchey, R. A., Benedict, W. S., Clough, S. A., Burch, D. E., Calfee, R. F., Fox, K., Rothman, L. S., and Garing, J. S. (1973). AFCRL atmospheric absorption line parameters compilation, *AFCRL-TR-T3-0096*.

McClenny, W. A., and Russwurm, G. M. (1978). Laser-based long path monitoring of ambient gases—analysis of two systems, *Atmos. Environ.*, *12*, 1443.

McCormick, M. P. (1982). Lidar measurements of Mount St. Helens effluents, *Opt. Eng.*, *21*, 340.

McCormick, M. P., and Swissler, T. J. (1983). Stratospheric aerosol mass and latitudinal distribution of the El Chichón eruption cloud for October 1982, *Geophys. Res. Lett.*, *10*, 877.

McCormick, M. P., Swissler, T. J., Fuller, W. H., Hunt, W. H., and Osborn, M. T. (1984). Airborne and ground-based lidar measurements of the El Chichon stratospheric aerosol from 90°N to 56°S, *Geof. Int. 23*, 187.

McDermid, I. S., Laudenslager, J. B., and Pacala, T. J. (1983). New technological developments for the remote detection of atmospheric hydroxyl radicals, *Appl. Opt.*, *22*, 2586.

McIlrath, T. J. (1980). Fluorescence lidar, *Opt. Eng.*, *19*, 494.

Measures, R. M. (1984). *Laser Remote Sensing: Fundamentals and Applications*, Wiley, New York.

Megie, G., and Menzies, R. T. (1980). Complementarity of UV and IR differential absorption lidar for global measurements of atmospheric species, *Appl. Opt.*, *19*, 1173.

Megie, G., Bos, F., Blamont, J. E., and Chanin, M. L. (1978). Simultaneous nighttime lidar measurements of atmospheric sodium and potassium, *Planet. Space Sci.*, *26*, 27.

Megie, G., Pelon, J., and Flamant, P. (1984). Spaceborne lidar applications to meteorology and environmental studies, in *Space Laser Applications and Technology*, European Space Agency, 8-10 rue Mario-Nikis, 75738 Paris 15, France, p. 53.

Megie, G. J., Ancellet, G., and Pelon, J. (1985). Lidar measurements of ozone vertical profiles, *Appl. Opt.*, *24*, 3454.

Melfi, S. H. (1972). Remote measurements of the atmosphere using Raman scattering, *Appl. Opt.*, *11*, 1605.

Melfi, S. H., and Whiteman, D. (1985). Observation of lower-atmospheric moisture structure and its evolution using a Raman lidar, *Bull. Am. Meteorol. Soc. 66*, 1288.

Melfi, S. H., Lawrence, J. D., Jr., and McCormick, M. P. (1969). Observation of Raman scattering by water vapor, *Opt. Lett.*, 5, 233.

Melfi, S. H., Spinhirne, J. D., Chou, S.-H., and Palm, S. P. (1985). Lidar observations of vertically organized convection in the planetary boundary layer over the ocean, *J. Clim. and Appl. Meteorol.*, *24*, 806.

Menyuk, N., and Iseler, G. W. (1979). Efficient frequency tripling of CO_2 laser radiation in tandem $CdGeAs_2$ crystals, *Opt. Lett.*, *4*, 55.

Menyuk, N., and Killinger, D. K. (1983). Assessment of relative error sources in IR DIAL measurement accuracy, *Appl. Opt.*, *22*, 2690.

Menyuk, N., Killinger, D. K., and Menyuk, C. R. (1982). Limitations of signal averaging due to temporal correlation in laser remote-sensing measurement, *Appl. Opt.*, *21*, 3377.

Menyuk, N., Killinger, D. K., and Menyuk, C. R. (1985). Error reduction in laser remote sensing: combined effects of cross correlation and signal averaging, *Appl. Opt.*, *24*, 118.

Menzies, R. T. (1976a). Ozone spectroscopy with a CO_2 waveguide laser, *Appl. Opt.*, *15*, 2597.

Menzies, R. T. (1976b). Laser heterodyne detection techniques, in *Laser Monitoring of the Atmosphere* (E. D. Hinkley, ed.), Springer-Verlag, Berlin, p. 297.

Menzies, R. T. (1983). A re-evaluation of laser heterodyne radiometer ClO measurements, *Geophys, Res. Lett.*, *10*, 729.

Menzies, R. T. (1985). A comparison of Doppler lidar wind sensors for earth-orbit global measurement applications, *The Symposium and Workshop on Global Wind Measurements*, Columbia, MD, sponsored by NASA, July 29 - Aug. 1, 1985.

Menzies, R. T., Rutledge, C. W., Zanteson, R. A., and Spears, D. L. (1981). Balloon-borne laser heterodyne radiometer for measurements of stratospheric trace species, *Appl. Opt.*, *20*, 536.

Menzies, R. T., Flamant, P. H., Kavaya, M. J., and Kuiper, E. N. (1984a). Tunable mode and line selection by injection in a TEA-CO_2 laser, *Appl. Opt.*, *23*, 3584.

Menzies, R. T., Kavaya, M. J., Flamant, P. H., and Haner, D. A. (1984b). Atmospheric aerosol backscatter measurements using a tunable coherent CO_2 lidar, *Appl. Opt.*, *23*, 2510.

Miyagi, M., Hongo, A., Aizawa, Y., and Kawakami, S. (1983). Fabrication of germanium-coated hollow waveguides for infrared transmission, *Appl. Phys. Lett.*, *43*, 430.

Molina, L. T., and Grant, W. B. (1984). FTIR spectrometer determined absorption coefficients of seven hydrazine fuel gases: implications for laser remote sensing, *Appl. Opt.*, *23*, 8389.

Moore, C. B. (1965). Gas-laser frequency selection by molecular absorption, *Appl. Opt.*, *4*, 252.

Mudd, H. T., Jr., Kruger, C. H., and Murray, E. R. (1982). Measurement of IR laser backscatter spectra from sulfuric acid and ammonium sulfate aerosols, *Appl. Opt.*, *21*, 1146.

Murray, E. R. (1978). Remote measurement of gases using differential absorption lidar, *Opt. Eng.*, *17*, 30.

Murray, E. R., Hake, R. D., Jr., van der Laan, J. E., and Hawley, J. G. (1976). Atmospheric water vapor measurements with an infrared ($10-\mu m$) differential-absorption lidar system, *Appl. Phys. Lett.*, *28*, 542.

Murray, E. R., Powell, D. D., and van der Laan, J. E. (1979). Measurement of average atmospheric temperature using a CO_2 laser radar, *Appl. Opt.*, *19*, 1794.

Nevitt, T. J., and Bohren, C. F. (1984). Infrared backscattering by irregularly shaped particles: a statistical approach, *J. Climate Appl. Meteorol.*, *23*, 1342.

Newley, K., Reichert, W. M., Andrade, J. D., and Benner, R. E. (1984). Remote spectroscopic sensing of chemical adsorption using a single multimode optical fiber, *Appl. Opt.*, *23*, 1812.

O'Neil, R. A., Buje-Bijunos, L., and Rayner, D. M. (1980). Field performance of a laser fluorosensor for the detection of oil spills, *Appl. Opt.*, *19*, 863.

Pacala, T. J., McDermid, I. S., and Laudenslager, J. B. (1982). A wavelength scannable XeCl oscillator-ring amplifier laser system, *Appl. Phys. Lett.*, *40*, 1.

Persson, U., Marthinsson, B., Johansson, J., and Eng, S. T. (1980). Temperature and pressure dependence of NH_3 and C_2H_4 absorption cross sections at CO_2 laser wavelengths, *Appl. Opt.*, *19*, 1711.

Petheram, J. C. (1981). Differential backscatter from the atmospheric aerosol: the implications for IR differential absorption lidar, *Appl. Opt.*, *20*, 3941.

Post, M. J. (1984). Aerosol backscatter profiles at CO_2 wavelengths: the NOAA data base, *Appl. Opt.*, *23*, 2507.

Poultney, S. K., Brumfield, M. L., and Siviter, J. H., Jr. (1977). Quantitative remote measurements of pollutants from stationary sources using Raman lidar, *Appl. Opt.*, *16*, 3180.

Rediker, R. H., Melngailis, I., and Mooradian, A. (1984). Lasers, their development and applications at M.I.T. Lincoln Laboratory, *IEEE J. Quantum Electron.*, *QE-20*, 602.

Reid, J., Sinclair, R. L., Grant, W. B., and Menzies, R. T. (1985). High sensitivity detection of trace gases at atmospheric pressure using tunable diode lasers, *Opt. Quantum Electron.*, *17*, 31.

Remsberg, E. E., and Gordley, L. L. (1978). Analysis of differential absorption lidar from the Space Shuttle, *Appl. Opt.*, *17*, 624.

Renault, D., Pourny, J. C., and Capitini, R. (1980). Daytime Raman-lidar measurements of water vapor, *Opt. Lett.*, *5*, 233.

Rosen, H., Robish, R., and Chamberlain, O. (1975). Remote detection of pollutants using resonance Raman scattering, *Appl. Opt.*, *14*, 2703.

Rothe, K. W. (1980). Monitoring of various atmospheric constituents using a CW chemical hydrogen/deuterium laser and a pulsed carbon dioxide laser, *Radio Electron. Eng.*, *50*, 567.

Rothe, K. W., Brinkmann, V., and Walther, H. (1974). Remote measurement of NO_2 emission from a chemical factory by the differential absorption technique, *Appl. Phys.*, *4*, 181.

Rothe, K. W., Walther, H., and Werner, J. (1983). Differential absorption measurements with fixed frequency IR and UV lasers, in *Optical and Laser Remote Sensing* (D. K. Killinger and A. Mooradian, eds.), Springer-Verlag, Berlin, p. 10.

Rothman, L. S., Gamache, R. R., Barbe, A., Goldman, A., Gillis, J. R., Brown, L. R., Toth, R. A., Flaud, J.-M., and Camy-Peyret, C. (1983a). AFGL atmospheric absorption line parameters compilation: 1982 edition, *Appl. Opt.*, *22*, 2247.

Rothman, L. S., Goldman, A., Gillis, J. R., Gamache, R. R., Pickett, H. M., Poynter, R. L., Husson, N., and Chedin, A. (1983b). AFGL trace gas compilation: 1982 version, *Appl. Opt.*, *22*, 1616.

Russell, P. B., and Morley, B. M. (1982). Orbiting lidar simulations: 2. Density, temperature, aerosol, and cloud measurements by a wavelength-combining technique, *Appl. Opt.*, *21*, 1554.

Russell, P. B., Morley, B. M., Livingston, J. M., Grams, G. W., and Patterson, E. M. (1982). Orbiting lidar simulations: 1. Aerosol and cloud measurements by an independent-wavelength technique, *Appl. Opt.*, *21*, 1541.

Sasano, Y., Browell, E. V., and Ismail, S. (1985). Error caused by using a constant extinction/backscattering ratio in the lidar solution, *Appl. Opt.* 24, 3929.

Schotland, R. M. (1966). Some observations of the vertical profile of water vapor by a laser optical radar, in *Proc. 4th Symp. Remote Sensing of Environ.*, University of Michigan, Ann Arbor, Mich., p. 273.

Schotland, R. M. (1974). Errors in the lidar measurement of atmospheric gases by differential absorption, *J. Appl. Meteorol. 13*, 71.

Schwiesow, R. L. (1983). Potential for a lidar-based portable 1 km meteorologic tower, *J. Appl. Meteorol.*, *22*, 881.

Shapiro, J. H. (1985). Correlation scales of laser speckle in heterodyne detection, *Appl. Opt. 24*, 1883.

Shipley, S. T., Tracy, D. H., Eloranta, E. W., Trauger, J. T., Sroga, J. T., Roesler, F. L., and Weinman, J. A. (1983). High spectral resolution lidar to measure optical scattering properties of atmospheric aerosols: 1. Theory and instrumentation, *Appl. Opt.*, *22*, 3716.

Shipley, S. T., Browell, E. V., McDougal, D. S., Orndorff, B. L., and Haagenson, P. (1984). Airborne lidar observations of long-range transport in the free troposphere, *Environ. Sci. Technol.*, *18*, 749.

Shumate, M. S., Lundqvist, S., Persson, U., and Eng, S. T. (1982). Differential reflectance of natural and man-made materials at CO_2 laser wavelengths, *Appl. Opt.*, *21*, 2386.

Shumate, M. S., Grant, W. B., and Menzies, R. T. (1983). Remote measurement of trace gases with the JPL laser absorption spectrometer, in *Optical and Laser Remote Sensing* (D. K. Killinger and A. Mooradian, eds.), Springer-Verlag, Berlin, p. 31.

Siegman, A. E. (1979). *An Introduction to Lasers and Masers*, McGraw-Hill, New York.

Spinhirne, J. D., Hansen, M. Z., and Caudill, L. O. (1982). Cloud top remote sensing by airborne lidar, *Appl. Opt.*, *21*, 1564.

Spinhirne, J. D., Hansen, M. Z., and Simpson, J. (1983). The structure and phase of cloud tops as observed by polarization lidar, *J. Climate Appl. Meteorol.*, *22*, 1319.

Sroga, J. T., Eloranta, E. W., Shipley, S. T., Roesler, F. L., and Tryon, P. J. (1983). High spectral resolution lidar to measure optical scattering properties of atmospheric aerosol: 2. Calibration and data analysis, *Appl. Opt.*, *22*, 3725.

Staehr, W., Lahmann, W., and Weitkamp, C. (1985). Range-resolved differential absorption lidar: optimization of range and sensitivity, *Appl. Opt.*, *24*, 1950.

Takeuchi, N., Baba, H., Sakurai, K., and Ueno, T. (1986). Diode-laser random-modulation CW lidar, *Appl. Opt.*, *25*, 63.

Tonnissen, A., Wanner, J., Rothe, K. W., and Walther, H. (1979). Application of a CW chemical laser for remote pollution monitoring and process control, *Appl. Phys.*, *18*, 297.

Uchino, O., Maeda, M., Yamamura, H., and Hirono, M. (1983a). Observation of stratospheric vertical ozone distribution by a XeCl lidar, *J. Geophys. Res.*, *88*, 5273.

Uchino, O., Tokunaga, M., Maeda, M., and Miyozoe, Y. (1983b). Differential-absorption-lidar measurement of tropospheric ozone with excimer-Raman hybrid laser, *Opt. Lett.*, *8*, 347.

Uthe, E. E. (1983). Application of surface based and airborne lidar systems for environmental monitoring, *J. Air Pollut. Control Assoc.*, *33*, 1149.

Uthe, E. E., Morley, B. M., and Nelson, N. B. (1982). Airborne lidar measurement of smoke plume distribution, vertical transmission and particle size, *Appl. Opt.*, *21*, 460.

Volz, F. E. (1983). Infrared optical constants of aerosols at some locations, *Appl. Opt.*, *22*, 3690.

Webster, C. R., and Menzies, R. T. (1984). *In-situ* measurement of stratospheric nitric oxide using a balloon-borne tunable diode laser spectrometer, *Appl. Opt.*, *23*, 1140.

Weinman, J. A. (1984). Tomography lidar to measure the extinction coefficients of atmospheric aerosols, *Appl. Opt.*, *23*, 3882.

Werner, J., Rothe, K. W., and Walther, H. (1983). Monitoring of the stratospheric ozone layer by laser radar, *Appl. Phys.*, *B32*, 113.

Werner, J., Rothe, K. W., and Walther, H. (1984). Measurements of the ozone profile up to 50 km altitude by differential absorption laser radar, *Proc. Quadrennial Ozone Symp.*, Halkidiki, Greece.

Wiesemann, W. and Lehmann, F. (1985). Reliability of airborne CO_2 DIAL measurements: Schemes for testing technical performance and reducing interference from differential reflectance, *Appl. Opt.*, *24*, 3481.

Wilkerson, T. D., and Schwemmer, G. K. (1982). Lidar techniques for humidity and temperature measurement, *Opt. Eng.*, *21*, 1022.

Wolfe, D. C., Jr., and Byer, R. L. (1982). Model studies of laser absorption computed tomography for remote air pollution measurement, *Appl. Opt.*, *21*, 1165.

Zuev, V. E. (1982). *Laser Beams in the Atmosphere*, Consultants Bureau, New York.

9

Applications of Laser-Induced Fluorescence Spectroscopy for Combustion and Plasma Diagnostics

ROBERT P. LUCHT *Combustion Research Facility, Sandia National Laboratories, Livermore, California*

9.1 INTRODUCTION

Laser-induced fluorescence spectroscopy is a sensitive and powerful technique for detecting molecules and atoms, measuring species concentrations and energy-level population distributions, and for probing energy transfer processes in molecules and atoms. In this chapter the principles of laser-induced fluorescence (LIF) are reviewed. Instrumentation for laser-induced fluorescence measurements is described, and applications of the technique in the areas of combustion and fusion plasma diagnostics are discussed.

In an LIF measurement, molecules (atoms) are typically excited from ground electronic levels to excited electronic levels by absorption of laser radiation. Once excited, molecules can decay back to the ground electronic level by spontaneous emission of a photon; the intensity of the spontaneous emission, or fluorescence, is proportional to the excited-state species number density. Fluorescence is in many cases a very sensitive species detection method. Fluorescence is also much more suited for spatially resolved "point" measurements than line-averaged techniques such as absorption.

However, LIF is an indirect method of measuring ground-state populations. Calculating ground-state populations from measured fluorescence intensities is complicated because molecules (or atoms) can be transferred out of the laser-excited upper level by collisions with other species, either to other excited levels or back to the ground electronic level. Consequently, both the fluorescence spectrum and fluorescence intensity are sensitive to the rate and types of energy transfer processes. The quantitative application of fluorescence to measurement of species concentrations is therefore dependent on the accurate modeling of the collisional and radiative transfer in laser-excited molecules.

A laser-induced fluorescence experiment is illustrated schematically in Fig. 9.1. A laser beam tuned to a species resonance is directed
into the medium of interest, exciting molecules along the length of the
beam. The resulting spontaneous emission is nearly isotropic and can
be collected by a lens system that images a probe volume onto an entrance aperture for the detection system. The spatial resolution of
the fluorescence measurement is determined by the geometry of the
collecting optics as well as the dimensions of the laser beam.

A monochromator or filter is usually used in a fluorescence experiment to discriminate against scattered laser light and other background luminescence (including fluorescence from species other than
the one of interest). The fluorescence signal is detected using a photomultiplier or other light-sensitive detector. The electrical signal
from the light detector is then measured by suitable signal-processing
electronics.

9.2 PRINCIPLES OF LASER-INDUCED FLUORES-
CENCE SPECTROSCOPY

The fundamental physics of laser-induced fluorescence spectroscopy
is addressed in a number of excellent books and review articles. The
interaction of laser radiation with molecular (atomic) resonances is

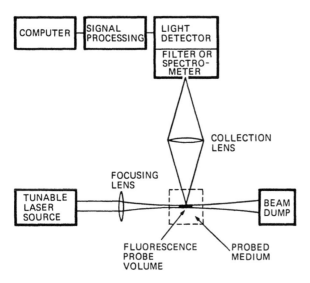

FIG. 9.1 Schematic diagram of a typical laser-induced fluorescence
experiment.

discussed in depth by Yariv (1975), Allen and Eberly (1975), and in Chapter 1 of this book. For the purposes of this chapter, a rate equation formulation will be used to model the interaction of the laser and resonance. Coherent effects, which cannot be accurately described via a rate-equation formulation, may be important when characteristic collisional times are much longer than the laser pulse length and thus may be significant in low-pressure flames and plasmas. However, as pointed out by Altkorn and Zare (1984), the multiaxial-mode nature of the exciting laser radiation and change in the axial-mode structure from laser shot to laser shot tend to obscure coherent effects.

The photon statistics, intensity, and spectrum of resonance fluorescence are discussed by Loudon (1983). The statistics of resonance fluorescence reduce to those of chaotic light in the limit that collisional dephasing is much faster than collisional transfer or spontaneous emission. In atomic fluorescence measurements in low-pressure flames and plasmas, the coupling of the resonance and the spontaneous emission field may be an important consideration. Polarization properties of resonance fluorescence are discussed by Altkorn and Zare (1984), Doherty and Crosley (1984), and Nieuwesteeg et al. (1983).

For many LIF experiments, especially molecular fluorescence, light is usually collected from numerous levels other than the level directly excited by the laser. Collisional and radiative transfer between energy levels in a laser-excited molecule or atom is usually the most important consideration for quantitative measurements of concentration or temperature. Laser-induced fluorescence is an indirect measurement of ground-state populations of atoms or molecules, and it is important to understand the population dynamics that occur when a species is excited by laser radiation.

The simplest case to analyze, the two-level model, is illustrated schematically in Fig. 9.2. The major assumption of the two-level model is that only those two levels which are coupled by laser radiation undergo population change during the laser pulse. The rate equation for the excited-state population N_2 (cm^{-3}) is

$$\frac{dN_2}{dt} = N_1(W_{12} + Q_{12}) - N_2(W_{21} + Q_{21} + A_{21}) \qquad (9.1)$$

where N_1 (cm^{-3}) is the population of ground level 1, W_{12} and W_{21} (s^{-1}) are the laser-induced absorption and stimulated emission rates, Q_{12} (s^{-1}) the collisional excitation rate, Q_{21} (s^{-1}) the collisional quenching rate, and A_{21} (s^{-1}) the spontaneous emission rate. Assuming that the total number density of the system is constant ($N_1 + N_2 = N_T$) and that the system has reached steady state ($dN_2/dt = dN_1/dt = 0$), we obtain

LEVEL 2

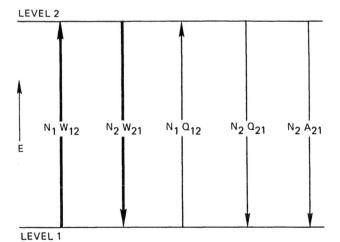

$N_1 W_{12}$ $N_2 W_{21}$ $N_1 Q_{12}$ $N_2 Q_{21}$ $N_2 A_{21}$

E

LEVEL 1

FIG. 9.2 Two-level model for laser excitation and fluorescence detection.

$$N_2 = \frac{N_T(W_{12} + Q_{12})}{Q_{21} + Q_{12} + A_{21} + W_{21} + W_{12}} \tag{9.2}$$

In many cases collisional excitation from the ground state to excited electronic states can be neglected in comparison to other rates. This is the case for virtually all species in flames. However, in plasmas, collisional excitation of excited electronic levels may be significant. This is discussed in more detail in Section 9.5; for the present we assume that $Q_{12} = 0$ and that the population of level 2 prior to laser excitation is negligible.

The laser-induced rates W_{12} and W_{21} are discussed in Chapter 1 [Eqs. (1.6)-(1.23)]. In the absence of coherence effects, W_{12} and W_{21} are approximately proportional to laser power for a given laser frequency spectrum and absorption line profile (Yariv, 1975). The absorption and stimulated emission rates are related by $g_1 W_{12} = g_2 W_{21}$, where g_1 and g_2 are level degeneracies. At low laser power ($W_{12} + W_{21} \ll Q_{21} + A_{21}$) the upper-level population is proportional to the laser-induced absorption rate,

$$N_2 = \frac{W_{12} N_T}{Q_{21} + A_{21}} \tag{9.3}$$

Measurements performed in this low-laser-power regime, where the excited-state number density is linearly proportional to laser power, are called linear fluorescence measurements. In linear fluorescence measurements the upper-level population is inversely proportional to the collisional quenching and spontaneous emission rates.

The observed fluorescence signal is proportional to the upper-level population N_2. The number of photons n_p (s^{-1}) incident on the light detector per unit time is given by

$$n_p = \frac{N_2 A_{21} V_c \epsilon \Omega_c}{4\pi} \tag{9.4}$$

where V_c (cm^3) is the fluorescence probe volume, ϵ is the efficiency of the optical collection system and includes such factors as the transmission of lenses and filters, and Ω_c is the solid angle subtended by the collection optics. In writing Eq. (9.4) it has been assumed that the fluorescence is isotropic. The fluorescence signal is given by

$$V_f = n_p h c \nu_{21} \phi_p G_p \tag{9.5}$$

where h (J-s) is Planck's constant, c (cm/s) the velocity of light, ν_{21} the fluorescence transition frequency (cm^{-1}), ϕ_p the quantum efficiency of the detector, and G_p (V/W) the detector gain.

For linear fluorescence measurements, the observed signal depends on the fluorescence yield ϕ_f, where

$$\phi_f = \frac{A_{21}}{A_{21} + Q_{21}} \tag{9.6}$$

as can be shown by substituting Eqs. (9.3) and (9.4) into Eq. (9.5). The dependence of the fluorescence yield on the quenching rate Q is a serious complication for probing media such as flames or plasmas using linear fluorescence. In such media, the variety of quenching partners and spatial inhomogeneities in medium properties make correction for the quenching rate difficult.

One method that has been suggested for reducing or eliminating the dependence of the fluorescence signal on the fluorescence yield is saturation of the laser-pumped transition (Piepmeier, 1972; Daily, 1977). At high laser powers, the absorption and stimulated emission rates can become much larger than the collisional quenching and spontaneous emission rates ($W_{12} + W_{21} \gg Q_{21} + A_{21}$). The upper-level population approaches the saturated limit in this case,

$$N_2 = \frac{g_2}{g_1 + g_2} N_T \tag{9.7}$$

At saturation the upper-level population is independent of both the laser power and the collisional quenching and spontaneous emission rate constants. It can be shown by substituting Eqs. (9.4) and (9.7) into Eq. (9.5) that the fluorescence signal is independent of the collisional quenching rate constant and proportional to the desired total number density at complete saturation.

The signal expressions obtained for the linear and saturated cases are summarized in Eqs. (9.8a) and (9.8b), respectively:

$$V_f(\text{linear}) = \frac{N_T W_{12}}{Q_{21} + A_{21}} hc\nu_{21} A_{21} \frac{V_c \epsilon_c \Omega_c}{4\pi} \Phi_p G_p \tag{9.8a}$$

$$V_f(\text{sat}) = \frac{N_T g_2}{g_1 + g_2} hc\nu_{21} A_{21} \frac{V_c \epsilon_c \Omega_c}{4\pi} \Phi_p G_p \tag{9.8b}$$

where the fluorescence signal has been written as V_f = (excited-state number density) (power emitted per excited-state species) (optical system collection efficiency) (detection electronics gain). This two-level model is valuable for elucidating the fluorescence process and the saturation behavior of fluorescence signals, but cannot be applied in general for real atoms and molecules.

In most cases population changes occur for numerous levels in addition to those directly coupled by laser radiation. Figure 9.3 illustrates the situation where three levels are involved in the laser excitation process. The three-level model is analyzed in Example 1 in the Appendix.

Radiative and collisional interactions which a molecule undergoes during laser excitation are illustrated schematically in Fig. 9.4. Molecules are excited from a single vibration-rotation level in the lower electronic level to a single vibration-rotation level in the excited electronic level by laser absorption. Whether or not an isolated transition is excited depends on the rotational line spacing, laser line width, and the laser power, if the electric field intensity is high enough that lines are Stark-broadened (Gross and McKenzie, 1983; Goldsmith, 1985).

Once excited, molecules can decay back to the original level or other levels in the ground state via spontaneous emission. Selection rules limit the number of vibration-rotation levels coupled by spontaneous emission, so that the fluorescence spectrum from a single upper rotational level consists of only a few lines. This is illustrated in Fig. 9.5a, where the fluorescence spectrum of OH in a 30-torr flame

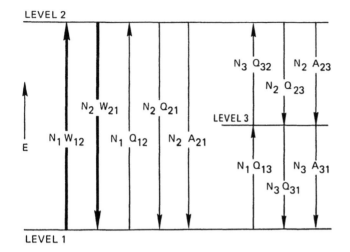

FIG. 9.3 Three-level model for laser excitation and fluorescence detection.

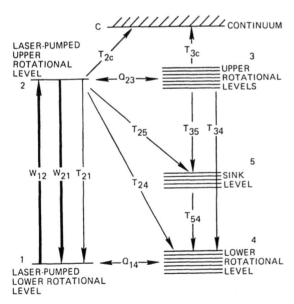

FIG. 9.4 Laser excitation dynamics in a molecule. The transfer rate $T_{ij} = Q_{ij} + A_{ij}$.

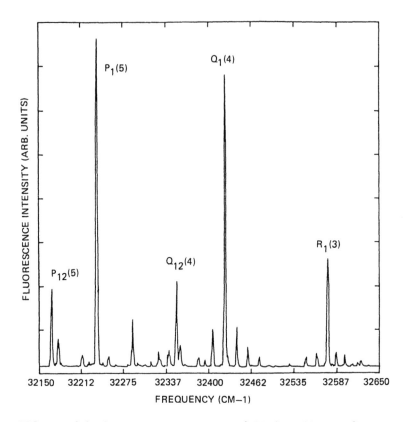

FIG. 9.5 (a) Fluorescence spectrum of OH in a 30-torr flame at two different time delays relative to the beginning of the laser (Lucht et al., 1986). The times after the beginning of the laser excitation were (a) 4 ns and (b) 10 ns. The $P_1(5)$ line of the $OHX^2\Pi$-$A^2\Sigma^+$ $(0,0)$ band was directly excited by the laser. The fluorescence lines labeled $P_{12}(5)$, $P_1(5)$, $Q_1(4)$, $Q_{12}(4)$, and $R_1(3)$ all have the same upper level (Dicke and Crosswhite, 1962).

is shown (Lucht et al., 1986). The fluorescence spectrum was time-resolved using a sampling oscilloscope and fast photomultiplier. The spectrum in Fig. 9.5a was recorded early in the laser pulse before significant collisional transfer occurred, and transitions other than those from the laser-pumped upper level are very weak.

Molecules can be transferred to other excited levels by rotational or vibrational energy transfer, or back to the ground electronic level by electronic quenching. Other rotational and vibrational levels in the upper electronic level are coupled to the ground electronic level by spontaneous emission. Thus, if significant rotational and/or vi-

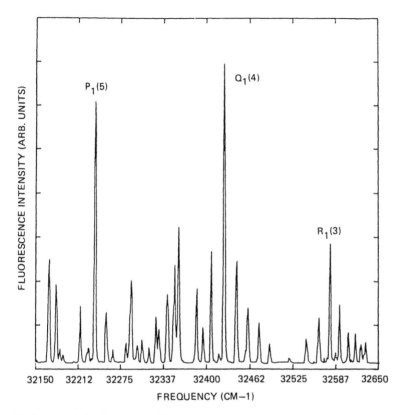

FIG. 9.5 (b)   (continued)

brational energy transfer occurs before molecules fluoresce, the fluorescence spectrum can become very complicated. The OH spectrum shown in Fig. 9.5b, for example, was recorded later in the laser pulse after significant rotational relaxation had occurred. Application of rate-equation models to molecular fluorescence as well as a signal-to-noise calculation for OH fluorescence are illustrated in Example 2 (see the Appendix).

9.3 LASER SOURCES AND DETECTORS FOR FLUORESCENCE MEASUREMENTS

9.3.1 Laser Sources

The development of the continuously tunable dye laser in the late 1960s and early 1970s (Soffer and McFarland, 1967; Hänsch, 1972) led to a surge of activity in fluorescence spectroscopy and technique development. Prior to the introduction of the dye laser, resonance lamps were the primary optical excitation sources used for fluorescence measurements. The use of lasers as excitation sources seemed to offer many advantages over the use of resonance lamps, including higher spectral intensity and better control over excitation line width and temporal pulse length. However, the limited tuning capability of laser sources prior to the dye laser severely restricted the applicability of lasers in fluorescence spectroscopy. The continuous tunability of dye lasers was quickly exploited for fluorescence measurements in analytical chemistry (Denton and Malmstadt, 1971; Fraser and Winefordner, 1971; Kuhl and Marowsky, 1971) and in plasma diagnostics (Dimock et al., 1969).

Since that time, developments in dye laser technology have continually led to development of new techniques and to improvements in existing techniques in fluorescence spectroscopy. For example, the signal-to-noise ratio of fluorescence measurements is directly related to the absorption rate. Thus, increases in dye laser spectral intensity allowed the application of fluorescence spectroscopy to combustion media and other systems where background luminosity is a potentially significant problem. The development of spectrally narrow, single-mode ring dye lasers holds the promise of significant applications in high-resolution fluorescence spectroscopy (Rea and Hanson, 1983). The development of dye lasers with pulse lengths of picoseconds or less has enabled the application of fluorescence spectroscopy in time-resolved measurements of ultrafast energy-transfer processes (Campillo and Shapiro, 1983).

One of the most significant developments in laser technology for LIF studies has been the introduction of commercial Nd:YAG laser-pumped dye laser systems. The high spectral intensity of these systems relative to flash-lamp-pumped and nitrogen laser-pumped dye lasers has greatly expanded the possible applications of fluorescence spectroscopy. Commercially available, Q-switched Nd:YAG lasers produce approximately 250 mJ of radiation at 1064 nm in a 10 to 20-ns pulse. The Nd:YAG radiation may pass through an amplifier external to the laser cavity to increase the pulse energy to almost 1 J. The fundamental radiation is typically frequency doubled (532 nm) or frequency tripled (355 nm) with efficiencies of approximately 40% and 20%, respectively. The frequency-doubled or frequency-tripled light is used to pump dye lasers, with conversion efficiencies as high as 35% for 560-nm light from Rhodamine 6G pumped by 532-nm radiation.

The wavelength of the dye laser radiation is determined by the laser dye used and by a grating which serves as a spectrally selective reflector at one end of the laser cavity. The line width of the dye laser radiation is determined by the grating resolution. Further narrowing of the dye laser line width is typically achieved by placing an etalon within the laser cavity. Radiation from dye lasers pumped by

Nd:YAG lasers has high enough intensity that frequency-conversion techniques such as frequency doubling and Raman shifting are very efficient. Commercially available Nd:YAG-pumped dye lasers have significant intensity over a wavelength range of 200 to 3000 nm. Recent developments in frequency tripling of dye laser radiation in mixtures of xenon and krypton have extended this wavelength range below 100 nm (Hilbig and Wallenstein, 1981).

9.3.2 Light-Detection Devices

The photomultiplier is the most common detection device used in fluorescence spectroscopy and is reviewed in numerous excellent reviews (Lytle, 1974; Engstrom, 1980). Photomultipliers have numerous advantages for fluorescence measurements, including their sensitivity, high gain, wide dynamic range, and fast temporal response.

The use of streak cameras in conjunction with picosecond dye lasers has opened up new possibilities for time-resolved fluorescence spectroscopy. Streak cameras with time resolution down to a picosecond are now commercially available. The use of the streak camera in fluorescence spectroscopy has been reviewed by Campillo and Shapiro (1983). The streak camera and mode-locked dye laser have proved especially useful for investigating energy transfer in biological processes.

An exciting new development has been the use of array detectors for one- and two-dimensional fluorescence imaging. The two main types of detectors which have been applied are the silicon-intensified-target (SIT) vidicon and the self-scanned photodiode (SSPD) array; these and other array detectors are discussed in detail by Chang and Long (1982). The SIT vidicon has a fiber-optic faceplate which is coated on the inside with a photocathode material. Electrons emitted by the photocathode are accelerated by an electrostatic field and strike the target, an n-type silicon wafer, where electron-hole pairs are created. The other side of the silicon wafer has an array of p-type silicon disks approximately 8 μm in diameter charged to a large negative potential. Holes created in the n-type silicon wafer can diffuse to the n-p junctions and discharge the p-type silicon disks; the amount of depleted charge is then read out by an electron scan beam. The chief disadvantage of the SIT vidicon is image lag; it takes several electron beam scans to completely recharge the p-type disks. The sensitivity of the SIT vidicon is sometimes enhanced by adding a second stage of electrostatic intensification; this configuration is given the acronym ISIT (intensified SIT vidicon).

The self-scanned photodiode array is an integrated circuit chip which includes both an array of silicon photodiodes and a shift register that scans the array (Chang and Long, 1982; Horlick and Codding, 1973; Snow, 1976). Unlike the vidicon, electron-hole pairs in the SSPF are generated by incident photons. The chief advantage of the SSPD over the SIT vidicon is that the SSPD has virtually no lag. The SSPD array can be coupled with a microchannel plate image intensifier (Csorba, 1980) to obtain sensitivity comparable to that of the ISIT.

9.4 APPLICATIONS OF LIF IN COMBUSTION FLOWS

One of the most successful areas of application of LIF diagnostics during the past decade has been in combustion. LIF is a powerful tool for studying the complex chemistry and structure of combustion flows. LIF has been used to measure both flame temperature and the concentrations of reactive intermediates and pollutant species. It has been applied in a variety of combustion systems, ranging from flat flames (laminar flames with uniform properties in the radial dimension) to turbulent diffusion flames. LIF possesses the necessary sensitivity and spatial resolution to accurately map radical concentrations in flames which have significant gradients on millimeter-length scales. In addition, it is a nonintrusive technique; the fluorescence measurement does not alter the flame chemistry (if used with care) or fluid dynamics. Such perturbations are always a possibility when a physical probe such as a thermocouple is used. In many cases physical probes cannot be used because they will not survive in the combustion flow of interest.

9.4.1 Concentration Measurements

Species that have been measured using LIF in combustion systems include:

Free radical species. LIF has been used extensively for the detection and measurement of transient free radical species such as OH, O and H, reactive intermediates which are important in flame chemistry. Free radicals play important roles in the decomposition of fuel molecules and in the formation of combustion products, including undesirable pollutant species such as CO and NO. In addition, transport of free radicals from flame reaction zones plays a crucial role in stabilization and ignition of flames.

Pollutant species. Concentration measurements of pollutant species such as NO, CO, NO_2, and SO_2 have obvious importance for monitoring and controlling pollutant emissions. Spatially and temporally resolved measurements could lead to new insights into the formation of these species during the combustion process. Integration of new detector and computer technologies has opened the door to potential control strategies for operation of combustion devices with minimal pollutant emission.

Metallic species. Metal atoms such as Na, K and V and compounds such as NaS are important in corrosion, fouling, and slagging in coal- and oil-fired combustors. LIF measurements of alkali metal atoms in flames have been performed for over a decade; indeed, sodium has served as a model species for the development of most LIF techniques. Laser-induced fluorescence techniques for alkali metal compounds, on the other hand, are much less well developed. LIF measurements of species such as NaS are handicapped by the lack of a well-developed spectroscopic data base (Gole, 1981).

9.4.1A. Linear Fluorescence Measurements

The quantitative application of fluorescence diagnostics to combustion systems is complicated by the collisional quenching dependence of the signal. In flames at atmospheric pressure, typical collisional transfer rate constants are on the order of 10^9 to 10^{10} s^{-1}, much faster than typical spontaneous emission rates of 10^5 to 10^8 s^{-1}. Therefore, the fluorescence yield is usually inversely proportional to the quenching rate constant. For most flame radicals, quenching data, even at room temperature, are completely lacking. Application of such data, if they exist, to flame measurements is often unreliable due to the wide range of temperatures and quenching species encountered in combustion systems. The large gradients in temperature and species concentration which can occur within a flame imply that similarly large gradients may occur in the quenching rate.

Various methods have been proposed for correcting linear fluorescence measurements for quenching effects. Daily (1976) proposed that the quenching rate be directly measured by time resolving the fluorescence signal. This method has been demonstrated for both OH (Stepowski and Cottereau, 1979) and CH (Cattolica et al., 1984) in low-pressure flames. Recently, Bergano et al. (1983) used a picosecond dye laser and a streak camera detector to perform time-resolved measurements of OH fluorescence in an atmospheric pressure flame.

Bechtel and Teets (1980) corrected linear fluorescence measurements by measuring major species concentrations using laser Raman scattering and calculating the quenching rate from measured room temperature cross sections. Recent measurements (Fairchild et al., 1983; Bergano et al., 1983) have shown that for OH, the cross section for electronic quenching is approximately a factor of 2 lower at flame temperatures than at room temperature. This temperature dependence complicates the data analysis using the technique of Bechtel and Teets (1980), but as the data base for quenching of fluorescence species expands, this technique will become more accurate.

Muller et al. (1979) proposed that variations in the fluorescence yield could be minimized by restricting the fluorescence detection bandwidth such that only the directly excited upper rotational level was observed. The effective quenching rate was then the sum of the rotational, vibrational, and electronic collisional transfer rates. Muller et al. (1979) used this technique to measure OH, SH, S_2, SO, and SO_2 in atmospheric pressure $H_2/O_2/N_2$ flames seeded with H_2S.

In many cases, conditions in a combustion flow are sufficiently uniform that the fluorescence yield changes only slightly over the spatial dimensions of the fluorescence measurement. In such cases, a one-point absorption calibration serves to transform relative fluorescence intensities into absolute number densities. This approach has been used extensively for measurements in the postflame regions of laminar flat flame burners, where temperatures and major species concentrations change very slowly as a function of distance above the burner. Postflame, linear fluorescence studies include measurements of OH and NH concentrations in methane-nitrous oxide flames (Anderson et al., 1982a), NH and OH concentrations in hydrogen-nitrous oxide flames (Cattolica et al., 1982), and OH, NO, CN, and NH measurements in ammonia-oxygen flames (Morley, 1981). Cattolica and Schefer (1983) measured OH concentrations in the boundary layer next to heated platinum (catalytic) and quartz (noncatalytic) plates using the same technique; flow conditions were nearly uniform within the boundary layer. Smith and Chandler (1986) performed CN and CH concentration measurements in low-pressure $H_2/O_2/Ar/HCN$ flames using a CW dye laser pumped by the ultraviolet emission of an argon laser. The signal-to-noise ratio of the LIF measurements was excellent due to the use of CW excitation and phase-locked detection, even though absorption due to CH and CN was barely detectable even in a multipass arrangement.

LIF has been used to demonstrate perturbations in flow fields caused by the use of physical probes. Stepowski et al. (1981) used LIF to measure the perturbation in the OH profile caused by a quartz probe used for mass spectrometric sampling. They found that the presence of the probe significantly decreased the OH concentration in the preheat region of a low-pressure flame. Smith and Chandler (1986) used LIF to measure CN concentration and rotational temperature in the vicinity of a mass spectrometer sampling probe. They found that CN concentrations were reduced in the vicinity of the probe, but that the rotational temperature was unaffected.

In a fluorescence experiment which had a major impact on the modeling of combustion processes which lead to unburned hydrocarbon emissions from internal combustion engines, Bechtel and Blint (1982) measured OH concentrations in the so-called "quench layer" of a flame next to a cold wall. The quench layer was proposed as a means of explaining unburned hydrocarbon emissions (Daniel, 1957). The drop in flame temperature near the walls of the combustion chamber was thought to inhibit complete hydrocarbon combustion. To test the quench-layer hypothesis, Bechtel and Blint (1980) and Clendening et al. (1981) used spontaneous Raman scattering to measure fuel concentrations in premixed flames near cooled walls. The fuel concentrations decreased much more rapidly than predicted by quench-layer theory. Similar doubts on the validity of the quench-layer hypothesis were raised by sampling measurements in an internal combustion engine (LoRusso et al., 1980). It is now thought that mass diffusion of fuel out of the quench layer and diffusion of radicals into the quench layer, thus leading to increased fuel oxidation rates, accounts for the rapid decrease in fuel concentrations in the quench layer.

Bechtel and Blint (1982) measured OH concentrations by laser-induced fluorescence in the same flame investigated earlier using laser Raman spectroscopy. The flame apparatus and fluorescence geometry are shown in Fig. 9.6. OH fluorescence was excited by tuning the output of a frequency-doubled, flash-lamp-pumped dye laser to various isolated transitions withint the $A^2\Sigma^+-X^2\Pi$ (0,0) band. The fluorescence signals were in the linear regime. The spectrometer bandpass was chosen such that fluorescence from the entire R branch was collected. It is important in a fluorescence experiment where broad-

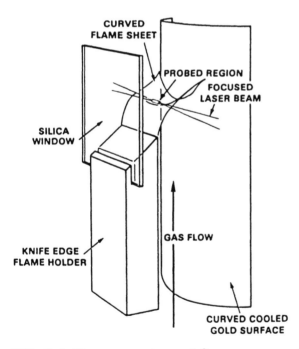

FIG. 9.6 Flame apparatus and fluorescence excitation and detection geometry for measurement of OH concentration near cooled surface. (Bechtel and Blint, 1982; reprinted with permission, copyright 1982, Society of Automotive Engineers, Inc.)

band detection is employed to avoid biases introduced when emission
from only a portion of the rotational transitions in an R, Q, or P
branch is collected; the effective fluorescence yield will change
anomalously as flame temperature varies or levels with different rotational quantum numbers are excited. For OH, collection of the R
branch is advantageous because of its spectral compactness relative to
the Q and P branches. Detection bandpass and therefore potential interferences from scattered laser light and flame emission can be minimized by monitoring R-branch fluorescence.

Absolute OH concentrations were calculated from the observed fluorescence intensities by performing an absorption calibration in the
postflame region, and correcting measurements at other flame positions
by calculating the collisional quenching rate using the technique of
Bechtel and Teets (1980). Typical results are shown in Fig. 9.7,
where OH concentrations along the cooled wall of the slot burner are

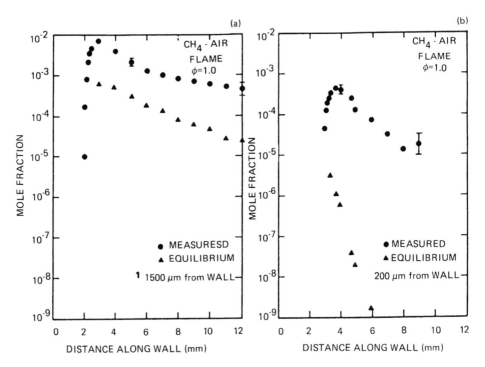

FIG. 9.7 Measured and calculated equilibrium OH concentration profiles as a function of distance along cooled wall. (Bechtel and Blint,
1982; reprinted with permission, copyright 1982, Society of Automotive Engineers, Inc.) The measurements were performed (a) 1.5 mm
and (b) 0.2 mm away from the cooled surface.

shown for distances of 1.5 and 0.2 mm from the wall, respectively.
At 1.5 mm from the wall, the OH profile peaks just after the flame
front and decreases slowly thereafter. Note that the OH concentration
is an order-of-magnitude higher than the equilibrium concentration,
calculated from the measured flame temperature and major species concentrations. The measurements at 0.2 mm are within the "quench layer." OH concentrations at 0.2 mm from the wall peak slightly later
and are more than an order-of-magnitude lower than OH concentrations at 1.5 mm from the cooled wall. However, at 0.2 mm from the
wall the measured OH concentrations exceed the calculated equilibrium
concentrations by many orders of magnitude. The large superequilibrium populations of radicals such as OH together with diffusion of
fuel molecules out of the quench layer account for the rapid decrease
in methane concentrations along the cooled wall, despite the low gas
temperatures near the wall. These results, together with the afore-
mentioned Raman scattering and sampling probe measurements, have
virtually eliminated the quench layer as a possible source of unburned
hydrocarbon (UHC) emissions in internal combustion engines.

9.4.1B. Saturated Fluorescence Measurements

Another approach to overcoming the collisional quenching dependence of the fluorescence signal is optical saturation of the absorption
transition, proposed by Piepmeier (1972). Daily (1977) suggested the
application of saturated fluorescence as a diagnostic for turbulent
flames, where the quenching rate can vary considerably from laser
shot to laser shot. The first saturated fluorescence measurements in
flames were performed by Omenetto et al. (1973), who saturated five
species, including sodium, using a nitrogen-pumped dye laser.
Numerous investigations of saturated sodium fluorescence have since
been reported (Kuhl et al., 1973; Sharp and Goldwasser, 1976; Smith
et al., 1977; Pasternack et al., 1978; Daily and Chan, 1978; Allen
et al., 1979; Van Calcar et al., 1979; and Muller et al., 1980). Sodium has served as a model species for the development of saturated
fluorescence techniques.

Saturated fluorescence has also been used to measure the flame
radicals CH (Bonczyk and Shirley, 1979; Verdieck and Bonczyk,
1981; Mailander, 1978), CN (Bonczyk and Shirley, 1979; Verdieck and
Boncyzk, 1981), C_2 (Baronavski and McDonald, 1977; Mailander,
1978), NH (Salmon et al., 1984), and OH (Alden et al., 1982a; Lucht
et al., 1983a). The quantitative application of laser-saturated fluorescence (LSF) to measurements of diatomic flame radicals has been
complicated by their vibrational-rotational structure (Lucht and Laurendeau, 1979; Berg and Shackleford, 1979; Lucht et al., 1980).

Lucht et al. (1983a) performed laser-saturated fluorescence measurements of the OH radical in flames at pressures of 30 to 242 torr,
and compared the fluorescence measurements with independent absorp-

tion measurements. They demonstrated that the collisional transfer dependence of the fluorescence signal was reduced by a factor of 5 to 10 when the laser-excited rotational transition was saturated and fluorescence was observed only from the laser-pumped upper rotational level. This fluorescence scheme, the balanced cross-rate model, was based on numerical analysis of the laser excitation dynamics of OH (Lucht et al., 1980). The balanced cross-rate methodology was recently applied successfully to measurements of NH in low-pressure flames (Salmon et al., 1984) and to OH measurements in sooting methane-air flames (Lucht et al., 1985).

This methodology was also applied to single-pulse OH measurements in a turbulent jet diffusion flame (Lucht et al., 1984). The OH measurements were performed in order to better understand the interaction between turbulence and flame chemistry; OH measurements in this flame were especially useful in this regard because the hydrogen-oxygen chemical reaction system is well understood. The jet diffusion flames had previously been characterized using single-shot spontaneous Raman scattering, from which probability density functions (PDFs) of flame temperature and major species concentration had been determined. Equilibrium OH concentration PDFs were calculated from the major species concentration and temperature PDFs, and compared to measured OH PDFs. Departures from equilibrium conditions are an indication that the rate of mixing of reactants is comparable to or faster than chemical reaction rates.

The experiment is illustrated schematically in Fig. 9.8. A frequency-doubled beam from a Nd:YAG laser-pumped dye laser was passed vertically through the jet diffusion flame. The laser beam had sufficient intensity for nearly completely saturation of the $Q_1(8)$ rotational transition in the $A^2\Sigma^+ - X^2\Pi$ (0,0) band. Fluorescence was collected at right angles to the laser beam and imaged onto the entrance slit of a 3/4-m spectrometer. Fluorescence was detected only from the laser-pumped upper rotational level; the spectrometer was tuned to the isolated $P_1(9)$ rotational transition. A 1P28 photomultiplier specially wired for fast response (2-ns FWHM) (Harris et al., 1976) was used to detect the light signal at the exit slit. The photomultiplier voltage at the peak of the fluorescence pulse was measured using a sampling oscilloscope with a 350-ps sampling window. The sampler voltages were recorded shot by shot on a digital oscilloscope interfaced to a minicomputer.

Probability density functions of OH concentration for a laminar calibration flame and in a Re = 8500 jet diffusion flame are shown in Fig. 9.9. The PDF in the laminar flame is a measure of the instrumental spread in single-shot measurements. The much broader PDF in the jet diffusion flame is due to variations in flame conditions at the measurement location (i.e., turbulence) from laser shot to laser shot. The maximum measured single-shot OH concentration in the turbulent

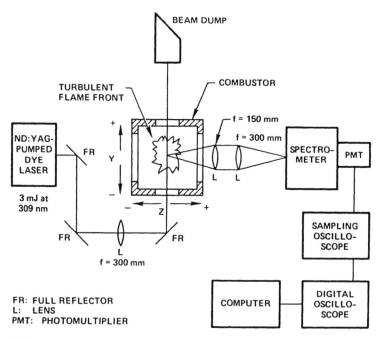

FIG. 9.8 Experimental schematic for single-pulse, laser-saturated fluorescence measurements of OH concentrations in a turbulent jet diffusion flame (Lucht et al., 1984). The jet diffusion flame rig is shown in cross section. It can be translated in three dimensions.

diffusion flame was approximately 6×10^{16} cm^{-3}, or three times the maximum adiabatic equilibrium concentration of 2.1×10^{16} for a stoichiometric hydrogen-air mixture.

Average OH number density profiles are shown in Fig. 9.10 at three different axial locations in the Re = 8500 jet diffusion flame. Close to the fuel nozzle ($x/d = 50$) fluid mechanical mixing rates are very high due to high shear rates, and the degree of superequilibrium is significant. Farther downstream the degree of superequilibrium decreases until at $x/d = 200$, the measured and equilibrium profiles are nearly identical.

The fluorescence results described above have already proven to be useful in turbulent flame modeling (Drake et al., 1984; Correa et al., 1984). Radical concentrations are a much more sensitive indication of chemical nonequilibrium conditions in the combustion flow than is the temperature; uncertainties in single-shot temperature measurements by Raman scattering are typically greater than predicted departures from equilibrium temperatures. The single-shot OH concentration measurements, on the other hand, provide a data base against which turbulent chemistry models can be compared in detail.

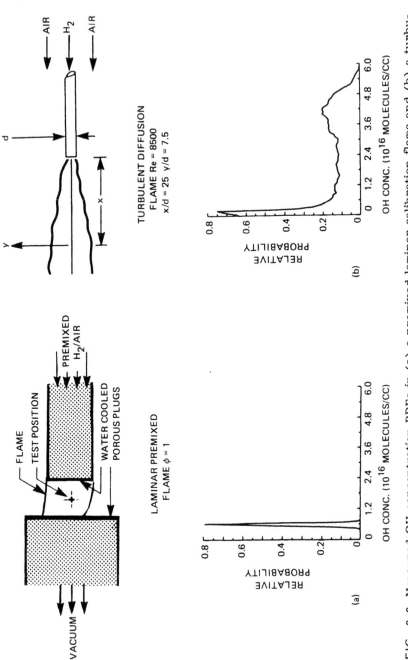

FIG. 9.9 Measured OH concentration PDFs in (a) a premixed laminar calibration flame and (b) a turbulent jet diffusion flame (Lucht et al., 1984). The flame and measurement point geometries are also indicated. The mean standard deviation of the laminar flame measurement is 7%. Each PDF consists of 1024 single-shot measurements.

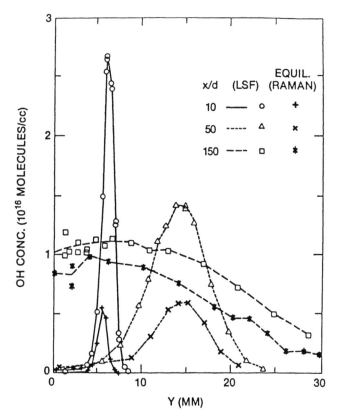

FIG. 9.10 Measured and calculated profiles of average OH concentration at various axial locations in the Re = 8500 turbulent jet diffusion flame (Drake et al., 1984).

9.4.1C. Two-Photon-Excited Fluorescence Measurements

For many important flame species fluorescence measurement is difficult because the first electronic resonance lies in the vacuum ultraviolet (VUV) region below 200 nm. In addition to the technical difficulty of generating laser radiation in this wavelength range, the penetration depth of VUV radiation in atmospheric-pressure flame gases is negligibly small. In recent years, however, the technique of two-photon-excited fluorescence (TPEF) has been demonstrated in flames, and species such as H, O, and CO have been measured for the first time by laser-induced fluorescence. The potential for such measurements was previously demonstrated by Hänsch et al. (1975) and Bokor et al. (1981), who measured H atoms by TPEF in a discharge flow cell, and by Bischel et al. (1982), who measured O and N atoms in a dis-

charge flow cell. Bischel et al. (1982) used Raman shifting of frequency-doubled dye laser radiation to generate sufficient laser intensity at the two-photon resonance wavelengths for O (226 nm) and N
(211 nm). The laser excitation and fluorescence detection scheme for
the oxygen atom is shown in Fig. 9.11. Note that although the laser
wavelength is deep in the ultraviolet, the fluorescence transitions are
in the near infrared.

Recently, both O and H have been measured in flames using
TPEF. Alden et al. (1982b) detected oxygen atoms in a flame using
the same Raman shifting scheme as that used by Bischel et al. (1982).
The same group recently demonstrated linear imaging of TPEF from
the O atom (Alden et al., 1984a). Lucht et al. (1983b) performed
TPEF measurements of the hydrogen atom in flames using the laser
system shown in Fig. 9.12 to generate approximately 70 µJ of energy

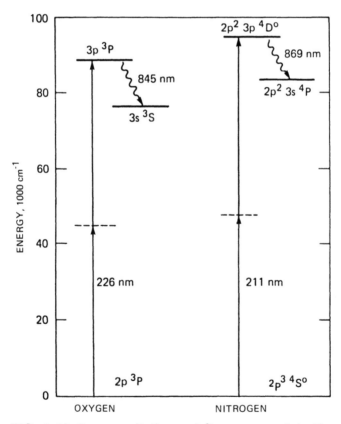

FIG. 9.11 Laser excitation and fluorescence detection schemes for the
oxygen and nitrogen atoms (Bischel et al., 1982).

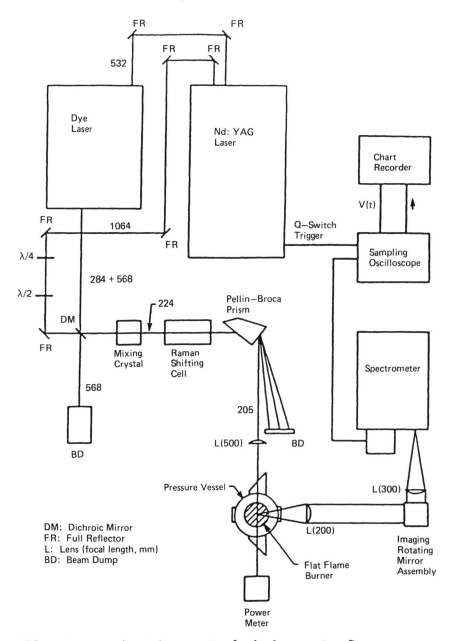

FIG. 9.12 Experimental apparatus for hydrogen atom fluorescence measurements (Lucht et al., 1983b).

at 205 nm. The n = 1 to n = 3 transition was pumped by the laser, and Balmer-α n = 3 to n = 2 fluorescence was detected. Alden et al. (1984b) recently demonstrated three-photon excitation of hydrogen atom fluorescence; the n = 4 electronic transition was pumped by the laser and n = 4 to n = 2 fluorescence was observed. The signal-to-noise ratio of the measurement was very low due to the weakness of the three-photon absorption cross section.

Recently, Goldsmith (1985) used a two-step excitation scheme to monitor hydrogen atom concentrations in a laminar hydrogen-air diffusion flame. Laser radiation at 243 nm was used to excite hydrogen atoms from n = 1 to n = 2. Excited-state atoms were then pumped from n = 2 to n = 3 by 656-nm laser light, and fluorescence back to n = 2 was monitored. The 243-nm beam was completely overlapped by the 656-nm beam, as shown in Fig. 9.13. The saturation of the n = 2 to n = 3 transition was essentially complete, as shown by the saturation curve in Fig. 9.13, because of the absence of laser beam edge effects (Daily, 1978; Salmon and Laurendeau, 1985) and because of the high

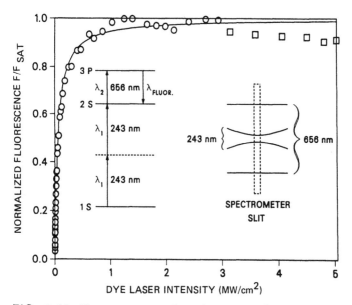

FIG. 9.13 Fluorescence saturation curve for two-step excitation of atomic hydrogen (Goldsmith, 1985). The fluorescence intensity is plotted versus the relative intensity of the 656-nm radiation. The decrease in fluorescence signal above a dye laser intensity of 3 MW/cm^2 is caused by the onset of Stark splitting of the Balmer-α transition. An energy-level schematic of the laser excitation and fluorescence and a schematic of the beam overlap geometry are also shown.

oscillator strength of the transition. The measurement signal-to-noise ratio was excellent, allowing linear imaging of hydrogen atom concentration profiles (Goldsmith and Anderson, 1986). The detectivity of this technique will probably be limited by scattered laser light because the detected fluorescence is at same wavelength as one of the laser beams.

Two-photon-excited fluorescence has also been used to detect NO in low-temperature flows (Gross and McKenzie, 1983), and OH (Crosley and Smith, 1983) and CO (Alden et al., 1984c) in flames. Although no quantitative number densities were reported for the CO measurements, the fluorescence signal was sufficiently strong to record single-shot line images in a bunsen burner flame using an intensified photodiode array.

9.4.1D. Fluorescence Imaging Measurements

One of the most significant developments over the last few years in combustion diagnostics has been the demonstration of fluorescence imaging techniques. The first line-imaging fluorescence measurement was reported by Alden et al. (1982a), who used a one-dimensional intensified diode array for OH fluorescence detection. Line-imaging measurements of O (Alden et al., 1984a; Goldsmith and Anderson, 1986), H (Goldsmith and Anderson, 1986), and CO (Alden et al., 1984b) have been reported.

The first two-dimensional imaging measurements were performed nearly simultaneously in 1982. Dyer and Crosley (1982) used a two-dimensional ISIT vidicon tube for detection of OH fluorescence in a bunsen burner flame, while Kychakoff et al. (1982) used a two-dimensional SSPD array for planar imaging of OH in a laminar flat flame burner and a turbulent premixed flame. Figure 9.14 shows the experimental apparatus of Kychakoff et al. (1982). Cylindrical lenses are used to form a sheet of laser light which induces fluorescence in the flame. The planar fluorescence is then imaged onto the intensified reticon photodiode array. The color filter is used to reject stray light, and the laser beam is polarized in the horizontal plane to minimize Rayleigh scattering into the detector. Kychakoff et al. have also performed planar fluorescence measurements of NO (1984a) in a laminar flame and OH in turbulent jet diffusion flames (1984b).

Cattolica and Stephenson (1983) performed two-line fluorescence temperature measurements in a plane by recording fluorescence images with two different excitation frequencies. Although the two different excitations were not performed simultaneously, the use of a two-dye-laser system similar to that of Gross and McKenzie (1983) would in principle allow single-shot, two-dimensional fluorescence temperature measurements.

Cattolica and Vosen (1984) performed fluorescence imaging measurements of OH concentration in transient flames in a constant-volume

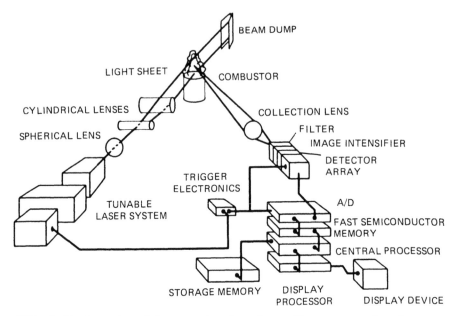

FIG. 9.14 Experimental apparatus for planar fluorescence imaging
measurements (Kychakoff et al., 1982).

combustion chamber. The constant-volume combustion chamber was
filled with a stoichiometric mixture of methane and air, and ignited by
a spark. The OH concentration field was mapped as a function of time
by triggering the laser at varying time delays relative to ignition.

Fluorescence imaging was also used to study the interaction of a
flame propagating out of a small cylindrical prechamber into a stag-
nant, stoichiometric methane-air mixture (Cattolica and Vosen, 1986).
The experimental apparatus is shown in Fig. 9.15. A portion of the
input beam was directed onto a diode array to correct for nonuniformi-
ties in laser intensity along the laser sheet. OH concentration fields
were measured outside the prechamber at various times after ignition,
providing important information on the interaction of fluid mechanics
and chemistry during flame propagation. The fluid mechanics of the
flame propagation were also studied using Schlieren photography.

9.4.2 LIF Temperature Measurements in Flames

LIF is attractive as a temperature diagnostic because of its nonintrusive nature, its excellent spatial resolution, and the strength of fluorescence signals, which makes single-shot temperature measurements
possible. LIF temperature measurements in flames, along with other
temperature-measurement techniques, were recently reviewed by Elder and Winefordner (1983).

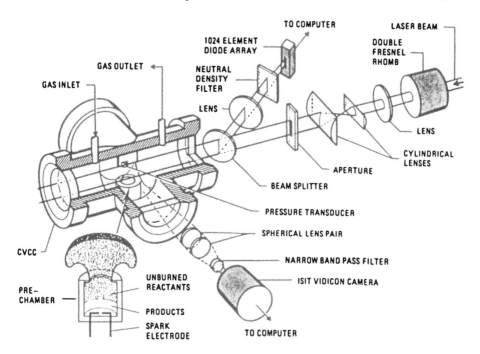

FIG. 9.15 Experimental apparatus for OH fluorescence imaging of OH in a constant-volume combustion chamber (Cattolica and Vosen, 1986). The ignition prechamber is shown at lower left.

OH is the most common species selected for laser-induced fluorescence temperature measurements for a number of reasons. OH is typically present in high concentrations in flames. The (0,0) and (1,0) vibrational bands of the $A^2\Sigma^+ - X^2\Pi$ electronic transition lie at the frequency-doubled wavelengths of high-efficiency rhodamine laser dyes. Consequently, OH LIF signals are very strong for a wide range of flame conditions. In addition, the frequency and radiative transition rates of OH rotational transitions are unusually well characterized (Dieke and Crosswhite, 1962; Dimpfl and Kinsey, 1979; Chidsey and Crosley, 1980), thus providing the spectroscopic data base needed for accurate temperature measurements.

LIF temperature measurement techniques may be classified according to whether the temperature is calculated from the ground-state population distribution (inferred from the fluorescence signal) or from an observed upper-level population distribution. For example, the ground-state population of OH is typically determined by scanning a narrowband dye laser across numerous OH rotational transitions and measuring the resultant fluorescence using a broadband detector

(Bechtel, 1979; Wang and Davis, 1974; Crosley and Smith, 1982; Anderson et al., 1982b); Chandler et al. (1985) used the same technique but with CN and CH LIF. Rotational temperatures are obtained from Boltzmann plots of fluorescence intensity versus the energy of the lower rotational level that is being probed. Rea and Hanson (1983) used a fast-scanning, frequency-doubled CW ring dye laser to scan across two rotational OH transitions at kilohertz rates, thus offering the possibility of collecting temperature data at kilohertz rates.

Systematic errors in calculation of ground-level populations can arise from collisional transfer effects (Crosley and Smith, 1982), absorption of the laser beam by flame gases, and fluorescence trapping. These effects can lead to large temperature errors because of the nonlinear dependence of level populations on temperature. Two-line fluorescence temperature measurement techniques which obviate some of these potential systematic errors either through fluorescence saturation (Lucht et al., 1982) to reduce the collisional transfer effects, or excitation of a single upper rotational level from two different ground rotational levels (Cattolica, 1981) have been demonstrated. In Cattolica's scheme, the same upper rotational level is excited from two different ground rotational levels. Because the same upper level is excited from two different lower levels, fluorescence trapping and collisional quenching effects are identical for the two excitation cases and do not influence the fluorescence temperature measurements.

Schemes have been proposed in which the temperature is deduced from observed rotational (Chan and Daily, 1980) or vibrational (Crosley and Smith, 1980) populations in the upper electronic state of laser-excited OH. The accuracy of such schemes depends on the availability of accurate collisional transfer models for excited states of OH.

In many cases OH is not present in high enough concentrations to obtain sufficient fluorescence signal for accurate temperature measurements. Seeding the flow with fluorescing species is an attractive alternative in some instances. Single-pulse, two-line temperature measurements in flows seeded with NO were recently demonstrated by Gross and McKenzie (1983). NO is relatively unreactive and can be used as a seed species for some flame stoichiometries and in nonreacting flows. Seeding the flow with NO could be especially useful for low-temperature regions where OH concentrations are very low. On the negative side, additional experimental complexity is necessary to ensure uniform seeding and the possibility of influencing flame chemistry exists.

Atomic LIF temperature measurement techniques are attractive because numerous possible seed atoms have strong electronic resonances in the visible, where dye lasers operate with the highest spectral intensity. The signal-to-noise ratio of such measurements is therefore potentially excellent. Atomic LIF temperature measurements were recently reviewed by Zizak et al. (1984). Bradshaw and coworkers (1980) have outlined five methods for measuring flame temperature using atomic fluorescence.

Haraguchi et al. (1977) demonstrated a two-line atomic LIF temperature measurement technique using both indium and thallium as seed materials. In this technique, the same upper electronic level is excited from two different lower levels. The possibility of using this two-line atomic LIF technique to perform temperature measurements at kilohertz data rates was addressed by Joklik and Daily (1982). They seeded a flame with indium and used two dye lasers in a preliminary two-line fluorescence measurement. Although the time resolution obtained in these initial experiments was only 10 s, the authors concluded that with improved equipment temperature measurements with a time resolution of less than 1 ms were feasible. Zizak and Winefordner (1982) measured the postflame temperature of a gasoline-air flame using a technique in which the temperature is calculated from the upper-level distribution of laser-excited atoms aspirated into the flame.

9.5 APPLICATION OF LIF TO FUSION PLASMA DIAGNOSTICS

The use of laser-induced fluorescence as a plasma diagnostic was first proposed by Measures (1968), who suggested fluorescence detection of neutral atoms such as potassium and ions such as barium (Ba II). Measures also suggested that plasma properties such as electron density and temperature, in addition to species number density, could be measured using LIF. LIF is especially well suited for measurements in low-density, high-temperature plasmas because it is a nonperturbing measurement and because of its excellent sensitivity.

There has been a great deal of activity in the last few years in LIF diagnostics of hydrogen fusion plasmas. Applications of LIF as well as other laser techniques for plasma diagnostics are reviewed in recent articles by Peacock and Burgess (1981) and Danielewicz et al. (1984). The application of LIF to actual fusion plasma devices is complicated by the size of the machines and the difficulty of obtaining optical access. In terms of the medium itself, calculation of number densities from fluorescence signals is complicated because excited electronic levels may be significantly populated at the high temperatures characteristic of fusion plasmas, and because thermodynamic equilibrium may not exist in the plasma.

Figure 9.16 illustrates typical energy transfer processes which are important for laser-excited species in a plasma. In contrast with flames, collisional excitation of excited electronic levels cannot be neglected; excited electronic levels may have significant population prior to laser excitation. In Fig. 9.16 the ground level 1 and excited

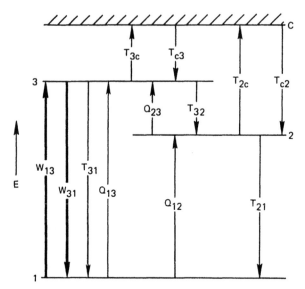

FIG. 9.16 Four-level model for laser excitation of atoms or ions in plasma. The upper electronic levels are for some species strongly coupled to the continuum either through radiative processes or electron collisions. The transfer rate $T_{ij} = Q_{ij} + A_{ij}$.

electronic level 3 are coupled by laser excitation; the rate equations for this case are analyzed in Example 3 (Appendix). In many cases, and in particular for the neutral hydrogen atom, acceptably large fluorescence signals may be obtained when atoms are pumped from one excited level to a higher-lying electronic level (e.g., level 2 to level 3).

The predominant energy transfer processes for laser-excited species in high-temperature, low-density plasmas are spontaneous emission and collisions with electrons. The effect of collisional transfer on LIF measurements of hydrogen atoms in plasmas has been addressed in detail by Gohil and Burgess (1983). Collisional transfer in the laser-excited state can be used to infer plasma properties. For example, Tsuchida et al. (1983) have used LIF measurements of helium atoms to measure electron density in a plasma. They measured a collision rate from the directly excited upper level to another upper level from the ratio of resonance fluorescence to collision-induced fluorescence from the nonresonant upper level. The electron temperature and collisional transfer cross sections were measured or calculated independently of the LIF measurement. Therefore, the electron density could be calculated from the measured collision rate.

Typically, collisions of laser-excited species with neutrals and ions are much less important than electron collisions because electron velocities are so much higher. However, charge-exchange reactions with heavy impurities such as iron atoms can play important roles in the overall plasma energy balance (see Sec. 9.5.2). Ionization and recombination processes are, of course, very important in plasmas, so transfer processes to and from the ionization continuum must be considered.

Only atomic and ionic species are of interest in most plasmas; molecules do not exist in appreciable quantities. Atomic transitions are usually easy to saturate in plasmas because of the high oscillator strengths of the transitions and the low collisional quenching rates. The large Doppler widths of atomic transitions in plasmas which have very high translational temperatures complicates analysis of the laser excitation and makes it difficult to excite all velocity classes efficiently. On the other hand, the translational temperature can be calculated from the observed fluorescence or laser excitation spectral profiles.

9.5.1 LIF Measurements of the Neutral Hydrogen Atom

Several groups have performed measurements of neutral hydrogen densities and velocities in magnetically confined plasmas. Neutral hydrogen is not confined by the magnetic field and thus is very important in the interaction of the plasma with the walls of the fusion device (Gohil and Burgess, 1983). Sputtering of impurities may be caused when high-energy neutrals, including hydrogen, hit the wall or limiter. In addition, neutral hydrogen plays an important role in charge-exchange reactions with impurities in the plasma (Peacock and Burgess, 1981). After charge exchange with neutral hydrogen, impurity species tend to be in highly excited electronic states which radiate strongly and can significantly affect the overall plasma energy balance.

Spatially resolved measurements of neutral hydrogen densities and velocities are thus of great value in characterizing plasma conditions. Laser-induced fluorescence measurements using both Lyman-α and Balmer-α excitation have been performed. The relative advantages of the two different excitation schemes have been analyzed in detail by Gohil and Burgess (1983).

The first reported LIF measurements of hydrogen atoms in a fusion plasma were performed by Burakov et al. (1977). They used Balmer-α excitation and measured hydrogen atom concentrations down to 10^9 cm^{-3} on the centerline of a tokamak plasma discharge. The intensity of the flash-lamp-pumped dye laser used to excite the hydrogen was sufficient to saturate the transition completely. The plasma had an electron number density n_e and temperature T_e of 10^{13} cm^{-3} and 300 to 350 eV, respectively, and a hydrogen gas pressure of 7×10^{-5}

torr. The relative populations of the n = 1, 2, and 3 levels prior to laser excitation were determined using the data of Johnson and Hinnov (1973) and the known values for n_e and T_e. Consequently, the total population of neutral hydrogen could be calculated from the observed laser-induced population change in the n = 3 level.

Razdobarin et al. (1979) upgraded the laser and detection system of Burakov et al. (1977) by increasing the duration of the laser pulse and adding a compensating circuit to suppress plasma emission noise caused by collective oscillation. The detection limit was lowered to 3×10^8 cm^{-3} because of the resulting increase in the signal-to-noise ratio.

Gohil et al. (1983) performed Balmer-α fluorescence measurements in the HBTX1A reverse-field pinch plasma. The laser excitation dynamics for hydrogen were analyzed in detail (Gohil and Burgess, 1983; Gohil et al., 1983) for a wide range of electron temperatures and densities. Based on this detailed analysis and independent measurement of the electron temperature and density, Gohil et al. (1983) determined that the ratio of the laser-induced emission to background emission at the Balmer-α wavelength was a factor of 4 higher than predicted.

Lyman-α excitation provides potentially much greater sensitivity for LIF measurements of neutral hydrogen densities because the great majority of hydrogen atoms occupy the ground state even at plasma temperatures. Bogen et al. (1982) frequency-tripled radiation from an excimer-pumped dye laser in a phase-matched krypton-argon mixture to obtain approximately 80 W of laser radiation at 121.6 nm in a 2-ns pulse. The frequency-tripled radiation was used to excite the Lyman-α transition for neutral hydrogen in the cleaning discharge plasmas of the ASDEX-tokamak. The experiment is depicted schematically in Fig. 9.17. Figure 9.18 shows an excitation scan of the Lyman-α line in the discharge. The FWHM of the profile, which is primarily Doppler broadened, corresponds to a temperature of 6000 K. However, the Gaussian fit to the excitation line is poor, especially in the wings. This is a possible indication of the presence of an excess of translationally hot atoms. The estimated detection limit of the fluorescence system was 10^9 cm^{-3}, limited primarily by scattered laser light.

Mertens and Bogen (1984) recently used the same Lyman-α LIF technique to measure the velocity distribution of deuterium atoms sputtered from niobium and vanadium targets saturated with deuterium. The wall and limiters of magnetically confined fusion reactors are expected to be saturated with hydrogen and the sputtering of hydrogen atoms from the walls might have a significant influence on plasma properties near the wall.

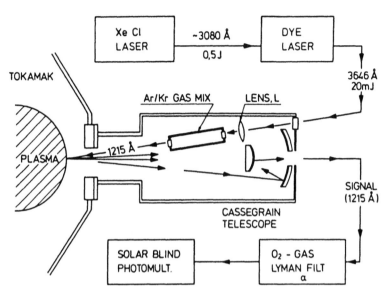

FIG. 9.17 Experimental apparatus used for Lyman-α LIF measurements of neutral hydrogen concentrations in a tokamak plasma (Bogen
et al., 1982).

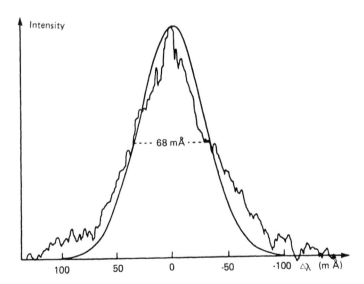

FIG. 9.18 Fluorescence intensity versus dye-laser wavelength (frequency-tripled) for the hydrogen atom Lyman-α resonance. The measurements were taken in a 600-A, 50-Hz cleaning discharge. The
laser line width is approximately 10 mA. The solid curve is a calculated Doppler profile at a temperature of 6000 K (Bogen et al., 1982).

9.5.2 LIF Measurements of Impurities in Fusion Plasmas

Metal impurities can seriously affect the overall energy balance of a
hydrogen fusion plasma because of the intense radiation from excited
electronic states of the metal. Laser-induced fluorescence has been
used to measure the concentrations and velocities of a number of metallic impurities in hydrogen plasmas. Such measurements have been
very useful in elucidating the mechanisms by which metal impurities
are introduced into the plasma.

The potential for measuring sputtered metal species by LIF was
demonstrated by Elbern et al. (1978). They performed LIF measurements of the density and velocity distribution of iron atoms sputtered
from a stainless steel surface following impact of high-energy deuterium, helium, or argon ions.

In the last few years a number of LIF measurements of impurity
concentrations within tokamak reactors have been reported. Schweer
et al. (1980) used LIF techniques developed by Elbern et al. (1978)
to measure neutral iron atom concentrations in the Impurity Study Experiment Tokamak (ISX-B). Schweer et al. (1980) successfully measured Fe I at the beginning and end of plasma discharges. They observed no measurable signal in the steady-state discharge, which they
attributed to ionization of Fe I due to the high electron number density at steady state. Muller and Burrell (1981) also performed LIF measurements of neutral iron and titanium densities in ISX-B for discharges which were heated ohmically or by neutral beams. From the
spatial distribution of the measured impurity densities and from the
relative concentrations for different plasma operating conditions, they
concluded that sputtering due to collisions of charge-exchange neutrals with the stainless steel vessel walls was the primary source of
impurities. Dullni et al. (1982) performed LIF measurements of aluminum concentration and velocity in the Elmo Bumpy Torus Scale
(EBT-S), a non-tokamak device. They correlated the observed aluminum flux from the walls with hydrogen pressure in the device.

Muller et al. (1982) reported LIF measurements of neutral titanium
and deuterium in the Doublet III tokamak. The experiment is depicted
schematically in Fig. 9.19. The $(a^3F_2$-$v^3F_2^0)$ Ti I transition at 294.20
nm was pumped by the dye laser and fluorescence from the $(v^3F_2^0$-
$b^3F_2)$ transition was monitored to avoid stray light interferences.
The laser energy was high enough to completely saturate the $(a^3F_2$-
$v^3F_2^0)$ transition. The laser was tuned by observing the optogalvanic
signal from a metal hollow-cathode lamp. The laser was fired at various delays relative to the start of the plasma discharge to obtain time
histories of titanium and deuterium number densities. Figure 9.20
shows the concentration history of Ti I for two similar discharges, one
of which had neutral beam injection during the discharge. As expected, the Ti I density increased during neutral beam injection. After
the end of neutral beam injection, the Ti I number density decayed

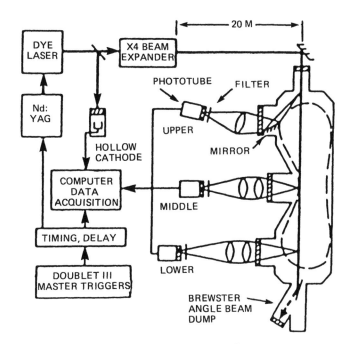

FIG. 9.19 Experimental apparatus for measurements of neutral titanium and deuterium concentration in the Doublet III tokamak (Muller et al., 1982).

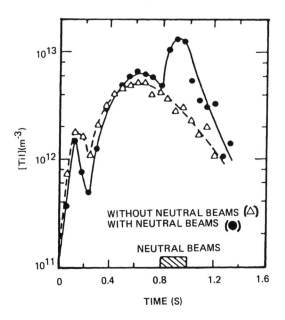

FIG. 9.20 Time-dependent titanium concentrations in the center of a discharge with and without neutral beam injection from 0.8 to 1.0 s after the beginning of the discharge. The plasma current was 500 kA, the electron density 6×10^{19} m^{-3}, and the strength of the magnetic field was 1.2 T (Muller et al., 1982)

with a time constant much closer to ion confinement times in the center
of the discharge rather than in the edge regions. This led the authors to conclude that charge-exchange reactions in the plasma center
created energetic neutrals which then sputtered titanium from the
vessel walls.

9.6 OTHER APPLICATIONS OF LIF

The two areas of LIF application which have been discussed thus far
are, of course, only a limited sampling of the areas where LIF measurements have proved useful; other such areas include analytical
chemistry, biochemistry, and chemical physics. LIF has been used
extensively in all these fields and only a few experiments and review
articles will be mentioned here. LIF applications in remote sensing
are discussed in Chapter 8.

9.6.1 Analytical Chemistry

Applications of LIF in analytical chemistry were recently reviewed by
Harris and Lytle (1983). Harris and Lytle (1983) discuss LIF measurements of atoms and molecules in solution, the use of LIF coupled
with liquid chromatography, and the use of time-resolved fluorimetry
for improved background rejection and species identification. Wehry
et al. (1983) review matrix-isolation LIF, in which large organic molecules are embedded in crystals cooled to cryogenic temperatures so as
to stimplify their spectra and enhance fluorescence selectivity.
Hirschfeld et al. (1983) recently described a technique in which fiber
optics were used for remote LIF detection of groundwater contaminants.

9.6.2 Biochemistry

Fluorescence may be used to study the structure of biomolecules and
to elucidate the pathways and kinetics of biochemical mechanisms.
LIF may be used to study the biological molecule directly, or another
molecule (such as a dye molecule) may be attached to the biomolecule
of interest to enhance the absorption and/or fluorescence yield. For
example, Anders (1983) performed LIF measurements of nucleic acids
(DNA) and bound complexes of DNA and dye molecules; the fluorescence yield is greatly enhanced by binding of the dye molecule.
    The use of picosecond lasers and streak cameras has enabled the
study of fast energy transfer processes in biological molecules. Shapiro et al. (1975) performed time-resolved measurements of exciton
transfer in DNA molecules by exciting the DNA with a 10-ps, fre-
quency-quadrupled Nd:YAG laser beam. Fluorescence from a bound

dye molecule was monitored using a streak camera. Pellegrino (1983) probed energy transfer processes which occur during photosynthesis in leaves. He performed time-resolved fluorescence measurements using a 6-ps frequency-doubled Nd:YAG laser and a streak camera.

Badley (1976) reviews LIF studies of the structure of biological membranes. LIF is especially useful for such studies because it is an in situ, real-time monitor of structural change in membranes. LIF measurements of the structure and reaction mechanisms of proteins and enzymes are reviewed by Churchich (1976).

9.6.3 Chemical Physics

LIF has found wide application in chemical physics as a monitor of specific quantum states of molecules (atoms) during energy transfer processes or chemical reactions. LIF has been widely used for measuring quantum-state distributions of reactant molecules in molecular beams (Zare and Dagdigian, 1974). Collisional transfer rates in laser-excited molecules can be determined from fluorescence measurements of the population distribution among quantum states in the excited electronic level (Lengel and Crosley, 1976). Fluorescence can also be used to monitor quantum state distributions following molecular photodissociation (Fotakis, 1983; references in Chapter 6).

9.7 SUMMARY

In this chapter applications of laser-induced fluorescence in combustion media and fusion plasmas are reviewed. In combustion, LIF has been used to obtain new insights into combustion processes in flames ranging in complexity from laminar flat flame burners to turbulent diffusion flames. Although LIF has not yet been applied in practical combustion devices, understanding gained from measurements in laboratory flames can be expected to influence design of practical devices.

In the near future several different thrusts in applications of LIF in combustion research can be anticipated. Fluorescence imaging will be an active area of both technique development and application. The main applications will be in the area of turbulent combustion, especially as techniques to acquire images at kilohertz rates are developed. Two-photon-excited fluorescence techniques are still in an early stage of development. Because of the importance of the species which are measurable using TPEF (O,H,CO), future applications in complex combustion flows should be numerous.

There is a great need for an improved collisional and radiative transfer rate data base in order to improve the absolute accuracy of LIF measurements of combustion species. Establishing such a data base for even a single fluorescent species is a time-consuming task because of the wide range of temperatures and collision partners that must be considered. Excellent work has been done in the last few years on the data base necessary for accurate LIF measurements of the OH radical; a great deal of work remains to be done, however, to improve the data base for other fluorescent flame species.

Application of LIF to practical combustion devices for monitoring of temperature, and radical and pollutant concentrations would provide valuable data for modeling of such devices. The two main barriers to such applications appear to be the high luminosity and high pressure which are frequently characteristic of such devices. LIF measurements in luminous flames have recently been demonstrated, and time gating and spectral filtering can drastically reduce the interferences due to background luminosity. Measurements in high-pressure flames certainly seem feasible, although no such measurements have yet been reported.

In contrast to LIF measurements in combustion, LIF has been applied as a fusion plasma diagnostic on actual experimental fusion devices. LIF has probably had its greatest impact in understanding the source and behavior of impurities such as iron in fusion devices. If optical access problems could be overcome, the application of fluorescence imaging techniques to impurity measurements in plasmas would be extremely interesting.

Hydrogen atom concentration and velocity measurements in operating fusion devices have been fairly successful. There is certainly a pressing need for higher-intensity sources at the Lyman-α wavelength. The application of two-step excitation techniques such as those of Bogen and Lie (1980) and Goldsmith (1985) offers the possibility of obtaining line images of hydrogen density.

The potential for measuring plasma properties such as electron density using LIF has been demonstrated. Such measurements typically require accurate knowledge of collisional and radiative transfer cross sections and other plasma parameters, such as electron temperature, and are thus applicable only in well-characterized plasmas.

APPENDIX

Example 1

Here we derive the rate equations for the three-level model shown in Fig. 9.3, neglecting collisional excitation ($Q_{12} = Q_{13} = 0$), and the expression for the population ratio N_3/N_2 at steady state. Then assuming that the laser-pumped transition is completely saturated, we show that N_2/N_T, where $N_T = N_1 + N_2 + N_3$, is no longer independent of collisional transfer rates, as is the case for a two-level model.

The rate and conservation equations for the three-level system are

$$\frac{dN_2}{dt} = N_1 W_{12} - N_2(W_{21} + Q_{21} + A_{21} + Q_{23} + A_{23}) + N_3 Q_{32} \qquad (9A.1)$$

$$\frac{dN_3}{dt} = N_2(Q_{23} + A_{23}) - N_3(Q_{31} + A_{31} + Q_{32}) \qquad (9A.2)$$

$$N_T = N_1 + N_2 + N_3 \qquad (9A.3)$$

At steady state $dN_2/dt = dN_3/dt = 0$, and the population ratio N_3/N_2 becomes

$$\frac{N_3}{N_2} = \frac{Q_{23} + A_{23}}{Q_{31} + A_{31} + Q_{32}} = \beta \qquad (9A.4)$$

Substituting Eqs. (9A.3) and (9A.4) into Eq. (9A.1) and assuming steady state, we obtain

$$\frac{N_2}{N_T} = \frac{W_{12}}{W_{12}(1 + \beta) + W_{21} + Q_{21} + A_{21} + Q_{23} + A_{23} - Q_{32}\beta} \qquad (9A.5)$$

At full saturation ($W_{12}, W_{21} \gg Q_{ij}, A_{ij}$) the population ratio N_2/N_T becomes

$$\frac{N_2}{N_T} = \frac{W_{12}}{W_{12}(1 + \beta) + W_{21}} \qquad (9A.6)$$

Equation (9A.6) reduces to the two-level result, Eq. (9A.7), only when β approaches zero. Consequently, in a three-level system the population fraction of the laser-excited upper level is not independent of collisional transfer rates, even when the transition is fully saturated.

Example 2

What is the minimum concentration of the OH radical which can be detected by LIF in flame gases at atmospheric pressure? Assume that the laser has a pulse energy of 0.01 mJ, a pulse duration of 1 μs, a focused diameter of 200 μm, and that the laser is single-frequency and tuned to the $Q_1(8)$ transition of the $A^2\Sigma^+-X^2\Pi$ (0,0) band. Assume that the electronic quenching rate is 1.0×10^9 s^{-1} and that the homogeneous width of the absorption line is 0.1 cm^{-1}. The fluorescence detection volume has a length of 0.1 cm, the collection solid angle is 0.012 sr (f/8 optics), the detection efficiency of the collection system is 0.1, and the photodetector quantum efficiency is 0.1. The OH detection limit is reached when the signal-to-noise ratio is equal to 1, assuming that detection is limited by photon statistics.

The laser-induced absorption rate W_{12} is given by (Yariv, 1975)

$$W_{12} = \frac{A_{21} c^2 I g(0)}{8 \pi n^2 h \nu^3} \tag{9A.7}$$

where $A_{21} = 6.40 \times 10^5$ s^{-1} (Dimpfl and Kinsey, 1979), $c = 2.998 \times 10^8$ m/s, $h = 6.626 \times 10^{-34}$ J-s, and $\nu = 9.69841 \times 10^{14}$ s^{-1} for the $Q_1(8)$ transition (Dieke and Crosswhite, 1962). The line-shape factor $g(\nu) = g(0) = 2T_2 = 1/\pi \Delta \nu$, assuming that the laser is tuned to the line center and neglecting Doppler broadening. Assuming that the laser output is a square pulse of 1 μs duration and focuses to a uniform-intensity 200-μm-diameter circle, the laser intensity $I = 3.2 \times 10^8$ W/m^2. Substituting these values into Eq. (9A.7) gives $W_{12} = 1.3 \times 10^8$ s^{-1}. Note that the fluorescence signal is not saturated, because $W_{12}/Q_{21} = 0.13$.

The optimum OH detection limit will be achieved by collecting fluorescence from the entire $A^2\Sigma^+ - X^2\Pi$ (0,0) band, for which $A_{ug} = 1.46 \times 10^6$ s^{-1}. Because of the long pulse length compared to characteristic collisional times, steady-state conditions can be assumed. The rotational structure of the two bands which are coupled by the laser radiation must, however, be considered. Assuming that quenching occurs directly to the ground vibration-rotation band, and not to some intermediate levels(s), the total populations N_u and N_g of the upper and ground sets of rotation-vibration levels are related by

$$N_g f_1 W_{12} = N_u (Q_{ug} + A_{ug} + W_{21} f_2) \tag{9A.8}$$

where f_1 and f_2 are the population fractions for the lower and upper rotational levels which are directly coupled by the laser radiation. (Referring to Fig. 9.4, $N_u = N_2 + N_3$ and $N_g = N_1 + N_4$). For weak excitation ($W_{12} \ll Q_{ug}$), f_1 will be equal to the Boltzmann fraction to a good approximation. The upper-level population fraction will probably be far from the rotational equilibrium value for both weak excitation and saturated conditions (Lucht and Laurendeau, 1978), but for weak excitation the ratio N_2/N_1 is not very sensitive to the value of f_2.

Under weak excitation conditions $N_g \approx N_{OH}$, and Eq. (9A.8) becomes

$$N_u = \frac{N_g W_{12} f_1}{Q_{ug} + A_{ug}} = \frac{N_{OH} W_{12} f_1}{Q_{ug} + A_{ug}}$$

The population fraction $f_1 \approx f_{1B}$, where f_{1B} is the Boltzmann fraction for level 1, and is given by

$$f_{1B} = \frac{(2J_1 + 1) \exp(-hcE_1/kT)}{Z_r Z_v Z_e} \qquad (9A.10)$$

where for $J_1 = 8.5$ and $E_1 = 1324.24$ cm^{-1} (Dieke and Crosswhite, 1962). The partition functions are given by

$$Z_r = \frac{kT}{hcB_0} \qquad Z_v = \left[1 - \exp\left(-\frac{hc\omega_e}{kT}\right)\right]^{-1} \qquad (9A.11)$$

where for OH $B_0 = 18.515$ cm^{-1}, $\omega_e = 3735.2$ cm^{-1}, and $Z_e = g_e = 4$ (Dieke and Crosswhite, 1962). At a temperature of 2000 K, $f_{1B}(e) = 0.0216$.

The detectability limit for a single laser pulse is reached when 10 photons are incident on the 0.1 quantum efficiency photocathode during the laser pulse. The number of photons incident on the photocathode during the laser pulse of duration Δt (seconds) is given by

$$n_p \, \Delta t = \frac{N_{OH} \, \Delta t \, f_{1B} W_{12} \Phi f_c V \epsilon \Omega_c}{4\pi} \qquad (9A.12)$$

where Eqs. (9A.6) and (9A.9) have been substituted into Eq. (9A.4). At the detection limit, $n_p \, \Delta t = 10$. The minimum detectable OH number density is thus given by

$$N_{OH} = \frac{40\pi}{f_{1B} W_{12} \, \Delta t \, \Phi_f V_c \epsilon \Omega_c} \qquad (9A.13)$$

Substituting the values of parameters in Eq. (9A.13) and solving the equation gives a minimum OH detectability of 7.8×10^{13} cm^{-3}, or 0.21 ppm at atmospheric pressure and 2000 K.

Example 3

(a) Here we derive the rate equations for the upper-level population N_3 for the four-level model shown in Fig. 9.16. Assuming weak excitation from the ground state ($N_1 \approx N_1^0$), we derive the expression for the steady-state upper-level population N_3, neglecting transfer to and from the ionization continuum ($T_{c2} = T_{c2} = T_{c3} = T_{3c} = 0$) and upward collisional transfer ($Q_{12} = Q_{13} = Q_{23} = 0$). (b) Considering the case where another laser beam is crossed in the probe volume and induces a photoionization process W_{3c}, we derive the dependence of N_3 on T_3 for $W_{3c} \gg T_3$.

(a) The rate equation for level 3 is given by

$$\frac{dN_3}{dt} = N_1 W_{13} + N_2 Q_{23} + N_c T_{c3} - N_3(W_{31} + T_{31} + T_{32} + T_{3c})$$

(9A.14)

Under weak excitation conditions, neglecting transfer to and from the continuum, and neglecting upward collisional transfer, the rate equation becomes

$$\frac{dN_3}{dt} = N_1^0 W_{13} - N_3(W_{31} + T_{31} + T_{32})$$

(9A.15)

At steady state the upper-level population N_3 is given by

$$N_3 = \frac{N_1^0 W_{13}}{W_{31} + T_3}$$

(9A.16)

where $T_3 = Q_{32} + Q_{31} + A_{32} + A_{31}$.

(b) For the case where a photoionizing beam is crossed in the same volume as the exciting laser beam, the rate equation becomes

$$\frac{dN_3}{dt} = N_1^0 W_{13} - N_3(W_{31} + T_3 + W_{3c})$$

(9A.17)

When $W_{3c} \gg W_{31}$, Eq. (9A.17) reduces to

$$N_3 = \frac{N_1^0 W_{13}}{W_{3c}}$$

(9A.18)

at steady state. The upper-level population, and consequently the fluorescence signal, is now independent of collisional transfer and spontaneous emission rates. The fluorescence signal is also reduced relative to the case where no photoionization beam is present. This technique has been proposed as a means of eliminating the quenching dependence of the fluorescence signal for hydrogen atom measurements (Lucht et al., 1983b).

ACKNOWLEDGMENTS

The author wishes to acknowledge his collaborations with Professors Normand M. Laurendeau and Galen B. King, and Dr. J. Thaddeus Salmon of the School of Mechanical Engineering, Purdue University, West Lafayette, Indiana, and with Dr. Donald W. Sweeney of Sandia National Laboratories, Livermore, California, on much of the fluorescence research described in this chapter. The author also wishes to thank Drs. John E. M. Goldsmith, Greg W. Foltz and Wen L. Hsu of Sandia National Laboratories, Livermore, California for reviewing the manuscript and for many helpful suggestions.

BIBLIOGRAPHY

Allen, L., and Eberly, J. H. (1975). *Optical Resonance and Two-Level Atoms*, Wiley, New York. Discusses the interaction of coherent radiation with two-level resonances. The semiclassical approach is emphasized, although the fully rate equation and quantum mechanical approaches are discussed.

Chang, R. K., and Long, M. B. (1982). Optical multichannel detection, in *Light Scattering in Solids* (M. Cardona and G. Guntherodt, eds.), Springer-Verlag, New York, p. 179. Reviews the physical principles and selected applications of array detectors.

Crosley, D. R., ed. (1980). *Laser Probes for Combustion Chemistry*, ACS Symp. Ser., Vol. 134, American Chemical Society, Washington, D.C. A broad review of work in combustion diagnostics, including applications of LIF and Raman spectroscopy.

Crosley, D. R. (1981). Collisional effects on laser-induced fluorescence flame measurements, *Opt. Eng.*, *20*, 511. The effects of electronic, vibrational, and rotational energy transfer on LIF measurements in combustion are discussed.

Danielewicz, E. J., Luhmann, N. C., Jr., and Peebles, W. A. (1984). Applications of lasers to magnetic confinement fusion plasmas, *Opt. Eng.*, *23*, 475. Reviews laser diagnostics of plasmas, including interferometry, Thomson scattering, and fluorescence.

Eckbreth, A. C., Bonczyk, P. A., and Verdieck, J. F. (1979). Combustion diagnostics by laser raman and fluorescence techniques, *Prog. Energy Combust. Sci.*, *5*, 253. Evaluates the applicability of laser diagnostics to practical combustion systems, discussing potential signal interferences in detail.

Klainer, S. M., ed. (1983). *Opt. Eng.*, *20*, 507. A special journal issue devoted to applications of fluorescence spectroscopy.

Loudon, R. (1983). *The Quantum Theory of Light*, 2nd ed., Clarendon Press, Oxford. Discusses resonance fluorescence in detail.

Peacock, N. J., and Burgess, D. D. (1981). New developments in
measurement techniques for high temperature plasmas, *Philos.
Trans. R. Soc. London, A 300*, 665. Reviews plasma characteristics and a variety of diagnostic techniques in addition to laser-
based techniques.

Schofield, K., and Steinberg, M. (1981). Quantitative atomic and
molecular fluorescence in the study of detailed combustion processes, *Opt. Eng.*, *20*, 501. Reviews LIF measurements in flames,
especially those oriented toward studying the chemical kinetics of
combustion reactions.

Wehry, E. L., ed. (1976). *Modern Fluorescence Spectroscopy*, Vols.
1 and 2, Plenum Press, New York. These volumes are a comprehensive survey of applications of fluorescence in analytical chemistry and biology.

Yariv, A. (1975). *Quantum Electronics*, Wiley, New York. A classic
text on lasers and the interactions of laser radiation with matter.

REFERENCES

Alden, M., Edner, H., Holmstedt, G., Svanberg, S., and Hogberg,
T. (1982a). Single-pulse laser-induced OH fluorescence in an
atmospheric pressure flame, spatially resolved with a diode array
detector, *Appl. Opt.*, *21*, 1236.

Alden, M., Edner, H., Grafstrom, P., and Svanberg, S. (1982b).
Two-photon excitation of atomic oxygen in a flame, *Opt. Commun.*,
42, 244.

Alden, M., Schawlow, A. L., Svanberg, S., Wendt, W., and Zhang,
P.-L. (1984a). Three-photon-excited fluorescence detection of
atomic hydrogen in an atmospheric pressure flame, *Opt. Lett.*,
9, 211.

Alden, M., Hertz, H. M., Svanberg, S., and Wallin, S. (1984b).
Imaging laser-induced fluorescence of oxygen atoms in a flame,
Appl. Opt., *23*, 3255.

Alden, M., Wallin, S., and Wendt, W. (1984c). Applications of two-
photon absorption for detection of CO in combustion gases, *Appl.
Phys.*, *33*, 205.

Allen, L., and Eberly, J. H. (1975). *Optical Resonance and Two-
Level Atoms*, Wiley, New York.

Allen, J. E., Jr., Anderson, W. R., Crosley, D. R., and Fansler, T. D. (1979). Energy transfer and quenching rates of laser-pumped electronically excited alkalis in flames, *Seventeenth Symp. (Int.) Combustion*, Combustion Institute, Pittsburgh, Pa., p. 797.

Altkorn, R., and Zare, R. N. (1984). Effects of saturation on laser-induced fluorescence measurements of population and polarization, *Annu. Rev. Phys. Chem.*, *35*, 265.

Anders, A. (1983). Laser fluorescence spectroscopy of biomolecules: nucleic acids, *Opt. Eng.*, *22*, 592.

Anderson, W. R., Decker, L. J., and Kotlar, A. J. (1982a). Concentration profiles of OH and NH in a stoichiometric CH_4/N_2O flame by laser excited fluorescence and absorption, *Combust. Flame*, *48*, 179.

Anderson, W. R., Decker, L. J., and Kotlar, A. J. (1982b). Temperature profile of a stoichiometric CH_4/N_2O flame from laser excited fluorescence measurements on OH, *Combust. Flame*, *48*, 163.

Azzazy, M., and Daily, J. W. (1983). Fluorescence measurements of OH in a turbulent flame, *AIAA J.*, *21*, 1100.

Badley, R. A. (1976). Fluorescent probing of dynamic and molecular organization of biological membranes, in *Modern Fluorescence Spectroscopy*, (E. L. Wehry, ed.), Plenum Press, New York, p. 91.

Baronavski, A. P., and McDonald, J. R. (1977). Application of saturation spectroscopy to the measurement of C_2, $^3\Pi_u$ concentrations in oxy-acetylene flames, *Appl. Opt.*, *16*, 1897.

Bechtel, J. H. (1979). Temperature measurements of the hydroxyl radical and molecular nitrogen in premixed, laminar flames by laser techniques, *Appl. Opt.*, *18*, 2100.

Bechtel, J. H., and Blint, R. J. (1980). Structure of a flame-wall interface by laser Raman scattering, *Appl. Phys. Lett.*, *37*, 576.

Bechtel, J. H., and Blint, R. J. (1982). Hydrocarbon combustion near a cooled wall, *Soc. Automot. Eng. Tech. Pap. Ser. 820063*.

Bechtel, J. H., and Teets, R. E. (1980). Hydroxyl and its concentration profile in methane-air flames, *Appl. Opt.*, *18*, 4138.

Berg, J. O., and Shackleford, W. L. (1979). Rotational redistribution effect on saturated laser-induced fluorescence, *Appl. Opt.*, *18*, 2093.

Bergano, N. S., Jaanimagi, P. A., Salour, M. M., and Bechtel, J. H. (1983). Picosecond laser-spectroscopy measurement of hydroxyl fluorescence lifetime in flames, *Opt. Lett.*, *8*, 443.

Bischel, W. K., Perry, B. E., and Crosley, D. R. (1982). Detection of fluorescence from O and N atoms induced by two-photon absorption, *Appl. Opt.*, *21*, 1419.

Bogen, P., and Lie, Y. T. (1980). Detection of atomic hydrogen by resonance fluorescence using two-step excitation with L_α and H_α radiation, *J. Nucl. Mater.*, *93/94*, 363.

Bogen, P., Dreyfus, R. W., Lie, Y. T., and Langer, H. (1982). Measurement of atomic hydrogen densities and velocities by laser-induced fluorescence at L_α, *J. Nucl. Mater.*, *111/112*, 75.

Bokor, J., Freeman, R. R., White, J. C., and Storz, R. H. (1981). Two-photon excitation of the n = 3 level in H and D atoms, *Phys. Rev.*, *A24*, 612.

Bonczyk, P. A., and Shirley, J. A. (1979). Measurement of CH and CN concentration in flames by laser-induced saturated fluorescence, *Combust. Flame*, *34*, 253.

Bradshaw, J. D., Omenetto, N., Zizak, G., Bower, J. N., and Winefordner, J. D. (1980). Five laser-excited fluorescence methods for measuring spatial flame temperatures: theoretical basis, *Appl. Opt.*, *19*, 2709.

Burakov, V. S., Misyakov, P. Y., Naumenko, P. A., Nechaev, S. V., Razdobarin, G. T., Semenov, V. V., Sokolova, L. V., and Folomkin, I. P. (1977). Use of the method of resonance fluorescence with a dye laser for plasma diagnostics in the FT-1 tokamak installation, *JETP Lett.*, *26*, 403.

Campillo, A. J., and Shapiro, S. L. (1983). Picosecond streak camera fluorometry—a review, *IEEE J. Quantum Electron.*, *QE-19*, 585.

Cattolica, R. J. (1981). OH rotational temperature from two-line laser-excited fluorescence, *Appl. Opt.*, *20*, 1156.

Cattolica, R. J., and Schefer, R. W. (1983). Laser fluorescence measurements of the OH concentration in a combustion boundary layer, *Combust. Sci. Technol.*, *30*, 205.

Cattolica, R. J., and Stephenson, D. A. (1984). Two-dimensional imaging of flame temperature using laser-induced fluorescence, in *Dynamics of Flames and Reactive Systems*, Progress in Astronautics and Aeronautics Series, 95, AIAA, New York, p. 714.

Cattolica, R. J., and Vosen, S. R. (1984). Two-dimensional measurements of the [OH] in a constant volume combustion chamber, *Twentieth Symp. (Int.) Combustion*, Combustion Institute, Pittsburgh, Pa., p. 1273.

Cattolica, R. J., and Vosen, S. R. (1986). Two-dimensional fluorescence imaging of a flame-vortex interaction, *Combust. Sci. Technol.* (in press).

Cattolica, R. J. Smooke, M. D., and Dean, A. M. (1982). A hydrogen-nitrous oxide flame study, Sandia National Laboratories Report, SAND82-8776.

Cattolica, R. J., and Stepowski, D., and Cottereau, M. (1984). Laser fluorescence measurements of the CH radical in a low pressure flame, *J. Quant. Spectrosc. Radiat. Transfer* *32*, 363.

Chan, C., and Daily, J. W. (1980). Measurement of temperature in flames using laser in duced fluorescence spectroscopy of OH, *Appl. Opt.*, *19*, 1963.

Chang, R. K., and Long, M. B. (1982). Optical multichannel detection, in *Light Scattering in Solids* (M. Cardona and G. Guntherodt, eds.), Springer-Verlag, New York, p. 179.

Chidsey, I. L., and Crosley, D. R. (1980). *J. Quant. Spectrosc. Radiat. Transfer*, *23*, 187.

Churchich, J. E. (1976). Fluorescent probe studies of binding sites in proteins and enzymes, in *Modern Fluorescence Spectroscopy* (E. L. Wehry, ed.), Plenum Press, New York, p. 91.

Clendening, C. W., Jr., Shackleford, W., and Hilyard, R. (1981). Raman scattering measurements in a side-wall quench layer, *Eighteenth Symp. (Int.) Combustion*, Combustion Institute, Pittsburgh, Pa., p. 1583.

Correa, S. M., Drake, M. C., Pitz, R. W., and Shyy, W. (1984). Prediction and measurement of a non-equilibrium turbulent diffusion flame, *Twentieth Symp. (Int.) Combustion*, Combustion Institute, Pittsburgh, Pa., p. 337.

Crosley, D. R. (1981). Collisional effects on laser-induced fluorescence flame measurements, *Opt. Eng.*, *20*, 511.

Crosley, D. R., and Smith, G. P. (1980). Vibrational energy transfer in laser-excited OH as a flame thermometer, *Appl. Opt.*, *19*, 517.

Crosley, D. R., and Smith, G. P. (1982). Rotational energy transfer and LIF temperature measurements, *Combust. Flame*, *44*, 27.

Crosley, D. R., and Smith, G. P. (1983). Two-photon spectroscopy of the $A^2\Sigma^+ - X^2\Pi_i$ system of OH, *J. Phys. Chem.*, *79*, 4764.

Csorba, I. P. (1980). Current gain parameters of microchannel plates, Appl. Opt., 19, 3863.

Daily, J. W. (1976). Pulsed resonance spectroscopy applied to turbulent combustion flows, Appl. Opt., 15, 955.

Daily, J. W. (1977). Saturation effects in laser induced fluorescence spectroscopy, Appl. Opt., 16, 568.

Daily, J. W., and Chan, C. (1978). Laser-induced fluorescence measurement of sodium in flames, Combust. Flame, 33, 47.

Daniel, W. A. (1957). Flame quenching at the wall of an internal combustion engine, Sixth Symp. (Int.) Combustion, Reinhold, New York, p. 886.

Danielewicz, E. J., Luhmann, N. C., Jr., and Peebles, W. A. (1984). Applications of lasers to magnetic confinement fusion plasmas, Opt. Eng., 23, 475.

Denton, M. B., and Malmstadt, H. V. (1971). Tunable organic dye laser as an excitation source for atomic flame fluorescence spectroscopy, Appl. Phys. Lett., 18, 485.

Dieke, G. H., and Crosswhite, H. M. (1962). J. Quant. Spectrose. Radiat. Transfer, 2, 97.

Dimock, D. Hinnov, E., and Johnson, L. C. (1969). Phys. Fluids, 12, 1730.

Dimpfl, W. L., and Kinsey, J. L. (1979). J. Quant. Spectrosc. Radiat. Transfer, 21, 233.

Doherty, P. M., and Crosley, D. R. (1984). Polarization of laser-induced fluorescence in OH in an atmospheric pressure flame, Appl. Opt., 23, 713.

Drake, M. C., Pitz, R. W., Lapp, M., Fenimore, C. P., Lucht, R. P., Sweeney, D. W., and Laurendeau, M. M. (1984). Measurements of superequilibrium hydroxyl concentrations in turbulent nonpremixed flames using saturated fluorescence, Twentieth Symp. (Int.) Combustion, Combustion Institute, Pittsburgh, Pa., p. 327.

Dullni, E., Hintz, E., Roberto, J. B., Colchin, R. J., and Richards, R. K. (1982). Measurement of the density and velocity distribution of sputtered Al in EBT-S by laser-induced fluorescence, J. Nucl. Mater., 111/112, 61.

Dyer, M. J., and Crosley, D. R. (1982). Two-dimensional imaging of OH laser-induced fluorescence in a flame, Opt. Lett., 7, 382.

Elbern, A., Hintz, E., and Schweer, B. (1978). Measurement of the velocity distribution of metal atoms sputtered by light and heavy particles, *J. Nucl. Mater.*, *76/77*, 143.

Elder, M. L., and Winefordner, J. D. (1983). Temperature measurements in flames—a review, *Prog. Anal. At. Spectrosc.*, *6*, 293.

Engstrom, R. W. (1980). *RCA Photomultiplier Handbook*, RCA Corporation, Lancaster, Pa.

Fairchild, P. W., Smith, G. P., and Crosley, D. R. (1983). Collisional quenching of $A^2\Sigma^+$ OH at elevated temperatures, *J. Phys. Chem.*, *79*, 1795.

Fotakis, C. (1983). Photofragment fluorescence following ultraviolet laser multiphoton excitation, *Opt. Eng.*, *22*, 554.

Fraser, I. M., and Winefordner, J. D. (1971). Laser-excited atomic fluorescence flame spectrometry, *Anal. Chem.*, *43*, 1693.

Gohil, P., and Burgess, D. D. (1983). A comparison between laser induced fluorescence at Balmer-alpha and at Lyman-alpha for the measurement of neutral hydrogen densities in magnetically contained fusion plasmas, *Plasma Phys.*, *25*, 1149.

Gohil, P., Kolbe, G., Forrest, M. J., Burgess, D. D., and Hu, B. Z. (1983). A measurement of the neutral hydrogen density determined from Balmer alpha fluorescence scattering in the HBTX1A reverse field pinch, *J. Phys.*, *D16*, 333.

Goldsmith, J. E. M. (1985). Two-step saturated fluorescence detection of atomic hydrogen in flames, *Opt. Lett. 10*, 116.

Goldsmith, J. E. M., and Anderson, R. J. M. (1985). Imaging of atomic hydrogen in flames with two-step saturated fluorescence detection, *Opt. Lett. 11*, 67.

Gole, J. L. (1981). Aspects of sparsely studied gas phase chemistry of import to energy technologies, *Opt. Eng.*, *20*, 546.

Gross, K. P., and McKenzie, R. L. (1983). Single-pulse gas thermometry at low temperatures using two-photon laser-induced fluorescence in NO-N_2 mixtures, *Opt. Lett.*, *8*, 368.

Hänsch, T. W. (1972). Repetitively pulsed dye laser for high resolution spectroscopy, *Appl. Opt.*, *11*, 895.

Hänsch, T. W., Lee, S. A., Wallenstein, R., and Wieman, C. (1975). Doppler-free two-photon spectroscopy of hydrogen 1S-2S, *Phys. Rev. Lett.*, *34*, 307.

Haraguchi, H., Smith, B., Weeks, S., Johnson, D. J., and Wine-fordner, J. D. (1977). Measurement of small volume flame temperatures by the two-line atomic fluorescence method, *Appl. Spec.*, *31*, 156.

Harris, T. D., and Lytle, F. E. (1983). Analytical applications of laser absorption and emission spectroscopy, in *Ultrasensitive Laser Spectroscopy* (D. S. Kliger, ed.), Academic Press, New York, p. 369.

Harris, J. M., Lytle, F. E., and McCain, T. C. (1976). Squirrel-cage photomultiplier base design for measurement of nanosecond fluorescence decays, *Anal. Chem.*, *48*, 2095.

Hilbig, R., and Wallenstein, R. (1981). Enhanced production of tunable VUV radiation by phase-matched frequency tripling in krypton and xenon, *IEEE J. Quantum Electron.*, *QE-17*, 1566.

Hirschfeld, T. Deaton, T., Milanovich, E., and Klainer, S. (1983). Feasibility of using fiber optics for monitoring groundwater contaminants, *Opt. Eng.*, *22*, 527.

Horlick, G., and Codding, E. G. (1973). Some characteristics and applications of self-scanning linear silicon photodiode arrays as detectors of spectral information, *Anal. Chem.*, *45*, 1490.

Johnson, L. C., and Hinnov, E. (1973). Ionization, recombination, and population of excited levels in hydrogen plasmas, *J. Quant. Spectrosc. Radiat. Transfer*, *13*, 333.

Joklik, R. G., and Daily, J. W. (1982). An experimental study of two-line atomic fluorescence temperature measurement in flames, paper 82-43, Spring Meeting of the Western States Section of the Combustion Institute, Salt Lake City, Utah.

Kowalik, R. M., and Kruger, C. H. (1979). Laser fluorescence temperature measurements, *Combust. Flame*, *34*, 135.

Kuhl, J., and Marowsky, G. (1971). Narrow-band dye laser as a light source for fluorescence analysis in the subnanogram range, *Opt. Commun.*, *4*, 125.

Kuhl, J., Neumann, S., and Kriese, M. (1973). Influence of saturation phenomena on laser-excited atomic fluorescence flame spectrometry, *Z. Naturforsch*, *A28*, 273.

Kychakoff, G., Howe, R. D., Hanson, R. K., and McDaniel, J. C. (1982). Quantitative visualization of combustion species in a plane, *Appl. Opt.*, *21*, 3225.

Kychakoff, G., Knapp, K., Howe, R. D., and Hanson, R. K. (1984a). Flow visualization in combustion gases using nitric oxide fluorescence, *AIAA J.*, *22*, 153.

Kychakoff, G., Howe, R. D., Hanson, R. K., Drake, M. C., Pitz, R. W., Lapp, M., and Penney, C. M. (1984b). Visualization of turbulent flame fronts with planar laser-induced fluorescence, *Science*, *224*, 382.

Lengel, R. K., and Crosley, D. R. (1976). Energy transfer in $A^2\Sigma^+$ OH: I. Rotational, *J. Chem. Phys.*, *67*, 2085.

LoRusso, J. A., Lavoie, G. A., and Kaiser, E. W. (1980). An electrohydraulic gas sampling valve with applications to hydrocarbon emissions studies, *Soc. Automot. Eng. Tech. Pap. Ser. 800045*.

Loudon, R. (1983). *The Quantum Theory of Light*, 2nd ed. Clarendon Press, Oxford.

Lucht, R. P., and Laurendeau, N. M. (1978). *Appl. Opt.*, *18*, 856.

Lucht, R. P., and Laurendeau, N. M. (1979). Two-level model for near saturated fluorescence in diatomic molecules, *Appl. Opt.*, *18*, 856.

Lucht, R. P., Sweeney, D. W., and Laurendeau, N. M. (1980). Balanced cross-rate model for saturated molecular fluorescence in flames using a nanosecond pulse length laser, *Appl. Opt.*, *19*, 3295.

Lucht, R. P., Laurendeau, N. M., and Sweeney, D. W. (1982). Temperature measurement by two-line laser-saturated OH fluorescence in flames, *Appl. Opt.*, *21*, 3729.

Lucht, R. P., Sweeney, D. W., and Laurendeau, N. M. (1983a). Laser-saturated fluorescence measurements of OH concentration flames, *Combust. Flame*, *50*, 189.

Lucht, R. P., Salmon, J. T., King, G. B., Sweeney, D. W., and Laurendeau, N. M. (1983b). Two-photon-excited fluorescence measurement of hydrogen atoms in flames, *Opt. Lett.*, *8*, 365.

Lucht, R. P., Sweeney, D. W., Laurendeau, N. M., Drake, M. C., Lapp, M., and Pitz, R. W. (1984). Single pulse, laser-saturated fluorescence measurements of OH in turbulent nonpremixed flames, *Opt. Lett.*, *9*, 90.

Lucht, R. P., Sweeney, D. W., and Laurendeau, N. M. (1985). Laser-saturated fluorescence measurements of OH in atmospheric pressure $CH_4/O_2/N_2$ flames under sooting and non-sooting conditions, *Combust. Sci. Technol.*, *42*, 259.

Lucht, R. P., Sweeney, D. W., and Laurendeau, N. M. (1986). Time-resolved fluorescence investigation of rotational transfer in $A^2\Sigma^+$ (v = 0) OH, *J. Chem. Phys.* (submitted).

Lytle, F. E. (1974). Optical signals: detectors, *Anal. Chem.*, *46*, 545A.

Mailander, M. (1978). Determination of absolute transition probabilities and particle densities by saturated fluorescence excitation, *J. Appl. Phys.*, *49*, 1256.

Measures, R. M. (1968). Selective excitation spectroscopy and some possible applications, *J. Appl. Phys.*, *39*, 5232.

Mertens, P., and Bogen, P. (1984). Velocity distribution of hydrogen atoms sputtered from metal hydrides, *J. Nucl. Mater.*, *128/129*, 551.

Morley, C. (1981). The mechanism of NO formation from nitrogen compounds in hydrogen flames studied by laser fluorescence, *Eighteenth Symp. (Int.) Combustion*, Combustion Institute, Pittsburgh, Pa., p. 23.

Muller, C. H., III, and Burrell, K. H. (1981). Time-dependent measurements of metal impurity densities in a tokamak discharge by use of laser-induced fluorescence, *Phys. Rev. Lett.*, *47*, 330.

Muller, C. H., III, Schofield, K., Steinberg, M., and Broida, H. P. (1979). Sulfur chemistry in flames, *Seventeenth Symp. (Int.) Combustion*, Combustion Institute, Pittsburgh, Pa., p. 867.

Muller, C. H., III, Schofield, K., and Steinberg, M. (1980). Laser induced flame chemistry of $Li(2^2P_{1/2,3/2})$ and $Na(3^2P_{1/2,3/2})$. Implications for other saturated mode measurements, *J. Chem. Phys.*, *72*, 6620.

Muller, C. H., III, Eames, D. R., Burrell, K. H., and Bates, S. C. (1982). Dye laser fluorescence spectroscopy on the doublet III tokamak, *J. Nucl. Mater.*, *111/112*, 56.

Nieuwesteeg, K. J., Hollander, T., and Alkemade, C. T. J. (1983). Collisional depolarization of the $Na-D_2$ fluorescence at line wing excitation in a N_2-diluted H_2/O_2 flame at 1 atm, *J. Quant. Spectrosc. Radiat. Transfer*, *30*, 97.

Omenetto, N., Benetti, P., Hart, L. P., Winefordner, J. D., and Alkemade, C. Th. J. (1973). *Spectrochim. Acta*, *28B*, 289.

Pasternack, E., Baronavski, A. P., and McDonald, J. R. (1978). Application of saturation spectroscopy for measurement of atomic Na and MgO in acetylene flames, *J. Chem. Phys.*, *69*, 4830.

Peacock, N. J., and Burgess, D. D. (1981). New developments in measurement techniques for high temperature plasmas, *Philos. Trans. R. Soc. London*, *A 300*, 665.

Pellegrino, F. (1983). Ultrafast energy transfer processes in photo-synthetic systems probed by picosecond fluorescence spectroscopy, *Opt. Eng.*, *22*, 508.

Piepmeier, E. H. (1972). Theory of laser saturated atomic resonance fluorescence, *Spectrochim. Acta*, *B27*, 431.

Razdobarin, G. T., Semenov, V. V., Sokolova, L. V., Folomkin, L. P., Burakov, V. S., Misakov, P. Y., Naumenkov, P. A., and Nechaev, S. V. (1979). An absolute measurement of the neutral density profile in the Tokamak plasma by resonance fluorescence on the H-α line, *Nucl. Fusion*, *19*, 1439.

Rea, E. C., and Hanson, R. K. (1983). Fully resolved absorption/fluorescence lineshape measurements of OH using a rapid-scanning ring dye laser, paper 83-66, Fall Meeting of the Western States Section of the Combustion Institute, Los Angeles, Calif., October.

Salmon, J. T., and Laurendeau, N. M. (1985). Analysis of probe volume effects associated with laser-saturated fluorescence measurements, *Appl. Opt.*, *24*, 1313.

Salmon, J. T., Lucht, R. P., Sweeney, D. W., and Laurendeau, N. M. (1984). Laser-saturated fluorescence measurements of NH in a premixed, subatmospheric $CH_4/N_2O/Ar$ flame, *Twentieth Symp. (Int.) Combustion*, Combustion Institute, Pittsburgh, Pa., p. 1187.

Schweer, B., Rusbuldt, D., Hintz, E., Roberto, J. B., and Husinsky, W. R. (1980). Measurement of the density and velocity distribution of neutral Fe in ISX-B by laser fluorescence spectroscopy, *J. Nucl. Mater.*, *93/93*, 357.

Shapiro, S. L., Campillo, A. J., Kollman, V. H., and Goad, W. B. (1975). Exciton transfer in DNA, *Opt. Commun.*, *15*, 308.

Sharp, B. L., and Goldwasser, A. (1976). Some studies of the laser excited fluorescence of sodium, *Spectrochim. Acta*, *B31*, 431.

Smith, O. I., and Chandler, D. W. (1986). An experimental study of probe distortions to the structure of one-dimensional flames, *Combust. Flame 63*, 19.

Smith, B., Winefordner, J. D., and Omenetto, N. (1977). Atomic fluorescence of sodium under continuous-wave laser excitation, *J. Appl. Phys.*, *48*, 2676.

Snow, E. H. (1976). Self-scanning photodiode arrays for spectroscopy, *Res./Dev.*, April, 18.

Soffer, B. H., and McFarland, B. B. (1967). Continuously tunable narrow-band organic dye lasers, *Appl. Phys. Lett.*, *10*, 266.

Stepowski, D., and Cottereau, M. J. (1979). Direct measurement of OH local concentration in a flame from the fluorescence induced by a single laser pulse, *Appl. Opt.*, *18*, 354.

Stepowski, D., Puechberty, D., and Cottereau, M. J. (1981). Use of laser-induced fluorescence to study the perturbation of a flame by a probe, *Eighteenth Symp. (Int.) Combustion*, Combustion Institute, Pittsburgh, Pa., p. 1567.

Tsuchida, K., Miyake, S., Kadota, K., and Fujita, J. (1983). Plasma electron density measurements by the laser- and collision-induced fluorescence method, *Plasma Phys.*, *25*, 991.

Van Calcar, R. A., Van de Ven, M. J. M., Uitert, B. K., Biewenga, K. J., Hollander, T., and Alkemade, C. T. J. (1979). Saturation of sodium fluorescence in a flame irradiated with a pulsed tunable dye laser, *J. Quant. Spectrosc. Radiat. Transfer*, *21*, 11.

Verdieck, J. F., and Bonczyk, P. A. (1981). Laser-induced saturated fluorescence investigations of CH, CN, and NO in flames, *Eighteenth Symp. (Int.) Combustion*, Combustion Institute, Pittsburgh, Pa., p. 1559.

Wang, C. C., and Davis, L. I. (1974). Ground state population distribution of OH determined with a tunable UV laser, *Appl. Phys. Lett.*, *25*, 34.

Wehry, E. L., Conrad, V. B., Hammons, J. L., Maple, J. R., and Perry, M. B. (1983). Characterization of complex samples by laser-excited matrix-isolation fluorescence spectrometry, *Opt. Eng.*, *22*, 558.

Yariv, A. (1975). *Quantum Electronics*, Wiley, New York.

Zare, R. N., and Dagdigian, P. J. (1974). Tunable laser fluorescence method for product state analysis, *Science*, *185*, 739.

Zizak, G., and Winefordner, J. D. (1982). Application of thermally assisted atomic fluorescence technique to the temperature measurement in a gasoline-air flame, *Combust. Flame*, *44*, 35.

Zizak, G., Omenetto, N., and Winefordner, J. D. (1984). Laser-excited atomic fluorescence techniques for temperature measurements in flames: a summary, *Opt. Eng.*, *23*, 749.

Index